THE HUMAN IMPACT READER

Readings and Case Studies

Edited by
Andrew Goudie
Professor of Geography, University of Oxford

Advisory Editors

David E. Alexander, University of Massachusetts,
Amherst Basil Gomez, Indiana State University
H. O. Slaymaker, University of British Columbia
Stanley W. Trimble, University of California, Los Angeles

BLACKWELL
Publishers

Editorial matter and organization copyright © Andrew Goudie 1997

First published 1997

2 4 6 8 10 9 7 5 3 1

Blackwell Publishers Ltd
108 Cowley Road
Oxford OX4 1JF
UK

Blackwell Publishers Inc.
350 Main Street
Malden, MA 02148
USA

British Library Cataloguing in Publication Data

A CIP catalogue record for this book is available from the British Library.

Library of Congress Cataloging-in-Publication Data

Goudie, Andrew.
    The human impact reader : readings and case studies / edited by Andrew Goudie : advisory editors, David E. Alexander . . . [et al.].
        p.   cm. — (Blackwell readers on the natural environment)
    Includes bibliographical references.
    ISBN 0–631–19979–9 (acid-free paper). — ISBN 0–631–19981–0 (pbk. : acid-free paper)
    1. Nature—Effect of human beings on—Case studies.
I. Alexander, David (David E.)   II. Title.   III. Series.
GF75.G685   1997
304.2'8—dc21   96–47638
                                                                                CIP

Printed in Great Britain by T. J. International Limited, Padstow, Cornwall

This book is printed on acid-free paper

THE HUMAN IMPACT READER

0

Blackwell Readers on the Natural Environment

Published
The Human Impact Reader
Edited by Andrew Goudie

The Environmental Management Reader
Edited by Lewis Owen and Tim Unwin

CONTENTS

Contributors

D. W. Anderson, Dept. of Wildlife Ecology, University of Wisconsin, Madison, Wisconsin, USA

M. O. Andreae, Depts of Atmospheric Chemistry and Biogeochemistry, Max Planck Institute for Chemistry, PO Box 3060, D-6500, Mainz, Germany

P. G. Appleby, Dept. of Applied Mathematics and Theoretical Physics, University of Liverpool, Liverpool L69 3BX, UK

M. A. Bari, Water Authority of Western Australia, PO Box 100, Leaderville, WA 6007, Australia

R. W. Battarbee, Palaeoecology Research Unit, University College, 26 Bedford Way, London WC1H 0AP, UK

K. Benson-Evans, Dept. of Plant Science, University College, PO Box 78, Cardiff CF1 1XL, UK

A. Binns, School of African and Asian Studies, University of Sussex, Falmer, Brighton, Sussex BN1 9QN, UK

J. Boardman, School of Geography, University of Oxford, Mansfield Road, Oxford OX1 3TB, UK

M. A. Bolshov, Institute of Spectroscopy, Russian Academy of Sciences, Troitzk, 142092, Moscow Region, Russia

C. F. Boutron, Laboratoire de Glaciologie et Géophysique de L'Environnement du CNRS, 56 rue Molière, Domaine Universitaire, BP 96, 38402 St Martin d'Hères Cedex, France

F. H. Bormann, Yale University, School of Forestry and Environmental Studies, 370 Prospect Street, New Haven, CT 06511, USA

T. P. Burt, Dept. of Geography, University of Durham, Durham City, DH1 3LE, UK

J. -P. Candelone, Laboratoire de Glaciologie et Géophysique de L'Environnement du CNRS, 56 rue Molière, Domaine Universitaire, BP 96, 38402 St Martin d'Hères Cedex, France

L. Carbognin, Division of Geology, Institute for the Study of the Dynamics of Large Masses, National Research Council, Venice, Italy

F. M. Chambers, Dept. of Geography and Geology, Cheltenham and Gloucester College of Higher Education, Cheltenham, GL50 4AZ, UK

J. M. Coleman, Coastal Studies Institute, Louisiana State University, Baton Rouge, Louisiana, USA

P. J. Crutzen, Depts of Atmospheric Chemistry and Biogeochemistry, Max Planck Institute for Chemistry, PO Box 3060, D-6500, Mainz, Germany

R. J. Delmas, Laboratoire de Glaciologie et Géophysique de L'Environnement du CNRS, 56 rue Molière, Domaine Universitaire, BP 96, 38402 St Martin d'Hères Cedex, France

A. Demitrack, 17 Dunnder Drive, Summit, New Jersey 07901, USA

A. P. Dobson, Dept. of Biology, Princeton University, Princeton, New Jersey, USA

R. Evans, Dept. of Geography, University of Cambridge, Cambridge CB2 3EN, UK

J. C. Farman, British Antarctic Survey, European Ozone Research Coordinating Unit, Madingley Road, Cambridge CB3 0ET, UK

R. J. Flower, Palaeoecology Research Unit, University College, 26 Bedford Way, London WC1H 0AP, UK

W. J. Freeland, Conservation Commission of the Northern Territory, PO Box 496,

Palmerston, Northern Territory, 0831, Australia

B. G. Gardiner, British Antarctic Survey, Natural Environment Research Council, High Cross, Madingley Road, Cambridge CB3 0ET, UK

U. Görlach, Laboratoire de Glaciologie et Géophysique de L'Environnement du CNRS, 56 rue Molière, Domaine Universitaire, BP 96, 38402 St Martin d'Hères Cedex, France

A. Grainger, School of Geography, University of Leeds, Leeds LS2 9JF, UK

J. Hansen, NASA Institute for Space Studies, Goddard Space Flight Center, New York, NY 10025, USA

R. Harriman, Freshwater Fisheries Laboratory, Pitlochry, Perthshire PH16 5LB, UK

N. E. Haycock, Quest Environmental, PO Box 45, Harpenden, Herts AL5 5JL, UK

J. J. Hickey, Dept. of Wildlife Ecology, University of Wisconsin, Madison, Wisconsin, USA

D. Johnson, NASA Institute for Space Studies, Goddard Space Flight Center, New York, NY 10025, USA

R. Jones, Dept. of Biology, Grant University, Peterborough, Ontario K9J 7B8, Canada

V. J. Jones, Dept. of Geography, University of Newcastle-u-Tyne, Newcastle-u-Tyne, NE1 7RU, UK

M. A. K. Khalil, Portland State University, Dept. of Physics, Portland, Oregon 97207, USA

A. Lacis, NASA Institute for Space Studies, Goddard Space Flight Center, New York, NY 10025, USA

S. Lebedeff, NASA Institute for Space Studies, Goddard Space Flight Center, New York, NY 10025, USA

P. Lee, NASA Institute for Space Studies, Goddard Space Flight Center, New York, NY 10025, USA

G. E. Likens, Dept. of Biological Sciences, Dartmouth College, Hanover, New Hampshire 03755, USA

S. W. Lund, US Army Corps of Engineers, Wilmington, North Carolina, USA

C. H. McLellan, Centre for Environmental Technology, Imperial College, Exhibition Rd, London SW7, UK

R. B. Meade, Dept. of Civil Engineering, Virginia Military Institute, Lexington, VA 24450-0304, USA

P. P. Micklin, Dept. of Geography, Western Michigan University, Kalamazoo, MI 49008, USA

N. J. Middleton, School of Geography, University of Oxford, Oxford OX1 3TB, UK

N. Myers, Upper Meadow, Old Road, Headington, Oxford OX3 8SZ, UK

C. A. Nobre, University of São Paulo, DCA, IAY, Cloade Univ, Rua Matao 1226, BR-05508000 São Paulo, Brazil

A. J. Peck, Division of Land Resources Management, CSIRO, Private Bag, PO, Wembley WA 6014, Australia

R. S. Pierce, Northeastern Forest Experimental Station, US Forest Service, Durham, New Hampshire 03755, USA

L. F. Pitelka, Electric Power Research Institute, PO Box 10412, Palo Alto, California, 94303, USA

R. A. Rasmussen, Institute of Atmospheric Sciences, Oregon Graduate Center of Science and Technology, 19600 N. W. Von Neumann Dr., Beaverton, Oregon 97006, USA

D. J. Raynal, College of Environmental Science and Forestry, State University of New York, Syracuse, New York 13210, USA

D. Rind, NASA Institute for Space Studies, Goddard Space Flight Center, New York, NY 10025, USA

B. Rippey, Limnology Laboratory, University of Ulster, Traad Point, Magherafeltll, Co Derry, BT 45 6LR, UK

H. H. Roberts, Coastal Studies Institute, Louisiana State University, Baton Rouge, Louisiana, USA

F. S. Rowland, University of California Irvine, Dept. of Chemistry, Irvine, CA 92697, USA

G. Russell, NASA Institute for Space Studies, Goddard Space Flight Center, New York, NY 10025, USA

E. Salati, University of São Paulo, Esalq, Dept. FIS Meteorol, C Postal 9, BR-13400, Piracicaba, Sp. Brazil

B. Salvat, University of Perpignan, Ephe, CNRS, URA 1453, F-66860 Perpignan, France

A. P. Schick, The Hebrew University, Jerusalem, Israel

N. J. Schofield, Water Authority of Western Australia, PO Box 100, Leaderville, WA 6007, Australia

J. D. Shanklin, British Antarctic Survey, Natural Environment Research Council, High Cross, Madingley Road, Cambridge CB3 0ET, UK

H. Sioli, Max-Planck Institut für Limnologie, Postfach 165, N-2320 PLON, Germany

A. C. Stevenson, Dept. of Geography, University of Newcastle-u-Tyne, Newcastle-u-Tyne, NE1 7RU, UK

D. S. G. Thomas, Dept. of Geography, University of Sheffield, Sheffield S10 2TN, UK

C. Tickell, Green College, Woodstock Road, Oxford OX2 6HG, UK

S. W. Trimble, Dept. of Geography, University of California, Los Angeles, California 90024, USA

R. S. Tye, Dept. of Marine Sciences, Louisiana State University, Baton Rouge, Louisiana, USA

T. H. Van Andel, Dept. of Earth Sciences, University of Cambridge, Cambridge CB2 3EQ, UK

H. J. Walker, Dept. of Geography and Anthropology, Louisiana State University, Baton Rouge, LA 70803, USA

T. M. L. Wigley, University Corporation for Atmospheric Research, National Center for Atmospheric Research, Boulder, Colorado, USA

D. S. Wilcove, Environmental Def. Fund, Washington D.C. 20009, USA

M. G. Wolman, The Johns Hopkins University, Baltimore, Maryland, USA

D. H. Yaalon, Hebrew University of Jerusalem, Institute of Earth Sciences, Jerusalem 91904, Israel

B. Yaron, Aro, Inst. Soils and Water, Dept. of Soil Organization and Residues Chem., Bet Dagan 50250, Israel

E. Zangger, Dept. of Earth Sciences, University of Cambridge, Cambridge CB2 3EQ, UK

PREFACE: THE HUMAN IMPACT – A DEVELOPING LITERATURE

Although various individual writers had discussed some of the possible ways in which humans could transform nature and landscapes it was not until 1864 that a wide-ranging synthesis appeared. This was George Perkins Marsh's *Man and Nature* – 'a prodigiously learned discussion of the modifications of the flora and fauna and of the destruction wrought on forests, waters, soils and sands' (Kates et al., 1990, p. 3).

The following extract illustrates the breadth of his interests and the ramifying connections he identified between human actions and environmental changes:

Vast forests have disappeared from mountain spurs and ridges; the vegetable earth accumulated beneath the trees by the decay of leaves and fallen trunks, the soil of the alpine pastures which skirted and indented the woods, and the mould of the upland fields, are washed away; meadows, once fertilised by irrigation, are waste and unproductive, because the cisterns and reservoirs that supplied the ancient canals are broken, or the springs that fed them dried up; rivers famous in history and song have shrunk to humble brooklets; the willows that ornamented and protected the banks of lesser water-courses are gone, and the rivulets have ceased to exist as perennial currents, because the little water that finds its way into their old channels is evaporated by the droughts of summer, or absorbed by the parched earth, before it reaches the lowlands; the beds of the brooks have

widened into broad expanses of pebbles and gravel, over which, though in the hot season passed dryshod, in winter sealike torrents thunder; the entrances of navigable streams are obstructed by sandbars and harbors, once marts of an extensive commerce, and are shoaled by the deposits of the rivers at whose mouths they lie; the elevation of the beds of estuaries, and the consequently diminished velocity of the streams which flow into them, have converted thousands of leagues of shallow sea and fertile lowland into unproductive and miasmatic morasses. (Marsh, 1965, p. 9)

More than a third of the book is concerned with 'the woods'; Marsh does not touch upon important themes like the modifications of mid-latitude grasslands, and he is much concerned with western civilization. Nevertheless, employing an eloquent style and copious footnotes, Marsh, the versatile Vermonter, stands as a landmark in the study of environment.

Marsh, however, was not totally pessimistic about the future role of humankind or entirely unimpressed by positive human achievements (1965, pp. 43–4).

New forests have been planted; inundations of flowing streams restrained by heavy walls of masonry and other constructions; torrents compelled to aid, by depositing the slime with which they are charged, in filling up lowlands, and raising the level of morasses which their own overflows had created; ground submerged by the encroachment of the ocean, or exposed to be covered by its tides, has been rescued from its dominion by diking; swamps and even lakes have been drained, and their beds brought within the domain of agricultural industry; drifting coast dunes have been checked and made productive by plantation; seas and inland waters have been repeopled with fish, and even the sands of the Sahara have been fertilised by artesian fountains. These achievements are far more glorious than the proudest triumphs of war . . .

Marsh, as Kates et al. (1990, p. 3) point out, straddled two of the great environmental philosophies of his time – the Arcadian/ Romantic notion of an harmonious nature and the Victorian notion of humans above nature.

In subsequent decades there were other important statements about human influences though many geographers were concerned with how environment influenced humans rather than vice versa. The concept of 'Geographical Determinism' was a central part of Geography. None the less, important figures included Reclus, Woeikof, Brunhes, and Friedrich in Europe (see Goudie, 1993, pp. 4–6) and Sauer in the United States. Sauer, in the 1920s and 1930s, reacted against any naïve Geographical Determinism. He decried the destructive occupation of the soil associated with colonial expansion, reintroduced Marsh to a wide public, recognized the ecological virtues of some allegedly 'primitive' peoples, concerned himself with the great theme of plant and animal domestication, concentrated on the landscape changes that resulted from human action, and gave clear and far-sighted warnings about the need for conservation (Sauer, 1938; Speth, 1977). Indeed many historical geographers, inspired by Sauer, recognized the importance of the evolution of the cultural landscape through such processes as the clearing of the woodland (Darby, 1956), the draining of wetlands (Williams, 1970) and the introduction of alien plants and animals (McKnight, 1959).

In 1956 these and other themes were explored in detail in a major volume arising from a symposium held in Princeton the previous year. The volume, edited by W. L. Thomas, Jr., was called *Man's Role in Changing the Face of the Earth*. Its impact is difficult to assess, for although it contained many brilliant individual contributions it did not really lead into a new phase of investigation into humanly induced environmental change. As Kates et al. (1990, p. 4) write:

In identifying the human forces that had changed the states or faces of the biosphere, *Man's Role* drew upon a much broader store of knowledge than did its predecessor, although systematic documentation remained sparse. Papers dealt at length or in passing with various components of the biosphere: population, land use, water management, climate, marine and terrestrial biota, forests, carbon and other elements, earth materials, water quality, and nuclear radiation. Several aspects of society responsible for changes in these were discussed directly or indirectly, especially technology, urbanization, and trade.

The impact of *Man's Role* is difficult to assess. The volume did not directly herald a new generation of environmental research, even among scholars within the two main disciplines involved, anthropology and geography. *Man's Role* seems at least to have anticipated the ecological movement of the 1960s, although direct links between the two have not been demonstrated. Its dispassionate, academic approach was certainly foreign to the style of that movement, a portion of which Sauer apparently labelled the ecological binge.

It was from about 1970 that the human impact on the natural environment became a central concern for many disciplines and many scientists. Words like 'ecology' and 'environment' became common currency, as later did 'sustainability'. It became increasingly clear that humans not only caused change at the local level (e.g. by accelerating rates of erosion in an area that had been stripped of forest) but also promoted change at the global level. Themes like biological magnification of pesticides through the food chain became of common concern following Rachel Carson's *Silent Spring* (1963).

Trans-boundary acid-rain pollution was seen to be degrading large spreads of forest. Tropical deforestation was seen to be threatening the richness or 'biodiversity' of earth's flora and fauna. A hole was identified in the protective layer of stratospheric ozone over Antarctica. The alleged spread of the world's deserts through 'desertification' became a central concern of various United Nations agencies. The view was put forward, and became the subject of a substantial level of consensus, that the world could be facing rapid and substantial warming because of the 'greenhouse effect'. In many parts of the world fish stocks were seen to be anything but finite. People questioned the consequences of intensive agriculture and recognized that large engineering schemes, such as the building of major dams and the diversion of rivers, could spawn tragedies like the desiccation of the Aral Sea.

These concerns have generated a huge literature in the form of new journals, new books, directories of environmental data, countless papers in scientific journals, and seemingly endless quantities of consultants' reports and government documents. Some of the more valuable journals are listed in table 0.1, some of the more valuable books in table 0.2, and some of the more accessible directories of environmental data in table 0.3.

The purpose of this volume of readings is to bring together a selection of papers that have appeared either as individual chapters in books or as substantive or review articles in scientific journals. The great majority of them have appeared over the last three decades and some of them are included not because they give the most up-to-date account of a particular phenomenon, but because in their time they were exceptionally influential, innovative, informative or important. The papers are grouped, not always easily, into a series of subject areas: geomorphological and surface impacts; soil impacts; water impacts; climatic and atmospheric impacts; and biospheric impacts. Each group is accompanied by a brief introduction and commentary, and references are provided to very recent literature that may serve to place some of the earlier papers in their context.

References

Carson, R. 1963, *Silent Spring*. London: Hamish Hamilton.

Darby, H. C. 1956, The clearing of the woodland in Europe. In W. L. Thomas (ed.), *Man's Role in Changing the Face of the Earth*. Chicago: University of Chicago Press, pp. 183–216.

Goudie, A. S. 1993, *The Human Impact on the Natural Environment* (4th edition). Oxford: Blackwell Publishers.

Kates, R. W., Turner, B. L. I., and Clark, W. C. 1990. The great transformation . In B. L. Turner, W. C. Clark, R W. Kates, J. F.

Table 0.1   Some key journals relating to the human impact on the environment.

Name of Journal	Publisher
Ambio	Royal Swedish Academy of Sciences
Biological Conservation	Elsevier
Environmental Conservation	Elsevier Sequoia
Global Environmental Change	Butterworth-Heinemann
Land Degradation and Rehabilitation	Wiley
Environmental Management	Springer Verlag
Science of the Total Environment	Elsevier
Journal of Environmental Quality	American Society of Agronomy Crop Science Society of America Soil Science Society of America

Table 0.2 Some key recent textbooks relating to general aspects of the human impact on the environment.

Author(s)	Title	Date	Publisher
Goudie, A.	*The Human Impact on the Natural Environment* (4th edition)	1993	Blackwell Publishers

A general textbook which has a similar structure to, and should be used in conjunction with, this book of readings.

| Goudie, A.S. and Viles, H.A. | *The Earth Transformed* | 1997 | Blackwell Publishers |

A new highly illustrated, wide-ranging survey of human activities and their environmental consequences. It is particularly rich in case studies from many parts of the world.

| Johnson, D.L. and Lewis, L.A. | *Land Degradation: Creation and Destruction* | 1995 | Blackwell Publishers |

A broadly-based study of intentional and unintentional causes of many aspects of land degradation.

| Kemp, D.O. | *Global Environmental Issues: A Climatological Approach* (2nd edition) | 1994 | Routledge |

An introductory and clear look at climatological and atmospheric changes. It is especially strong on matters like global warming and ozone depletion.

| Mackenzie, F.T. and Mackenzie, J.A. | *Our Changing Planet* | 1995 | Prentice Hall |

This book considers both natural and human-induced global change and approaches it with a strong systems flavour.

| Mannion, A.M. | *Global Environmental Change* | 1991 | Longman |

A geographical approach to understanding global change at all time scales. It considers natural and anthropogenic changes.

| Mannion, A.M. | *Agriculture and Environmental Change* | 1995 | Wiley |

Agriculture is a dominating cause of environmental changes in many parts of the world. This is a comprehensive and well-referenced survey.

| Mannion, A.M and Bowlby, S.R. (eds) | *Environmental Issues in the 1990s* | 1992 | Wiley |

An edited series of review essays, some of them better than others, which enables one to appreciate the diversity of environmental issues we face today.

| Meyer, W.B. and Turner, B.L. (eds) | *Changes in Land Use and Land Cover: A Global Perspective* | 1994 | Cambridge University Press |

An extremely useful series of edited essays that looks at the causes, trends and consequences of land-use and land-cover changes. It tackles issues like afforestation, deforestation and grasslands.

Author(s)	Title	Date	Publisher
Meyer, W.B.	*Human Impact on the Earth*	1996	Cambridge University Press

A wonderful point-of-entry to the literature. This book brims over with thought-provoking epigrams.

Middleton, N.	*The Global Casino: An Introduction to Environmental Issues*	1995	Edward Arnold

An issues-based approach to environmental change which outlines both the workings of the physical environment and the political, economic and social frameworks in which the issues occur.

Nisbet, E.G.	*Leaving Eden: To Protect and Manage the Earth*	1991	Cambridge University Press

A book that not only reviews the causes of environmental problems but also provides a discussion of some of the ways in which the environment can be protected and restored.

O'Riordan, T. (ed.)	*Environmental Science for Environmental Management*	1995	Longman Scientific

This volume concentrates not so much on environmental issues *per se* but on the ways in which the environment can be protected and restored.

Roberts, N. (ed.)	*The Changing Global Environment*	1994	Blackwell Publishers

A particularly useful series of edited essays because of its low-latitude concerns and its case studies.

Simmons, I.G.	*Changing the Face of the Earth*	1989	Blackwell Publishers

An historical tour de force of the way in which through time and with different levels of technology humans have transformed their environment.

Simmons, I.G.	*Environmental History: A Concise Introduction*	1993	Blackwell

A delightful 'history of the world in only five chapters', provides an introduction to the long history of the interactions between nature and human culture.

Tolba, M.K. and El-Kholy, O.A.	*The World Environment 1972–1992*	1992	Chapman & Hall

A compendium of environmental and social trends since the 1970s, with a useful commentary.

Turner, B.L. et al. (eds)	*The Earth as Transformed by Human Action*	1990	Cambridge University Press

A great international compendium of tremendous significance that considers the main changes that have taken place in population and society, the transformations that have been achieved on land, water, ocean, atmosphere, the biota and chemical cycles. It has a well-judged series of regional case studies.

Table 0.3   Some important annual directories of environmental data and trends.

Name	Publisher
The Green Globe Yearbook	Fridtjof Nansen Institute with Oxford University Press
Vital Signs	Worldwatch Institute with Earthscan Publications
Environmental Data Report	United Nations Environment Programme with Blackwell Reference
World Resources	World Resources Institute, United Nations Environment Programme, United Nations Development Programme with Oxford University Press
State of the World	Worldwatch Institute with W.W. Norton

Richards, J. T. Matthews and W. B. Meyer (eds), *The Earth as Transformed by Human Action*. Cambridge: Cambridge University Press, pp. 1–17.

McKnight, T. L. 1959, The feral horse in Anglo-America. *Geographical Review* 49:506–25.

Sauer, C. O. 1938, Destructive exploitation in modern colonial expansion. *International Geographical Congress, Amsterdam*, vol. III, sect. IIIC, 494–9.

Speth, W. W. 1977, Carl Ortwin Sauer on destructive exploitation. *Biological Conservation* 11: 145–60.

Thomas, W. L. (ed.) 1956, *Man's Role in Changing the Face of the Earth*. Chicago: University of Chicago Press.

Williams, M. 1970, *The Draining of the Somerset Levels*. Cambridge: Cambridge University Press.

ACKNOWLEDGEMENTS

I am grateful to the authors and publishers of the papers used in this collection for their permission to reproduce their material. I would also like to thank Dr Adrian Parker for assisting in the task of collecting the papers, and Amanda Wallace and Jan Magee for typing some of the material.

Andrew Goudie

The editor and publishers wish to thank the following for permission to use copyright material:

Academic Press Ltd, London, for Thomas, D.S.G. and Middleton, N.J. (1993) 'Salinization: new perspectives on a major Desertification issue', *Journal of Arid Environments*, 24, 95–105;

American Association for the Advancement of Science for Hickey, J.J. and Anderson D.W. (1968) 'Chlorinated hydrocarbons and eggshell changes in raptorial and fish-eating birds', *Science*, 162, 271–2. Copyright © 1968 American Association for the Advancement of Science; Bormann, F.H., Likens, G.E., Fisher, D.W. and Pierce, R.S. (1968) 'Nutrient loss accelerated by clear-cutting of a forest ecosystem', *Science*, 158, 882–4. Copyright © 1968 American Association for the Advancement of Science; Landsberg, H.E. (1970) 'Man-made climatic changes', *Science*, 170, 1265–8. Copyright © American Association for the Advancement of Science; and Micklin, P.P. (1988) 'Desiccation of the Aral Sea: a water management disaster in the Soviet Union', *Science*, 241, 1170–5. Copyright © 1988 American Association for the Advancement of Science; Hansen, J., Johnson, D., Lacis, A., Lebedeff, S., Lee, P., Rind, D. and Russell, G. (1981) 'Climatic impact of increasing atmospheric carbon dioxide', *Science*, 213, 957–66. Copyright © 1981 American Association for the Advancement of Science; Crutzen, P. J. and Andreae, M. O. (1990) 'Biomass burning in the tropics: impacts on atmospheric chemistry and biogeochemical cycles', *Science*, 250, 1669–78. Copyright © 1990 American Association for the Advancement of Science; and Likens, G. E. and Bormann, F. H. (1974) 'Acid rain: a serious regional environmental problem', *Science*, 184, 1176–9. Copyright © 1974 American Association for the Advancement of Science;

American Geophysical Union for Wolman, M.G. and Schick, A.P. (1967) 'Effects of construction on fluvial sediment, urban and suburban areas of Maryland', *Water Resources Research*, 3, 451–64. Copyright © by the American Geophysical Union;

Blackwell Science Ltd for Freeland, W.J. (1990) 'Large herbivorous mammals: exotic species in northern Australia', *Journal of Biogeography*, 17, 445–9;

CAB International for Boardman, J. and Evans, R. (1994) 'Soil erosion in Britain: a review' in *Conserving Soil Resources: European Perspectives*, ed., R.J. Rickson, 3–10;

Chapman & Hall for Myers, N. (1990) 'The biodiversity challenge: expanded hot-spots analysis', *The Environmentalist*, 10(4), 243–56;

CSIRO for Peck, A.J. (1978) 'Salinization of non-irrigated soils and associated streams: a review', *Australian Journal of Soil Research*, 16, 157–68;

The Ecological Society of America for Pitelka, L.F. and Raynal, D.J. (1989) 'Forest decline and acidic deposition', *Ecology*, 70, 2–10;

Elsevier Science Ltd for Salvat, B. (1992) 'Coral reefs – a challenging ecosystem for human societies', *Global Environmental Change*, 2, 12–18; Walker, H.J. (1984) 'Man's impact on shorelines and nearshore environments: a geomorphological perspective', *Geoforum*, 15, 395–417; and Khalil, M.A.K. and Rasmussen, R.A. (1987) 'Atmospheric methane: trends over the last 10,000 years', *Atmospheric Environment*, 21, 2445–52;

Elsevier Science NL for Meade, R.B. (1991) 'Reservoirs and earthquakes', *Engineering Geology*, 30, 245–62; and Bari, M.A. and Schofield, N.J. (1992) 'Lowering of a shallow, saline water table by extensive eucalypt reforestation', *Journal of Hydrology*, 133, 273–91;

The Geographical Association for Binns, T. (1990) 'Is desertification a myth?', *Geography*, 75, 106–13; Hollis, G.E. (1988) 'Rain, roads, roofs and runoff: hydrology in cities', *Geography*, 73, 9–18; and Burt, T.P. and Haycock, N.E. (1990) 'Farming and the nitrate problem', *Geography*, 76, 60–3;

Geografiska Annaler for Walker, H.J., Coleman, J.M., Roberts, H.H. and Tye, R.S. (1987) 'Wetland loss in Louisiana', *Geografiska Annaler*, 69A, 189–200;

Journal of Field Archaeology and the Trustees of Boston University for Van Andel, T.H., Zangger, E. and Demitrack, A. (1990) 'Land use and soil erosion in prehistorical and historical Greece', *Journal of Field Archaeology*, 17, 379–96;

Kluwer Academic Publishers for Salati, E. and Nobre, C.A. (1991) 'Possible climatic impacts of tropical deforestation', *Climatic Change*, 9, 177–96;

Nature for Boutron, C.F., Görlach, U., Candelone, J.-P., Bolshov, M.A. and Delmas, R.J. (1991) 'Decrease in anthropogenic lead, cadmium and zinc in Greenland snows since the late 1960s', *Nature*, 353, 153–6. Copyright © 1991 Macmillan Magazines Ltd; Battarbee, R.W., Flower, R.J., Stevenson, A.C. and Rippey, B. (1985) 'Lake acidification in Galloway: a palaeoecological test of competing hypotheses', *Nature*, 314, 350–2. Copyright © 1985 Macmillan Magazines Ltd; Battarbee, R.W., Flower, R.J., Stevenson, A.C., Jones, V.J., Harriman, R. and Appleby, P.G. (1988) 'Diatom and chemical evidence for reversibility of acidification of Scottish lochs', *Nature*, 332, 530–2. Copyright © 1988 Macmillan Magazines Ltd; Wigley, T.M.L. (1989) 'Possible climate change due to sulphur dioxide-derived cloud condensation nuclei', *Nature*, 339, 365–7. Copyright © 1989 Macmillan Magazines Ltd; and Farman, J.C., Gardiner, B.G. and Shanklin, J.D. (1985) 'Large losses of total ozone in Antarctica reveal seasonal ClO_x/NO_x interactions', *Nature*, 315, 207–10. Copyright © 1985 Macmillan Magazines Ltd;

Royal Geographical Society for Sioli, H. (1985) 'The effects of deforestation in Amazonia', *Geographical Journal*, 151, 197–203; Grainger, A. (1993) 'Rates of deforestation in the humid tropics: estimates and measurements', *Geographical Journal*, 159, 33–44; and Tickell, C. (1993) 'The human species: a suicidal success?', *Geographical Journal*, 159, 219–26;

Scientific Research Society for Rowland, F.S. (1989) 'Chlorofluorocarbons and the depletion of stratospheric ozone', *American Scientist*, 77, 36–45;

Soil and Water Conservation Society for Trimble, S.W. and Lund, S.W. (1982) 'Soil conservation in the Coon Creek Basin, Wisconsin', *Journal of Soil and Water Conservation*, 37, 355–6;

UNESCO Publishing for Carbognin, L. (1985) 'Land subsidence: a worldwide environmental hazard', *Nature and Resources*, 21, 2–12. Coypright © 1985 UNESCO;

Waverly for Yaalon, D.H. and Yaron, B. (1996) 'Framework for man-made soil changes – an outline of metapedogenesis', *Soil Science*, 102, 272–7. Copyright © Williams & Wilkins;

John Wiley & Sons Ltd for Jones, R., Benson-Evans, K. and Chambers, F.M. (1985)

'Human influence upon sedimentation in Llangorse Lake, Wales', *Earth Surface Processes and Landforms*, 10, 227–35;

Andrew Dobson for D.S. Wilcove, McLellan, C.H. and Dobson, A.P. (1986) 'Habitat fragmentation in the temperate zone' in *Conservation Biology: The Science of Scarcity and Diversity*, ed. M.E. Soule, Sinuaer, pp. 237–56.

Every effort has been made to trace the copyright holders but if any have been inadvertently overlooked the publishers will be pleased to make the necessary arrangement at the first opportunity.

Dedicated to Heather

PART I

Geomorphological and Surface Impacts

Introduction

In the late eighteenth century, as Clarence Glacken related so well (Glacken, 1967), a number of scientists were intrigued by some of the consequences of deforestation in the Alps of Europe: accelerated flooding, erosion, sediment deposition and channel braiding. This was one of the first modern stages in the study of how humans could transform the geomorphological environment, an endeavour to which the term 'anthropogeomorphology' is often applied (Goudie, 1993). However, although individual geologists recognized the effects of tree felling and land drainage activities on the rate of operation of geomorphological processes, such actions were in general thought to be relatively insignificant in comparison with the power of natural forces exercised by volcanoes, great rivers, pounding breakers and the like.

A major change of emphasis came in the middle of the nineteenth century with the publication of George Perkins Marsh's *Man and Nature* (1864), a clarion call for the nascent conservation movement. Marsh pointed to the many ways in which nature was being transformed by human activities. In particular he pointed to the multifarious and ramifying consequences of the clearance of the woodlands for a cascade of geomorphological processes extending down from mountain slopes to estuaries, sandbars and coastal dunes.

Following Marsh there was a third stage of anthropogeomorphology in which a group of scientists (that included names like Shaler, McGee, Gilbert, Horton and Bennett) explored the impacts that European settlement and development had had on the landscape of North America. Other scientists considered the sometimes destructive effects of European colonial activities in Africa and elsewhere. However, with the lone exception of Sherlock's little monograph in the 1920s (Sherlock, 1922) there were few attempts to gain an overall view of the geomorphological power of humans until the environmentally conscious decades, starting in the 1970s, when a series of studies started to appear. Environmental conscientiousness was not, however, the only stimulus to this phase of work, which is exemplified, inter alia, by the works of Gregory and Walling (1979), Goudie (1981) and Nir (1983). As the relative importance of denudation chronology, which was concerned with reconstructing the erosional history of landscapes over long (pre-human) time-spans, became less central to geomorphology, and as geomorphologists concerned themselves with shorter (human) time-scales and operation of processes, so the influence of human actions became more evident.

The chapters in Part I are selected because they provide a range of reviews of geomorphological impacts which cover a range of processes and environments: the world's coastlines (chapter 1), which are modified by the very large proportion of the Earth's population that lives in close proximity to the sea; the acceleration of ground subsidence because of such processes as fluid extraction and mineral mining (chapter 2); and the acceleration of seismic activity (earthquakes) brought about by the construction of increasingly large reservoirs (chapter 3).

There is also a small second section of chapters which are included to demonstrate that human modifications of landforms and land-forming processes has a history and that some of the undesirable effects of human activities can be dealt with through careful management. Lake cores are a particularly fine way of reconstructing long- and medium-term changes in land use, erosion in catchments, sedimentation in basins and water chemistry (see also chapters 17 and 18). Many studies could have been selected, but the one that has been included, from central Wales (chapter 4), combines, succinctly and concisely, a description of methods and results. Old maps and air photos are another potent source of information on geomorphological change and these, together with field surveys and the location of old human artefacts (fences, dams, bridges, etc.), have been used to good effect over many years by Trimble and his co-workers to reconstruct the history of erosion and sedimentation in various parts of the United States in response to changes in land management (chapter 5).

Because of the intimate relationships between geomorphological processes and forms and the other components of the environment (e.g. climate, hydrology, vegetation and soil) there are various chapters elsewhere in this volume (e.g. chapters 10, 13 and 32) that serve to illustrate further the role that humans have played in this area.

Finally, it is worth mentioning that in the late 1980s, a further stage of anthropogeomorphology developed as the significance of global change and particularly of global warming became appreciated (see Goudie, 1989 for an early review). Geomorphologists have been active in contributing to the deliberations of the Intergovernmental Panel on Climate Change (IPCC). This is likely to be a major area for research into the twenty-first century and will serve to sharpen the predictive capabilities of geomorphologists. It is, however, a minefield of uncertainties and controversies (see also the introduction to Part VI).

References

Glacken, C. J. 1967, *Traces on the Rhodian Shore: Nature and Culture in Western Thought from Ancient Times to the end of the Eighteenth Century*. Berkeley: University of California Press.

Goudie, A. S. 1981, *The Human Impact* (1st edition). Oxford: Basil Blackwell.

Goudie, A. S. 1989, The global geomorphological future. *Zeitschrift für Geomorphologie Supplementband*, 79: 51–62.

Goudie, A. S. 1993, Human influence in geomorphology. *Geomorphology* 7:37–59.

Gregory, K. J., and Walling, D. (eds) 1979, *Man and Environmental Processes*. Folkestone: Dawson.

Marsh, G. P. 1864, *Man and Nature*. New York: Scribner.

Nir, D. 1983, *Man, a Geomorphological Agent, an Introduction to Anthropogenic Geomorphology*. Jerusalem: Keter.

Sherlock, R. L. 1922, *Man as a Geological Agent*. London: Witherby.

1

MAN'S IMPACT ON SHORELINES AND NEARSHORE ENVIRONMENTS: A GEOMORPHOLOGICAL PERSPECTIVE*

H. J. Walker

Introduction

The first shorelines on Earth came into existence when land and sea became distinct entities and, from that day, they have been among the most changeable of surface features. Like all earthly forms, shorelines and their bordering environments are reflections of the materials from which they have been made, and of the processes to which they have been subjected. Through time, they have undergone continuous change in length, position and form, while at the same time the processes involved in their evolution have increased in number and varied in relative importance.

The contemporary 450,000-km long shoreline is more than twice the length it was when the last major round of rifting began 200 million years ago, and its position relative to sea-level on a world-wide basis is about 150 m higher than it was only 18,000 years ago, when the continental ice-sheets

began their last major retreat. Physical and chemical agents, which have been operative upon shore materials and forms from the first were joined by biological agents once life began to evolve. Man, as an agent of coastal modification, is only the most recent (albeit the most versatile) of these biological agents (Walker, 1978).

The Natural Shoreline

The gross form of the Earth's coasts is primarily determined by plate tectonics. The Earth's crust is composed of a number of relatively rigid but mobile plates, the boundaries of only some of which correspond with shorelines. Coasts that occur at the edges of plates tend to be hilly or mountainous and may be either continental [e.g. the coast of Western South America (destructive margin) or the Red Sea (constructive margin)] or insular (the Aleutian Island Arc); those embedded within plates are

* Originally published in *Geoforum*, 1984, vol. 15, pp. 395–417.

tectonically relatively quiet and vary from hilly plateau types (much of the African coast) to generally low-lying types, such as most of the eastern seaboard of the Americas (Inman and Nordstrom, 1971).

The tectonic activity associated with plate movement is also largely responsible for the arrangement and lithology of continental drainage basins and, therefore, along with climate and wave action, for the type and amount of sediments that reach any particular shoreline. Nearly two-thirds of the world's largest rivers carry sediment to the Atlantic Ocean and its marginal seas. Of the nearly 15 km^3 of sediment contributed by land to the ocean each year, less than 1% (0.12 km^3) is derived directly from shoreline cliff erosion (Kuenen, 1950). Most of the rest, some 14.2 km^3 per year, is carried to the coast by rivers and streams and especially by large rivers. The load of large rivers tends to be fine-textured; it has been estimated that 98% of the sediment that the Mississippi River delivers to the Gulf of Mexico is silt and clay. The limited amount of sand contributed to shorelines each year via short rivers (especially those draining the leading edge of plates) and by cliff erosion is augmented from offshore by wave action, and from onshore by wind transport, biotic production and chemical precipitation.

The present-day appearance of shorelines depends not only on the character of the primary tectonic forms and origin of contributed sediments, but also on the way they have been modified by erosion, transportation and deposition by a variety of processes. Especially important are the modifications caused by water movement. Waves, the main cause of coastal change, are aided in a number of ways by currents, tides, surges (storm, typhoon, tsunami. . .) and sea-level change. In addition, some coastal modification is caused by wind, ice, plants and animals, and a variety of chemical processes. Indeed, along certain coastal sectors these other processes may dominate.

Given such great variation in materials and processes, it should not be surprising that the types and distributions of shoreline forms are also great. Although variety is almost unlimited, there are nonetheless a number of basic shoreline types, including: cliffs, dunes, beaches and barriers, deltas, estuaries and lagoons, and those primarily biotic in composition.

Cliffs, predominant erosional, are a reflection of both subaerial and oceanic processes. They are likely to be steep when removal of material at the cliff base is rapid (either in relative or absolute terms) and gentle when slumping predominates. Many cliffed shorelines are bordered by platforms, although some plunge directly into the sea to well below low tide level and show little evidence of change since the sea-level reached its present position 5000–6000 years ago. Dunes, formed mainly by subaerial processes, are an integral, even if often fragile, part of many shore systems. They can exist only when there is an adequate supply of sand (usually the beach) and when winds are sufficient to transport the material inland. Today dunes are viewed as flexible barriers that serve both as a protector of the environments behind them and as a reserve of sand available for natural nourishment during beach attack.

The creation of depositional forms such as beaches, spits and barrier islands is only possible when material is available for their growth and maintenance; material that may be inorganic or organic and that may range in size from ooze to boulders. Most beaches are composed mainly of sand, although along some coasts shingle is common. Beach size and shape ranges from small arcuate pockets separated by headlands to linear forms that may stretch over 200 km. Barrier islands and spits differ from mainland beaches in that they have a double shoreline, one that faces the open ocean and another that faces a lagoon, bay or estuary. Some 12 or 13% of the mainland shorelines of the world are protected by such natural barriers (Fox, 1983).

Deltas, which occupy less than 2% of the world's coastline, vary in shape and size in relation to the quantity and type of sediment supplied by rivers and the ability of oceanic processes to rework and redistribute them (Coleman and Wright, 1971). The relative importance of river and ocean processes in deltaic development changes with delta modification; wave action, for example, varies as the subaqueous portion of the delta grows seaward. Deltas are composed of a number of sub environmental forms such as natural levees (both subaerial and subaqueous), distributary channels, lakes, marshes, beaches, spits and barriers (Russell, 1967) which, when combined, provide an areal mosaic that contrasts with most other major coastal forms.

Estuaries, lagoons and marshes comprise a group of entities, many of which owe their existence to sea-level rise and the relative stillstand that followed. Sea-level rise in the Post-glacial, or Flandrian, brought about submergence of the lower portion of many valleys fashioned by fluvial, glacial and tectonic processes, often extending into tributaries to create highly irregular shorelines such as that displayed by Chesapeake Bay. Since stillstand, deposition has predominated in such relatively protected coastal environments. Nonetheless, there have been great variations, e.g. estuarine fill has been relatively great in coastal plain estuaries and relatively small in most fjords. Along many shorelines sedimentation has resulted in the creation of the barrier islands backed by shallow lagoons normally connected by tidal inlets with the sea. In the process of filling, estuaries and lagoons are converted to mudflats, marshes and swamps, and thus display some of the characteristics common in deltas.

Organic reefs occur in intertidal and subtidal environments along many tropical and some temperate shorelines. Coral reefs dominate in the tropics, where they range from linear forms, such as the Great Barrier Reef, to the cirular shapes of some coral atolls. Non-coral reefs are formed by a variety of organisms including vermetid gastropods, red algae, oysters and polychaete worms. Organic reefs may help stabilize shorelines, reduce erosion and aid in sediment trapping (Jablonski, 1982).

There are also shorelines dominated by plants. Mangroves are common along tropical coasts, especially in areas protected from strong waves such as bays, lagoons and estuaries. Today one of the most serious problems along many tropical coasts is the utilization of the mangrove vegetation for charcoal, firewood, and lumber (UNESCO, 1979). Along protected shorelines in non-mangrove areas, saltwater marsh grasses frequently dominate. They, like mangroves, are indicative of sediment accumulation and an advancing shoreline and, like mangroves, are easily destroyed by human activity.

In addition to these major forms, shorelines display a vast range of small forms, including cliff-foot notches, ripple marks, mudlumps, washover fans in dunes, etc. Most are ephemeral forms and can be eliminated or altered by human impact. However, in this paper it is the impact on the major forms – cliffs, beaches, barriers, deltas, etc. – that is emphasized.

Although the forms described above are natural, they are anything but static. Slight changes in sediment source, weather patterns or oceanic circulation, for example, may result in major modifications in shoreline expression. It is precisely because of this 'fragile nature' of form and 'delicate balance' of process that the introduction of a new (potentially universal) agent – in this case, man – into the system can have such a distinctive impact.

Man – Discovery and Initial Utilization

Although we do not know the thoughts man may have had when he first encountered the sea shore, we do know that, on a geological timescale, it did not take him long to become a dominant factor in its transformation. Nonetheless, early in man's history, the most important effects may have been more ecological than morphological. Some sections of the coast were convenient as sources of food and material resources, and must have been capitalized upon early. Middens, dating from at least 70,000 BC (Volman, 1978), attest to some utilization. Although over-exploitation may have occurred, it is likely that permanent ecologic damage was rare (Steward, 1948).

Just how extensive these early shoreline shell middens were is unknown because most of the sites were subsequently destroyed, or at least buried, during sea-level rise. Somewhat better knowledge exists about midden formation since sea-level reached its present position. Along many shorelines they are numerous and attest to very active exploitation of shellfish during the past 3000–4000 years. For example, along the shoreline of San Francisco Bay over 400 middens have been documented (Hedgpeth, 1975). None-theless, our knowledge of their extent remains limited despite the fact that most should be conspicuous features, largely because they have been mined intensively for a variety of purposes including road construction.

Despite the fact that middens represent localized changes in shore material composition and the addition of new forms at the sea's edge, they can hardly be considered morphologically significant, at least from today's viewpoint (Walker, 1981b).

Such low-level morphologic activity prevailed until at least the last few millennia. Indeed, one might even go further, as suggested above, and write that, even today, many human shoreline activities have little more significance, morphologically, than those of earlier times. As the author once wrote elsewhere (Walker, 1981a):

Swimming in the surf, walking along the waterline, building sand castles, or surf fishing have little more effect on the beach than grunion "running" with the tide to spawn, turtles burying their eggs, or the seagulls dropping clams from the air to crack them for food.

Man – an Agent of Shoreline Modification

The thought above was concluded by noting that

once the swimmer is backed by parking lots, hot-dog stands, and condominiums, the walker is joined by offroad vehicles (ORVs), the sandcastle builder is replaced by a bulldozer, and the fisherman gives way to a mariculture enterprise, another dimension in shore utilization has appeared.

Shell middens are the product of human activity and may well represent the earliest anthropogenic land-forms along the shore-line. Simple as they may be, they nonetheless reflect humans as an agent capable of destruction (erosion), removal (trans-portation) and construction (deposition), the same three aspects of shoreline modi-fication that are critical today. In the case of the shell midden these 'geomorphological processes' were used to satisfy one objective: the acquisition of food by a collecting-gathering society. From such an elementary (instinctive?) beginning, man has developed into one of the most important exogenic agents in terms of the destruction, alteration and creation of shoreline landforms.

The nature of coastal change wrought by man depends on several variables, including: (1) the reasons for (objective of) change; (2) the techniques, equipment, and materials used in effecting change; (3) the amount of energy (money) expended in their change; (4) the level of scientific and engineering knowledge involved; and (5) whether the change is intentional or un-intentional and whether direct or indirect.

The reasons for human modification of shorelines and neighbouring environments are almost unlimited in number. Virtually every endeavor of modern man is involved to some extent, a fact that should not be too surprising when one notes that the contem-porary rate of population increase in the coastal zone of many countries is several times that occurring elsewhere. Shorelines are being modified in the name of esthetics, agriculture, commerce, energy production, fisheries, housing, industry, mariculture, mining, shoreline protection, recreation, transportation, waste disposal and so on.

Equally as varied are the intensities with which such modifications are applied, ranging from those on an individual basis to schemes of national (and international) scope. For example, in his State of the Union Address on 25 January 1984, President Reagan of the United States included the cleaning of the Chesapeake Bay as a project of national importance right alongside the establishment of a permanently manned space station.

Because of the degree of contrast among the human objectives listed above, the type, location and extent of impact varies greatly. Two examples suffice to illustrate this point. First, a shoreline modified for commercial purposes is likely to be located near a city whereas that for agriculture will normally be more remote and generally more extensive. Second, a protective structure placed at the base of an eroding cliff in front of a private house will usually be made of inexpensive materials (tyres, logs, country rock. . .); one around a nuclear-power plant will consist of an expensive concrete seawall faced with prefabricated armour units. These two 'protectors' not only vary in terms of the material used and form created, but also in their effect on the operation of natural processes.

Human modification may involve changes in material, process and form, singly or in combination. The alteration of one almost always affects the other two, a condition that applies to natural changes as well as those engineered by man. Theoretically, there are eight possible combinations, ranging between the purely natural in all three parameters to the purely artificial in all three. Today, most of the world's shoreline still exhibits natural materials, processes and forms, even though human-generated flotsam and jetsam are present along much of it. At the opposite extreme is the shoreline that has been eliminated; where what was formerly a shoreline is now replaced by what can, at best, be only considered a part of the 'nearshore' environment.

Shoreline Modification – Setting the Stage

Indirect and unintentional modification

The first major morphologic impact of man on the shoreline was indirect and unintentional and, for that matter, it was almost certainly the result of non-coastal activities by non-coastal peoples (Walker, 1981a). Although there is debate as to where agriculture originated (Solheim, 1972), there is no doubt that it, together with animal husbandry, were being practiced in the Near East, while the sea-level was still on the rise. Certain agricultural practices and overgrazing increased soil erosion, which led to increased amounts of sediment being transported to the Persian Gulf, thereby helping slow the rate by which the Gulf was deepened (today it averages only 31 m in depth) and accentuated the growth of the Tigris–Euphrates delta. Since stillstand was reached, the Tigris–Euphrates delta has advanced some 300 km into the Persian Gulf, at a rate of about 1 km every 15 years. With only slightly later starts, similar results occurred in Pakistan and China (especially at the changing mouths of the Hwang Ho) and subsequently elsewhere (Davis, 1956).

As in the case of shell middens, which are being built along some shorelines even today, the indirect unintentional increase in sediment load because of agriculture, animal husbandry and deforestation is still occurring. One might ask how much of the contemporary loads of such major rivers as the Yangtze and Mississippi are the result of human activity. And, for that matter, the same question can be posed for any river or stream that drains a basin that has been altered by man. The opposite effect, i.e. the reduction in contributed load because of dams and mining, is treated later.

Direct and intentional modification

Indirect unintentional activities led to the earliest direct and intentional modifications of the shorelines which manifested themselves in two major but highly different ways. The first was associated with the implementation of agricultural practices on the newly formed deltaic plains. Man must have realized early that diking and canalization could be used to advantage. Reclamation was born and, as a major coastal modifier, has continued to the present.

Intensification of agricultural practices made possible the development of permanent settlements which, in turn, led to occupational specialization, social and political organization, commercial development, and external trade. This trade, at least around the Mediterranean Sea, early became dependent upon the sea; navigators rapidly developed their art and engineers soon learned to protect harbors, build breakwaters and drain swamps. Thus, harbor construction was probably the second major intentional shoreline modification: archaeological evidence of such endeavors is plentiful.

As is the case today, some of the early engineering works had unanticipated results. Strabo (*c*.7 BC) noted that the harbor of Ephesus was "made narrow by engineers, but they, along with the king that ordered it, were deceived as to the result, . . . silt, made the whole of the harbor, . . . more shallow". Nonetheless, some of the feats of these early engineers were very effective. For example, the "Phoenicians developed a 'continuous self-flushing' harbor at Tyre" (Inman, 1974).

Shoreline Modification – Today's Manifestation

In order to categorize the impact of modern man on the shoreline and nearshore environments, the following discussion is divided into two parts; the first part deals with those activities and forms associated with major displacements of the shoreline, the second with those associated with its stabilization. Major displacement occurs when the shoreline is moved either seaward (as in land reclamation) or landward (as in water reclamation). Stabilization, as used here, includes not only the protection of the shoreline and adjacent environments, but also other localized shoreline modification.

Shoreline displacement

Land reclamation. As Chapman (1982b) writes: "in the coastal zone, reclamation usually signifies exclusion of marine or estuarine waters from littoral or riparian lands formerly periodically or permanently inundated". Exclusion of unwanted water can be accomplished by raising the level of the area to be reclaimed or by damming off the area and draining it. As one of the first conscious efforts of man to modify the shoreline, land reclamation often capitalized on the natural (and humanly aggravated) sediment growth at the mouths of rivers. By building dikes

(embankments, seawalls, dams, levees) sequentially as deposition occurred a number of agricultural societies kept pace with nature.

Utilizing natural processes during reclamation is practiced at the present time in South China. The people of the Pearl River delta, having observed the natural sequence of events in deltaic growth, describe it in the phrase: "yii you, lu po, huo li, cao pu" which means "fish swim, rudders touch bottom, cranes stand, grass spreads". They select those areas in which this sequence is especially rapid and attempt to accelerate the processes involved. In 1979 a long dike was constructed off the shore from Ai Nan near the mouth of the western-most branch of the Pearl River delta. This dike, part of a structure that will eventually encircle 51 km² of reclaimed land, was built so that it is submerged at high tide in order to trap sediment during ebbing flow. Trapping is to be allowed to continue until the late 1980s, when the dike will be raised and the remainder of the fill needed trucked in from a land source (Walker, 1980).

Although reclamation projects are older in Asia than elsewhere, it is in the Netherlands that reclamation has had its most dramatic development. The peoples of what is now the Low Countries lived on mounds (*terpen*) they built in the marshes prior to the time (about 1000 years ago) that they began to construct dikes for protection from the sea. Subsequently, diking and its companion activity, polderizing, made possible the utilization of vast areas, most of which are below sea-level. Van Veen (1962) noted that in the Netherlands over 3800 km² of land had been gained by these practices along the seashore since 1200 AD. In the process the shoreline, which has been moved gradually seaward, has been shortened to less than one-quarter of its original length.

Not all reclamation is directly involved with the open sea coast. Many lagoons, bays and estuaries have been dammed and reclaimed. In the Netherlands poldering of the Zuiderzee will, when completed, provide an additional 2250 km² of agriculture land; in Japan the largest lagoon, Hachirogata, was reclaimed by placing a

52-km long dike around it. Volker (1982) notes that the coastal polders in the Ganges delta occupy some 30,000 km² making them probably the largest in the world. However, calculations of the amount of 'created land' along coasts vary greatly; for example, one estimate places the total area impoldered in the world (including non-coastal polders) between 100,000 and 500,000 km² (Volker, 1982).

Most large reclamation projects are for agriculture and aquaculture, and are often remotely located. Reclamation for housing, industry and commerce is usually adjacent to an urban area. Expansion into the sea (at the head of a bay or in an estuary, lagoon or marsh) is often technologically relatively easy. The list of major cities that have added to their area in this fashion is almost endless and includes virtually all of those large cities located on coasts. Today over 40% of Japan's industry is located on reclaimed land. In Singapore, where reclamation has become almost a way of life, the shoreline has been displaced to such an extent that the country is now 10% larger than when founded (Wong, 1984).

Recent 'new' types of reclamation include those associated with the construction of airports and of offshore drilling platforms. Airports may be on fill peninsulas extending out from land (e.g. Hong Kong and Hawaii) or on completely separated islands (e.g. the new airport in Osaka Bay, Japan). Although offshore islands have for long been constructed in the Pacific Ocean near Long Beach, CA to support oil wells, the most recent islands are along the coast of the Beaufort Sea off Alaska, where sea ice action is of major significance.

All reclamation projects, from the smallest man-made island to the largest polder, involve not only the elimination of the natural shoreline, but the construction of a new, artificial one. It may be in the form of a short dam across the mouth of a lagoon or of a long dike across a mudflat. These structures, intended to protect the reclaimed area from flooding by the sea, are variable in terms of the construction techniques employed and materials used, and range from in-situ sediment bulldozed into

a dike to concrete-poured seawalls.

These barriers have either continuous or intermittent contact with water and thus affect (and are affected by) tides, nearshore waves and currents, and in turn rates of erosion, sediment transport and deposition. In Singapore "the velocity of currents has increased as a result of the straightening of the coastline" (Wong, 1984). In Hong Kong reclamation projects have been responsible for disrupting littoral drift in some locations as well as speeding up longshore currents in others (So, personal communication). Although over 500 km² of mudflats have been reclaimed in South Korea since 1945, the total area of contemporary mudflats has not decreased because "new tidal flats are formed again in front of the dykes whenever tidal flats are newly reclaimed" (Park, 1984).

Water reclamation. Although the term reclamation is usually taken to mean 'conversion to usable land', the same idea prevails when land, swamp or marsh is converted to a water-body for a particular purpose. Exclusion of 'unwanted' land (including vegetation) can be accomplished by lowering the level of the area to be developed. In this type of 'reclamation' the shoreline is moved landward. Protection of the shoreline of the new water-body may be provided by artificial structures if excavation has been in unconsolidated materials, or natural if in consolidated.

One of the best examples of this type of 'reclamation' is provided by the new harbors recently constructed in Japan. In order to aid in decentralization, during the 1960s the Japanese began to construct large ports in non-metropolitan locations such as Tomakomai, Hokkaido and Kashima, Honshu. The beach at Kashima is backed by sand dunes into which a Y-shaped harbor was cut. The harbor itself is protected from ocean waves by the dune ridge, although an extensive breakwater system had to be built to protect the entrance.

By extension any lagoon, estuary, marsh or other wetland area that is converted into water (intentionally or unintentionally) fits this category. The famous canals of Venice and the infamous ones of Louisiana, U.S.A. are examples. Much of the 32,000 km² of marshland in Louisiana has become open water, either by direct canalization or indirectly because of it.

One of the most significant recent direct impacts on the coastal zone results from the combination of both land and water reclamation. The rapid rise in the importance of the shoreline for recreation has led to the expansion of elaborately developed marinas, particularly in developed countries.

Unintentional displacement. Shoreline displacement is often incidental to other, frequently remote, activities. Increased rates of sediment deposition can result from a range of activities, in addition to agriculture, grazing and deforestation. For example, gold mining in California during the last century was responsible for adding over 10 million m^3 of sediment to San Francisco Bay (Hedgpeth, 1975), and the dumping of spoil and other wastes into the nearshore waters adjacent to most coastal cities is, or has the potential of becoming, a major shoreline modifier.

Even more conspicuous, although areally not so important, are those displacements resulting from the reduction of sediment supply to the shore. Responsible for extensive erosion along many shorelines, such reduction is caused by river damming or diversion and riverbed mining. Possibly the most publicized example is the erosion caused at the front of the Nile delta by the construction of the Aswan Dam in Egypt. However, one of the most illuminating examples of shoreline displacement is at Niigata, Japan. Situated on a long stretch of beach on the Japan Sea coast, the shoreline received much of its supply of sand via the Shinano River. This river over the years was leveed, mined, dammed and gradually diverted away from its outlet in the city of Niigata. Prior to these actions the beach was wide and backed by at least three dune ridges. Erosion has been so severe that today only part of the innermost dune ridge remains and has to be heavily protected by seawalls and groins (Kayane, personal communication). Even though the present shoreline is protected, the offshore area is still affected because of sediment starvation.

Horikawa (1978) wrote that "hydrographic charts indicated that the 8 m and 16 m depth contour lines are approaching shore at a speed of 6 to 7 m/year and 4 to 5 m/year, respectively". Incidentally, the load of the Shinano River now being delivered to another location on the Japan Sea coast has built 30 km^2 of new land.

Unintentional shoreline displacement also occurs because of human-induced subsidence. The extraction of water, gas and oil from subterranean sources has resulted in the subsidence of parts of a number of coastal cities in Japan, Italy and the United States, among others. Many of these subsided locations are now protected by extensive dikes.

Shoreline Stabilization

Man, by tradition (and necessity?), adjusts to, or adapts for his own purpose, a particular set of environmental conditions and, because of a limited temporal and spatial perspective, then attempts to prevent changes in those natural or artificially modified conditions. Because the shoreline is one of the most dynamic of environments, it is little wonder that man often finds himself at odds with nature along it. His natural (apparently culturally conditioned) reaction is to stabilize and protect.

Protection of the shoreline is often incidental to other objectives, as noted above. In this section the major objective is to examine the ways human stabilization alters material, process and form along the basic shoreline types.

Cliffy shorelines

Cliffs, found along some 40% of the world's coasts (Coleman and Murray, 1976), have been relatively exempt from human modification. Today the construction of highways along cliff coasts, partly to take advantage of their scenic attributes, as in Italy and Japan, is of major significance. Such cliffs must be considered as scarred, and some have artificial coatings placed on them to prevent rock fall. More important, especially along low cliffs in unconsolidated material, is the prevention of cliff retreat. Although there are several reasons for trying to prevent cliff retreat, the most important is to eliminate the potential loss of property at the top of the cliff. The rate of retreat, and therefore change in cliff form, is often increased by human activities such as the removal of rock debris from the base of the cliff, increasing the rate of undercutting the removal of protective vegetation from the cliff face, the alteration of ground water tables and the improper location of artificial structures along the shore.

Possibly the most common method used to prevent cliff erosion is the addition of new materials at the cliff base. The range in type of material used is great and includes almost anything that can be declared dispensable, including old automobiles. Generally such randomly selected materials have little beneficial effect on erosional processes and may even aggravate them. Somewhat more substantial structures are seawalls of various type placed at the cliff base.

Cliff-base structures are geomorphically very significant. First, they add an artificial form in front of the natural cliff; second, they eliminate the sediment that normally would have been supplied to the beach below; and third, they alter the shoreline wave and current conditions. Such structures are often responsible for starvation of downdrift beaches. Clayton (1980) stated that the artificial protection of the Norfolk (England) cliffs threatens beach stability for up to 50 km downdrift.

Coastal dunes

The 'flexible barrier' nature of dunes suggests that they possess a protective attribute along with a degree of instability. Dunes are easily changed and, because they normally lie between the beach and human habitation, they are often impacted heavily. They may be mined, scarred by foot and vehicular traffic, devegetated by man and animals, and excavated for building sites. These activities often lead to further dune destruction by increasing the amount of wind erosion. Blowouts are common in such dunal areas. Dunes may also suffer from starvation if separated from the beach by seawalls and summer homes, or by other

alteration of the sand source region.

Although it has been known for several hundred years that dune stabilization can be achieved by revegetation, the value of dunes in the coastal setting has only recently begun to receive much attention. Coastal dune management takes several forms, including such drastic methods as the use of bulldozers in reshaping eroded dunes. The most common procedure used in dune stabilization is revegetation; i.e. the planting of native or exotic grasses, shrubs or trees in order to trap and anchor the sand. Sand fences are frequently used and may be made of bamboo, wooden slats, fabric or other materials. Whatever material is used must have a degree of porosity, for a solid barrier causes turbulence and scouring (Woodhouse, 1978). Temporary protection is sometimes provided by covering dune surfaces with brush or even by coating them with bitumen (Chapman, 1982a).

Beaches and barriers

Coastal erosion along shorelines has elicited much concern and expenditure of human energy and funds. Funds usually have gone to pay for artificial structures that are designed to protect the shoreline and the human property associated with it. In some instances the objective has been to rebuild a beach, or even create one, and in others it has been to provide harbor protection. Each of the many structures placed on the shore by man is unique in that each affects shoreline processes in a distinctive way.

The seawall is one of the most common structures used along eroding shores. Placed parallel to the shoreline, it is usually massive, being designed to protect the backshore. The face of the seawall is made more or less vertical, although various concave designs have been devised to reduce wave impact. The substitution of a low, impermeable cliff for a gently sloping, permeable sand or shingle beach concentrates wave energy. Scour is aggravated, water depth is increased and large waves may result. Undermining and eventual failure of the seawall is not uncommon. Because of such potential damage to seawalls, they are frequently protected by riprap or fabricated armor blocks. These blocks, often referred to generally as tetrapods, are now made in a variety of forms. Because of their shape some have been given descriptive names such as 'turtle' and 'igloo' (Walker, 1976).

In addition to the change in profile caused by seawalls, longitudinal changes can also occur. Zenkovich (1973) reported that long-shore flow velocities of up to 8 m/sec developed because of seawalls along the northern Black Sea coast, resulting in the movement of larger sized sediments than along similar shorelines without seawalls.

The groin, constructed normal or nearly normal to the shoreline, is used to retard shore erosion and/or trap littoral drift. The first effect of the groin is morphologic, for it compartmentalizes and lengthens the shore-line, unlike the seawall, which has the opposite effect. Second, the groin, by inter-cepting and diverting a longshore current, traps sediment on its updrift side and thus deprives downdrift positions of sediment they might otherwise have received. This deprivation (starvation) often leads to erosion and the necessity for more groin construction.

The type, length, spacing and number of groins affect not only the appearance of the shoreline but also their effectiveness in achieving the desired results. If permeable, they allow some water and sediment to move through; if impermeable, sediment not deposited is carried around their ends and often lost in deep water.

Groins are one of the most common artifi-cial structures to be found along shorelines. For example, in 1977 Japan possessed 9393 groins, a 49-km stretch on the United States shore of Lake Erie has over 200 and a 12-km section of Yucatan, Mexico has 159 (Meyer-Arendt, personal communication).

The breakwater, as the term implies, is a structure that is designed to reduce the impact of waves. When parallel to the shore, breakwaters behave much like seawalls; when at an angle, like groins. Some are attached to the shore, others separated from

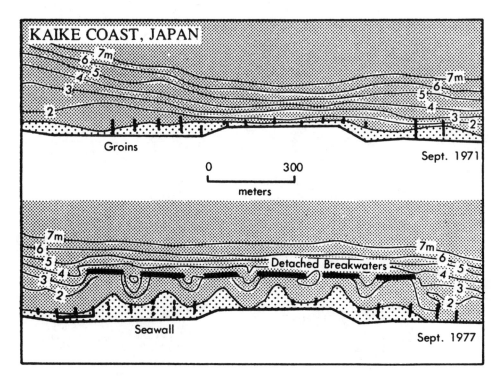

Figure 1.1 [*orig. figure 18*]   Changes in shoreline following the construction of detached breakwaters at Kaike, Japan (from Toyoshima, 1983).

it. Most are exposed but some are submerged during all or part of a tidal cycle. Materials used in breakwater construction are variable but usually are chosen according to the wave energy level to which they will be subjected. Because of their wide use in connection with harbors, breakwaters are probably the best known of all artificial shoreline structures. Further, with such great variation in material, position and shape (both longitudinal and cross-sectional), they add a wide range of forms to the shoreline. Yet their *raison d'etre* is their ability to alter oceanic processes and thereby shelter water-bodies from waves, protect beaches from erosion, prevent silting of harbor entrances and cause beach accretion (Rosen and Vajda, 1983).

Most breakwaters, especially those that are stationary, act primarily as wave reflectors, although refraction, diffraction and dissipation are also involved. Floating breakwaters, which may be desirable because of water depth, poor bottom conditions or seasonal fluctuations in water level, are mainly dissipators of wave energy (Harms, 1981).

Unique effects are generated by detached, segmented breakwaters. When sand is present it will be deposited usually in the form of sand spits or tombolos. In 1971 a project to build detached breakwaters at Kaike, Japan, was initiated. By 1977 six had been completed with beneficial results (figure 1.1). A report by Toyoshima (1983) states that, by the time the series of 11 breakwaters was completed in 1981, 284,900 m^3 of sand had been deposited between the breakwaters and the shore and 221,300 km^3 just outside the breakwaters. Further, the bathymetric lines offshore had become smoother.

Sometimes detached breakwaters and groins are used together. It has been found that if groins are constructed behind the breakwater, the shoreline of each compartment becomes concave, similar to those deposits shown in figure 1.1. On the other

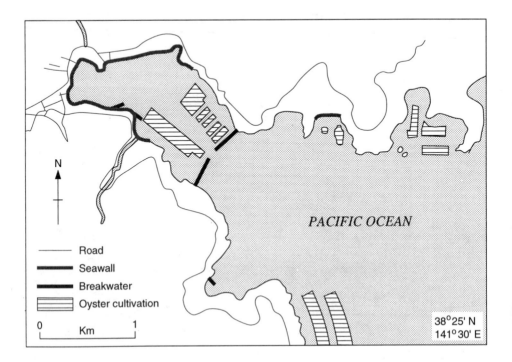

Figure 1.2 [*orig. figure 19*]   The use of bay-mouth breakwaters as protection against tsunami, Sanriku coast, Japan.

hand, when the groins are built between the segmented breakwaters, the shoreline becomes convex (Sato and Tanaka, 1981).

Although detached breakwaters have been used in the United States, Israel, Italy and many other countries, it is in Japan where they have been most popular. Between 1962 and 1981 the Japanese increased the number of detached breakwaters along their shoreline from 205 to 2305, or more than 11 times.

A specialized type of breakwater has been constructed at the mouths of some bays in Japan (figure 1.2). They are designed to reduce the impact of tsunami within the bay. Another type of storm-surge barrier, located at the mouths of rivers or lagoon inlets, can be mechanically closed during storm surges (Horner, 1979).

Instead of, or in conjunction with, the 'passive' techniques of shoreline protection or modification discussed above, there are what have been referred to as 'soft' or 'geomorphically compatible dynamic' methods. In this approach, the "forms are allowed to be freely worked by waves, currents, winds, or biological processes" (Nordstrom and Allen, 1980). Included are beach fill, offshore mounds, sand dunes and vegetation, the latter two of which have been discussed earlier. Beach fill or artificial beach nourishment is becoming more favored as a method of combatting shoreline recession. Many of the changes that occur as a result of the addition of artificial fill depend upon the compatability of the sediment introduced into the system and how and where it is placed.

Emplacement on the beach itself steepens the profile leading to accelerated erosion, the creation of a scarp (at least at first) and subsequent movement offshore of sizeable amounts of the fill. Material will also usually be moved longitudinally. Sorting, which accompanies all of these profile adjustments, depends mainly on the nature of the wave climate and the foreignness of the fill.

In order to take advantage of natural littoral processes, procedures used today occasionally involve the dumping of large amounts of sand at selected locations on the beach (known as 'feeder beaches') or in

mounds offshore. The sand is then transported by wave and current to the desired locations. The mounds created offshore behave much like submerged breakwaters in that they will alter water movement in their vicinity (Walker, 1981a).

Deltas

Stabilization in deltas is one of man's oldest endeavors at stabilization because it has been associated with reclamation. Artificial levees, although primarily designed to prevent flooding of the adjacent floodplains, have had many other geomorphological ramifications.

At the mouth of rivers they may be responsible for two major modifications in natural delta growth. First they, and their jettied extensions, tend to 'shoot' sediment into the deeper water offshore. In the case of the Mississippi River much of the sediment load is transported over the edge of the continental shelf and lost. Second, because lesser amounts of sediment are now available for distribution by longshore currents, adjacent shorelines are subject to erosion.

In addition to levees, delta areas are impacted by many other human activities and structures. Seawalls, revetments, bulkheads, groins, breakwaters and jetties are all present and all are indicative of attempts at stabilizing deltaic shorelines.

Lagoons and estuaries

A major difference between lagoons and estuaries and the open ocean is that the erosive processes are less severe. Therefore, stabilization practices are generally less substantial and less expensive. However, because such shorelines are also impacted by man, and often more intensively than on the open coast, structures are common.

Recent attention has focused on the ecological impact on these water-bodies rather than the morphological impact. However, human activities easily disrupt the natural sedimentation rates, and disturbances such as waves created by power boats cause shoreline erosion and structural response.

Nonetheless, it is at the tidal inlets in lagoons and at the mouths of estuaries that

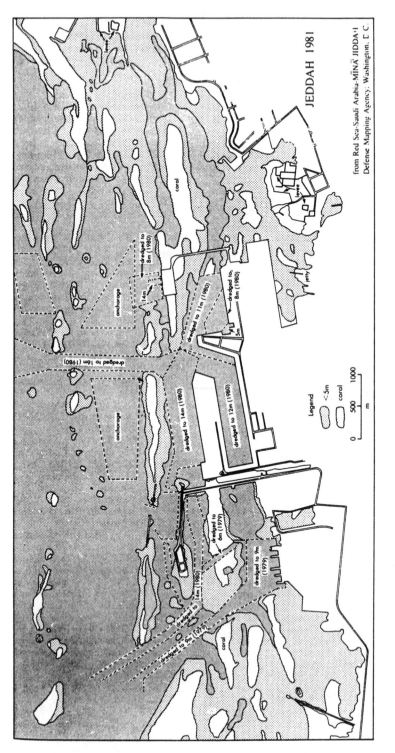

Figure 1.3 [*orig. figure* 23] Jeddah harbor, Saudi Arabia, partially excavated in Red Sea coral.

major human impacts occur. Because of the desire of man to move freely between lagoon or estuary and the ocean, jettying and dredging are common at their juncture.

The basic functions of a jetty are to prevent sediment from drifting into the entrance channel, to serve as a training wall for the tidal stream or river, and to reduce wave action at the entrance. Although jetties perform much like groins, they differ from them in three major ways: they are normally longer, they are usually single or at most paired, and they are adjacent to an entrance way on one side.

Like groins, they trap sand on the updrift side and can be responsible for erosion on a downdrift coast because of sediment loss. This reduction of sediment supply may also affect the lagoon by depriving it of the sediment that formerly entered the inlet. In many locations sand accumulates to such an extent that it begins to bypass the jetty. If the jetty extends into water of sufficient depth the sediment will be lost to the shore zone; if, on the other hand, the sediment is not lost, the channel entrance will begin to shoal, necessitating dredging.

Dredging is frequently used in jettied situations, especially at river mouths. Although the major objective of dredging is to maintain a channel sufficiently deep to handle boat traffic, a secondary objective is often to provide sand for beach nourishment projects.

Jetties may also have an effect along shorelines where there is zero net drift. Resulting alterations involve "shoreline advances adjacent to the jetties to fill 'embayments' formed between the jetties and the prejetty shoreline, with erosion at greater distances from the jetties to supply that sand. The changes are transient, however, usually being complete within 10 to 20 years" (Komar, 1983).

Biotic shorelines

Many of the world's biotic shorelines are also being impacted by man. However, in these cases the impact, with few exceptions, is not related to any attempts at stabilization. Coral reefs have been, and are being mined in many parts of the tropics. Along the shorelines, as in Sri Lanka, mining has destroyed the natural reef to such an extent that the inner shore now needs protection (Amarasinghe and Wickremeratne, 1983). The need for harbors along coral reef shorelines has prompted excavation of channels in the coral, as at the harbor of Jeddah, Saudi Arabia (figure 1.3).

Similar results have occurred along the Gulf Coast of the United States, where natural shell (especially oyster) reefs have been excessively mined for lime and construction materials. Laws are now being formulated that will require replacement with artificial 'reefs'.

The impact on coasts dominated by vegetation has been even more severe. Many coastal marshes have been completely eliminated, as noted in the section on reclamation. There is a major concern today for mangrove shorelines because many are being destroyed. Stabilization of such shorelines is probably best thought of in terms of a 'hands-off policy' rather than in terms of artificial structures.

Conclusions and the Future

Although man has only recently attained significance as a geomorphologic agent, he has nonetheless become one of the most important. There are few types of environments that show his abilities at transformation better than the shoreline. Many human-induced changes recognized along the shoreline today occurred without man even knowing they were taking place. Others, such as the vast reclamation projects that were begun several millennia ago, were largely predictable in terms of their shoreline impact.

Once man began to give thought to changing the shoreline into harbors or stabilizing it for the protection of life and property, his efforts became conspicuous. Today artificial structures are dominant along many coastlines of the world. Japan, for example, has nearly 4000 harbors and over 7900 km of dikes and seawalls along its 32,000-km long shoreline.

The intensity of coastal modification depends upon many factors – political, economic, social, scientific and technologic, among others. Many structures are made in response to natural processes and changes such as tsunami breakwaters and typhoon seawalls; many others are necessitated because of human-induced changes such as subsidence, beach-sand mining and canalization.

During much of human history, expediency dictated the nature of shoreline use and modification. The finiteness of the coastal zone, the limit of its resources and the fragileness of its ecosystems generally went unrecognized. Indeed, it has been only within the past couple of decades that much attention has been directed at the rational management of this unique environment. Although presently advocated in only a few of the more developed coastal countries, the management concept is spreading. Mitchell (1982) optimistically notes that "Because coastal management generally relies on sophisticated scientific research and data gathering, the existence of a national program underscores a public commitment to the use of science in resource management decisions". Of course, just how man will impact the shoreline in the future is unknown. However, it is known that he is becoming an ever more potent force and that, even with today's technology, he has the ability to make major, almost instantaneous, modifications. If Mitchell (1982) is correct, such modifications will be made only after due consideration of the consequences.

Especially significant in the long run will be man's response to a change in sea-level that is aggravated by human-induced climatic changes. Hansen et al. (1981) argue that the anthropogenic carbon dioxide warming in evidence today will continue and that during the 21st century, there will be "erosion of the West Antarctic ice sheet with a consequent worldwide rise in sea level, . . .". With the world's major cities (and therefore industry, commerce, capital, etc.) located on the shoreline, any significant rise in sea-level will trigger a major geomorphologic response in man.

References

Adams, R. D., Barrett, B. B., Blackmon, J. H., Gane, G. W. and McIntire, W. G. (1976) *Barataria Basin: Geological Processes and Framework*. Sea Grant, LSU, Baton Rouge, LA.

Amarasinghe, S. R. and Wickremeratne, H. J. M. (1983) The evolution and implementation of legislation for coastal zone management in a developing country – the Sri Lankan experience, In: *Coastal Zone 80*, pp. 2822–2841, O. T. Magoon (Ed.). New York.

Carson, R. L. (1955) *The Edge of the Sea*. Houghton Mifflin, Boston, MA.

Chapman, D. M. (1982a) Dune stabilization, In: *The Encyclopedia of Beaches and Coastal Environments*, pp. 379–380, M. L. Schwartz (Ed.). Hutchinson Ross, Stroudsburg.

Chapman, D. M. (1982b) Land reclamation, In: *The Encyclopedia of Beaches and Coastal Environments*, pp. 513–516, M. L. Schwartz (Ed.). Hutchinson Ross, Stroudsburg.

Clayton, K. M. (1980) Coastal protection along the East Anglian Coast, U. K., *Z. Geomorph. Suppl.*, 34, 165–172.

Coleman, J. M. and Murray, S. P. (1976) Coastal sciences: recent advances and future outlook, In: *Science, Technology, and the Modern Navy*, pp. 346–370. Office of Naval Research, Arlington, VA.

Coleman, J. M. and Wright, L. D. (1971) *Analysis of Major River Systems and Their Deltas: Procedures and Rationale, with Two Examples*. Coastal Studies Institute, LSU, Baton Rouge, LA.

Davis, J. H. (1956) Influences of man upon coastlines, In: *Man's Role in Changing the Face of the Earth*, pp. 504–521, W. L. Thomas (Ed.). Chicago, IL.

Fox, W. T. (1983) *At the Sea's Edge*. Prentice-Hall, Englewood Cliffs, NJ.

Hansen, J., Johnson, D., Lacis, A., Lebedeff, S., Lee, P., Rind, D. and Russell, G. (1981) Climate impact of increasing atmospheric carbon dioxide, *Science*, 213, 957–966.

Harms, V. W. (1981) Floating breakwater performance comparison, *Coastal Engng*, 17, 2137–2158.

Hedgpeth, J. W. (1975) San Francisco Bay, In: *Coastal Resources*, Vol. 12, pp. 23–30, H. J. Walker (Ed.). LSU, Baton Rouge, LA.

Horikawa, K. (1978) *Coastal Engineering*. Wiley, New York.

Horner, R. W. (1979) The Thames Barrier Project,

Geogrl. Jl, 145, 242–253.

Housing and Development Board (1980) *Annual Report 1979/80*. Singapore.

Inman, D. L. (1974) Ancient and modern harbors: a repeating phylogeny, *Coastal Engng*, 15, 2049–2067.

Inman, D. L. and Nordstrom, C. E. (1971) On the tectonic and morphologic classification of coasts, *J. Geol.*, 79, 1–21.

Jablonski, D. (1982) Reefs, noncoral, In: *The Encyclopedia of Beaches and Coastal Environments*, pp. 679–681, M. L. Schwartz (Ed.). Hutchinson Ross, Stroudsburg.

Komar, P. D. (1983) Coastal erosion in response to the construction of jetties and breakwaters, In: *CRC Handbook of Coastal Processes and Erosion*, pp. 191–204, P. D. Komar (Ed.). CRC Press, Boca Raton, FL.

Kuenen, P. H. (1950) *Marine Geology*. Wiley, New York.

Macgregor, A. R. (1968) *Fife and Angus Geology*. Blackwood, London.

Meyer-Arendt, K. and Davis, D. (1984) Louisiana, In: *Artificial Structures*. H. J. Walker (Ed.). In press.

Mitchell, J. K. (1982) Coastal zone management: a comparative analysis of national programs, In: *Ocean Yearbook 3*, pp. 258–319, E. M. Borgese and H. Ginzburg (Eds.) The University of Chicago Press, Chicago IL.

Nordstrom, K. F. and Allen, J. R. (1980) Geomorphically compatible solutions to beach erosion, *Z. Geomorph, Suppl.*, 34, 142–154.

Park, D. W. (1984) South Korea, In: *Artificial Structures*. H. J. Walker (Ed.). In press.

Pirazzoli, P. A. (1983) Flooding ("Acqua Alta") in Venice (Italy): a worsening phenomenon, In: *Coastal Problems in the Mediterranean Sea*, E. C. F. Bird and F. Fabbri (Ed.). IGU/CCE, Bologna.

Rosen, D. S. and Vajda, M. (1983) Sedimentological influences of detached breakwaters, *Coastal Engng*, 18, 1930–1949.

Russell, R. J. (1967) Aspects of coastal morphology, *Geog. Ann., Ser. A.* , 24, 299–309.

Sato, S. and Tanaka, N. (1981) Artificial resort beach protected by offshore breakwaters and groins, *Coastal Engng*, 17, 2003–2022.

Shen, J. Y. (1980) The evolution of the Yangtze River and its delta, Manuscript.

Singer, C. et al. (1956) *A History of Technology, Vol II*. Oxford University Press, Oxford.

Solheim, W. G. II (1972) An earlier agricultural revolution, *Scient. Am.*, 226, 34–41.

Steward, J. H. (1948) *Handbook of South American Indians, Vol. I*. U.S. Government Printing Office. Washington, DC.

Strabo (c.7 BC) *Geography*, 14.1.24, In: *The Geography of Strabo.*, pp. 229–230, H. L. Jones (trans. 1929). Heinemann, London.

Toyoshima, O. (1983) Variation of foreshore due to detached breakwaters, *Coastal Engng*, 18, 1873–1892.

UNESCO Reports in Marine Sciences (1979) *The Mangrove Ecosystem: Scientific Aspects and Human Impact*. Division of Marine Sciences. UNESCO, Paris.

Van Veen, J. (1962) *Dredge Drain Reclaim*. Drukkenj Trio, The Hague.

Volker, A. (1982) Polders: an ancient approach to land reclamation, *Nat. Resources*, 18, 2–13.

Volman, T. P. (1978) Early archeological evidence for shellfish collecting, *Science*, 201, 911–913.

Walker, H. J. (1976) Shoreline protection in Japan. In: *Time-Stressed Coastal Environments*, pp. 269–278, E. L. Pruitt (Ed.). The Coastal Society, Washington. DC.

Walker, H. J. (1978) Research in coastal geomorphology: basic and applied, In: *Geomorphology: Present Problems and Future Prospects*, pp. 203–223. C. Embleton et al. (Eds.). Oxford University Press, Oxford.

Walker, H. J. (1980) The Pearl River delta, *Scient. Bull.*, 5, 1–6.

Walker, H. J. (1981a) Man and shoreline modification, In: *Coastal Dynamics and Scientific Sites*, pp. 55–90. Komazawa University Press, Tokyo.

Walker, H. J. (1981b) The peopling of the coast, In: *The Environment, Chinese and American Views*, pp. 90–105, L. J. C. Ma and A. G. Noble (Eds.). Methuen, London.

Wong, P. P. (1984) Singapore, In: *Artificial Structures*, H. J. Walker (Ed.). In press.

Woodhouse, W. W., Jr. (1978) *Dune Building and Stabilization with Vegetation*. Coastal Engineering Research Center, Fort Belvoir, VA.

Zenkovich, V. P. (1973) Geomorphological problems of protecting the Caucasian Black Sea coast, *Geogrl Jl*, 139, 460–466.

2

LAND SUBSIDENCE: A WORLDWIDE ENVIRONMENTAL HAZARD*

L. Carbognin

Land subsidence is the surface sign, and the last step, of a variety of subsurface displacement processes. The word 'subsidence', in its common use, merely indicates vertical, downward movement independent from the causal mechanism of its occurrence, areal extent, or rate of movement.

Land subsidence is associated with natural causes that constitute the geological history of the affected area (geological or natural subsidence) but it can be accelerated or even set in motion, by human intervention (man-induced subsidence).

Among the most common natural causes which lead to subsidence, we can enumerate:

1   Deep-seated tectonic movement, volcanic activity, earthquakes, and isostasy.
2   Compaction (i.e. the decrease in thickness) of recent, fine-grained deposits subjected to the loading of the overlying sediments (overburden), or by vibration during earthquakes.
3   Drying-up of lacustrine basins due to natural evolution of the environment, and oxidation of highly organic soils.

The most common man-made subsidence results from:

1   Removal of fluids from the subsoil, i.e.:
    (a) withdrawal of groundwater, thermal water and gas bearing water, causing a decline of the water head;
    (b) pumping of crude oil, natural gas, oil and associated gas, causing underground pressure depletion.
2   Compaction of fine sediments induced by:
    (a) application of water. Some unconsolidated dry deposits collapse when wetted: this process is referred to as hydrocompaction or shallow subsidence;
    (b) land drainage and reclamation and biochemical oxidation of peat and organic soils;
    (c) loading by buildings or other engineering structures and vibration on sediments.
3   Mineral mining.

The largest and most remarkable instances of man-induced subsidence have been caused by the pumping of subsurface fluids.

* Originally published in *Nature and Resources*, 1985, vol. 21, pp. 2–12.

Such a phenomenon was first observed in Galveston Bay, Texas, in the early 1920s, where subsidence occurred above the Goose Creek oil field. The pumping of petroleum from the underlying strata caused the land to sink simultaneously, and cracks at the ground surface appeared quite suddenly.

But now pockets and regions of land sinking attributable to the same basic cause are a far more common phenomenon over the earth. In fact, in the last decades, with the growing need for water and energy to accommodate industrial, urban and agricultural developments, man has come to rely increasingly on underground resources, often extracting them imprudently.

It must, however, be mentioned that fluid exploitations will produce land subsidence only under certain geological conditions, in general, where the deposits involved are mostly composed of unconsolidated late Cenozoic and Neozoic sediments of high initial porosity. Almost all the subsiding areas where strata are tapped, are characterized by underlying, semi-confined or confined aquifers made up of sand/or gravel of high permeability and low compressibility, interbedded with layers of clay and/or silt of low vertical permeability and high compressibility (aquitards). The aquifers and aquitards are of variable thicknesses.

The Subsoil and Mechanism of Subsidence

Even though there are many differences between the extraction of water from that of other fluids, the principles involved in the mechanics of land subsidence are nonetheless the same. The variables of the system, among which are the size of the reservoirs, the lithologic nature of the deposits, their depth and physical characteristics, the effective tension and its increase and the time of subjection to the increase, determine the 'nature' of the subsidence.

Since water is by far the most common fluid met with in the subsoil, and the most exploited, let us see what happens after tapping the aquifers.

To understand the mechanism of land subsidence better, it is essential to have some idea of the structure of the subsoil. The types of sediment common to the entire zones of subsidence are:

- *Sand and gravel*. A loose non-coherent material consisting of small rock and mineral particles, distinguishable by the naked eye. Gravels are different from sands, having larger particles. Layers of these materials are the source of productive aquifers, due to their high permeability. When subjected to loadings their behaviour is essentially elastic.

- *Silt*. Rock fragment, mineral, or detrital particles with little or no plasticity. The silts, that are not very permeable, are an intermediate aggregate between sands and clay, i.e. with a diameter smaller than fine sand and larger than coarse clay.

- *Clay*. An aggregate of microscopic particles which originated from the chemical degradation of rock components. It is formed primarily of silicon, aluminium, water, iron, alkalies, etc. It develops plasticity when a limited amount of water is present. The permeability of clay is very, very low.

- *Peat*. A dark brown or black aggregate of macroscopic fragments produced by the decomposition and disintegration of any vegetative organic matter (moss, trees, etc.). It is highly compressible.

The particles of all these materials have dimensions that vary from a pebble to a large molecule. In particular, coarse soils are composed of granules of about 0.06 mm in diameter, still visible with a magnifying glass; silt contains granules between 0.06 mm and 2μ ($1\mu = 0.001$ mm) that can only be observed with a microscope, and clay is composed of granules smaller than 2μ. In the subsoil it is difficult to find an aggregate in a pure form; generally there are mixed types such as silty-sand, clayey-silt, sandy-clay, etc.

In subsidence the basic physical characteristics which the behaviour of the subsoil depends on are permeability and consolidation.

The *permeability* of a material is its capacity to transmit water (or other fluids) under pressure; it is dependent on the size and shape and interconnection of the spaces (pores) in the porous medium. The degree of permeability of a soil is defined in relation to the facility of water to flow through these pores. This is expressed by the permeability coefficient, commonly evaluated in the laboratory by observing the rate of movement of fluid through a determined sample of material. Indeed, the coefficient is higher for the coarse material and lower for fine-grained sediments.

Consolidation is the gradual reduction of the soil volume when underground equilibrium conditions are disturbed. Strictly speaking consolidation is the gradual reduction in water content of a soil, as a result of an increase in load, while compaction is the decrease in the volume as a result of the same cause. Compaction, as used by engineers, is synonymous with 'one-dimensional consolidation'. In a broad sense, the terms 'consolidation' and 'compaction' are used with the same meaning, and 'compressibility' means only the aptitude of a soil to be deformed.

When a new pressure is exerted on layers, for example by the overloading of new sediments or any external load, the particles are pushed together, the water content of the sediment is squeezed out, the volume decreases and the deposit undergoes compaction. The same process occurs when the hydrostatic pressure of the water percolating in the pores of sediments is reduced after pumping begins.

To evaluate compaction, a laboratory test called the oedometric test is usually performed on undisturbed samples of representative fine-grained deposits. The coefficient of consolidation expresses the rate of consolidation under a load increase. In the process of compaction, part of the energy is spent in the breakdown of the 'flow' structures of the particles and part is absorbed by the work of elastic deformation. This means that if the hydrostatic pressure is increased, one may observe that the water content and volume also begin to increase. In sand, where the flow structure is absent,

the greatest part of the energy is spent in elastic deformation which is reversible; in clay, on the contrary, the greatest part of the energy is spent in the permanent breakdown of the flocculation structure, and the deformation is almost completely irreversible.

While the law of permeability of a porous medium had already been formulated in 1856, on the basis of the experiments by the French hydraulic engineer Henry Darcy, the 'reaction' of an aquifer system to any, more or less rapid, perturbation could only be completely understood when Karl Terzaghi developed his theory of vertical consolidation in 1923, introducing the fundamental principle of 'effective stress'. This principle, valid for any depth, states that the total overburden load (geostatic pressure or total stress), σc, of a vertical column of unit cross section is sustained partly by the hydrostatic pressure (neutral stress), p, of the water in the pores of the porous medium, and partly by the intergranular pressure (effective stress), σz, that is by the forces which the grains exchange among themselves on contact. The formula is quite simple:

$$\sigma c = p + \sigma z$$

but the concept it expresses is of paramount importance. The total load per unit area, σc, is constant for each determined depth; when a variation in subsurface flow occurs, it is accompanied by a variation in p, then σz must vary. That is, if the pressure in a water-bearing stratum decreases, the intergranular stress must increase. Experience has demonstrated that while the variations of p have negligible effects on the deformation of the subsoil, an increase of σz causes the beds to compact.

From a practical point of view one can say: a pumping well produces a disturbance that propagates its effect, namely, the pressure head decline, in space and time, through the hydrogeological system (figure 2.1(a)). The reduction of the pore pressure at any point in the subsurface system increases the effective intergranular stress (figure 2.1(b)), under whose influence the formations compact – the amount of compaction being proportional to the compressibility of the compacting unit – and, last step, the ground

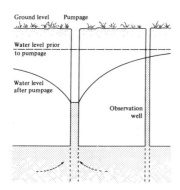

Figure 2.1(a) [*orig. figure 1(a)*]   Water withdrawal from a pumping well creates a disturbance in the pre-existing hydrodynamic equilibrium and generates a cone of depression.

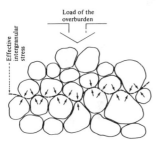

Figure 2.1(b) [*orig. figure 1(b)*]   The decrease of the water pressure produces an increase of the effective intergranular stress under whose influence soils compact.

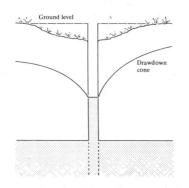

Figure 2.1(c) [*orig. figure 1(c)*]   The underground compaction induces a ground surface settlement, which is proportional to the decrease in water pressure and follows the shape of the drawdown cone.

surface sinks taking the shape of a bowl centred above the point of maximum withdrawal (figure 2.1(c)).

This is valid for poorly consolidated soils; if on the contrary, the formation is well-consolidated, i.e. if the grains are strongly held together, it will retain its structure after the water has been withdrawn, without any subsidence.

What this means then is that because layers of sand and gravel (aquifers) compact quite elastically, their compaction is very negligible owing to their low compressibility. On the other hand, the clayey and silty layers (aquitards) that separate aquifers are highly compressible and will undergo significant compaction which is mostly non-recoverable.

Field Measurements of Subsidence

As previously pointed out, land subsidence at the surface is the additive result of the subsurface compaction of the various formations.

To measure the deformation of any individual clay layer, compaction recorders (extensometers) are installed in wells at different depths. Two types are generally used: pipe and anchored-cable both of which are illustrated in figure 2.2. For the operation of either type of equipment, the well should be drilled as nearly vertical as possible and, for the anchor and cable installations the inside diameter of the well casing should be at least 20 cm to decrease the cable-casing friction.

Even if subsidence processes are concealed below ground, it is common practice to directly keep the ground surface elevation under control by surveying a number of firmly anchored concrete posts (bench marks). Bench marks are set up according to a network which covers the area known as, or suspected of being, subsident and extends into a broader regional network connected to reference bench marks assumed to be stable.

In every country there exists a national levelling network but an additional number of bench marks are established locally to provide a denser grid over areas of special

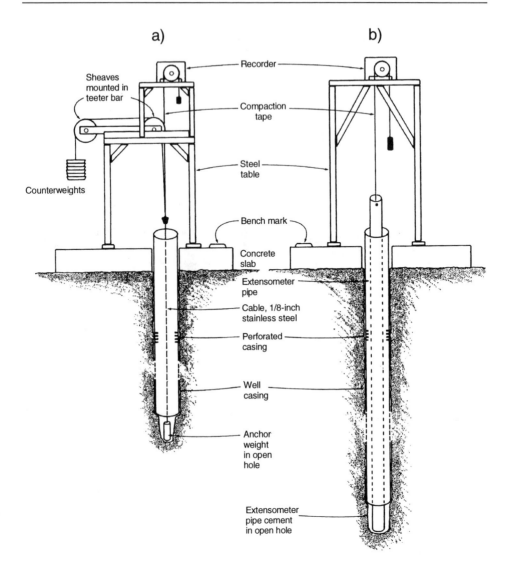

Figure 2.2 [*orig. figure 2*]   Recording extensometer installations. A. cable assembly; B. pipe assembly (after Poland, 1977).

interest. Periodically surveyed by precise levelling procedures, bench marks will show whether, where and how much ground surface lowering has occurred. Maps and/or profiles of land subsidence are prepared from measured changes in the elevation of bench marks (figure 2.3).

Some Case Histories of Natural Subsidence

Ground surface displacements resulting from tectonic movement, isostasy, and natural compaction involve almost the whole earth's crust.

Tectonic subsidence occurs on a very long-term scale and does not cause concern since it has the smallest impact on human

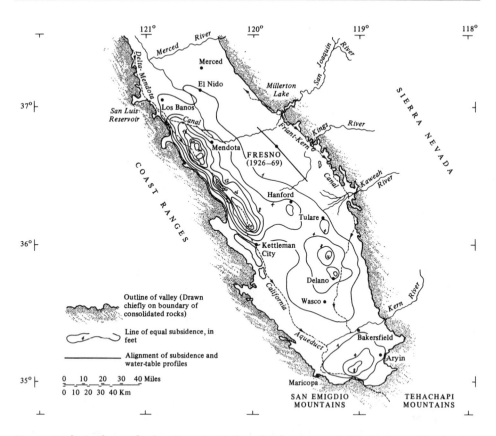

Figure 2.3 [*orig. figure 3*] San Joaquin Valley, California: areal distribution of land subsidence, 1926–70, due to groundwater withdrawal (after J. F. Poland, Lofgren, Ireland and Pugh, 1975. 'Land Subsidence in the San Joaquin Valley as of 1972'. U.S. Geol. Survey, Prof. Paper 437-H).

activities. But tectonic activity may generate seismic and volcanic events responsible for both vertical and horizontal movements, even on a vast scale. In this connection it is worth mentioning that the 1959 Hegben earthquake in Montana caused an area of about 1,500 sq. km to subside asymmetrically with a maximum of 6.6 metres; the 1964 Alaska earthquake provoked vast lowering of a regional area of 800 by 160 km, and the vibrations then induced on the deposits led to a compaction responsible for an additional notable subsidence.

Roman ruins

A very interesting example of vertical downward movement, which today has reversed its trend, is given by the Italian seismic-volcanic area of Pozzuoli, close to Naples. Historically the ground of the 'Campi Flegrei' area was highest in the Roman times; nowadays the structures of the ancient metropolis, including the entire port, lie submerged in the Gulf of Pozzuoli. Worth mentioning are the ruins of the Roman Imperial Palace at Baia recently found underwater at a depth of 7 m; indeed, it is believed that the subsidence responsible for its disappearance was of about 14 m (Cotecchia, 1984).

The temple of Jupiter Serapis, at Pozzuoli, is the most singular monument of the region both for its geological as well as its archaeological interest. For 2,000 years it has represented the most significant available metric index for assessing the up and down land displacements that have occurred. In particular, the columns of the temple are marked by the burrows of litodomi, marine perforating mollusks which live at the surface of the water, indicators of the past levels of submergence.

The very, very long period characterized by a low irrelevant subsidence stopped in 1969. Since then an irregular upward movement has been observed, some with the alarming rates of 5 mm/day. A total ground raising of 2.60 m has been recorded from 1970 to 1983.

Surface loading over large areas of the earth is another very large-scale phenomenon leading to ground surface movements. For example the crust moves under the weight of ice in an ice age, as a consequence of melting (uplift) or freezing of ice (downlift). This process, called isostasy, is of importance in regions of Scandinavia and the eastern North American continent.

Natural compaction of fine-grained deposits due to the loading of the overburden is accompanied by land subsidence. This type of subsidence chiefly occurs in Quaternary soils and it is most apparent where a great thickness of sediments accumulates rapidly; typical examples are the areas of river deltas. One can in particular mention the modern Mississippi River Delta where 400 to 500 million tons of sediments are deposited each year and the levee deposits on the lower delta have subsided more than 6 m.

Examples of Man-induced Subsidence

It is impossible in this article to list and analyze all the areas affected by man-made subsidence, but the case histories described here are notable for the magnitude of subsidence or the interest in their induced effects, or causes.

The most numerous and spectacular instances of man-made subsidence are those due to rash fluid exploitation.

California agriculture

Numerous land sinkings have been recorded in California (United States) where vast amounts of water are withdrawn for irrigation, although also for domestic and industrial uses. Groundwater withdrawals in California are chiefly from intermountain basins in which the valley fill of late Tertiary and Quaternary age consists mainly of alluvial deposits, but also in places of lacustrine and shallow marine origin. In many of the basins, the aquifers being pumped are semiconfined or confined below depths ranging from 30 to 200 m, with much of the withdrawal coming from confined aquifer systems.

Water withdrawals have caused a piezometric decline in the aquifer of tens of metres, with recorded measurements of up to 153 m in the San Joaquin Valley where one of the greatest and most extensive man-induced subsidences in the world occurred with a magnitude of 9 m. Groundwater overdraft has prevailed in much of the valley since the 1930s. Subsidence apparently began concurrently, and became of widespread concern in the late 1940s.

Pumping extractions increased especially after the Second World War; during the 1950s an amount of more than one-quarter of all groundwater pumped for irrigation in the United States was used in the San Joaquin Valley alone. In the early 1960s pumping lifts frequently exceeded 150 m. The close correlation between water level decline, compaction of the aquifer systems and land subsidence became apparent.

Figure 2.3 has shown the curves of equal subsidence for 1926–70, but a more striking demonstration is visible on a utility pole showing the ground heights recorded in 1925, 1955 and 1977. In fifty years, 20 billion m^3 of land have disappeared, corresponding to the volume of consolidation caused by the extraction of about 70 billion m^3 of groundwater. Since that period, subsidence has slowed in most of the San Joaquin Valley as a result of the drastic reduction of groundwater pumping with the large volume of surface water being substituted. As of 1976 artesian heads of the aquifers have recovered toward pre-subsidence levels and subsidence rates of the surface have decreased to near zero in much of the valley.

Mexico City: a sinking capital

Mexico City, Mexico, groundwater withdrawal has also produced 9 metres of land subsidence. But while in the San Joaquin Valley a large number of aquitards were subjected to compaction, in Mexico City most of the 9 m of compaction has occurred chiefly in the top 50 m below the land surface, especially in two very highly compressible silty-clay beds, 25–30 m and 5–10 m thick. This subsidence began in the last century, when deep wells were drilled all over the city to supply water for different uses.

In 1925 land surface sinking was confirmed through comparison of two precision levellings of 1877 and 1924, and in 1948 it was demonstrated that the subsidence of the city was due to groundwater exploitation. The rate of subsidence up to 1938 was about 4 cm/year; from 1938 to 1948 the yearly rate was about 15 cm and increased to 30 cm/year from 1948 to 1952. After 1952 the rate decreased gradually reaching between 1970 and 1973 a value equal to 5cm/year. Today land subsidence is not a serious problem. But in 1959 land subsidence had already exceeded 4 m beneath all the old city and was as much as 7.5 m in the northeast part (reaching about 9 m late in the 1970s). Protrusion of well casings was a common occurrence in the subsiding area as a demonstration of the compaction of the sediments within the casing depth.

The great and long continued settlement of Mexico City has caused many problems in water transport, drainage, in the construction of buildings and other engineering structures of the city. Pumping of underground water in the central part of the city has been prohibited since the 1970s and locally water injections have been carried out to improve the piezometric head.

Japan's water for industry

Japan offers other interesting examples of man-induced subsidence. Investigations and surveys have recognized 59 places affected by land subsidence by 1981, most of them concentrated in industrial regions on the coast. The total sinking areas covered 9,520 sq. km, accounting for about 12 per cent of habitable land of the nation. According to many investigations the principal cause of subsidence in almost all the 59 places is the over-exploitation of groundwater from confined aquifers located in alluvial and shallow marine deposits.

Even though the groundwater is chiefly used for industrial purposes, large withdrawals are also made for domestic uses, fish farming, and irrigation. Among the subsiding areas, the well-known ones are those of Niigata, Chiba, Osaka, Taipai Basin with an overall subsidence (1.5 to 3 m), less than that of Tokyo (about 4.50 m).

Let us examine the latter case. The history of land subsidence has been described in detail for some areas of Tokyo for this century (figure 2.4). In general the rate of land sinking has annually increased in Tokyo since 1950; by 1961 about 74 sq. km were subsiding (around) 10 to 15 cm/year. The maximum subsidence of 4.57 m was recorded in the period 1920–75 in the Koto ward of Tokyo, an area in the northeastern section of the city, where the groundwater head had dropped to about 60 m because of the intensive pumping from various aquifers. As a consequence of this subsidence the potential danger of tidal flood under a typhoon has been increasing year by year. Dikes had to be built along the edge of Tokyo Bay to prevent the city from flooding. Concurrently a series of restrictions in groundwater was taken. As a result of these legal interventions, the water levels recovered quickly all over the area, the subsidence first decreased and then stopped. A land rebound of a few centimetres was also measured at various places.

Venice and its lagoon

Another example of land subsidence due to groundwater withdrawals is that of Venice. The interest in this case is due not so much to the amount of subsidence, which is relatively small, but to the special prestige of the city of Venice which runs the risk of disappearing into the lagoon. We must in fact remember that Venice was built in the water and it is with respect to water that it must come to terms. The subsidence of Venice

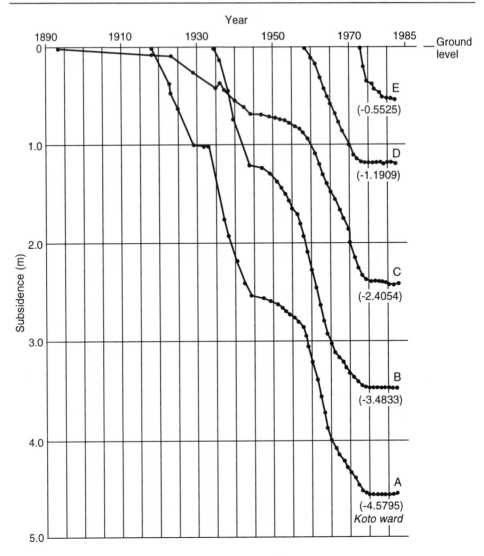

Figure 2.4 [*orig. figure 7*]   Total subsidence of several bench marks in the lowland of Tokyo since 1890 (updated after Y. I. Inaba et al. 'Reviews of land subsidence researches in Tokyo'. *Proceedings of the First International Symposium on Land Subsidence*, 1969, IAHS Publ. No. 89, Vol. I, p. 87–98).

relative to the mean sea level has been quantified as 22 cm from the beginning of the century, most of which occurred in the twenty years from 1950–70. During this period in fact, the events of the 'acqua alta'

(high water) increased (figure 2.5) and the problem aroused worldwide attention since the existence of the city and its historical monuments were threatened.[1]

For the same reason the subsidence of

---

[1] For a more comprehensive treatment of the subject, see Augusto Ghetti and Michel Batisse, 'The Overall Protection of Venice and its Lagoon'. *Nature and Resources*. Vol. XIX, No. 4, October–December 1983.

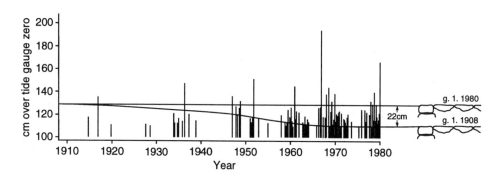

Figure 2.5 [*orig. figure 8*]   The 'aqua alta' occurrences shown here as vertical lines and revealing an increase with time dependent on the loss in ground surface elevation (P. Gatto and L. Carbognin, 'The Lagoon of Venice: Natural Environmental Trend and Man-induced Modification', *Hydrological Sciences Bulletin*, 1981, Vol. 26, No. 4).

Ravenna, a coastal city south of Venice, has created a very great concern. Ravenna is known all over the world for the splendour and uniqueness of its mosaics, and for its monuments spanning the fifth to the fifteenth century.

The subsidence, whose maximum has been of about 1.20 m, endangers the entire territory and influences the hydrologic balance of reclamation and the river network. As a consequence of land settlement, a regression of the shoreline, in some places even more than 130 m, has been observed, which in turn caused the destruction and submersion of the famous pine forest. The harbour installations sank below sea level, as well as building foundations in the historical centre where the existence of several important monuments is seriously jeopardized.

Asian cities

Other cases of land subsidence due to groundwater withdrawals have been reported recently from the People's Republic of China. The most impressive among them refers to Shanghai which sank 2.63 m between 1921 and 1965, and to the Tianjin area, which is located in the northern China Plain, where a maximum cumulative amount of settlement equal to 2.15 m has been observed since 1950.

In Thailand, groundwater is pumped from unconsolidated deposits of sand, gravel and clay for the industrial and domestic needs of Bangkok. Starting in 1955, the water level declined by 50 m in 28 years, with as much as 88 cm of surface subsidence between 1978 and 1983 in the area east-southeast of Bangkok. Flooding of the city has become a problem.

Due to the relatively recent exploitation of geothermal fluids the consequence of subsidence has so far received only minor attention, although it may be of significant magnitude. Evidence of this is given in New Zealand where the Wairakei field, the most extensively developed one, has experienced a ground surface settlement up to nearly 5 m from 1956 to 1974. Other areas indicate a much lower rate of subsidence, but demonstrate the need for further investigations in this field.

Pumping of gas-bearing water has also caused subsidence at different places. Two examples worth mentioning are those of Niigata, Japan, and the Po River Delta, Italy.

At Niigata a piezometric head decline of about 50 m in the period 1950–68 induced a subsidence up to 2.60 m. In the Po Delta area more than 3 m of land sinking was recorded between 1951 and 1965 following a piezometric decline also of 50 m. The two areas are very similar in their subsurface geology: the deposits involved in the process belong chiefly to the Pleistocene and consist of lagoonal and marine sediments. The main difference between the two cases is the thickness of the sediments subject to compaction, 0–1,000 m at Niigata and 100–600 m at the Po Delta. Since both zones are close to the sea, the consequences are quite the same: coastal facilities have been severely damaged; some river and road embankments disappeared; and farms and houses, perpetually flooded, have had to be abandoned.

Oil and gas withdrawal: effects and remedies

For subsidence due to oil and gas withdrawal, since the productive fields are generally considerably deeper than aquifers (they can reach a depth of 3,000 m), the surface effects are not always so apparent. In fact, among the thousands of oil fields existing in the world, the most serious effects of subsidence have been felt in the Los Angeles–Long Beach area of California. Above the extremely active Wilmington oil field the ground surface subsided about 9 m between 1926 and 1968; the maximum annual rate of 71 cm was reached at the centre of the bowl in 1952. The geologic structure of this area is a broad, asymmetrical anticline broken by a series of transverse normal faults. The seven major producing zones range in age from Lower Pliocene to Upper Miocene, spanning a vertical section of about 1,500 m. In general most of the compaction took place in the oil zones between 600 and 1,200 m.

This case is particularly noteworthy for three reasons: (a) the vertical settlement of 9 m at the centre, one of the greatest subsidences to occur in the world; (b) many fields yielding oil from deposits of similar age and from an equivalent depth range of approximately similar lithology, and with comparable decrease in fluid pressure have not experienced subsidence of more than a few feet; and (c) corrective remedial action taken to stop subsidence.

A new process of salt-water injection was successfully used to control the subsidence which even accomplished some elastic rebound of land surface and raised the field's oil production.

Venezuela offers other troublesome instances of oil-produced subsidence; worth mentioning is the 3.4 m of subsidence over the Lagumillas field measured in the period 1926–54.

Considering gas production, there is evidence of several tens of centimetres of subsidence in the gas fields of Rio Vista and River Island, United States. For the gas extraction in Groningen, Netherlands, a few mm/year of land sinking are occurring today, but measurements and mathematical models have suggested the expectation of 1 m of subsidence up to the year 2056. This gas field is perhaps the largest in the world. It was discovered in 1959 and its large size of approximately 900 sq. km was completely identified in 1963. The sandstone reservoir is situated at a depth of nearly 2,900 m and has an average thickness of about 150 m.

Compaction: English fens and Russian plains

A peculiar process of compaction which occurs for reasons quite different to water exploitation is that called *hydrocompaction* as mentioned in the introduction. The soils subjected to this type of subsidence, namely loessial deposits, can support relatively large loads when they are dry, without compacting. When they are wet, the bond between individual grains is reduced, rapid compaction occurs, followed by a subsidence of the land surface. Hydrocompaction usually begins near the land surface taking place soon after the deposit becomes wet, then progresses downward with the advancing water front.

Reports of subsidence due to hydrocompaction refer principally to the United States, but in vast areas in western Europe and west-central Asia where loess deposits are present, subsidences due to application of water are not uncommon. Nearly 350 sq. km of farm land are irrigated in western and southern San Joaquin Valley with a resulting land settlement of nearly 3 m occurring. In several areas of the Ukraine, a region of the USSR, about 65 per cent of which is covered with loess soils with a thickness from 20 to 40 m, subsidence may range from 1.0 to 1.5 m.

Subsidence due to hydrocompaction causes serious concern either in the design or maintenance of aqueducts, irrigation canals, buildings and other major engineering structures. Surface cracks, sunken ditches and other evidence of hydro-compaction are readily apparent even to the casual observer.

Compaction of recent deposits often occurs as one of the results of drainage and reclamation works, carried out on wetlands (e.g. swamp, mudflat, etc.) in order to transform them into more profitable areas. Generally the settlement rate is proportional to the drainage rate: the lower the water table, the greater the subsidence; but the amount of subsidence is always larger in the presence of peat and organic soils because they undergo irreversible oxidation from biochemical action on drying that reduces the volume. For example in the Fens of Cambridgeshire and Lincolnshire, United Kingdom, a ground surface settlement ranging between 5 and 10 m has been observed since 1800. The same process occurred in the reclaimed plains of the USSR and in the Florida Everglades in the United States. The latter represents the most extensive deposits of organic soils in the world (8,000 sq. km). About 1.6 m of subsidence has occurred, with different sinking patterns since 1913.

With regard to land subsidence in reclaimed areas, perhaps the Netherland polders, in Europe, offer the best example, and in particular the Yssel Lake area. Here the recently reclaimed 'Flevoland' subsided within fifteen to twenty-five years after reclamation, from a few centimetres in the shallow, sandy layer, to a little over a metre in the deposits with a peat soil.[2]

In Israel the swamp area of about 4,000 ha reclaimed in the Hula Valley in 1958 had experienced in 1980 a subsidence ranging from 8 cm/year in the high organic peat soil to 2.8 cm/year in the sedimentary lake deposit soil.

New York's sinking airport

Loading of the surface due to man-made structures has induced soil compaction and therefore subsidence. Intensive urbanization has been one of the biggest responsible factors of this in many areas. For example the area of New York's La Guardia Airport had subsided by artificial loading more than 2 m after twenty-five years of operation.

In addition, urbanization may induce further compaction by vibration produced by heavy traffic, and a variety of man-made sources.

Land subsidence can also result from the extraction of solids such as coal, mineral ores, salt and sulphur, either in the solid state or as a result of solution mining techniques.

The cavities produced by extraction will always cause stress and deformation in the adjacent soils. The resulting displacements at ground surface mostly depend on the geological and geotechnical conditions of the deposits, the amount of material removed and the size of the underground opening. The cavities formed in the subsoil often progress upwards through the soil overburden; when the underground opening is sufficiently close to the land surface a sudden collapse of the surface ('sinkhole') occurs. In regions underlain by limestone these events are a common occurrence.

One must remember at this point that in the Karst terrains sinkholes may result from natural causes. Nevertheless, land subsidence due to natural sinkholes generally develops in geologic time and it is not regarded as so serious a problem as subsidence resulting from man-induced sinkholes. The incidence of sinkhole development can be increased considerably when natural equilibrium conditions are changed by man's activities, in particular those that change groundwater levels or increase infiltration. These are widespread in the southeastern part of the United States, in particular in Georgia, Alabama and Florida. Their diameter ranges from 1 to 90 m, and their depth from 0.3 to 30 m.

Coal mining operations have caused subsidence in Germany and in Great Britain,

---

[2] See also A. Voler, 'Polders: an Ancient Approach to Land Reclamation', *Nature and Resources*. Vol. XVIII, No. 4, October–December 1982.

in Nigeria and other places in South Africa and in Australia. Two localities of instances of subsidence due to extraction of salt are the town of Tuzca, Yugoslavia, and the districts of Cheshire in England. However, in general, surface deformations caused by mining activities are confined to relatively small areas immediately overlying the mines and do not create great concern.

The remedies

Land subsidence has always produced severe damage on the environment and particularly in coastal-plain regions, where land can sink below sea level. Disappearance of stretches of beach, increase in flooding, shore erosion, pollution of fresh water from salt water intrusion and spoiling of harbour efficiency are the quite common detrimental effects. In many areas land subsidence has ruined buildings, wrecked bridges, cracked city pavements, twisted railroad tracks, etc. It is evident that all these devastating consequences represent an incalculable cost for mankind.

What can be done to avoid or reverse subsidence? It is obvious that subsidence due to natural causes is unavoidable and remedial measures can only be taken against man-induced subsidence.

A number of countermeasures to prevent damages or to reverse subsidence have been adopted. They can be summarized as follows: (a) cessation or regulation of subsurface resource extractions, legal action, or replacement with substitute supply; (b) construction of dikes in areas susceptible to flooding; (c) repressuring by water injection. This method, highly successfully applied in the Wilmington oil field, cannot be adopted everywhere because it can produce very serious problems which stem from the quality of water injected, especially in an aquifer system.

Since subsidence is a mostly irreversible occurrence, the environment will never recover fully its original condition.

Conclusion

In the past, man was the 'victim' of subsidence because he lacked the knowledge to either predict or to control the phenomenon. Today, however, research on the causes of subsidence and the possibility of forecasting physical changes through mathematical models, should help planners to avoid, or at least minimize, the dangers when building urban and industrial complexes.

3

RESERVOIRS AND EARTHQUAKES*

R. B. Meade

Introduction

The practice of suspending judgement until the evidence can be evaluated is common to most scientific investigation. On the contrary, studies of reservoir-induced seismicity have routinely assumed a cause and effect relationship between the reservoir operation and earthquakes. In such a case the details are not evaluated but instead, form the basis for the characterization of reservoir-induced seismicity.

The idea that stress changes due to reservoirs could trigger earthquakes is plausible. Reservoirs impose stresses of significant magnitudes on crustal rocks at depths rarely equalled by any other man-made work. The ability to prove a cause and effect relationship between reservoir activity and earthquakes is beyond the capabilities of scientists and engineers because of the severely limited ability to measure stress below depths of several kilometers.

Despite the lack of direct evidence for evaluating a cause and effect relationship, the question of the role of reservoir activity in triggering earthquakes is sufficiently important to justify the serious review and evaluation of the available circumstantial

evidence. About 1981, the US Army Corps of Engineers sponsored a study to evaluate the evidence of reservoir-induced earthquakes critically (Meade, 1982). This study took an in-depth look at all the cases of damaging earthquakes that had been linked to reservoir activity in published reports prior to 1980. A listing of these cases appears as table 3.1.

Table 3.1 [*orig. table I*]   Cases reviewed in the 1982 study.

Reservoir	Location	Largest Earthquake
Hoover (Lake Mead)	USA	$M_L = 5.0$
Kariba	Zambia	$m_b = 5.8$
Kremasta	Greece	$M_s = 6.3$
Koyna	India	$M_s = 6.5$
Kurobe	Japan	$M_s = 4.9$
Manic 3	Canada	$m_b = 4.3$
Hsinfengkiang	China	$M_s = 6.1$
Nurek	USSR	$M_s = 4.5$

The study used reservoir water level data and earthquake catalog data to judge the correlation between reservoir-induced

* Originally published in *Engineering Geology*, 1991, vol. 30, pp. 245–62.

stress changes and the occurrence of earth-quakes. The correlation was measured by examining the number of earthquakes occurring during certain water level stages.

In this paper the results of the study are briefly reviewed and the methods applied to data from Koyna and Aswan that appeared after publication of the study.

The Method

Two questions form the starting point of the investigation. First, has the occurrence of earthquakes increased since impoundment of the reservoir? Second, do the earthquakes occur at times when the reservoir is weakening the crust?

In cases where the annual number of earthquakes has not been altered by filling of the reservior, charges of induced seismicity rest on changes in location of earthquake activity or a close relationship in the time of occurrence of earthquakes with water level changes.

Any active fault that is cut by a reservoir will be influenced by the changes in water level. The correlation of shallow focus microearthquake activity on active faults within reservoirs should always be detectable. Studies of active faults use microearthquake data to refine fault dimensions and location because the high rate of occurrence of microearthquakes provides a large amount of data after a few weeks of observation. The assumption that all microearthquakes represent a precursor of a large earthquake is false.

Microearthquakes release so little energy that changes in the water level could account for all of the energy released (Banks and Meade, 1982). Microearthquakes found near plutons, deep mines, and volcanoes are examples of seismicity generated by local stress changes. These small earthquakes are not related to large tectonically spawned earthquakes. In contrast to small shocks, the energy released by damaging earthquakes is so large that the stress changes caused by the reservoir are minor. Although the reservoir could possibly trigger a large shock, the energy changes caused by the reservoir should not increase the size of the maximum earthquake that could take place on a fault.

An earthquake catalog of the area of influence of the reservoir provides the data to answer the first question. Earthquakes large enough to be felt, not microearthquakes, were taken as data. Only felt earthquakes were considered in order to screen possible non-tectonic seismicity from the data. The presence of microearthquakes without a history of larger earthquakes should not influence the siting or operation of reservoirs. The rarity of damaging earthquakes occurring near reservoirs supports the decision to reject data that contains only microearthquakes.

The region of the crust within 20 km of the reservoir boundary was adopted as the zone of influence for the reservoir. This zone should be considered to be three-dimensional, although in the study the depth dimension was not explicitly used. This limit was assigned based on evaluation of three cases of induced seismicity due to fluid injection. Two of the cases were in the USA. One was at Rocky Mountain Arsenal, in Denver, Colorado, and the second, in the Rangely Oil Fields, also in Colorado. The third case was in Matsushiro, Japan.

The question of increased earthquake activity after impoundment is difficult to answer for areas normally considered earthquake country such as Greece, Japan, and central Asia. In seismically active areas, the earthquakes occurring in the reservoir area may be unrelated to the reservoir. None of the literature summarizing reservoir-induced seismicity has attempted to separate coincidental earthquake activity from induced activity. These studies, that listed all reported cases of induced seismicity, assumed that all earthquakes near reservoirs were induced. This assumption is unwarranted and misleading.

For example, the reputation of Kremasta as a case of reservoir-induced seismicity was based solely on coincidence. At Kremasta, the evaluation of the pre-impoundment and post-impoundment seismicity taken with the relatively deep focus of the one large earthquake in the reservoir zone of influence implies that triggering by the reservoir was unlikely. The Kremasta data are presented later in this paper.

Water level records and an earthquake catalog for the reservoir area form the raw data to address the second question, that of correlation of the reservoir operation to earthquakes. Changes in water level are interpreted as affecting the crust in the reservoir area in one of two ways. The water level change can move the crust toward failure or not. The analysis of water level is based on the regional stress conditions as evaluated from fault type or fault plane solutions of earthquakes. The interpretation of water level effects on stress in the crust are not complex but a familiarity with the principle of effective stress is essential.

When a reservoir water level rises the stresses due to the weight of the water rise. The water pressure within the crustal rocks is changed in two ways. One effect is caused by the change in shearing stresses due to water weight. This effect occurs immediately but dissipates completely after some time. The other change is due to the change in water level elevation (head). This change occurs slowly and is permanent.

The water loading is applied within the reservoir boundaries. This limited loading area creates shearing stresses within the crust as well as a change in compressive stresses. Rock at depths of several kilometers or more will always experience an increase in pore water pressure with an increase in shear stress. This change in pore water pressure is temporary and diminishes as water flows out of the crustal rocks. The time for flow to take place depends on permeability of the rocks and the variability in permeability within the crust. This variability is not known with confidence at earthquake focal depths. An assumption must be made regarding the rate of dissipation of temporary changes due to shearing stresses associated with water level changes.

The rising water level can directly change the water pressure within the crustal rocks just as pumping water into or out of porous rock changes the water pressure within the pores of the rock. The change in water pressure is not temporary but acts to change the steady-state water pressure. Of course,

if water levels continually fluctuate no constant steady-state pressure can be achieved. A reservoir may produce effects similar to an injection well for the part of the crust adjacent to the reservoir. For the crust directly below the reservoir the changes in total stress due to water weight rather than pore water pressure will dominate.

The reservoir water level will form the basis of long term water pressure if the water in the crustal rock and the reservoir water form part of the same flow regime. It is possible that some impervious layer creates a boundary between the reservoir water and the water at focal depths so that the reservoir resembles a bathtub rather than a standpipe recording the water pressure in the crust. The unity of the reservoir water and the crustal water has been described as a case with fluid communication and a bathtub situation has been described as a case with no fluid communication. Fluid communication was assumed in each case.

The water level data is used to assign a stability state to the crust. When changes in the water level would weaken the crust a stability state of T is assigned, meaning triggering is possible. When the water level is

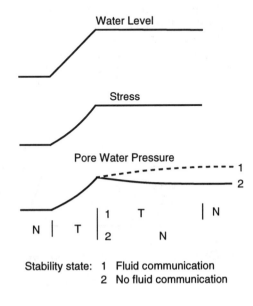

Figure 3.1 [*orig. figure 1*] Effects of changing water level.

changing in a manner that would not weaken the crust a state of N is assigned, meaning that triggering is not possible. The effects of water level on the stability state are shown in figure 3.1.

The earthquake catalog data is examined with respect to the stability state at the time that an earthquake occurred. The numbers of earthquakes occurring during each stability state show the correlation of the reservoir and the earthquakes. In cases where the seismicity is not triggered by the reservoir, earthquakes should take place during state N as often as during state T. If most of the earthquakes take place during state T or if little earthquake activity is recorded during state N, then the reservoir is triggering the activity.

A complete description of the method is given in the Corps of Engineers study (Meade, 1982). The method is simple in concept, but sometimes the earthquake catalog data is too sparse to apply the method confidently. The absence of accurate pre-impoundment data is frequent and often the locations of the epicenter and focus have a significant measurement error.

Another problem is that the seasonal cycles of filling make it difficult to assign state N except in times when the water level falls below the previous year's low. The tendency to assign a state of T when conditions are uncertain due to the time lag in raising pore pressures biases the evaluation in favor of triggering.

Results of the 1982 Study

A judgement based on circumstantial evidence is rarely conclusive. The conclusions drawn in the 1982 study are reasoned opinions or beliefs. The author of the study depicted his beliefs as shown in figure 3.2.

The two-question method focussed attention on the quantity and quality of the data. The chief limitation of the data was the incompleteness of the earthquake catalogs for the reservoir area of influence. The quality of the pre-impoundment catalog was high for only one case, Nurek. The other seven cases had inconsistent or incomplete catalog data. For these cases the evaluation of the seismicity in the reservoir areas was a major effort. Data was drawn from a wide variety of published sources. Incompleteness of the catalog data was one of the principal reasons for placing cases as undecided in the results of the study.

Nurek (USSR)

Nurek Dam is a 315 m high earth dam on the Vakhish River in Central Tadyikistan in the Soviet Union. The seismicity of the region has been studied since the 1930's. Detailed studies began in 1955 when a high sensitivity network was installed. By the mid 1960's shocks greater than magnitude 1.7 (Soviet energy class $K=7$) could be located to within an accuracy of 5 km.

Filling began in 1967 but the first substantial increase in water level took place in

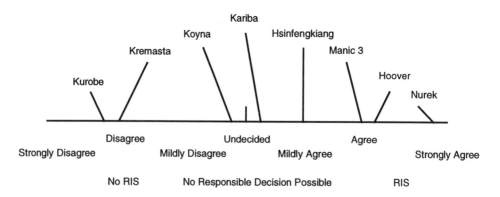

Figure 3.2 [orig. figure 2] Decision diagram for Corps of Engineers Study (Meade, 1982). RIS = reservoir-induced seismicity.

1971. From 1960 to 1971 the average number of earthquakes per quarter (3 months) was 26. In 1971 the average number was 40 and in the last quarter of 1972 133 shocks were recorded. These data are shown in figure 3.3. This catalog is the most complete of any case examined in the study. The quality of the data permitted use of a statistical test. A two-sample Wilcoxon test was performed to test the hypothesis that the median 3 month average before impoundment was the same as after impoundment. The alternate hypothesis was that the median after filling was larger. The null hypothesis was rejected at a 99% level of confidence (P value = 0.0003).

Nurek was also unique among these cases for its thrust fault setting. Simpson and Negmatullaev (1978) suggested that drops in water level cause bursts of seismic acitivity. During the 1976–77 filling cycle the testing of a power tunnel caused a rapid 3 m drop in the water level. A burst of shocks occurred. Simpson inferred that the rapid drop caused the shocks. At the time of the 3 m drop the water depth was about 160 m.

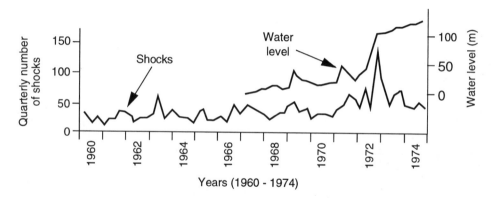

Figure 3.3 [orig. figure 3]   Quarterly number of shocks (K>=7) at Nurek (after Soboleva and Mamadaliev, 1976).

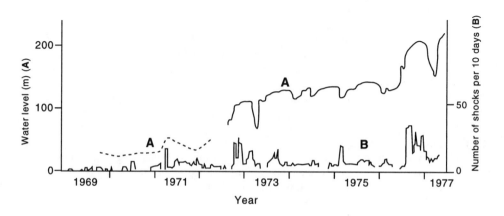

Figure 3.4 [orig. figure 4]   Correlation data for Nurek (after Simpson and Negamatullaev, 1978).

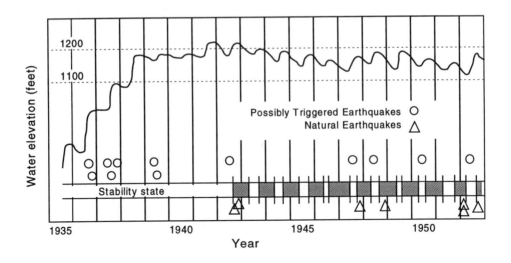

Figure 3.5 [*orig. figure 5*]   Correlation data for Lake Mead. Stability state bar open = triggering possible; stability state bar shaded = triggering impossible.

The water level data for that period is shown in figure 3.4.

Hoover (USA)

Located on the Colorado River southeast of Las Vegas, Nevada, the Hoover Dam impounded Lake Mead. In 1935 the reservoir was the largest in the world until surpassed by Kariba about 1958. No earthquakes (M_L>4.5) had occurred in the reservoir area for 15 yr prior to impoundment. Twelve earthquakes (M_L>4.5) occurred in the reservoir area in the 29 yr after impoundment. Eight of the twelve occurred at times when the reservoir was weakening the crust (stability state T). These data are shown in table 3.2 and figure 3.5 (Meade, 1985). Since 1964 no earthquakes (M_L>4.5) have occurred in the reservoir area.

Manic 3 (Canada)

Hydro-Quebec began the Manicougan–Outardes hydroelectric project in 1959. Located north of the St. Lawrence River in eastern Quebec the project used nineteen dams to create seven reservoirs on the Manicougan and Aux–Outardes Rivers. Three of these reservoirs, Manic 3, Manic 5, and Aux–Outardes 4, are large reservoirs

with dams over 100 m high. The entire project area is vast, as shown in figure 3.6.

The seismicity of the area is low to moderate. At the southern terminus of the project near Baie Cameau there is moderate seismicity associated with the St. Lawrence River valley. Proceeding north towards the huge impact crater north of Manic 5 the seismicity becomes low. The seismic history of eastern Canada is given by Basham et al. (1979). The catalog of the region is complete for shocks of magnitude 4 and above. A list of shocks in the project area (until the end of 1981) is shown in table 3.3. The last of the shocks in 1975 took place near Manic 3.

One sensitive instrument was installed to monitor the project area in December 1974 (Le Blanc and Anglin, 1978). The instrument could detect microearthquakes as small as M = 1 in a 200 km radius. From January 1975 to mid-September 1975, no microearthquake activity was observed in the Manic 3 area. The filling of Manic 3 began in August 1975. In mid-September a few microearthquakes were recorded from the Manic 3 area. On October 20th portable instrumentation was sent to Manic 3 to locate the activity. Three days later a magnitude 4.1 shock took place near the dam. The portable array had not

Table 3.2 [*orig. table II*] Earthquakes at Lake Mead.

Date	Stability state
09/07/36	T
09/20/36	T
04/28/37	T
06/18/37	T
11/12/37	T
05/04/39	T
06/11/39	T
06/04/42	T
08/11/42	N
09/09/42	N
07/30/47	T
09/30/47	N
07/30/47	T
09/30/47	N
05/09/48	T
11/02/48	N
05/07/40	T
02/08/52	N
02/20/52	N
05/24/52	T
10/20/52	N
04/19/58	N
03/25/63	N
04/23/63	N
09/23/64	N

Stability state codes: T = triggering possible; N = triggering not possible.
Source for catalog data: Rogers and Lee (1976).
Note: All earthquakes had magnitudes larger than 4.0 or MMI greater than or equal to V.

been installed before the shock but the instruments were operating within 5 h after the shock. Filling was temporarily halted in response to the shock. Aftershock studies indicated that the events were located in a small area about 8 km upstream of the dam. Filling was resumed and the water level rose slowly to maximum pool and remained at that level regulated by the discharge from Manic 5. No other shocks have been felt.

Hsinfengkiang (China)

This reservoir is located in southern China in an area described as unstable by the Chinese. The reservoir area is well populated. In the 25 yr prior to filling, four earthquakes, intensity V–VI were reported. Small shocks (M_S = 2 or 3) were felt in the

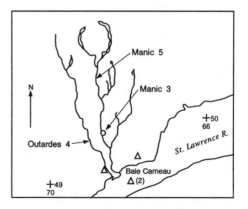

Figure 3.6 [*orig. figure 6*] Regional seismicity near Manic 3. Triangles = pre-filling earthquakes; pentagon = postfilling earthquake.

Table 3.3 [*orig. table III*] Regional earthquakes in the vicinity of Manic 3, 49–50°N, 67.5–69°W.

Date	Latitude (°N)	Longitude (°W)	Magnitude
06/12/17	49.0	68.0	4.0
05/17/38	49.0	68.0	4.0
06/23/44	49.4	67.8	5.0
10/21/58	49.2	68.5	4.0
Filling of Manic 3			
10/23/75	49.8	68.6	4.0

Note: Prior to 1963 only earthquakes of magnitude 4 or higher could have been recorded for this area.

reservoir area during October 1959, one month after filling began. An array of instruments was installed in July 1961 and recorded about five shocks per month in July. This rate increased to eleven per month by February 1962.

A large shock (M_S = 6.1) occurred on March 19, 1962. A large earthquake may have foreshocks and aftershocks that are components of the one large quake. The foreshocks and aftershocks can distort the catalog data. The evaluation of seismicity should use independent events as the raw data for any discussion of correlation of water level and activity. The occurrence of one large event makes the data ambiguous. Foreshocks and aftershocks should be culled out of the catalog data but no clear rules are available. Almost all events were located within 5 km of the reservoir. Data were published by Shen et al. (1974) and are shown in figure 3.7. If microearthquakes are discounted by drawing a line at 10 Joules then a few significant periods of seismicity can be seen. Stability decreased from October 1959 to mid 1963. In late 1963 there

was an increase in activity that was not due to the reservoir. During 1964 the water level rose and seismicity increased and included a magnitude 5.3 (M_S) shock. During 1967 the lake level was relatively low for the year. In 1968 the lake level rose abruptly and remained high for about a year with no increase in seismicity. After 1965 nearly all of the events were located in the narrow canyon near the dam and seem unrelated to reservoir level.

Kariba (Zambia)

Little pre-impoundment data was available. There were active faults in the reservoir area but they had produced no large quakes. Gough and Gough (1970) report that a tremor was felt near Binga in 1956 (see figure 3.8). A visitor to the area observed that the locals were not alarmed by the shock. In 1958 Lake Kariba was the largest sustained loading that man had created. Filling occurred slowly over several years (1958–1963). Seismic instruments were installed in 1959 after filling had begun. The array began recording small shocks within

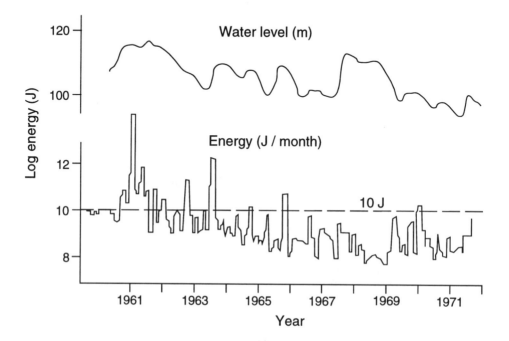

Figure 3.7 [*orig. figure 7*] Correlation data for Hsinfengkiang (after Shen et al., 1974).

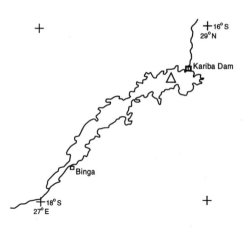

Figure 3.8 [*orig. figure 8*]   Epicenters at Kariba. A triangle marks the vicinity of large earthquakes, M > 5.

weeks of installation. In 1963 several large quakes occurred near the deepest portion of the reservoir. Most of the activity took place in a burst of seven large quakes in a 3-day period of September 23 through 25th and two more large quakes in October and November. No large earthquakes occurred again until 1966 when they struck in April 1966, April 1967, and June 1968. During these years the water level varied little. The activity in 1963 occurred during filling but the later events seem unrelated to the reservoir.

Koyna (India)

The seismicity at Koyna was examined in detail in the original study. Koyna is located in India near the central west coast in a region of lava flows known as the Deccan plateau.

The pre-impoundment seismicity ($M_L > 4$) was negligible based upon the records of the Benioff seismograph installed about 100 km away at Poona. The seismograph at Poona had been operating for about 12 yr before filling began in 1962. Gupta and Rastogi (1976) state that earthquakes as large as magnitude 4 would have been recorded at Poona. Mild tremors began with filling and the size of the earthquakes increased, capped by a magnitude 5.5 earthquake in September 1967 and a magnitude 6 earth-

quake in December of 1967. No nearby instrumentation was available until 1964. The earthquake epicenters were downstream of the dam within 20 km of the reservoir as shown in figure 3.9. The post-impoundment seismicity is substantial, so the answer to the first question regarding the increase in earthquake activity is yes.

The answer to the question of correlation is more difficult. The reservoir completed four complete seasonal cycles from 1963 to 1967 before the largest earthquake occurred as shown in figure 3.10. The lake level was highest in 1965. During the first loading in 1963, the lake level reached elevation 2145 ft and did not fall below 2040 ft. This 30 m (100 ft) of seasonal fluctuation could raise pore pressures a maximum of about 3 bars. The typical cycle was a rapid rise from of about 30 m within 8 to 16 weeks followed by a slow decline during the remaining 36 to 44 weeks each year. The 1967 filling cycle was similar to previous years but the maximum level was maintained slightly longer than usual, potentially allowing for an increase in pore pressure above that of previous years. Assignment of a stability state of T for all periods of rising water level was considered. The assignment of a stability state of N during all cases of declining water levels was not justified. In those years where the water level declined below the previous year's minimum a state of N was imposed until the water rose past the previous year's minimum. In June, July and August of 1966 and from April to July of 1967 the circumstances permitted an assignment of state N.

A catalog of earthquakes was assembled by Guha et al. (1974). The magnitude of located earthquakes ranged from 2.1 to 3.9 plus the magnitude 7 earthquake of December 10, 1967. No earthquakes larger than magnitude 3.9 were located until the largest earthquake occurred. Of these small earthquakes recorded from 1965, 1966, and 1967, most took place in the brief 2-day episodes during the months of November and December. These periods of earthquake activity take place when the water level has been declining for weeks. Any triggering theory had to postulate a delay effect to

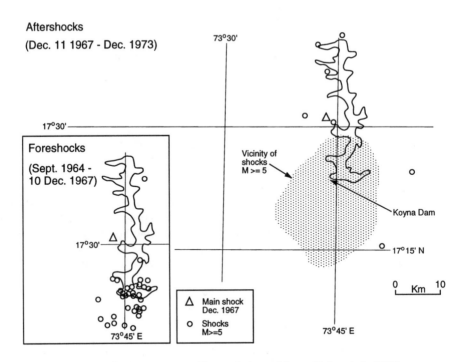

Figure 3.9 [*orig. figure 9*]   Epicenters at Koyna (adapted from Guha et al., 1974).

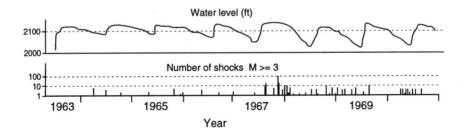

Figure 3.10 [*orig. figure 10*]   Correlation data at Koyna 1963–1970 (adapted from Guha et al., 1974).

make this sequence plausible. These shocks are unrelated to the reservoir.

Kremasta

Located in central Greece, Kremasta is within earthquake country. The dam was built in 1964 and upon filling small shocks were felt near the dam. On February 5, 1966, a large shock ($M_S = 6.3$) took place within the reservoir zone of influence. This large quake was located outside of the reservoir boundary and had a focal depth probably in excess of 20 km. The large quake fits easily into the normal seismicity pattern of central Greece as shown in figure 3.11. Small quakes felt at the dam (Therianos, 1974) may have been induced but they were unrelated to the quake of February 5, 1966.

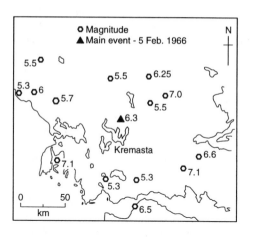

Figure 3.11 [*orig. figure 11*]   Regional seismicity near Kremasta.

Kurobe (Japan)

Kurobe Dam was begun in 1956 in a mountainous region of the central island of Honshu. On August 1961 a magnitude 4.9 earthquake occurred within 10 km of the dam. Several days later on August 21 a magnitude 4 shock took place at nearly the same location. The earthquakes at Kurobe were consistent with the regional seismicity. Three quakes ($M > 3.8$) were associated with the reservoir. The two largest quakes occurred in Agust 1961 within 20 km of epicenters of large ($M = 6$) historic quakes as shown in figure 3.12. The first quake on August 19, 1961 took place just 8 h after the Kita Mino earthquake ($M = 7.2$). The epicenter of the Kita Mino quake was about 100 km from the dam but the earthquake caused "rather strong" shaking in the Kurobe area as shown by the isoseismals (figure 3.13).

Additional Results

Koyna (India)

Intermittent activity has continued as reported by Gupta (1985). In the years 1975 through 1980, 27 earthquakes of magnitude 4 or larger have occurred. Of these 27, eight

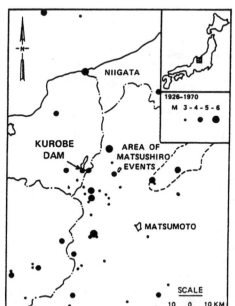

Figure 3.12 [*orig. figure 12*]   Regional seismicity near Kurobe (from Hagiwara and Ohtake, 1972).

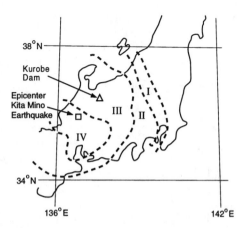

Figure 3.13 [*orig. figure 13*]   Isoseismals of the Kita Mino earthquake (adapted from Hagiwara and Kayano, 1961), isoseismals on Japan Meterological intensity scale.

occurred on September 20, 1980. These eight plus three others took place in August and September during or immediately after a period of summer filling. The one other earthquake of 1980 occurred during declining water level. Of the 15 earthquakes that occurred from 1975 through 1979, five occurred during or immediately after summer filling and 10 occurred during the slow water level decline that occurs during most of the year. This data is summarized in table 3.4 and figure 3.14.

In a typical year (1975–1980), most of the earthquakes took place during a non-triggering stability state. Any claim that the reservoir is triggering earthquakes looks selectively at the 1980 experience and ignores the bulk of the data.

The recent data supports the conclusion of the original evaluation (Meade, 1982) that the Koyna seismicity was unrelated to the reservoir.

Aswan (Egypt)

The rapids at Aswan have been dammed since 1932 when a 52 m high dam was raised. In 1970, a high dam located 7 km upstream of the old dam was built to a height of 111 m. The reservoir level began to rise in 1970 and by 1975 the reservoir depth was 93 m. The typical variation in water level since 1975 was 6 m/yr with a maximum of 10 m (Gibowicz et al., 1983).

The historical seismicity of southern Egypt is low, but damaging earthquakes are not unknown in the region. Early records of a large earthquake in the area in 1210 BC caused cracks in a temple at Abu-Simbel (Gibowicz et al., 1983). Seventeen years after filling, a magnitude 5.5 earthquake took place on the Kalabsha fault on November 14, 1981. This fault runs through an arm of the reservoir near Abu-Simbel (figure 3.15). The Kalabsha Wadi area was within the reservoir of the old 1932 dam but the water depth was rather shallow and varied seasonally. Aftershock activity continued during the next year, the largest being a magnitude 4.5 shock on August 20, 1982.

The single episode of 1981 was not accompanied by any unique reservoir activity. The reservoir had been filled initially 17 yr before the earthquake and the mild seasonal fluctuations of 10 m or less had been

Table 3.4 [*orig. table IV*] Koyna earthquake data 1975–1980.

Total number of earthquakes $M_L > = 4$	27
Number occurring in 1980	12
Number occurring on September 20, 1980	8
Number during 1975 through 1979	15
Number occurring 1975 through 1979 during stability state N	10
Number occurring 1975 through 1979 during stability state T	5

Data adapted from Gupta (1985).

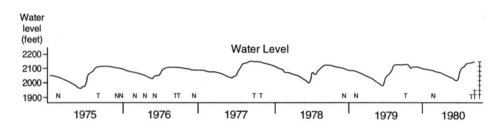

Figure 3.14 [*orig. figure 14*] Correlation data at Koyna, 1975–1980 (adapted from Gupta, 1985). N and T = earthquakes with magnitude > 4; N = triggering impossible; T = triggering possible.

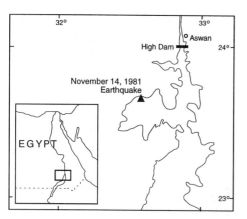

Figure 3.15 [*orig. figure 15*]   Epicenter near Aswan of the November 14, 1981 earthquake.

repeated for years as well. This earthquake was not triggered by filling. The initial rise in water occurred more than 10 yr prior to the earthquake. The seasonal fluctuations were modest, compared with Koyna for example. The reservoir operations have continued without any large earthquakes since 1982.

The continued presence of micro-earthquakes on an active fault do not provide proof of a relationship between the reservoir and the Kalabsha fault. The micro-seismic activity of the Kalabsha fault was not investigated prior to impoundment, so it is not possible to conclude that the seismicity

has changed in location due to impoundment.

The lack of earthquakes triggered by filling in 1975 and the continued lack of large earthquakes since the one episode in 1981 show that the reservoir at Aswan is not an example of induced seismicity.

Conclusions

Koyna and Aswan have been put forth as cases of induced seismicity (Gupta, 1985; Simpson et al., 1985). The evidence of induced seismicity is lacking in each of these examples. The merits of these two cases relative to the findings in the first study is shown in figure 3.16. Use of these cases as examples of induced seismicity has been built on the false assumption that every earthquake occurring in or near a reservoir is induced by the reservoir.

References

Banks, D. C. and Meade, R. B., 1982. Slope Stability; Control and Remedial Measures; and Reservoir Induced Seismicity: Considerations at Natural and Man-Made Lakes, Proc., 4th Int. Congr. Eng. Geol., New Delhi.

Basham, P. W., Weichert, D. H. and Berry, M. J., 1979. Regional assessment of seismic risk in Eastern Canada. *Bull. Seismol. Soc. Am.*, 69 (5): 1567–1602.

Gibowicz, S. J., Droste, Z., Kebeasy, R. M., Ibrahim, E. M. and Albert, R. N. H., 1983. A microearthquake survey in the Abu-Simbel area in Egypt. *Eng. Geol.*, 19 (2): 95–109.

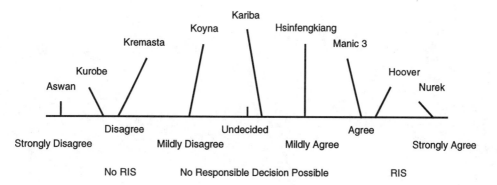

Figure 3.16 [*orig. figure 16*]   Revised decision diagram (1990). RIS = reservoir-induced seismicity.

Gough, D. I. and Gough, W. I., 1970. Load induced earthquakes at Kariba – II. *Geophys. J. R. Astron. Soc.*, 21: 211–217.

Guha, S. K., Gosavi, P. D., Nand, K., Padale, J. G. and Marwadi, S. C., 1974. Koyna Earthquakes (October 1962 to December 1973). Report of the Central Water and Power Research Station, Khadakwasha (South), Poona-24, India, 340 pp.

Gupta, H. K., 1985. The present status of reservoir induced seismicity investigations with special emphasis on Koyna earthquakes. In: S. J. Duda and J. Vaněk (Editors), Quantification of Earthquakes. *Tectonophysics*, 118 (3/4): 257–279.

Gupta, H. K. and Rastogi, B. K., 1976. *Dams and Earthquakes*. Elsevier, New York.

Hagiwara, T. and Kayano, I., 1961. Seismological observations of the Kita Mino earthquake, August 19, 1961 and its aftershocks. *Bull. Earthquake Res. Inst.*, 39: 873–880 (in Japanese).

Hagiwara, T. and Ohtake, M., 1972. Seismic activity associated with the filling of the reservoir behind Kurobe Dam, Japan, 1963–1970. *Tectonophysics*, 15: 241-254.

Le Blanc, G. and Anglin, F., 1978. Induced seismicity at the Manic 3 reservoir, Quebec. *Bull. Seismol. Soc. Am.*, 68 (5): 1469–1485.

Meade, R. B., 1982. The evidence for reservoir-induced macroearthquakes. Rep. 19, State-of-the-art for assessing earthquake hazards in the United States. Misc. Pap. S-73-1. US Army Corps of Engineers, Waterways Experiment Station, Vicksburg, Miss., 192 pp.

Meade, R. B., 1985. The Hoover Dam earthquakes reconsidered. Proc., Water Power '85, Las Vegas, NV. pp. 1308–1315.

Rogers, A. M. and Lee, W. H. K., 1976. Seismic studies of earthquakes in the Lake Mead, Nevada–Arizona region. *Bull. Seismol. Soc. Am.*, 66 (5): 1657–1681.

Shen, C., Chen, H., Chang, C., Huang, L., Li, T., Yang, C., Wang, T. and Lo. H., 1974. Earthquakes induced by reservoir impounding and their effect on the Hsinfengkiang Dam. *Sci. Sinica*, 17: 239–272.

Simpson, D. W. and Negmatullaev, S., Kh., 1978. Induced seismicity studies in Soviet Central Asia. *Earthquake Inf. Bull.*, 10 (6): 208–213.

Simpson, D. W., Kebeasy, R. M., Maamoun, M., Ibrahim, E. M. and Albert, R. N., 1985. Induced seismicity around Aswan Lake (abstract). In: S. J. Duda and J. Vaněk (Editors), Quantification of Earthquakes. *Tectonophysics*, 118 (3/4): 281–282.

Soboleva, O. V. and Mamadaliev, U. A., 1976. The influence of Nurek Reservoir on local earthquake activity. *Eng. Geol.*, 10 (2–4): 293–305.

Therianos, A., 1974. The seismic activity of The Kremasta area – Greece – between 1967 and 1972. *Eng. Geol.*, 8: 49–52.

4

HUMAN INFLUENCE UPON SEDIMENTATION IN LLANGORSE LAKE, WALES*

R. Jones, K. Benson-Evans and F. M. Chambers

Introduction

The Severn Basin has been adopted as the reference area in Britain for Project 158 (The Palaeohydrology of the Temperate Zone in the Last 15000 Years) in the International Geological Correlation Programme (IGCP) (Gregory, 1983). The Severn Basin comprises the catchments of two major rivers – the Severn and the Wye; their sources are located within a few kilometres on the eastern slopes of Plynlimon in Mid Wales from which the rivers diverge markedly, with the Wye taking the more southerly route, until their confluence in the Severn Estuary. Palaeolimnological studies under Subproject 158 B (Lake and Mire Environments) in the Severn Basin have so far been concentrated in an area of the Wye catchment at Llangorse Lake, the largest natural body of freshwater in the southern half of Wales (figure 4.1(a)), and the refer-

ence site adopted for the South Wales region under Subproject 158 B.

Llangorse Lake is situated at 155 m O. D. in the Brecon Beacons National Park at National Grid Reference SO 13 26. It is approximately 150 ha in extent and predominantly shallow but for two localized troughs where the water depth exceeds 7 m (figure 4.1 (b)). The major inflow is from the south via the Afon Llynfi; small spring-fed streams debouching on the north and east shores have minimal flow in summer. A more detailed description of the lake is given by Jones and Benson-Evans (1974). Aquatic macrophytes and lakeside vegetation have been described by Seddon (1964) and by Guile (1965). Modern nutrient and phytoplankton studies have been conducted by Jones and Benson-Evans (1974) and by Tai (1975).

Palaeoecological investigations at Llangorse were initiated by Dr. Seddon in 1972

* Originally published in *Earth Surface Processes and Landforms*, 1985, vol. 10, pp. 227–35.

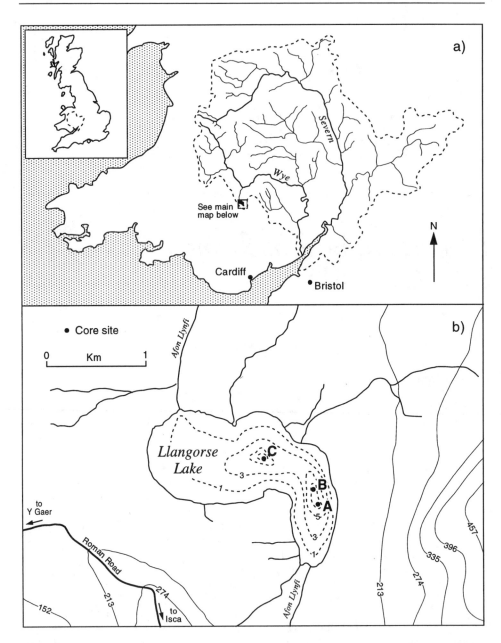

Figure 4.1 [*orig. figure 1*]   (a) Map showing location of Llangorse Lake within IGCP Project 158 Reference Area: the Severn Basin. (b) Map showing Llangorse Lake with inflow (from south) and outflow stream Afon Llynfi, and location of core sites A (12.35 m core), B and C from south to north. All contours in metres.

and continued more recently by Jones, Benson-Evans, and Chambers. The result of biological and chemical studies on sediments from short Mackereth cores collected by Seddon (site B) and by Jones (site C) were reported by Jones et al. (1978). Cores from both troughs in the lake exhibited a marked change in stratigraphy from a (lower) dark brown nekron mud to a red–brown silty clay over an interval of 6 to 10 cm. Radiocarbon dating of cores taken by Dr. Seddon indicated that the dramatic change in stratigraphy corresponded with Romano-British times. The change to predominantly allochthonous sedimentation was tentatively attributed to an intensification of arable agriculture and increased soil erosion in the Llynfi catchment following pacification of the Silures by Roman troops led by Frontinus in the campaign of AD 74–78 (Jones et al., 1978).

The present paper is concerned with evidence for human influence upon sedimentation in the Post-glacial as indicated by biological (pollen; diatom) and chemical (multielement) analyses on a 12.35 m core (site A) collected in 1980. Particular emphasis is placed on the section of the core in which the influence of man is first apparent. This core was extracted using a Livingstone sampler, since previous attempts using Mackereth corers failed to penetrate deeper than 3.5 m of sediment.

Methods

In essence the techniques applied and the general methods adopted for analyses followed the guidelines for Subproject B of IGCP 158, detailed in Berglund (1979, 1983).

The 1980 core was collected in 1 m increments employing a Livingstone sampler from a floating platform located above one of the two deep-water troughs in the lake (figure 4.1 (b)). The core was sectioned in the laboratory into 1 cm segments and subsampled for chemical, physical, pollen, and diatom analyses for sediment-based ecological study (Oldfield, 1977).

Analyses were conducted on samples at varying depth intervals to ensure representation of all units of sediment stratigraphy.

Water content (% wet wt) and dry wt per ml of fresh sediment were determined by drying sediment at 110° C for 48 hours). Oven-dried (110° C for 48 hours) samples were ground with an agate mortar and pestle for multielement determination by radio-frequency analysis (Fassel and Knisely, 1974) after digestion in a hydrofluoric/perchloric/nitric acid mixture. Loss on ignition was determined by muffling ground, oven-dried samples at 550° C for one hour and then carbonates were determined by wt loss between 550° C and 950° C (Wetzel, 1970).

Samples were prepared for pollen analysis after Barber (1976), with silicone fluid as mounting medium. Spore tablets (*sensu* Stockmarr, 1971) were added to each sample to permit later calculation of absolute pollen frequency. Samples were initially prepared at 16 cm intervals; samples were then interdigitated until the behaviour of the curves became apparent; contiguous samples were analysed over the horizons of the elm (*Ulmus*) decline. The relative pollen diagram (figure 4.3) shows trees, shrubs, and a summary curve for non-arboreal taxa (NAP); spores and obligate aquatics were excluded from the pollen sum of 500 total land pollen (TLP) at each horizon.

Methods for preparing subsamples for diatom counts were detailed in Jones et al. (1978).

Samples for radiocarbon dating were submitted to East Kilbride (Core A and Core B) and to Cardiff (Core C) Radiocarbon Laboratories. Quoted dates are uncorrected, in conventional radiocarbon years BP.

Results

The stratigraphy of the 12.35 m core is depicted in figure 4.2. The core shows three major stratigraphic units. An upper red-brown silty clay of *c.*4 m corresponds with similar sediments in two short cores in which the base of the silty clay dates from *c.*1800 BP. The uppermost organic sediments have been radiocarbon-dated to 1980 ± 80 BP (SRR – 2378) in this core, whilst the lowermost organic deposits date from 8720 ± 80 BP (SRR – 2384); hence the organic section from

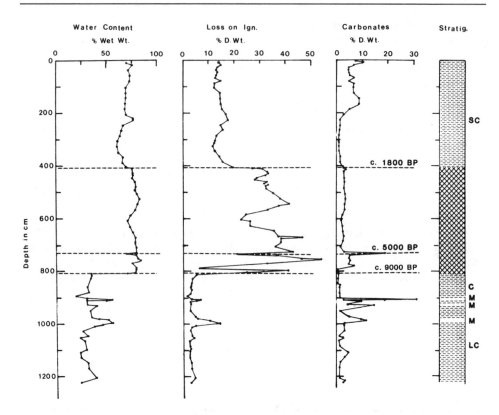

Figure 4.2 [*orig. figure* 2] Depth profiles of water content, loss on ignition, carbonates and stratigraphy of 12.35 m core A. LC: laminated clay; M: marl; C: clay; SC: silty clay.

8 to 4 m depth apparently corresponds with some 7000 years of deposition in the Post-glacial. Predominantly inorganic sediments, some of which are laminated, and bands of marl are found below the organic section of the core. Results of determinations of water content, loss on ignition, and carbonates for the whole core are presented in figure 4.2.

This paper is concerned with evidence for human influence upon sedimentation in the Post-glacial. Detailed results from the organic section of the core from *c*.8 m to 6.75 m depth are considered first. A relative pollen diagram is presented from 6.75 to *c*.8 m depth in figure 4.3. No diatom frustules were encountered in sediments below the elm decline horizon (737–736 cm); hence, diatom data are not presented here. Density, water content, loss on ignition, carbonates, and phosphorus determinations for the

interval 6.75 to 8.25 m depth are presented in figure 4.4, depth profiles for Ca, Mg, K, Ti, Al, Fe, Mn, and Fe : Mn ratio in figure 4.5.

The water content and loss on ignition (figure 4.2) reflect the gross stratigraphy described earlier. A substantial increase of water content and organic matter (loss on ignition is considered to reflect organic matter since above 6 per cent C, loss on ignition is proportional to organic carbon: Mackereth, 1966) occurred at the lower transition and a decrease of these parameters occurred at the upper transition between the organic and silty clay sediments.

High carbonate concentrations were associated with the bands of marl in the lower sediments and with a lighter coloured, more compact segment of the core between 730 and 736 cm (figure 4.2). Reference to figure 4.3 shows that this later peak in

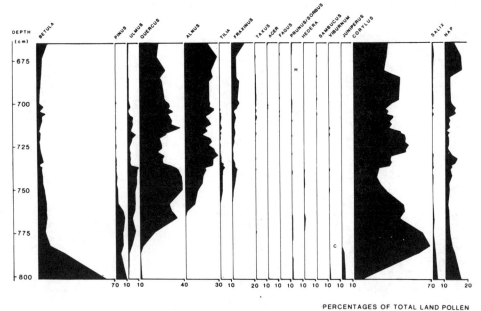

PERCENTAGES OF TOTAL LAND POLLEN

Figure 4.3 [*orig. figure 3*]   Pollen diagram for core A showing individual tree and shrub, plus summary curve for non-arboreal (NAP) pollen types from *c*.6.75 to 8 m depth. C: *Cornus*, H: *Crataegus*-type. The primary elm (*Ulmus*) decline horizon is from 737–736 cm depth.

Table 4.1 [*orig. table I*]   Radiocarbon dates on sediments from core A.

Depth (cm)	Sediment	¹⁴C Age (yr BP)	Lab no.	Comment
302–311	Red–brown silty clay	2530±90	SRR–2376	Too old: major contamination by 'old' carbon
376–386	Red–brown silty clay	2230±80	SRR–2377	Too old: contaminated by 'old' carbon
406–416	Transition	1980±80	SRR–2378	In close agreement with date from core B
451–461	Brown/black organic mud	3080±80	SRR–2379	
527–537	Brown/black organic mud	3240±70	SRR–2380	
604–614	Brown/black organic mud	4380±80	SRR–2381	
677–687	Brown/black organic mud	4140±80	SRR–2382	
759–769	Brown/black organic mud	8150±90	SRR–2383	
797–805	Brown/black organic mud	8720±80	SRR–2384	

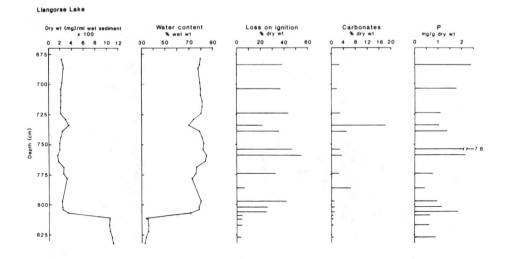

Figure 4.4 [*orig. figure 4*]   Depth profiles of density, water content, loss on ignition, carbonate and phosphorus from 6.75 to 8.25 m depth in core A.

carbonate concentrations immediately post-dates the elm decline horizon at which elm pollen representation declines from 10 per cent to 3 per cent of TLP. This is the earliest horizon with strong evidence for human impact on the Llangorse sediments. Radiocarbon dates have not been obtained for this specific horizon on this core but the indications from dated horizons above and below (table 4.1) are such that the date of this elm decline horizon is not inconsistent with an expected age of *c*.5000 BP or just before, (cf. Smith and Pilcher, 1973).

The part of the core between 825 and 675 cm exhibited changes in stratigraphy which correlated with sediment chemistry. Declines of organic matter above the inorganic/organic transition were due to high carbonates in the sediments. Organic matter is highest in the sediments of the Atlantic period and is associated with the highest concentration of P in the core (figure 4.4). The highest concentrations of Ca (figure 4.5)

were not associated with the highest concentrations of Mg and K in this part of the core, presumably because Ca is primarily associated in lake sediments with biological activity (Mackereth, 1966) whereas Mg and K are associated with the mineral products of erosion in lake sediments. Concentrations of Fe and Mn were low in the sediments with high carbonate content and the concentrations of both elements increased in the Atlantic period. The concentration of iron increased in the sediments after the elm decline but that of manganese did not (figure 4.5).

Discussion

The decline of *Pinus* pollen and increase of *Alnus* pollen representation at a depth of *c*.760 cm (figure 4.3) represents the Boreal/Atlantic transition of *c*.7500 BP; some 45 cm of sediment accumulated in the Boreal period from *c*.9000–7500 BP. The rate of

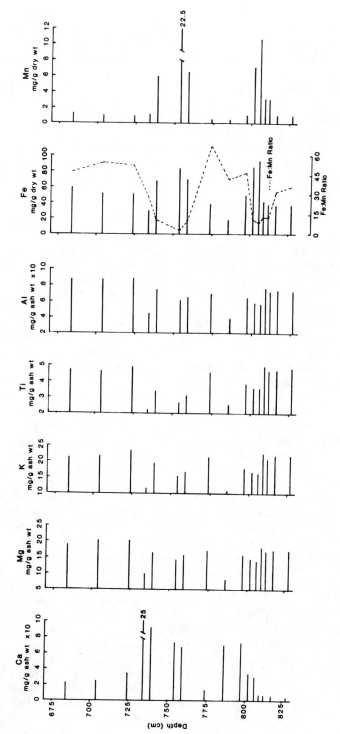

Figure 4.5 [*orig. figure 5*] Depth profiles of calcium, magnesium, potassium, titanium, aluminium, iron, and manganese, with Fe : Mn ratio from 6.75 to 8.25 m depth in core A.

sediment accumulation from 7500 BP to the Atlantic/Sub-Boreal transition at c.5000 BP (indicated by the decline of elm pollen) was slower since only c.25 cm of sediments accumulated during this 2500 year period. This period from 7500–5000 BP is considered by many to be the time of the Post-glacial climatic optimum. At this time, the rate of transfer of particles from catchments to British lakes was generally low (Pennington, 1981) with leaching of catchment soils postulated as the main process by which nutrients entered lakes (Mackereth, 1966). Evidence for leaching of nutrients from Llangorse catchment soils in the Atlantic period is the decline in the sediments of K and Mg, two elements postulated by Mackereth (1966) to be largely associated with the mineral products of erosion. The concentration of titanium, considered to be an essentially conservative element by Cowgill and Hutchinson (1970) and which appears to be associated with the clay fraction of soils (Goldschmidt, 1954), also declined through the Boreal/Atlantic transition, further suggesting a mainly leaching input of nutrients. The increase of Mn relative to Fe in the lake sediments and consequent fall of the Fe : Mn ratio during the Atlantic period suggests reducing conditions in catchment soils, such that Mn was mobilized largely in preference to Fe (Mackereth, 1966). Some iron, however, was also being mobilized because the sediment iron content in this period is higher than in the preceding and immediately following periods. It is also apparent that the deeper waters of the lake were sufficiently oxidizing to prevent mobilization of Mn from the sediment. It appears that aluminium was mobilized to some extent from catchment soils because its concentration in Atlantic period sediments did not decline as did K, Mg, and Ti. The increase of Ca and P in the sediments might be explained by their accumulation in sediments following leaching from catchment soils.

The high concentrations of Ca and the band of marl in the Boreal and Atlantic sediments suggests that base-demanding species of diatoms were probably present in the phytoplankton at that time but no frustules were found in the organic sediments which accumulated in the period up to 5000 BP. Since it is unlikely that diatoms were absent from the lake prior to the elm decline, it is postulated that the pH of bottom sediments was sufficiently high to induce dissolution of the frustules (Werner, 1977; Engstrom and Wright, 1984).

Some British lake sediments contain evidence for a general perturbation of the environment c.5000 BP which is detailed by Pennington (1981). There is often a significant decline of elm pollen and sometimes a change in lake sediment composition (Smith, 1981, pp. 147–9), the extreme of which is a change from organic mud to pink clay in the sediments of Barfield Tarn in the English Lake District. This change in stratigraphy is believed to have been due to ploughing of slopes around the Tarn for agricultural purposes (Pennington, 1981). The changes in Llangorse lake sediments at the time of the elm decline are also consistent with an environmental perturbation. There was an increase in dry weight density of the sediments due to a three-fold increase in carbonate content (figure 4.4). Calcium, too, increased but other elements (Mg, K, Ti, Al, Fe, and Mn) declined in concentration. Following the elm decline, however, the concentration of these elements, with the exception of Ca and Mn, increased in the sediments. These increases, particularly of K, Mg, and Ti suggest increased erosional input into the lake (Mackereth, 1966). Pennington (1981) postulated downcutting of stream channels into mineral soils as the cause of increased erosional input in lakes in northern England. She suggested that increased streamflow could be caused either by an increase in the precipitation/ evaporation ratio or by the clearance of primary forest by man. The decline in arboreal pollen and increase in non-arboreal pollen (NAP) representation in the Sub-Boreal (figure 4.3) at Llangorse Lake support the forest clearance hypothesis.

It appears that sediment accumulation in the period 5000 to 2000 BP increased substantially. Radiocarbon dates on organic horizons above the elm decline, however, are not sequential (table 4.1), suggesting input of 'old' carbon into samples yielding older-than-expected dates. Whilst sediment slumping or lake shore erosion (Pennington, 1980) are possible pathways and both could contribute to apparent increased profundal sedimentation rates (Dearing, 1982), there is no clear evidence of lake sediment reworking from the pollen record. Hence, in the 3000 years following the elm decline (i.e. during the Neolithic, Bronze, and Iron Ages), some 325 cm of largely organic sediments accumulated at Site A.

There followed an abrupt change in sediment stratigraphy through an interval of less than 10 cm which had been observed also by Jones et al. (1978) in cores collected from sites B and C (figure 4.1(b)). In core B this change in stratigraphy was radiocarbon-dated at 1790 + 60 yr BP (SRR – 129). (The uppermost organic sediments in core A were dated to 1980 + 80 BP (SRR – 2378).)

During the past 1800 years, c.4 m of silty clay has accumulated at site A while at sites B and C the accumulation has been 2.4 and 2.72 m respectively. The greater accumulation rate at site A is undoubtedly due to its being closer to the inflowing Afon Llynfi (figure 4.1(b)). The erosion of catchment soils is reflected in the radiocarbon dates on samples above the sediment transition which yielded significantly older-than-true radiocarbon ages (table 4.1).

Lead-210 dating of core A, as part of a study of heavy metals in Llangorse Lake sediments, provided a date of BP 1840 + 12 yr for sediments at a depth of 65 cm (Jones, in press). This suggests an intensification of the rate of removal of catchment soils to the lake during the past c.140 years. This intensifica-

tion probably began at the beginning of the 19th century when, in response to large increases in the price of cereals during the Napoleonic Wars and technological increases in agricultural practices, marginal lands and land further up hill slopes were brought into cultivation (Howells, 1977).

Today, in winter and spring, the suspended sediment load of inflowing streams is such that water entering Llangorse Lake is often red–brown in colour. However, this is not merely a recent phenomenon, for in AD 1188 the lake was reported by Giraldus Cambrensis to be at times tinged with red, as if blood flowed partially through veins and small channels (Rhys, 1908). Conceivably, this could be a reference to suspended sediment being carried into the lake by streams. A summary of the accumulation of sediment at site A is presented in table 4.2 where it can be seen that human activity has caused an increasing rate of transfer of catchment soils into Llangorse Lake during the past 5000 years.

Acknowledgements

Thanks are due to Dr. H. J. B. Birks for the loan of a Livingstone sampler and floating platform. Funds for chemical analyses were provided by Trent University. Radiocarbon dates on Core A were funded by N. E. R. C. Scientific Support Services. Mrs M. Patrick drew figure 4.1 [orig. figure 1].

References

Barber, K. E. 1976. 'History of vegetation', in Chapman, S. B. (Ed.), *Methods in Plant Ecology.* Blackwell Scientific Publications, Oxford, 5–83.

Table 4.2 [orig. table II]   Sediment accumulation rates at site A based on conventional pollen zones, ^{14}C and ^{210}Pb dating.

Pollen zone	Age (yr BP)	Depth (cm)	Accumulation rate (cm/100yr)
	AD1840–present*	65–0	59.0
VIII Sub-Atlantic	2800–AD 1840	c.445–65	14.1
VIIb Sub-Boreal	5000–2800	736–c.445	13.2
VIIa Atlantic	7500–5000	760–736	1.0
V, VI Boreal	9000–7500	c.805–760	3.0

* Based on ^{210}Pb date. May be underestimate of age of this horizon.

Berglund, B. E. (Ed.) 1979. *Palaeohydrological Changes in the Temperate Zone in the Last 15000 Years, Subproject B. Lake and Mire Environments Project Guide, Vols. I and II.* Lund University, Sweden.

Berglund, B. E. 1983. 'Palaeohydrological studies in lakes and mires – a palaeoecological research strategy', in Gregory, K. J. (Ed.), *Background to Palaeohydrology*, Wiley, Chichester, 237–254.

Cowgill, H. M. and Hutchinson, G. E. 1970. 'VII Chemistry and mineralogy of the sediments and their source materials', in Hutchinson, G. E. (Ed.), 'Ianula: An Account of the History and Development of Lago di Monterosi, Latrum, Italy'. *Trans, Americ. Phil. Soc.*, 60, 37–101.

Dearing, J. 1982. 'Core correlation and total sediment influx', in Berglund, B. E. (Ed.), *Palaeohydrological Changes in the Temperate Zone in the Last 15000 Years, Subproject B. Lake and Mire Environments Project Guide, Vol. III*, Lund University, Sweden, 1–23.

Engstrom, D. R. and Wright, H. E. 1984. 'Chemical stratigraphy of lake sediments as a record of environmental change', in Haworth, E. Y. and Lund, J. W. G. (Eds), *Lake Sediments and Environmental History*, University Press, Leicester, 11–67.

Fassel, V. A. and Knisely, R. N. 1974. 'Inductively coupled plasma – optical emission spectroscopy', *Anal. Chemistry*, 46, 1110–1120.

Goldschmidt, V. M. 1954. *Geochemistry*, Clarendon Press, Oxford.

Gregory, K. J. (Ed.) 1983. *Background to Palaeohydrology*, Wiley, Chichester, 486 pp.

Guile, D. P. M. 1965. *The Vegetation of Brecon Beacons National Park*, unpublished Ph.D. thesis, University of Wales.

Howells, B. 1977. 'Modern history', in Thomas, D. (Ed.), *Wales – A New Study*. David and Charles, London, 94–120.

Jones, R. in press. 'Heavy metals in the sediments of Llangorse Lake, Wales, since Celtic–Roman times', *Verh. Internat. Verein. Limnol.*, 22.

Jones, R. and Benson-Evans, K. 1974. 'Nutrient and phytoplankton studies of Llangorse Lake, a eutrophic lake in the Brecon Beacons National Park, Wales', *Field Studies*, 4, 61–75.

Jones, R., Benson-Evans, K., Chambers, F. M., Seddon, B. Abell, and Tai, Y. C. 1978. 'Biological and chemical studies of sediments from Llangorse Lake, Wales', *Verh. Internat. Verein. Limnol.*, 20, 642–648.

Mackereth, F. J. H. 1966. 'Some chemical observations on Post-glacial lake sediments.' *Phil. Trans. Roy. Soc. Land.* B 250, 165–213.

Oldfield, F. 1977. 'Lakes and their drainage basins as units of sediment-based ecological study', *Progress in Physical Geography*, 1, 460–504.

Pennington, W. 1980. 'The origin of pollen in lake sediments: an enclosed lake compared with one receiving inflow streams', *New Phytol.*, 83, 189–213.

Pennington, W. 1981. 'Records of a lake's life in time: the sediments', *Hydrobiologia*, 79, 197–219.

Rhys, E. 1908. *The Itinerary Through Wales and the Description of Wales by Giraldus Cambrensis*, J. M. Dent & Co., London.

Seddon, B. 1964. 'Some results of a lake flora survey of Wales', *Bot. Soc. Brit. Isles, Welsh Region Bull.*, 1, 3–6.

Smith, A. G. 1981. 'The Neolithic', in Simmons, I. G. and Tooley, M. J. (Eds), *The Environment in British Prehistory*. Duckworth, London, 125–209.

Smith, A. G. and Pilcher, J. R. 1973. 'Radiocarbon dates and vegetational history of the British Isles', *New Phytol.*, 72, 903–914.

Stockmarr, J. 1971. 'Tablets with spores used in absolute pollen analysis', *Pollen Spores*, 13, 615–621.

Tai, Y. C. 1975. *Phytoplankton Studies of Llangorse Lake, Breconshire*, unpublished Ph.D. thesis, University of Wales.

Werner, D. 1977. 'Silicate metabolism', in Werner, D. (Ed.), *The Biology of Diatoms*, Blackwell Scientific, Oxford, 110–149.

Wetzel, R. G. 1970. 'Recent and Post-glacial production rates of a marl lake', *Limnol. Oceanog.*, 15, 491–503.

5

SOIL CONSERVATION IN THE COON CREEK BASIN, WISCONSIN[*]

S. W. Trimble and S. W. Lund

Use of conservation practices after the 1930s reduced soil erosion and sedimentation rates significantly in much of the Driftless Area in the upper Midwest. Wisconsin's Coon Creek Basin, a 360-square-kilometer (139-square-mile) region southeast of LaCrosse, mirrored this striking transformation. Eroding slopes, many of them gullied, received improved management measures, such as contour plowing, contour strip-cropping, long rotation, crop-residue management, cover crops, improved fertilization, and controlled grazing (figure 5.1) (*3, 9*).

We compared sheet and rill erosion in 1934 and 1975 using the universal soil loss equation (*4*). Analysis of 10 sub-basins, comprising 9 percent of the Coon Creek Basin, indicated that average erosion rates in 1975 on upland fields [720 metric tons/square kilometer/year (3.2 tons/acre/year)] were about one-fourth those in 1934 [3,000 metric tons/square kilometer/year (13.4 tons/acre/year)] (*3*).

Our determination of sedimentation rates involved two methods: accumulation rates in small reservoirs and deposition rates in

the main valley. We surveyed and analyzed 10 small reservoirs for the period 1962–1975 and 5 for various periods between 1936 and

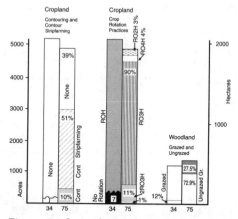

Figure 5.1 [*orig. figure 2*]   Land in various conservation practices, 10 sample sub-basins in the Coon Creek Basin, 1934 and 1975. For crop rotation practices, RO3H means 1 year of row crops, 1 year of oats or some small grain, and the following three years in hay. [Compiled in part from Lund (*3*), Tables 8 and 9.]

* Originally published in the *Journal of Soil and Water Conservation*, November–December 1982, vol. 37, pp. 355–6.

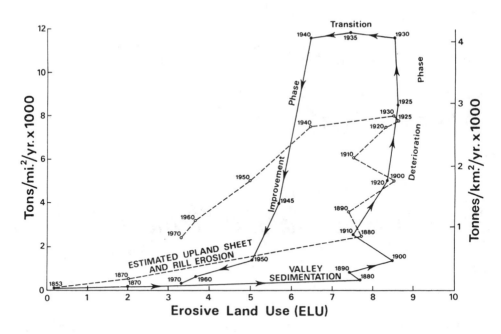

Figure 5.2 [*orig. figure 3*] Relation of erosion and sedimentation to erosive land use (ELU), Coon Creek Basin, Wisconsin, 1853–1975. ELU is an index created by weighting land use by the product of the cropping (C) and conservation practices (P) factors in the universal soil loss equation (table 5.1). The combined factors are multiplied by 100 to give whole numbers.

1945. The average annual sediment accumulation rate declined from about 4,800 metric tons per square kilometer (21.4 tons/acre) of drainage area in 1936–1945 to 55 metric tons per square kilometer (0.25 tons/acre) in 1962–1975.

In valleys we estimated deposition rates by detailed measurements (6), and distributed the sedimentation rates over time on the basis of deposition rates (7, 8) in the main valley. This analysis indicated that net annual deposition rates declined from about 3,700 metric tons per square kilometer (16.5 tons/acre) of drainage area for the 1930s to about 35 to 70 metric tons per square kilometer (0.16–0.31 tons/acre) in more recent time. Sedimentation data for both reservoirs and valleys indicated that later sedimentation rates were about 1 to 2 percent of those in the 1930s.

These reductions in erosion and sedimentation rates were caused principally by improvements in land management and to a lesser degree by changes in land use. Total

crop acreage did not change greatly, but conservation management of that cropland improved greatly (figure 5.1).

To describe the erosive effect of land use, we used a dimensionless, composite index termed erosive land use (ELU); land use is weighted by land treatment, which is the product of the vegetative cover factor (C) and the conservation practices factor (P) in the universal soil loss equation (5, table 5.1, 10). Although erosion and sedimentation rates were functions of ELU, a change in ELU did not immediately produce a commensurate change in erosion and sedimentation: both erosion and sedimentation rates rates lagged changes in ELU by 10 to 40 years (figure 5.2).

This lag increased with distance downstream and during the deterioration phase appeared to be caused by delayed reductions in soil infiltration capacities and delayed formation and growth of adequate upland and tributary channels to conduct larger storm-runoff peaks and sediment

Table 5.1 [*orig. table 1*]   Land use weighting factors to estimate erosive land use (ELU), 1853–1975. Factors are the product of the vegetative cover factor (C) and the conservation practices factor (P) in the universal soil loss equation.

	Weighting Factors			
Land Use	1853–1930	1940	1950	1960–Present
Row crops	.45	.32	.26	.20
Small grains	.20	.15	.13	.10
Hay and pasture	.04	.022	.014	.005
Farmstead and roads	.20	.15	.13	.10
Abandoned	.05	.045	.043	.04
Grazed forest	.03	.025	.023	.02
Ungrazed forest	.002	.002	.002	.002
Urban	.05	.05	.05	.05

load. Such channels, especially hillside and tributary gullies, were themselves sources of sediment and accounted for sedimentation rates that were higher than estimated upland erosion during the period from 1920 to 1940 (figure 5.2).

An additional factor also delayed increases in sedimentation rates in the larger stream valleys. Considerable sediment had been stored in small tributary valleys during the period from 1860 to 1920. These valleys were later trenched as runoff peaks became larger (7, 8).

During the improvement phase (figure 5.2), the lag of about 10 years between decreasing ELU and rates of erosion and sedimentation was caused by delayed restoration of soil infiltration capacities and healing and frequent occlusion of upland channels and gullies. These processes explain the disparity in 1975 between the estimated upland erosion and measured reservoir sedimentation rates. Sediment was not as easily conveyed downhill; more than 90 percent was lost as colluvium (upland deposition) on more level areas of cultivated fields, in grassed strips, along fence rows, and other locations where the velocity of overland water flow was reduced. In the 1930s, sedimentation rates were higher than upland erosion rates (figure 5.2). Therefore, as soil conservation measures became more effective, a smaller proportion of eroded material was conveyed from fields to streams.

The universal soil loss equation estimates how much soil moves; it does not predict how far that soil moves. Reports of alarmingly high erosion rates, as estimated by the universal soil loss equation, sometimes give the impression that soil loss and sedimentation are worse than they are (1).

We analyzed climatic change, an additional variable, by looking at the distribution of large rainstorms over time. Wet, stormy years came in cycles, but generally were not in phase with rates of erosion and sedimentation (figure 5.2). The worst erosion and sedimentation in the 1930s occurred during a dry period with few large storms, while the years in the "improvement phase" were marked by several wet, stormy periods.

Although the results of our study encompass one basin only, we believe this basin is representative of much of the Driftless Area. For example, the practice of contour stripcropping has spread from the Coon Creek Basin to the adjacent countryside (2). Other practices are similarly widespread. Although not as conclusive as this study, there is evidence too that other regions of the United States have benefited greatly from soil conservation (5).

References Cited

1 Brink, R. A., J. W. Densmore, and G. A. Hall. 1977. Soil deterioration and the growing world demand for food. *Science* 197: 625–630.

2 Johansen, H. E. 1969. Spatial diffusion of

contour strip cropping in Wisconsin. M. S. thesis. Univ. Wisc., Madison.

3 Lund, S. W. 1977. Conservation and soil lost through fluvial processes, Coon Creek watershed, Wisconsin, 1934–1975. M. S. thesis. Univ. Wisc., Milwaukee.

4 Soil Conservation Service, U.S. Department of Agriculture. 1975. Procedure for computing sheet and rill erosion on project areas. Tech. Release 51. Washington, D.C.

5 Trimble, S. W. 1974. Man-induced soil erosion on the Southern Piedmont, 1700–1970. *Soil Cons. Soc. Am.*, Ankeny, Iowa.

6 Trimble, S. W. 1976. Sedimentation in Coon Creek Valley, Wisconsin. In Proc., Third Federal Interagency Sedimentation Conf., Water Resources Council, Washington, D.C.

7 Trimble, S. W. 1981. Changes in sediment storage in the Coon Creek Basin. Driftless Area, Wisconsin, 1853 to 1975. *Science* 214: 181–183.

8 Trimble, S. W. 1982. A sediment budget for Coon Creek Basin in the Driftless Area, Wisconsin, 1853–1977. *Am. J. Sci.* (in press).

9 Trimble, S. W., and S. W. Lund. 1982. Soil conservation and the reduction of erosion and sedimentation in the Coon Creek Basin, Wisconsin. Prof. Paper 1234. U.S. Geol. Surv., Madison, Wisc.

10 Wischmeier, W. H., and D. D. Smith. 1965. Predicting rainfall erosion losses from cropland east of the Rocky Mountains. Agr. Handbk. 282. U.S. Dept. Agr., Washington, D.C.

PART II

Soil Impacts

INTRODUCTION

Soils are an integral and vital part of our environment. They interact with climate, vegetation, surface materials, water bodies and organisms. They provide food and fibre, support buildings and roads, and help to convert solar radiation into useable forms of energy and other resources. They tend to be thin, to take a long time to form, and to be prone to profound modifications in the face of human pressures. Such modification has been termed 'metapedogenesis' by Yaalon and Yaron (1966) in chapter 6. Another useful general review of soil metapedogenesis was that of Bidwell and Hole (1965) who used Jenny's (1941) range of soil forming factors (parent material, topography, climate, organisms and time) to assess whether the modifications achieved by humans were essentially beneficial or detrimental. The following summary is derived from Goudie (1993, p. 138):

1   *Parent Material*
    Beneficial: adding mineral fertilizers; accumulating shells and bones; accumulating ash locally; removing excess amounts of substances such as salts.
    Detrimental: removing through harvest more plant and animal nutrients than are replaced; adding materials in amounts toxic to plants or animals; altering soil constituents in a way to depress plant growth.
2   *Topography*
    Beneficial: checking erosion through surface roughening, land forming and structure building; raising land level by accumulation of material; land levelling.

Detrimental: causing subsidence by drainage of wetlands and by mining; accelerating erosion; excavating.
3   *Climate*
    Beneficial: adding water by irrigation; rainmaking by seeding clouds; removing water by drainage; diverting winds, etc.
    Detrimental: subjecting soil to excessive insolation, to extended frost action, to wind, etc.
4   *Organisms*
    Beneficial: introducing and controlling populations of plants and animals; adding organic matter including 'night-soil'; loosening soil by ploughing to admit more oxygen; fallowing; removing pathogenic organisms, e.g. by controlled burning.
    Detrimental: removing plants and animals; reducing organic content of soil through burning, ploughing, overgrazing, harvesting, etc.; adding or fostering pathogenic organisms; adding radioactive substances.
5   *Time*
    Beneficial: rejuvenating the soil by adding fresh parent material or through exposure of local parent material by soil erosion; reclaiming land from under water.
    Detrimental: degrading the soil by accelerated removal of nutrients from soil and vegetation cover; burying soil under solid fill or water.

It is not possible to provide chapters on all of these very many manifestations of the beneficial and detrimental effects of soil transformation by humans. However, no apology is needed for providing two chapters on different aspects of accelerated salinization, for this is one of the most unfortunate consequences of human activities of irrigation and deforestation. As Thomas and Middleton point out (chapter 7), salinization (or, more particularly, accelerated or secondary salinization) is a major process of land degradation and is arguably a type of desertification. However, the precise extent of salinization and associated waterlogging is not easy to determine, and it may have been exaggerated in the past. None the less, some more recent figures (table 7.1) do confirm that it is a problem of very widespread extent.

However, accelerated salinization, while it may often result from the expansion of irrigation in arid lands, does not solely result from that mechanism. As Peck points out in the second chapter on salinization (chapter 8), land clearance for agriculture can cause profound hydrological changes which lead to insidious saline seepage, as it does in large tracts of Australia.

Another important but often neglected process of soil transformation is soil drainage, achieved by surface ditching or by underdrainage with tile pipes and the like. Soil drainage has a long history and has been in many respects a successful way of increasing agricultural productivity or of allowing afforestation of badly drained areas. It has, however, had both detrimental and beneficial effects. The beneficial effects include the following: by leading water away, the water-table is lowered and stabilized, providing greater depth for the root zone. Moreover, well-drained soils warm up earlier in the spring and thus permit earlier planting and germination of crops. Farming is easier if the soil is not too wet, since the damage to crops by winter freezing may be minimized, undesirable salts carried away, and the general physical condition of the soil imrpoved. In addition, drained land may have certain inherent virtues: tending to be flat it is less prone to erosion and more amenable to mechanical cultivation. It will also be less prone to drought risk than certain other types of land. Paradoxically, by

reducing the area of saturated ground, drainage can alleviate flood risk in some situations by limiting the extent of a drainage basin that generates saturation excess overland flow.

On the other hand, there are a number of detrimental consequences. For example, some drainage systems, by raising drainage densities, can increase flood risk by reducing the distance over which unconcentrated overland flow (which is relatively slow) has to travel before reaching a channel (where flow is relatively fast).

Drainage can also cause long-term damage to soil quality. A fall in water level in organic soils can lead to the oxidation and eventual disappearance of peaty materials, which in the early stages of post-drainage use may be highly productive. This has occurred in the English Fenland, and in the Everglades of Florida (see chapter 2).

Some of the most contentious effects of drainage are those associated with the reduction of wetland wildlife habitats. In Britain this has become an important political issue, especially in the context of the Somerset Levels and the Halvergate Marshes.

The other theme that is developed in part II is soil erosion, though this was also addressed in chapters 4 and 5 in part I. Soil erosion is an insidious process, whether caused by wind or by water, and has recently been well reviewed by Morgan (1995). The chapter by van Andel, Zangger and Demitrack (chapter 9) demonstrates that accelerated soil erosion is not a new phenomenon and that it was both a significant problem in itself and a cause of valley sedimentation in prehistoric and historical Greece. There has, as they note, been some debate in the literature as to whether spasms of erosion are controlled by natural climatic fluctuations or by land-use changes. Chapter 10, by Boardman and Evans, illustrates that soil erosion is a current problem even in the relatively gentle climatic environment of the British Isles.

References

Bidwell, O. W., and Hole, F. D. 1965, Man as a factor of soil formation. *Soil Science*, 99: 65–72.

Goudie, A. S. 1993, *The Human Impact on the Environment* (4th edition). Oxford: Blackwell Publishers.

Jenny, H. 1941, *Factors of Soil Formation*. New York: McGraw Hill.

Morgan, R. P. C. 1995, *Soil Erosion and Conservation* (2nd edition). Harlow: Longman.

6

FRAMEWORK FOR MAN-MADE SOIL CHANGES – AN OUTLINE OF METAPEDOGENESIS*

D. H. Yaalon and B. Yaron

The vast activities of man as a soil modifier have not been as yet fully incorporated into the body of pedological investigations or into its general conceptual framework. Only occasionally do they serve as an object of a pedological study. More often than not the genetic processes altered by man's intervention are looked upon merely as deviations from the normal or genetical soil type, which do not concern the pedologist desirous of producing a soil map.

The investigations of these man-induced contemporary processes are pursued largely by the disciplines of soil chemistry, soil physics, soil microbiology, and soil technology, which, with the ever more refined modern methods of instrumentation, are able to follow in great detail a single or partial process active in soil formation. The study of these man-made changes in the soil body in its entity and in their interrelation with the environmental factors acting in the landscape has been neglected, often to the detriment of the applied agricultural and technological sciences.

The rate of change of the man-induced processes is relatively rapid, and present-day technology provides us with the means to modify all the agents of soil formation. The fundamental properties of a soil, which have been acquired under the influence of the environmental soil formers, are thus changed quantitatively within a short time, giving us a new soil with different characteristics. Often the man-made changes can be made reversible, provided a suitable input of energy is available to reverse the induced processes.

Studies of the influence of man on the soil fall into several groups. Some of the pedological aspects have been reviewed and discussed by Ganssen (5), Mückenhausen (13), and Scheffer (15), the first of these suggesting that the cultivated and cropped soils altered by secondary processes, caused by the activity of man, be treated as a separate entity from natural soils. Interesting historical outlines on the effect of man's cultural practices on soil conditions have been presented by Edelman (3, 4), Jacks (7) and others. Specific studies of profile changes have been made lately in several regions. They have treated mainly the effect of biovegetative changes and cultivation on

* Originally published in *Soil Science*, 1966, vol. 102, pp. 272–7.

soil profile properties (2, 10, 11, 12, 14, 16). Most recently Bidwell and Hole (1) have reviewed the impact of man on the soil and listed in detail the various man-created conditions which affect the factors and processes of soil formation.

It is a useful conceptual device to recognize models or frameworks, refer observations and experiments to them, and make them serve as a basis for predictions. The present exposition aims to show how man-induced changes in soil properties can be treated in a systematic way by utilizing the process-response model.

Man's manipulation of the soil does not affect only one or a few properties, but results in a simultaneous variation and an interrelated chain of changes of many soil parameters. We strongly urge, therefore, that the effects of man's economic activities be measured on the soil body in its entirety, and suggest experimental investigations in the field and in the laboratory to further this

objective. As man's activities extend over an ever larger part of the landscape and the effects of these activities become more drastic, it becomes more and more important to predict with a small margin of error the changes brought about by this technological impact on the soil.

A Systematic Framework for Man-made Soil Changes

We propose to study the man-made changes within a new reference system, independent of historical or environmental pedogenesis. The natural or normal soil serves as parent material and initial state of the man-directed functions. We call the man-induced processes and changes in the soil profile "metapedogenesis".

In the following we shall outline the parameters of the man-imposed soil-forming factors, which need not be analogous to the five classical pedogenetic factors and which

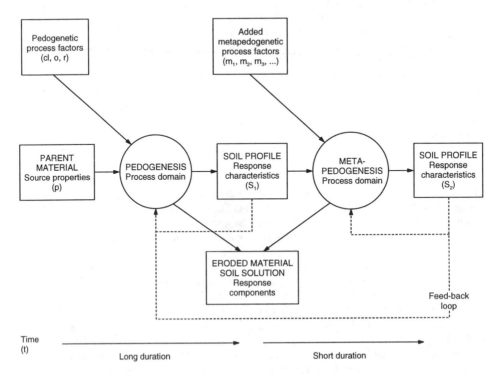

Figure 6.1 [orig. figure 1] A schematic diagram for a metapedogenetic process-response model. Feed-back is the effect of some response characteristics or properties on the processes.

are flexible enough to include any future technological factors which man may devise.

We start with a soil, whose development – without human intervention – is governed by the classical five pedogenetic factors. Let us consider that the soil is at the steady-state condition, with imperceptible changes taking place within a brief span of time. This soil (S_1) is the parent material at a defined time zero. The changes imposed by man starting at time zero are due to the new metapedogenetical factors (figure 6.1).

While in the natural process of soil formation the pedogenetical factors act slowly over a prolonged period of time, the metapedogenetical factors act on the soil properties with a strong and usually rapid effect. The metapedogenetical processes are mostly reversible. In mathematical language the following function can be written:

$$S_2 = f(S_1, m_1, m_2, m_3, \ldots)$$

where S_2 is the new soil produced by the action of the metapedogenetical factors (m_1, m_2, $m_3, \ldots$), acting singly or in combination, and which we group under the following headings: topographical factors; hydrological factors; chemical factors; and cultivation and cropping factors.

S_1 and S_2 represent an unspecified number of physical, chemical, and mineralogical properties of the soil profile. The metapedogenetical factors can be expressed in parameters which are defined and controlled by man. They define the soil system only in conjunction with the initial state of the soil S_1. Process-response models can be established for various partial processes, but because of feed-backs and interactions the process factor cannot be linked to any definite properties. A good example of feed-back effects is clay movement due to intensified leaching, for example irrigation. The illuviated clay decreases the soil's permeability which in turn affects infiltration and hence clay migration, and in turn the rate of leaching. The initial response becomes thus a process factor in the continuing sequence of process-response phenomena. By considering the total response of S_1 transforming into S_2

the composite of the many property functions and the whole sequence of events with time as variable is considered.

The transformation from S_1 to S_2 is sequential. Hence metapedogenetic functions can be established either by varying the intensity of the acting factors or by increasing the time. In order to obtain reversibility of the obtained phenomena it is usually not sufficient to remove the acting factor, it becomes necessary to apply an additional opposing force. Thus, while saline soils can be desalinized by drainage and leaching, and a destruction of the drainage system may again bring about the salinization of the soils, the redistribution of ions and clay produced by irrigation will not be immediately restored when irrigation is stopped.

Each of the exhaustive lists of the anthropogenic soil factors listed by Bidwell and Hole (1) can induce metapedogenesis. Selected examples of such soil changes have been assembled in table 6.1. In general, only changes at the level of soil family, as defined by the 7th Approximation (17), or higher levels have been considered as being of greatest interest in the present context. Changes at the level of soil series or soil type can, of course, also be studied within the framework of the metapedogenetical reference system.

It is not intended to propose a special classification system for the man-affected soils, though in certain cases supplemental groups may be necessary for some man-made soils in order to accommodate them in existing classification systems. Various attempts have been made to devise separate systems of classification for cultivated soil (5, 14), but this does not seem necessary, especially where the more modern morphogenetic system of soil classifications is used. The new United States soil classification system is especially comprehensive and encompasses a distinct effort to include man-induced properties in its category assignment (17).

Table 6.1 [*orig. table 1*]   Soil formation resulting from metapedogenetic processes.

Manipulation	Principal Processes Observed in the Soil	Soil Formation	
		Initial soil	Resulting soil
Topographical features			
Terracing or land leveling	Reduction of erosion; humus content increase; rejuvenation of pedogenetic processes; catenary slope differentiation altered	Lithosolic terra rossa (Lithic rhodustalfs)	Terra roma (Rhodustalfs)
		Rendzina (Rendolls)	Brown rendzina (Rendolls)
Dam construction on flood plains	Stopping of sedimentation and leaching; water table rise; salt accumulation;	Alluvial soil (Entisols)	Solonchak (Natrustent)
Hydrological factors			
Drainage, lowering of water table	Improved oxidation; structure formation; permeability change	Pseudogley (Aqualfs)	Parabraunerde (Udalfs)
Planting of wind breaks	Change of moisture regime; base saturation altered; carbonate leaching	Chernozem (Ustolls)	Leached chernozem (Altalfa)
Flooding of paddy fields	Hydromorphic water regime; reduced oxidation; gleying	Alluvium (Entisols)	Gley (Aquepts)
Chemical factors			
Irrigation with sodic water	Adsorption of sodium; structural degradation; decrease in permeability	Brunizem or Chestnut (Udolls)	Solonetz (Natrustalfs)
Clay marling or warping	Textural change in upper horizons; moisture regime change; base saturation altered	Regosol (Entisols)	Brown earth (Inceptisols)
Cultivation and cropping factors			
Deforestation and plowing in temperate areas	Mixing of upper horizons; change in pH; retardation of podsolization	Podsol (Spodosols)	Acid brown forest (Ochrepts)
		Lessive; grey-brown podsolic (Alfisols)	Brown earth (Ustalfs)
Deforestation and shifting cultivation in tropical areas	Erosion; dehydration of iron oxides	Latosol; ferrallitic soils (Oxisols)	Ferrallitic crusts (Acrox)
Over-grazing	Destruction of natural vegetation; erosion of surface horizons; reduction in infiltration	Various mountain soils (Ustent, Udents)	Lithosols; calcareous crusts

The table is not exhaustive but lists only selected examples, well documented in the literature, which show the variegated effects of the metapedogenetic processes in transforming soils.
* For readers' convenience, 7th Approximation equivalents are included.

Discussion

Selection of the systematic framework

There are three possible approaches (*a, b, c,* below) that could be used to form a suitable framework or reference system for the study of man-made soil changes.

(*a*) Man can be considered as an independent soil-forming factor on equal footing with, or in addition to, the five classical soil-forming factors. Jenny (8, 9) included the human factor among the three groups of organisms acting as pedologically independent variables. This approach has been followed more or less implicitly by many workers. Most soil books and regional surveys contain a brief chapter listing the changes, usually destructive, brought about by man's activities in deforestation, burning, irrigation, and cultivation. Specific studies of the changes in soil properties caused by man's activities are usually carried out by comparing unaffected virgin sites with cultivated ones. This approach enables the elucidation of the dynamics of the man-induced processes, especially if several points on a functional time sequence can be established.

One of the requirements of Jenny's functional approach and analysis is that in order to ascertain the role played by one of the soil-forming factors, it is necessary to demonstrate that all remaining factors are kept constant. Because of the relatively brief span of time over which man has acted as soil former, he has, in fact, in most cases acted as a modifier on already developed soils which have attained a certain dynamic equilibrium with the environmental conditions. These often cease to exert their influence with the appearance of man's activity, a point often neglected in such studies.

(*b*) It is possible to include all the man-induced processes and their conditioning factors within the classical framework of the five independent soil-forming factors, which according to Jenny (8) completely describe the soil system. Thus irrigation can be considered as a climatic factor, which changes the soil climate to a more humid one; marling, liming, and manuring are parent material factors similar to dust deposition on cumulative soils; land leveling and drainage would be topographical factors; burning and cropping biological factors, and so on. Bidwell and Hole (1) have grouped 38 such conditions according to the five factors.

We have not chosen this approach mainly because there are significant differences in the rates of reaction and reversibility between the environmental and man-induced processes that make it desirable to treat each separately. Also, many of the cultural changes fall within the domain of more than one of the factors and cannot be separated.

(*c*) According to Jenny (8, 9), whenever one of the variables that determine the properties of the soil system change, it is necessary to define a new zero time and a new parent material as the initial state of the system. The clarity of Jenny's functional system and the fact that it leads itself to experimental and quantitative evaluation, makes it particularly suitable for the study of the effect of man's activities on soil development. We suggested, therefore, the creation of a new reference system, with the natural or normal soil as the initial state of the man-directed functions.

Suggestions for metapedogenetic research

Genetic pedology uses largely descriptive methods in its research. It does not have the possibility of controlling quantitatively the influence of the factors that determine the formation of the soil. To a large extent it makes use of inductive reasoning and interpretations. Only the response is measured and identified and attributed to a likely set of processes.

In the studies of the metapedogenetic processes the descriptive methodology can be replaced by an experimental approach, which makes it possible to investigate quantitatively the processes occurring in the soil and to verify the processes postulated by inductive associations (6).

The metapedogenetical studies may be realized in two ways; (a) model laboratory studies on columns and undisturbed cores, and (b) field studies on experimental plots or lysimeters. In both cases, the controlled input of the metagenetical factor enables a quantitative solution of the equation shown above. Thus a laboratory study on models may give a quantitative appreciation on the chemical and hydrological factors and processes that accompany irrigation with saline water and that may lead to the formation of a secondary alkali soil. Field plots can be used in various ways and can give a quantitative appreciation of the new soil formed by, for example, cultivation and marling of a podsol.

Instead, considering only the simple partial processes, the changes, for example, in pH, base status, or salinity, the pedologist will evaluate the changes not only in the plow layer but in the whole sequence of soil horizons of the profile. Metapedogenetic changes have to be evaluated, not on the basis of the individual properties of a single horizon but according to the sum of the effects on the entire soil profile.

Today the pedologist gives the agronomist information on the nature of the soil and its inherent properties. This information helps the agronomist as does similar information furnished by the geologist, hydrologist, or meteorologist. By studying the metapedogenetical processes experimentally, the pedologist will be able to give to the agronomist a picture of the soil in the future, that is, to predict its response to various management and agrotechnical measures.

In science usefulness is in part measured by the ability to forecast and to predict relations that will obtain under given modified conditions. Metapedogenesis thus draws pedology closer to the needs of modern agriculture and technology. It extends its scope from the domain of a descriptive natural science to that of a quantitative applied science.

Summary

It is suggested that the man-induced changes in soil processes be studied in a systematic framework of a process-response model with the natural soil as parent material. The man-induced processes and soil profile changes are called metapedogenesis. Such a metapedogenetic concept provides a suitable framework within which all relevant factors can be marshalled and which can serve as a basis for the prediction of expected soil changes.

In general, the behavior of a metapedogenetic system, that is the resulting soil, depends on the intensity of the particular topographical, hydrological, chemical, or cultivation factor and on the capacity for adjustment of the initial soil. It is important that the total response on the soil profile as a whole be measured and studied, both in the field and in the laboratory.

References

1 Bidwell, O. W., and Hole, F. D. 1965 Man as a factor of soil formation. *Soil Sci.* 99: 65–72.
2 Damaska, J., and Nemecek, J. 1964 Contribution to the problem of genesis and fertility of soils in Czechoslovakia. *Rostlina Vyroba* 5–6: 540–555.
3 Edelman, C. H. 1950 *Soils of the Netherlands.* Holland Publishing Co., Amsterdam.
4 Edelman, C. H. 1954 L'importance de la pédologie pour la production agricole. Trans. Intern. Congr. Soil Sci. 5th Congr., Léopoldville I: 1–12.
5 Ganssen, R. 1959 Versuch einer vorläufigen genetischen Gliederung von Kultur und Wirtschaftsböden. *Z. Pflanzenernähr. Düng. Bodenk.* 87: 201–212.
6 Hallsworth, E. G., and Crawford, D. V., eds. 1965 *Experimental Pedology.* Butterworths, London.
7 Jacks, G. V. 1956 The influence of man on soil fertility. *Advance Sci.* No. 50, pp. 137–145.
8 Jenny, H. 1941 *Factors of Soil Formation.* McGraw-Hill Book Co., New York.
9 Jenny, H., and Ryachaudhouri, S. P. 1960 "Effect of Climate and Cultivation on Nitrogen and Organic Matter Reserves in

Indian Soils." Indian Council of Agricultural Research, New Delhi.

10 Lieberoth, I. 1962 The effect of cultivation on soil formation in the loess region of Saxony. *Albrecht-Thaer-Arch.* 6: 3–30.

11 Manil, G. 1956 Aspects dynamique du profile pedologique. *Trans. Intern. Congr. Soil Sci. 6th Congr.*, Paris V: 439–441.

12 Manil, G. 1963 Profil chimique, solum biodynamique et autres charactéristiques écologique du profil pedologique. *Science du Sol.* 1: 31–47.

13 Mückenhausen, E. 1961 Der Aufbau des Bodens und dessen Veränderung durch den landwirtschaftlichen Pflanzenbau. Vortragsreihe der 15. Hochschultagung Landwirtech, Fakult. Univ. Bonn: 1–28.

14 Nadezhdin, B. V. 1960 The problem of the principles of classification of anthropogenic soils. *Pochvovedenie* No. 1, pp. 64–71.

15 Scheffer, F. 1963 Das Transformationsvermögen der Böden als Grundlage zu ihrer Bewertung. *Landwirtsch. Forsch. Sonderh.* 17: 49–59.

16 Tavernier, R., and Pecrot, A. 1957 L'homme et l'evolution du sol en Belgique. *Pedologie* 7: 226–231.

17 U.S. Soil Survey Staff 1960 "Soil Classification, A Comprehensive System, 7th Approximation." U.S. Government Printing Office, Washington, D.C.

7

SALINIZATION: NEW PERSPECTIVES ON A MAJOR DESERTIFICATION ISSUE*

D. S. G. Thomas and N. J. Middleton

Introduction

Soil salinization, the concentration of salts in the surface or near-surface zones of soils, is a major process of land degradation, leading to falling crop yields and the loss of land from production in a range of environments. Human-induced salinization, often referred to as 'secondary salinization' to distinguish it from naturally salt-affected soils, is often cited as a major contributor to desertification processes in the world's drylands (e.g. Grainger, 1990), but has also been identified as a problem in more temperate environments (e.g. Szabolcs, 1974). It is a problem particularly associated with irrigation schemes in the world's drylands, reducing productivity in key food and cash crop growing regions. In Africa, for example, where short term food crises often overshadow more significant long term food production problems (McMillan and Hansen, 1986), it is regarded as limiting the effectiveness of large scale aid-funded irrigation projects (e.g. Barrow, 1987; Lewis and Berry, 1988). Despite being seen as better documented than other types of degrada-

tion (FAO/UNEP, 1984), it is difficult to establish the scale of salinization as a global and continental problem, leading to widely differing estimates of its extent and degree.

The purpose of this paper is 2-fold. Firstly, it provides a review of the nature of salinization and past attempts to evaluate the scale of its occurrence. Second, it introduces a new perspective on the issue, utilizing data from a new United Nations Environment Programme (UNEP) GIS database on land degradation (FAO/UNEP, 1984; UNEP-DC/PAC, 1990; Middleton and Thomas, 1992). This allows an assessment to be made of the scale of secondary salinization, including an evaluation of its importance within the broader land degradation category of chemical deterioration with particular reference to Africa.

The Nature of Salinization

Salinization is widely seen as a consequence of the misuse of irrigation (e.g. Vincent, 1990), cited by Kassas (1987) as one of the 'seven paths to desertification'. The problem is often viewed as being most acute in, but

* Originally published in *Journal of Arid Environments*, 1993, vol. 24, pp. 95–105.

not confined to, drylands. In such regions there is a natural tendency for the at- or near-surface accumulation of soluble salts, due to high evaporation rates favouring capillary rise combined with low absolute precipitation amounts and strongly seasonal regimes which together limit 'flushing' effects.

Salt accumulation can be exacerbated in specific locations when waters are added to the soil through agricultural practices or where natural vegetation communities are removed, and appears to be a problem particularly found in conjunction with irrigated agriculture. It is not a new problem however; though often considered a consequence of 'development' agricultural schemes in Lesser Developed Countries (e.g. Speece and Wilkinson, 1982; Grainger, 1990), it can equally apply to developed nations such as the USA (Berg et al., 1991) and has often been cited as a contributory factor in the fall of the Mesopotamian culture 6000 years ago (Jacobsen and Adams, 1958).

Causes of salinization

Discussions of the causes of secondary soil salinization are well represented in the literature (e.g. Goudie, 1986; Rhoades and Loveday, 1990; Rhoades, 1991; Barrow, 1991). Causes can be categorized and briefly summarized in five groups: salinization directly due to poor cultivation techniques; salinization due to indirect effects of irrigation schemes; salinization due to vegetation change; salinization due to sea water incursion, and salinization by disposal of saline wastes.

Irrigation is the cultivation technique widely associated with salinization (Lewis and Berry, 1988), and the FAO and UNEP (UNEP, 1992) note that the area of irrigated land world-wide has increased from *c*.8 million ha in 1800 to 250 million ha today. The quality as much as the quantity of irrigation water can affect the success of a project (Ayers and Westcot, 1985). This notwithstanding, there are perhaps four reasons why irrigation can cause salinization (Barrow, 1991): water leakage from supply canals, which for example raised the water table in parts of the Punjab by 7–9 m

over 10 years (El-Hinnawi and Hashmi, 1982: 196); over-application of water; poor drainage, and insufficient water application to leach salts away. In practice, making distinctions between the last three is often difficult, but it is an unfortunate paradox that in the water deficient drylands, over-application and bad drainage appear to be factors of considerable import (UNEP, 1992).

A number of indirect causes of salinization can also occur due to side effects of irrigation schemes. River barrages and dams can result in salinization of soils on valley terraces above a dam after its construction, due to changes in local water tables following the flooding of a reservoir. Irrigation schemes can also increase salinity in soils downstream of the water extraction point: the creation of about 30,000 km² of solonchak badlands on the shores of a rapidly shrinking Aral Sea between 1960 and 1989 is directly due to the expansion of irrigated lands from the waters of the Amu Darya and Syr Darya in Central Asia (Kotlyakov, 1991).

The direct effect of replacing a natural vegetation community with an anthropogenic one can be significant for salinization and may operate independently of, or in combination with, the direct effects of irrigation. In the former instance, it is reported to be a particular problem in parts of the drylands of the U.S.A. (Halvorson and Black, 1974; Brown et al., 1983; Berg et al., 1991) and Australia (Malcolm, 1983; Bettanay, 1986), leading to the development of saline seeps. The removal of natural vegetation and replacement with crops (or grassland for grazing) reduces evapotranspiration, particularly during the fallow period. The resultant soil water build up, below the root zone of shallow crops or grasses, leads to the down slope seepage of unsaturated groundwater, which picks up salts as it passes through permeable sediments. This eventually builds up saline groundwater over less permeable strata in lower slope positions (Halvorson and Black, 1974) over periods in excess of 20 years (Berg et al., 1991). The problem is greatest where water tables are shallow (Malcolm, 1983).

Soil salinization through sea water

incursion is a well known problem in the Middle East where excessive groundwater pumping has allowed the landward penetration of brine (Speece and Wilkinson, 1982). A further problem has been noted in West Africa, where drought and the extraction of water for irrigation have reduced the flow of the Senegal River to the sea, permitting considerable up-channel movement of ocean water (Anon, 1991). A similar problem has been created in the Nile delta, due to a reduced Nile discharge since the Aswan High Dam was completed in 1964 (Kishk, 1986).

Soil salinization can also occur when saline wastes from industrial and municipal sources are dumped directly on to soils. Petroleum wells, coal mines and industrial plants dispose of brackish water in this way, and emissions of municipal sludges and wastes may also carry high salt or alkali loads (Kovda, 1980).

Effects of salinization

From an ecological and agricultural perspective, salinization is a problem because it reduces productivity. The effects of salinization have been recently reviewed by Rhoades (1991) and can be viewed from the perspective of direct effects on the soil and direct effects on plants themselves.

In soils, the accumulation of salts reduces pore space and the ability to hold soil air and nutrients. Specifically, cations, which effectively embrace the salts (chlorides and sulphates) and carbonates of minerals such as magnesium, calcium and sodium, are attracted to clays by negative electrical charges. In this way, the presence of salts also contributes to the aggregation, and sealing, of soils. Salinity is measured in terms of electrical conductivity, with the exchangeable sodium percentage (ESP) or sodium adsorbtion ratio (SAR) and pH delimited.

Soil salinization also embraces the problem of alkalinization, the excess accumulation of sodium. The sodium hazard is complex, as the hazard associated with any SAR value increases as the degree of salinity rises. Wilcox (1955) identified sixteen classes of irrigation water on the basis of the combined relationship between salinity and alkalinity hazard, of which all bar one are damaging to soils and plants.

Salinization generally reduces the growth rate of plants. Energy that would otherwise contribute to growth is physiologically diverted to the process of overcoming the increased stress of extracting moisture from the affected soil (Rhoades, 1991). Although different species have different salt tolerances, with some indigenous dryland plants especially well adapted, some individual salts are especially toxic to crops even at extremely low concentrations, notably boron (Maas, 1986). Susceptibility is often greatest at the seedling stage (Norlyn and Epstein, 1984; Kumar and Tarafdar, 1989).

In addition to these impacts on soil productivity and crop yields, salinization also affects other aspects of the biosphere, with consequent impacts on human societies. Contamination of drinking water, toxic effects on soil micro-organisms and the role of waterlogged saline ground as a breeding ground for parasites and diseases are among the other adverse affects of salinization mentioned by Szabolcs (1987).

The extent of salinization: past views

According to estimates made by FAO and UNESCO, as much as half of all the world's existing irrigation schemes are more or less under the influence of secondary salinization, alkalization and waterlogging. In a recent FAO Soils Bulletin, for example, it is stated that 'Nearly 50 per cent of the irrigated land in the (world's) arid and semi-arid regions have some degree of soil salinization problems' (Abrol et al, 1988, page 1). About 10 million hectares of irrigated land are abandoned each year because of the adverse effects of irrigation, mainly secondary salinization and alkalization, according to Szabolcs (1987).

A reasonable estimate of the global extent of salt-affected soils has only recently been derived, largely based on the FAO/UNESCO Soil Map of the World. Szabolcs (1987) presents estimates for major world regions which are shown in table 7.1, indicating that some 952 million hectares are affected world-wide. Szabolcs also includes

Table 7.1 [*orig. table 1*]   Extent of salt-affected* soils by continent and sub-continent (after Szabolcs, 1979).

Region	Millions of Hectares
North America	15.7
Mexico & Central America	2.0
South America	129.2
Africa	80.5
South Asia	84.8
North & Central Asia	211.7
South-east Asia	20.0
Australia	357.3
Europe	50.8
Total	952.0

* *Note*: "salt-affected" includes both saline soils and alkali soils.

a number of regional maps of salt-affected soils, of which the Africa map is reproduced in figure 7.1.

It should be noted, however, that the data presented here on salt-affected soils includes both soils that are naturally salt-affected and those made saline or alkaline by human action. To date, there are no estimates at the continental scale of the extent and distribution of soils affected by human-induced or secondary salinization.

New Data on Salinization

Recently published data (UNEP/ISRIC, 1990; UNEP, 1992) have provided a new database on the overall extent and severity of human-induced soil degradation. These studies classify degradation into four categories, one of which, chemical deterioration, includes data that can be used to assess the extent and degree of secondary salinization. Before this can be considered in its own right, however, it is important to explain the way in which the data have been gathered.

GLASOD

Recognizing the problems in gaining reliable data on the extent and degree of processes of land degradation, UNEP commenced a programme for the Global

Assessment of Soil Degradation – GLASOD – in 1987, in conjunction with the International Soil Reference and Information Centre, or ISRIC (Oldeman, 1988; Oldeman et al., 1990). The GLASOD programme, with data stored and analysed on UNEP's Global Resources Information Database (GRID), allows evaluation of land degradation at different scales, for different regions, and by different causes (UNEP/ISRIC, 1990; UNEP, 1992).

For logistic, administrative and political reasons, the GLASOD project divided the world into 21 regions, four of which were in Africa. For each region, a coordinator gathered existing published and unpublished data on land degradation, and consulted local experts to re-evaluate and refine existing data and to provide new data using a systematic and standardised methodology. In all, over 250 local environmental and soil scientists were consulted (UNEP, 1992).

The assessment of soil degradation in each region was carried out and presented using a system of mapping units or 'polygons', each representing a discrete land area based on natural physiographic features (Oldeman et al., 1990). In addition to the global database, a second, more detailed, database was compiled for Africa. In the global database Africa was divided into 383 polygons, while in the specific African database the continent is divided into 898 polygons.

For each polygon, human-induced soil degradation was considered under the headings of water erosion, wind erosion, physical deterioration and chemical deterioration, the first two being categories of degradation processes by displacement of soil material, the second two categories of degradation processes by internal soil deterioration (Oldeman, 1988). For each of these, degradation was quantified according to its degree (none plus four classes of light, moderate, strong and extreme) and extent (five classes) which ranged from infrequent, representing up to 5 percent of the polygon affected by the degradation type, to dominant, representing more than 50 percent affected (Oldeman et al., 1990). Degree and

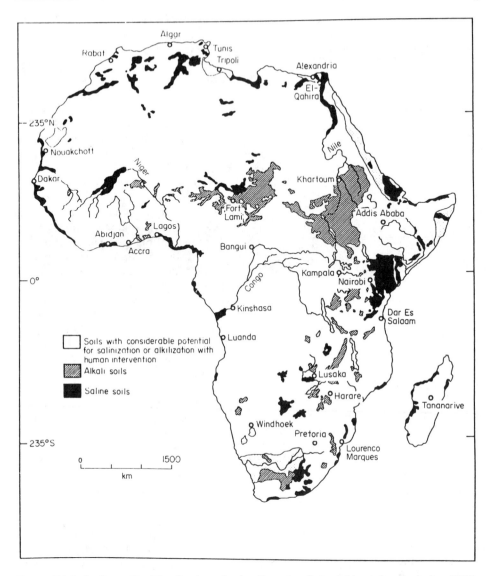

Figure 7.1 [*orig. figure 1*]   Distribution of salt-affected soils in Africa (after Szabolcs, 1979).

extent were combined in a twenty-cell matrix (figure 7.2), subdivided to establish the overall severity of degradation (Oldeman et al., 1990). Degradation severity is divided into four classes – low, medium, high and very high – as the basis for subsequent mapping, as utilised in UNEP/ISRIC (1990) and UNEP (1992). Overall, GLASOD is thought to provide a consistent and replicable database for the current and future assessment of soil degradation caused by human activities.

The GLASOD database also permits the extraction of numerical data on degradation, such as the size of actual areas affected by a particular process. Such data are a useful addition to the map, as the mapping procedures only plot information at the

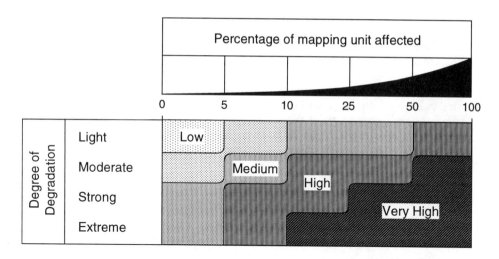

Figure 7.2 [*orig. figure 2*]   Severity classes of human-induced soil degradation used in GLASOD.

polygon-scale. Thus a map can appear visually to overestimate affected areas, because even if only a small part of a polygon is affected by a particular type of degradation, the classification calculated from the severity matrix will be plotted for the full polygon (UNEP, 1992).

GLASOD and salinization

The GLASOD database recognises several processes within the chemical deterioration category: salinization, nutrient depletion, acidification and pollution. While the recent global reports of overall soil degradation (UNEP/ISRIC, 1990; UNEP, 1992) do not distinguish between these processes, the original database allows them to be extracted individually, permitting an assessment of the secondary salinization problem to be made.

The regional contributions to the overall study characterized soil salinity by electrical conductivity (EC), ESP and pH as follows: non-saline soils: EC < 5 mS/cm, ESP < 15%, pH < 8.5; slightly saline soils: EC 5–8 mS/cm, ESP < 15%, pH < 8.5; moderately saline soils: EC 9–16 mS/cm, ESP < 15%, pH < 8.5; severely saline soils: EC > 16 mS/cm, ESP < 15%, pH < 8.5. The degree of human-induced salinization represents the following changes over a period of up to 50 years: Light salinization:

salinity increase of one class; moderate salinization: increase of two classes; severe salinization: increase of three classes; extreme salinization: increase of four classes. This polygon-scale information has then been applied to a revised matrix, providing an assessment of salinization severity from low to very high.

An estimate of the extent of human-induced salinization in Africa

The distribution of soils degraded by human-induced salinization in Africa is shown in figure 7.3. The initial impression of this map is that secondary salinization is very largely a dryland problem and of limited extent in Africa, particularly when considering that the apparently large areas of low severity salinization in Algeria and Libya represent isolated oases in otherwise large mapping polygons. Riverine irrigation schemes such as those on the Senegal River, Hadejia River in Nigeria, Logone River (part of the Chari system) in Cameroon, the Egyptian Nile, Awash River in Ethiopia, Shabelle River in Somalia, Limpopo River in Mozambique and the Orange River in South Africa, are all indicated as suffering from secondary salinization. Other notable areas where soils are degraded by human-induced salinization include oases south of the Atlas Mountains in Morocco and western Algeria, parts of north-east Algeria, central and southern Tunisia and Libya, where the widespread occurrence of sabkhas or 'Chotts' indicate saline surfaces and groundwater, the delta of the Nile and the mouth of Senegal's River Casamance, affected by marine intrusion, stretches of the Nigerian and Benin coast, part of Tanzania, South Africa's Cape and the western coast of Madagascar. Notable by their absence are such areas as the inland delta of the Niger

and Sudan's widespread Nile irrigation schemes.

Comparison with the map of salt-affected soils in figure 7.1, suggests that the salt problems of soils in much of coastal West Africa, northern Nigeria/southern Niger, Chad, Sudan, Kenya, Zambia, Zimbabwe and Botswana are due to naturally occurring salts rather than any human mismanagement.

Table 7.2 shows the extent of soils

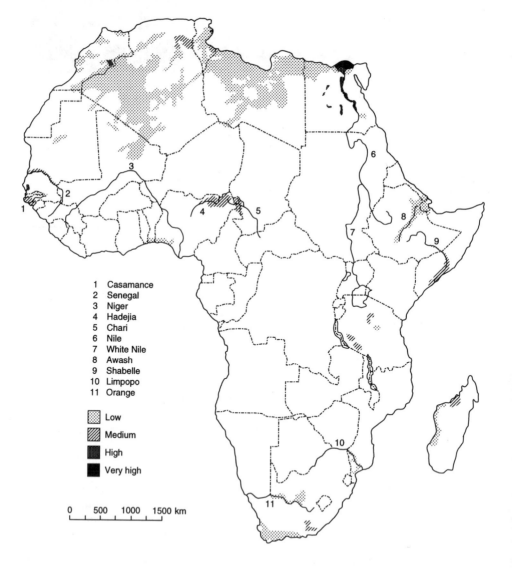

1   Casamance
2   Senegal
3   Niger
4   Hadejia
5   Chari
6   Nile
7   White Nile
8   Awash
9   Shabelle
10 Limpopo
11 Orange

Low
Medium
High
Very high

0   500   1000   1500 km

Figure 7.3 [*orig. figure 3*]   Distribution of soils degraded by human-induced salinization in Africa.

Table 7.2 [*orig. table 2*]   Extent of soils degraded by human-induced salinization in Africa by climatic zone (millions of hectares).

Degree	Arid	Semi-arid	Dry-subhumid	Total
Light	2.2	1.2	0.2	3.6
Moderate	1.2	0.8	0.1	2.1
Strong	0.0	0.0	0.0	0.0
Extreme	0.0	0.0	0.0	0.0
Total	3.4	2.0	0.3	5.7

degraded by secondary salinization in Africa's drylands, indicating that just 5.7 million hectares are affected. However, this figure does not include all areas shown in figure 7.3, since according to the climatic delimitations adopted by UNEP (1992) the salinized areas in most of Egypt, southern Senegal and Guinea-Bissau, coastal Benin, most of Tanzania and much of coastal Madagascar are designated as outside the so-called 'susceptible drylands' and therefore not areas where desertification can occur.

Even when taking this into account, the implication is clear. Less than ten percent of Africa's 80.5 million hectares of salt-affected soils are so affected because of human action.

Discussion

In the broader context of desertification, the information shown in figure 7.3 and table 7.2 is significant for a number of reasons. The very status of desertification as a major world environmental problem has been called into question in recent years (e.g. Nelson, 1988; Binns, 1990). One of the key problems researchers have come up against is the lack of standardised, reliable data concerning the phenomenon, and in this department alone, the GLASOD data presented here for soil degradation by salinization in Africa are of considerable value. Another area in which this study helps to clarify matters in the desertification debate is in the arena of the "man or nature?" question. The GLASOD data refer specifically to human-induced soil degradation, and thus, while naturally occurring salt-affected soils undoubtedly present problems for the

human user, the distinction between natural and human-induced problems of drylands is an important one in terms of the reclamation and future use of such areas.

A third area in which the GLASOD data presented here are of importance is in the true identification of the nature of problems in specific drylands, in this case Africa, widely regarded as the continent most critically affected by desertification and allied difficulties. As UNEP (1992) points out, correct identification of the problem is essential if appropriate measures to combat it are to have any chance of success. While some authorities have questioned the global approach of international agencies such as UNEP (e.g. Warren and Agnew, 1988), there is still a case to be made for such an approach in terms of assessing the scale of problems, prioritising those most in need of attention, and generating the political will to implement solutions. This being the case, the map shown in figure 7.4 is revealing. It shows human-induced soil degradation by loss of nutrients, another of the chemical deterioration processes assessed by GLASOD, which has been identified by a decline in organic matter, P, Ca, Mg or K, usually due to low-input agricultural practices. Figure 7.4 clearly indicates that soil degradation by nutrient loss inside, and indeed outside, African drylands is a much more widespread problem than that of secondary salinization. Table 7.3 shows that human-induced nutrient loss is responsible for soil degradation on more than seven times the land area degraded by secondary salinization in African drylands.

In terms of efforts to combat desertification, therefore, a concentration on improving nutrient inputs into, or reducing

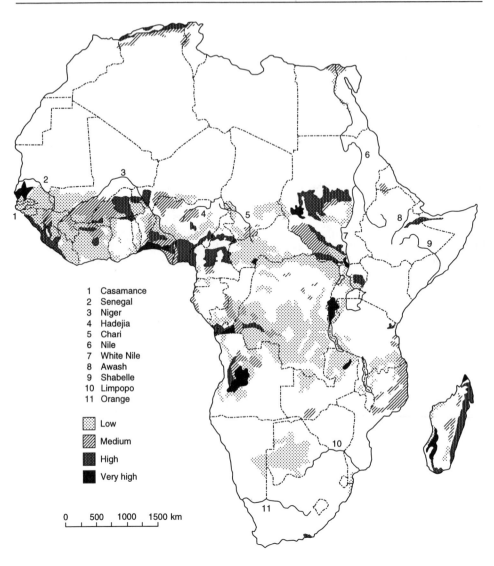

1 Casamance
2 Senegal
3 Niger
4 Hadejia
5 Chari
6 Nile
7 White Nile
8 Awash
9 Shabelle
10 Limpopo
11 Orange

Low
Medium
High
Very high

0   500   1000   1500 km

Figure 7.4 [*orig. figure 4*]   Distribution of soils degraded by human-induced loss of nutrients in Africa.

Table 7.3 [*orig. table 3*]   Extent of soils degraded by human-induced loss of nutrients in Africa by climatic zone (millions of hectares).

Degree	Arid	Semi-arid	Dry-subhumid	Total
Light	0.7	16.6	9.8	27.1
Moderate	2.3	4.3	2.1	8.7
Strong	0.6	4.2	1.2	6.0
Extreme	0.0	0.0	0.0	0.0
Total	3.6	25.1	13.1	41.8

soil nutrient losses caused by, African agriculture would pay more dividends than a concentration on reversing human-induced salinization. Such an effort would also have more effect on rural farmers than a concentration on large-scale irrigation schemes, although nutrient loss is also implicated in soil degradation on dryland irrigation areas such as on the inland delta of the River Niger in Mali, the Egyptian Nile and delta, the Sudanese White Nile, and Chad's Chari River (figure 7.4). In terms of the politics and economics of Africa's dryland countries, it is often the prestigious and foreign exchange earning large scale irrigation schemes that take priority, and since the FAO (1986) estimate that Africa has 9.5 million hectares of cropland under irrigation (6.1 million hectares of modern schemes plus 3.4 million hectares of small-scale, traditional flood, swamp and low-lift irrigation) the GLASOD data suggest that as much as half of all Africa's irrigated cropland may be suffering from secondary salinization. Hence, we do not suggest that human-induced salinization is not a serious problem for African dryland agriculture, but in terms of area and numbers of people affected, international and bilateral efforts at combating desertification would do well to focus on nutrient loss as the key component of soil degradation by chemical deterioration.

Conclusion

Salinization, or more importantly, secondary salinization, is widely quoted as a key process leading to soil degradation and desertification in the world's drylands. While salt-affected soils certainly provide problems for plant growth and crop yields, data from GLASOD suggest that in Africa human-induced salinization accounts for less than ten percent of all salt-affected soils. By contrast, human-induced loss of nutrients is a much more widespread problem for Africa's dryland agriculture, and as such warrants much greater attention from both the academic and international communities interested in relieving the problems of desertification.

References

Abrol, I. P., Yadav, J. S. P. and Massoud, F. I. (1988). Salt-affected soils and their management. *FAO Soils Bulletin*, 39. FAO, Rome.

Anon (1991). Senegal's saline soils have a future – despite the drought. *Spore*, 31: 5.

Ayers, R. S. and Westcot, D. W. (1985). Water quality for agriculture. *FAO Irrigation and Drainage Paper*, 29. FAO, Rome.

Barrow, C. J. (1991). *Land degradation*. Cambridge University Press, Cambridge, England.

Berg, W. A., Naney, J. W. and Smith, S. J. (1991). Salinity, nitrate and water in rangeland and terraced wheatland above saline seeps. *Journal of Environmental Quality*, 20: 8–11.

Bettenay, E. (1986). Salt affected soils in Australia. *Reclamation and Revegetation Research*, 5: 167–179.

Binns, T. (1990). Is desertification a myth? *Geography*, 75: 106–113.

Brown, P. L., Halvorson, A. D., Siddoway, F. H., Mayland, F. H. and Miller, M. R. (1983). Saline seeps diagnosis, control and reclamation. *U.S. Department of Agriculture Conservation Research Report* 30.

El-Hinnawi, E. and Hashmi, M. (1982). *Global environmental issues*. UNEP/Tycooly, Dublin.

FAO (1986). *African agriculture: the next 25 years. Annex IV Irrigation and water control*. FAO, Rome.

FAO/UNEP (1984). *Provisional methodology for assessment and mapping of desertification*. FAO/UNEP, Rome.

Goudie, A. S. (1985). *The human impact*. Blackwell, Oxford.

Grainger, A. (1990). *The threatening desert*. Earthscan, London.

Halvorson, A. D. and Black, A. L. (1974). Saline-seep development in dryland soils of northeastern Montana. *Journal of Soil and Water Conservation*, 29: 77–81.

Jacobsen, T. and Adams, R. M. (1958). Salt and silt in ancient Mesopotamian agriculture. *Science*, 128: 1251–1258.

Kassas, M. (1987). Seven paths to desertification. *Desertification Control Bulletin*, 15: 24–6.

Kishk, M. A. (1986). Land degradation in the Nile Valley. *Ambio*, 15: 226–230.

Kotlyakov, V. M. (1991). The Aral Sea basin. *Environment*, 33: 4–9, 36–38.

Kovda, V. A. (1980). *The problem of combatting salinization of irrigated soils*. UNEP, Nairobi.

Kumar, D. and Tarafdar, J. C. (1989). Genetic variation of salt tolerance in seedling emergence, early growth characters and phosphatase activity of sunflowers growing on arid soils. *Journal of Arid Environments*, 16: 263–269.

Lewis, L. A. and Berry, L. (1988). *African environments and resources*. Unwin Hyman, Boston.

McMillan, D. E. and Hansen, A. (1986). Overview: food in Sub-Saharan Africa. In Hansen, A. and McMillan, D. E. (Eds) *Food in Sub-Saharan Africa*. Riener, Boulder Colorado.

Maas, E. V. (1986). Salt tolerance of plants. *Applied Agricultural Research*, 1: 12–26.

Malcolm, C. V. (1983). Wheatbelt salinity, a review of the salt land problem in South-western Australia. *Western Australia Department of Agriculture, Technical Bulletin 52*.

Middleton, N. J. and Thomas, D. S. G. (1992). Text in UNEP (1992) op. cit.

Nelson, R. (1988). Dryland management – the desertification problem. *Environmental Department Working Paper No. 8*. World Bank, Washington.

Norlyn, J. D. and Epstein, E. (1984). Variability in salt tolerance of four triticale lines at germination and emergence. *Crop Science*, 24: 1090–1092.

Oldeman, L. R. (1988). Guidelines for general assessment of the status of human-induced soil degradation. *ISRIC Working Paper 88/4*.

Oldeman, L. R., Hakkeling, R. T. A. and Sombroek, W. G. (1990). *World map of the status of human-induced soil degradation. An explanatory note*. ISRIC/UNEP.

Rhoades, J. D. (1990). Soil salinity – causes and controls. In Goudie, A. S. (Ed.) *Techniques for desert reclamation*. Wiley, Chichester.

Rhoades, J. D. and Loveday, J. (1990). Salinity in irrigated crops. In Stewart, B. A. and Nielsen, D. R. (Eds) *Irrigation of Agricultural crops*. American Soil Association Monograph.

Speece, M. and Wilkinson, M. J. (1982). Environmental degradation and development of arid lands. *Desertification Control Bulletin*, 7: 2–9.

Szabolcs, I. (1974). *Salt affected soils in Europe*. Martinus Nijhoff, The Hague and Research Institute for Soil Science and Agricultural Chemistry of the Hungarian Academy of Sciences, Budapest.

Szabolcs, I. (1979). Review of research on salt-affected soils. *Natural Resources Research XV*. UNESCO, Paris.

Szabolcs, I. (1987). The global problems of salt-affected soils. *Acta Agronomica Hungarica*, 36: 159–172.

UNEP (1992). *World atlas of desertification*. Edward Arnold, London.

UNEP-DC/PAC (1990). *Desertification: a new assessment 1990–1992*. UNEP-DC/PAC, Nairobi.

UNEP/ISRIC (1990). *World map on status of human-induced soil degradation*. UNEP, Nairobi.

Vincent, L. (1990). Sustainable small-scale irrigation development. *Water Resource Development*, 6: 250–559.

Warren, A. and Agnew, C. (1988). An assessment of desertification and land degradation in arid and semi-arid areas. *Paper No. 2, Drylands Programme*, International Institute for Environment and Development, London.

Wilcox, L. V. (1955). Classification and use of irrigation waters. *U.S. Department of Agriculture, Circular 969*.

8

SALINIZATION OF NON-IRRIGATED SOILS AND ASSOCIATED STREAMS: A REVIEW*

A. J. Peck

Introduction

For thousands of years mankind has faced the problem of increasing salinity of irrigated soil (Jacobsen and Adams 1958). There is continuing research in many countries on this topic and the associated problem of increased salinity of streams due to irrigation return flows (U.S. Environmental Protection Agency 1972; Kovda et al. 1973). Within the last 100 years, similar problems have been recognized in non-irrigated areas too. These appear to be restricted to regions of Australia and North America which have been developed for agriculture in relatively recent times. Peck (1977) has briefly reviewed salinity in non-irrigated lands with special reference to south-western Australia. A research program has commenced to gather reliable data and to develop a detailed understanding of the effects of clearing native vegetation on the translocation of salts and the salinity of stream waters in this region. A major part of the work involves observations in several

small catchment areas. This paper reviews the magnitude of the problem and the results of previous investigations. Details of the current experiments will be presented for publication in the future.

The salinity problems which are discussed here are particular examples of solute and water transport in soils and porous rocks. The fundamentals of these transport processes have been reviewed in recent years (Gardner 1956; Biggar and Nielsen 1967; Philip 1970; Fried and Combarnous 1971; Hillel 1971; Raats and Gardner 1974). A recent monograph on agricultural drainage (van Schilfgaarde 1974) includes further material on these topics. Readers are referred to these papers so that the present work can concentrate on water and solute transport in specific landscape situations.

Magnitude of Salinity Problems in Non-Irrigated Land

Increasing salinity of streams and soils in Western Australia was first recognized by

* Originally published in *Australian Journal of Soil Research*, 1978, vol. 16, pp. 157–68.

railroad engineers more than 60 years ago (Bleazby 1917; Wood 1924). Malcolm and Stoneman (1976) estimate that 167000 ha of previously productive farmland in this State is now too saline to grow the preferred crops and pastures. An earlier survey indicated an average rate of increase of the saltland of about 3% per year, although there were decreases in some Shires (Lightfoot et al. 1964).

Little more than 1% of the total area of cultivated land in south-western Australia has become saline (Malcolm and Stoneman 1976), although the proportion is much larger on some farms. Affected areas are reputed to have been of above average productivity, but they now require special management. In many cases farm water supplies have become excessively saline too, and it has been necessary to construct new earth tanks and water supply systems. Bare, salt-affected soils are focal points for erosion.

Part of the salt which accumulates in surface soils is washed into streams by run-off. Associated discharges of saline groundwater in springs and directly into streams and rivers add further to the salt load in surface waters. The most recent data available show that in south-western Australia total dissolved solids (TDS) concentrations exceed 1000 mg l^{-1} in 33% of the average annual discharge from rivers, while a further 36% of the flow carries TDS in the range 500–1000mg l^{-1} (Anon. 1976). Moreover the salinity of many streams and rivers in this region is believed to be increasing. A compilation of data for the largest river (the Blackwood at Bridgetown) shows a threefold increase in the concentration of TDS from the 1910s to the 1960s (Peck et al. 1973), and chloride balances (Peck and Hurle 1973) suggest larger changes in many smaller streams. Detailed records of salinity in a major reservoir (Wellington Dam on the Collie River) show an increasing trend, although there is partial recovery in years of exceptionally large streamflow (Loh and Hewer 1977).

Northcote and Skene (1972) have mapped the extent of saline soils in Australia. They estimate that 197000 ha of land which once produced satisfactory crops and pastures is now saline. The problem occurs in Queensland (Pauli 1972), New South Wales (Wagner 1957; Logan 1958; Van Dijk 1969) and Tasmania (Colclough 1973), but estimates of the areas affected in those states are not known. Cope (1958) estimated the area affected in Victoria as more than 4000 ha, while Northcote and Skene (1972) quote a later estimate of 5000 ha in that State. Rowan (1971) estimated the rate of growth of salt pans in north-western Victoria as 1.5–4.2% per year. The problem also occurs in South Australia, where Matheson (1968) estimated that 14000 ha of land was affected. This area is believed to be increasing (Hartley 1976).

There is little information available on the effects of dryland agriculture on surface water salinity outside of Western Australia. Venables (1970) considered that streams in the Western Districts of Victoria have become more saline since European settlement, and Pauli (1972) indicates that a main effect of land clearing for agriculture in Queensland is an increase in the salinity of water. It is thought that some streams on the Eyre Peninsula in South Australia became saline after clearing of native vegetation, but no documentation is known.

Saline soils have developed in contiguous areas of non-irrigated farmland in the United States and Canada. McCracken (1973) suggests an area of 20000 ha in Manitoba, while Miller et al. (1976) estimate areas of 100000 ha in Alberta, 56000 ha in Montana, and 20000 to 40000 ha in North Dakota. They report substantial areas in Saskatchewan and South Dakota too. In Montana and Canada the affected area was estimated to be increasing at an average 10% a year for several years (Miller et al. 1976; Sommerfeldt 1977). Bahls and Miller (1973), Miller (1973) and Botz (1976) have discussed associated increases of salinity in streams and ponds, but this aspect of the problem in North America is not well documented.

From experience in Australia and North America, it is tempting to suggest that similar salinity problems may occur in other regions where large areas of land have been brought under cultivation during the last 100 years. However, no documentation of similar problems in other regions has been found by the author.

Terminology

Salinity problems in non-irrigated lands have been examined almost independently in different regions, and this had led to an undesirable profusion of terminology. Before proceeding further in this review, the relationships between some of these terms are examined.

A soil is usually considered to be saline when the electrical conductivity of the saturation extract exceeds $0.4\,S\,m^{-1}$ (United States Salinity Laboratory Staff 1954). However, due to the relatively high proportions of sodium and chloride ions in Australian soils, Northcote and Skene (1972) used the limits of 0.1% (loams and coarser soils) or 0.2% (clay loams and clays) of sodium chloride in the surface layer as criteria for salinity.

The term *secondary salinity* was used by Northcote and Skene (1972) to refer to areas where soluble salts have accumulated as a consequence of irrigation, agricultural practices, or the clearing of native vegetation. They subdivided secondary salinity into areas of *dryland salting* where irrigation is not practiced, and *secondary salinity of irrigated soils*. Conacher and Murray (1973) have used the term *salt scald* with (apparently) the same definition as dryland salting.

Teakle and Burvill (1938) recognized the development of saline soils without irrigation in three topographic situations which are described below.

(i)   In valleys where a saline water table has developed within 2 or 3 m of the soil surface. This has been referred to as *wet-pan salting* (Cope 1958; Colclough 1973), *valley salting* (Smith 1962), *lowland salinity* (Matheson 1968), and *Type A or C Salting* (van Dijk 1969) in different parts of Australia.

(ii) On slopes where a seasonal or permanent seepage of saline water has developed. This situation has been described as *seepage salting* (Cope 1958; Colclough 1973), *hillside seepage* (Smith 1962; Matheson 1968), and *Type B Salting* (Van Dijk 1969).

(iii) On certain soils where there is no evidence of a seasonal or permanent water table at the surface or within a considerable depth of the profile. This appears to be a less common situation which is referred to as *dryland salt* (Smith 1962), or *highland salt patch* (Matheson 1968). (Note that Northcote and Skene (1972) have used the term dryland salting to refer, in general, to the development of saline soil in non-irrigated land.)

McCracken (1973) and others in Canada have used the expression *dryland seepage and salinity*, and defined this as 'a condition caused by natural groundwater discharge, resulting in formerly productive soil becoming unproductive'. Smith (1975) has defined terms in connection with non-irrigated salinity problems in the United States. He uses the term *saline seep* which is defined by Brown (1976) as 'a recently developed wet salty area in non-irrigated soil on which crop production is reduced or eliminated. The soil surface is intermittently or continuously wet and white salt crusts are often present.' Salty water may flow out of the soil in a saline seep, or salts may accumulate at the soil surface by evaporation of water which moves upwards by capillarity from a shallow aquifer.

The use of the term 'recent' in Brown's definition of a saline seep is indefinite. For example, there are seeps in Western Australia which are believed to have developed on previously productive farmland more than 40 years ago (E. Bettenay, personal communication 1977). This may not be considered as 'recent'. On the other hand, there are substantial areas of soils in this region which were saline before the introduction of agriculture by European man during the last 150 years.

A detailed discussion of the relative merits of different terms is not within the objectives of this paper. It will be apparent that the definitions of dryland seepage and salinity and saline seepage are essentially the same, and they include the valley and hillside situations referred to above. My preference is for the simple expression salt seep and its derivatives to describe situations where surface salinity recently increased significantly and a seasonal or permanent water table occurs within 2 or 3 m of the ground surface.

In certain soils significant rates of accumulation of salts at the surface can result from a water table at substantially greater depth (Talsma 1963; Peck 1978). It is uncertain whether the third type of saltland situation referred to above occurs in these exceptional soils, or whether salts redistribute within these soils with essentially no loss of moisture or solute below the plant root zone. Other soils are saline as a result of natural processes without man's intervention.

General Theory of Saline Seepage

The first recorded observations of saline seepage appear to have been made by Wood (1924). From observations and measurements in South Australia and Western Australia over a period of 30 years, he hypothesized that the removal of native vegetation allows a greater penetration of rainfall into deeper soil layers which contain brackish or saline water. When the amount of water in the saline zone is increased greatly it will rise to the surface near the water course bringing salts with it. In broad terms this hypothesis has been supported by observations in Australia (Teakle and Burvill 1938; Cope 1958; Smith 1962; Bettenay et al. 1964; Rowan 1971) and North America (Greenlee et al. 1968; Clark 1971; Sommerfeldt 1977). The essential features of the problem are an accumulation of soluble salts in the soil profile, and some hydrologic disturbance which causes an accelerated redistribution of salt in the landscape, and discharge of saline water into drainage systems.

Solute Accumulation and Ionic Composition

By extraction of dissolved and adsorbed salts from soil cores, and integration over the profile depth, soluble salt storages in the range 17–95 kg m^{-2} have been determined in the deeply weathered (average 20 m) soils developed on granitic and gneissic rocks in south-western Australia (Dimmock et al. 1974). These soils are described in more detail in the following section of this review. Comparable data for other regions are not known, but there has been extensive sampling of soils to depths up to 3 m in many investigations of saline seepage (Teakle and Burvill 1938; Cope 1958; Bettenay et al. 1964; Greenlee et al. 1968; Rowan 1971; Halvorson and Black 1974). Ferguson (1976) concluded that the source of salts in seeps in Montana is the top 3 m of soil in nearby areas of groundwater recharge. Similarly Teakle and Burvill (1938) found that removal of native vegetation lead to a reduction of salts by about 3 kg m^{-2} in the top 1 m of soil in the area contributing to a seep in Western Australia.

The dominant ions in saline seeps and associated soils in Australia are Na^+ and Cl^- (Teakle and Burvill 1938; Cope 1958; Matheson 1968; Rowan 1971; Colclough 1973). Such evidence as there is (Wood 1924; Bettenay et al. 1964; Peck and Hurle 1973) suggests that the most important source of these solutes is the small quantity of oceanic salts found in rainfall (Hingston and Galaitis 1976).

In North America $Na^+ SO_4^{2-}$ are the dominant ions in saline seeps (Greenlee et al. 1968; Miller 1971, 1973; Halvorson and Black 1974). These ions have been released by glaciation and weathering of marine sediments from which the soils associated with seeps have been formed.

This comparison shows that the composition and primary source of the solutes is unimportant in saline seep problems relative to the presence of abundant soluble salt in permeable strata of the soil profile.

Saline Seepage in Relation to Soil Type

Saline seeps have developed in a variety of soil types in Australia. In the south-west they are usually associated with deeply weathered lateritic soils developed on granitic and gneissic parent rocks (Smith 1962; Bettenay et al. 1964). The A horizon of these soils is typically sand or gravelly sand and much more permeable than the underlying mottled clays and sandy clay of the weathered zone. Ephemeral perched water tables are often found in the A horizon material during the wet (winter) season. There is evidence (Smith 1962; Bettenay et al. 1964) of extensive permanent aquifers at depth, at least in areas which have been farmed for many years. The basement rock is believed to be essentially impermeable.

Cope (1958) emphasizes the importance of soil morphology in connection with saline seeps in central Victoria. In that area affected soils have a permeable A horizon with marked profile differentiation to a clay B horizon with poor structure, which is often sodic and relatively impermeable at depth.

The most common seeps in north-western Victoria are found in the swales, lower slopes, and nearby flats of sandy dunes which overlie much less permeable clay soils (Rowan 1971). Other seeps in this region have developed in lower topographic positions on clay soils of the broad plains. The type A (valley) seeps described by Van Dijk (1969) from his observations in the Southern Tablelands of New South Wales are associated with dense subsoils of low permeability which may include more permeable layers. His Type B (hillside) seeps occur in shallow (0.3 m) stony loams overlying an irregular thickness of dense clay and partly decomposed rock. The seeps occur in areas where the clay layer is very thin or non-existent. van Dijk's Type C (valley) seep is found in strongly differentiated silty-loam surface overlying dense clay subsoils. Surface runoff and seepage waters from adjacent higher land spread over the surface of the Type C seep.

Salt seeps in North America develop in permeable glacial till soils which overlie relatively dense and impermeable clays and shales (Greenlee et al. 1968; Clark 1973; Miller 1973; Nielsen 1973; Brun et al. 1976). Till thickness is variable, with seeps developing where it is less than 10–15 m (Bahls and Miller 1973; McCracken 1973). Lenses of sandstone, siltstone, lignite and sandy loam are found within the strata, and undoubtedly they contribute to lateral water movement at some sites (Miller 1971; Halvorson and Black 1974).

Comparison of the soil profiles and descriptions of the seeps in these different regions suggest the following conclusions.

(i) Salt accumulation may occur as a result of essentially surface movement of saline water to a lower-lying site with relatively impermeable subsoil, where a perched water table develops in some seasons and substantial evaporation takes place. It is a weakness of terminology to refer to this situation as saline seepage, since it is a surface problem.

(ii) In the second situation lateral movement of water from recharge to discharge areas occurs in an emphemeral or permanent unconfined aquifer in the more permeable surface horizons of a strongly differentiated soil profile.

(iii) Thirdly, lenses or layers of more permeable material within an otherwise relatively impermeable profile, such as clay, can serve as confined but leaky aquifers which transmit both hydraulic pressure and water from recharge to discharge areas.

The data presented earlier in this paper show that the total area affected, and the area of many individual saline seeps, is continuing to increase in Australia and North America. It should be recognized that the processes contributing to a particular seep may vary through time, both seasonally and from year to year. For example, a seep may first develop due to accumulation of surface water which forms a perched water table with some slow loss of water downwards through the perching layer. Subsequently, lateral subsurface water movement may reverse the direction of flow through the perching layer, bringing up salt from a leaky confined aquifer.

Change of Water Budget

Nearly 90 years ago Abbott (1880) reported increased flow of water in streams after killing or clearing native forest and woodland vegetation in New South Wales. He suggested that when the timber is dead the large proportion of the rainfall which was formerly taken up by the roots of the growing trees and evaporated from their leaves is allowed to find its way to the creeks and rivers. Later Wood (1924) referred to the role of cultivation, which, he suggested, increases the amount of rain entering the soil and thereby contributes to salt seeps.

Observations in many countries, including Australia (Holmes and Colville 1968; Boughton 1970), have confirmed that forests use more water than do grasslands under the same conditions. The difference in evapotranspiration is primarily attributed to relative rooting depths, but in south-western Australia another factor is undoubtedly that the grasses (pasture or crop) are annuals which grow in winter–spring, whereas the evergreen native forests continue to transpire through the summer and autumn. Differences in several other plant characteristics (albedo, leaf area index, canopy dimensions and geometry, plant-water relations, root density and distribution) may also contribute to the greater water use by forest than by grassland.

Although salt seeps in Australia have developed after native forest woodland or scrub has been removed, the pristine vegetation over some affected areas in North America was perennial grassland (prairie). The accumulation of soil moisture in those areas is believed to occur under snow packs, or when land is left under fallow (Greenlee et al. 1968; Clark 1971), evaporation from bare soil being even less than that from grassland or crop.

Soil Water and Groundwater Response

Following a reduction in evapotranspiration, soil moisture may accumulate at various points in the profile. Cope (1958), Rowan (1971) and Conacher (1975) concluded that ephemeral unconfined aquifers developed on top of a clay B horizon or plough layer, and most lateral flow of water takes place in this zone of saturation. Similarly Halvorson and Black (1974) found an unconfined aquifer, but this developed in glacial till soil overlying dense clay at a depth of about 6 m. Doering and Sandoval (1976) emphasize the importance of lateral water movement in bedded layers of lignite which are much more permeable than other strata in the profile. From their description, it seems likely that the lignite layers behave as confined (but probably leaky) aquifers. Similarly, Smith (1962, 1966) and Bettenay et al. (1964) in Western Australia, and van Dijk (1969) in New South Wales, report observations of confined aquifer systems which they associate with salt seeps.

Quite clearly, local properties of the soil profile will influence the depth or depths at which effects of a net increase of soil moisture will be transmitted laterally to a seepage area. Moreover, interactions between leaky confined and unconfined aquifers can be so complex that it may be impossible to attribute a particular seep to water movement in either one of these systems alone. Selim et al. (1975) have analysed steady saturated flow in a soil slab with a sloping surface for different relative hydraulic conductivities of three layers. These results suggest the possibility that, when the salinity of water differs between the layers, most of the water appearing at the seep may have passed through one aquifer, while most of the salt is released from another.

Australian studies of salt seeps have rarely attempted to define the extent and location of associated water intake areas. Smith (1962) considered that higher level soils with sandy or lateritic surfaces would be good intake areas in his study region, while Bettenay et al. (1964) suggested that run-off from large areas of exposed rock into surrounding coarse-textured soils was the major source in another region in Western Australia. Similarly, seeps in North America are believed to be part of local groundwater systems (Bahls and Miller 1973; Krogman 1973; Miller 1973; Halvorson and Black 1974; Doering and Sandoval 1976). That is, recharge takes place at or near a local topographic high, and discharge at an adjacent topographic low. Analysis of groundwater systems (Freeze and Witherspoon 1967, 1968) and groundwater discharge (Toth 1971) in relation to hydrogeologic settings and water table configurations can provide valuable insights into the spatial relationships between groundwater recharge and the development of seepage areas.

The contribution of local flooding to the water table depth in salt-affected valleys in Western Australia was considered by Smith (1966). He concluded that flooding does not contribute to salinity because groundwater gradients were always directed into the valley (there was no evidence of a mound beneath flooded land), and in some areas the potentiometric surface stood above ground level.

There appears to be a failure in the literature to consider the importance of salt-tolerant vegetation as an outlet for groundwater. When this vegetation is removed, the water table will rise owing to reduced discharge which is effectively the same as an increase of local recharge.

Dynamics and Reclamation of Saline Seeps

Although the total area affected by salt seeps is increasing in all regions, as discussed earlier in this report, the size of some seeps is decreasing and others have disappeared (Smith 1962; P. D. Brown, personal communication 1975; A. D. Halvorson, personal communication 1975). That is, there are important dynamic aspects to the problem. Further support for this was provided by Peck and Hurle (1973), who reported chloride balances of partly farmed catchments in south-western Australia. They showed that there were substantial net losses of chloride from these catchments and estimated characteristic times for reduction of the salt losses. These varied between 30 and 400 years with the larger times in areas of lower rainfall.

Variations of seep areas are commonly observed over periods of a few years (Halvorson 1973). These are attributed to fluctuations of yearly precipitation, and suggest that appropriate reclamation procedures should have a reasonably rapid effect.

The Western Australian Department of Agriculture (Anon. 1962) has recommended that saline seepages should be revegetated. Persistent cultivation and resowing with normal plants is said to be successful on mildly affected land, but only salt tolerant plants can be grown on severely affected or very wet sites (Smith 1961). The drainage of saline seeps is considered to be uneconomic in Western Australia (Anon. 1962), and in any case disposal of drainage water may add to the salt load in streams and rivers.

Conacher and Murray (1973) and Conacher (1974, 1975) have concluded that lateral water movement in shallow surface soil is the cause of saline seepages which they have studied in Western Australia. This lateral flow can be intercepted by banks which are run on, or close to the contour at intervals down slopes. The banks are formed by scraping a ditch down to and slightly into the clayey subsoil. Some of this material is included in the bank which reduces downslope leakage from the ditch. A small number of farmers in Western Australia have claimed successful rehabilitation of saltland by this method, which also serves to control surface water movement.

In a report on salt seepage in Western Australia, Pennefather (1950) recommended studies of dewatering plants and drainage to control surface and subsurface water movement. Other workers in this area (Smith 1962; Peck 1976) have also suggested that dewatering plants should be useful, although they have not agreed on the area which would be needed to transpire the water lost from a saline seep. There is no evidence that trees have been responsible for the reclamation of any seep, but where they have been planted the areas have probably been too small to expect transpiration at a rate approaching the loss of water from seeps.

The recommendations for salt seep reclamation in Victoria (Cope 1958; Rowan 1971), South Australia (Matheson 1968) and Tasmania (Colclough 1973) are generally similar to those described above, but with greater emphasis on the benefits of surface and subsurface water management by drainage, the use of deep-rooted plants to dry out water intake areas, and pasture management to maintain high transpiration rates. Wagner (1957) discussed reclamation of salt seeps in New South Wales. Several of his recommendations are the same as those above, but he also suggests deep cultivation on the contour to disperse the salt, which is normally concentrated at the surface, through the soil before reseeding with salt-tolerant species.

In North America there have been extensive studies into methods for early detection of potential seeps and their prevention or reclamation. The Wenner earth resistivity technique has been adapted as a survey tool to map electrical conductivity of the soil *in situ*, and this is related to the conductivity of the saturation extract (Halvorson and Rhoades 1976; Rhoades and Halvorson 1977). Water usage by various crops and management systems in the recharge areas associated with saline seeps in Montana has been studied, particularly by Brown and his colleagues (Brown and Ferguson 1973; Smith 1973; Brown et al. 1976). Considerable success in the reclamation of experimental areas has been achieved by appropriate management of vegetation in recharge areas (Halvorson and Reule 1976). Other investigators have studied the performance of drainage systems to reclaim salt seeps (Sommerfeldt 1976, 1977; Doering and Sandoval 1976; Berringer 1977; Stretch and Vulcan 1977; McCoy 1977), although they suggest that this method is probably uneconomic in most cases.

A novel approach to salt seep reclamation was described by Oosterveld (1977). He has investigated methods for enhancing evaporation of saline water so that the salt accumulation is confined to only a part of the original seepage area. The remainder of the seep is then returned to production. This technique is designed for the situation where there is no outlet for saline drainage effluent.

Lyles and Allen (1966) describe the results of an experiment in which ground surface topography was modified by machinery to encourage the leaching of salts in areas of more saline soil. In 2 years, 75–85% of the salt load in the uppermost 1.8 m of the profile was leached below that depth. They suggest that larger scale trials would be needed before changes in the salt balance of an entire soil unit could be detected.

While these various methods of reclamation have been studied experimentally, none of them has found widespread application.

References

Abbott, W. E. (1880). Ringbarking and its effects. *J.R. Soc. N.S.W.* 14, 97–102.

Anon. (1962). Flooding and salt problems in the wheatbelt. *J. Agric. West. Aust.* 3, 773–8.

Anon. (1976). Review of Australia's Water Resources, 1975. (Australian Government Publishing Service: Canberra.)

Bahls, L. E. and Miller, M. R. (1973). Saline seep in Montana. *In* 'Second Annual Report'. pp. 35–44. (Montana Environmental Quality Council: Helena, Mont.).

Berringer, R. (1977). 1974 Dryland salinity report. *In*, 'Dryland Salinity and Seepage in Alberta'. (Alberta Dryland Salinity Committee: Lethbridge, Alberta.)

Bettenay, E., Blackmore, A. V. and Hingston, F. J. (1964). Aspects of the hydrologic cycle and related salinity in the Belka valley, Western Australia. *Aust. J. Soil. Res.* 2. 187–210.

Biggar, J. W. and Nielsen, D. R. (1967). Miscible displacement and leaching phenomena. *In* 'Irrigation of Agricultural Lands.' Agronomy No. 11. pp. 254–74. (Am. Soc. Agron: Madison, Wisc.).

Bleazby, R. (1917) Railway water supplies in Western Australia: difficulties caused by salt in soil. *J. Inst. Civ. Eng.* 203. 394–400.

Botz, M. K. (1976). Salinity in hydrological systems in Montana. *In* 'Regional Saline Seep Control Symposium Proceeding'. Bull. No. 1132, pp. 91–8. (Montana State University: Bozeman, Mont.)

Boughton, W. C. (1970). Effect of land management on quantity and quality of available water: a review. Wat. Res. Lab. Rep. No. 120. (University of New South Wales: Sydney).

Brown, P. L. (1976). Saline-seep detection by visual observations. *In* 'Regional Saline Seep Control Symposium Proceedings'. Bull. No. 1132, pp. 59–61. (Montana State University: Bozeman, Mont.)

Brown, P. L. and Ferguson, H. (1973). Crop and soil management for possible control of saline seeps in Montana. *In* 'Governor's Saline Seep

Emergency Meeting Proceedings'. pp. 12–22. (Montana State University: Bozeman, Mont.)

Brown, P. L., Cleary, E. C., and Miller, M. R. (1976). Water use and rooting depths of crops for saline seep control. In 'Regional Saline Seep Control Symposium Proceedings'. Bull. No. 1132, pp. 125–36. (Montana State University: Bozeman, Mont.)

Brun, L. J., Deutsch, R. L., and Worcester, B. K. (1976). Preliminary soil water balance data for Stark County North Dakota. In 'Regional Saline Seep Control Symposium'. Bull. No. 1132, pp. 137–44. (Montana State University: Bozeman, Mont.)

Clark, C. O. (1971). Saline-seep development on non-irrigated cropland. In 'Saline Seep-Fallow Workshop Proceedings'. (Highwood Alkali Control Assn: Highwood, Mont.)

Clark, C. O. (1973). Saline seeps in Montana soils. In 'Governor's Saline Emergency Meeting'. pp. 29–30. (Montana State University: Bozeman, Mont.)

Colclough, J. D. (1973). Salt. Tasmanian J. Agric. 44, 171–80.

Conacher, A. J. (1974). Salt scald: a W. A. case study in rehabilitation Sci. Technol. (Surrey Hills, Vic.) 11, 14–6.

Conacher, A. J. (1975). Throughflow as a mechanism responsible for excessive soil salinisation in non-irrigated, previously arable lands in the Western Australian wheatbelt: a field study. Catena 2, 31–67.

Conacher, A. J., and Murray, I. D. (1973). Implications and causes of salinity problems in the Western Australia wheatbelt: The York–Mawson area. Aust. Geogr. Stud. 11, 40–61.

Cope, F. (1958). Catchment salting in Victoria. (Soil Conservation Authority of Victoria: Melbourne.)

van Dijk, D. C. (1969). Relict salt, a major cause of recent land damage in the Yass valley, Southern Tablelands, N. S. W. Aust. Geogr. 11, 13–21.

Dimmock, G. M., Bettenay, E., and Mulcahy, M. J. (1974). Salt content of lateritic profiles in the Darling Range, Western Australia. Aust. J. Soil Res. 12, 63–9.

Doering, E. J. and Sandoval, F. M. (1976). Hydrologic aspects of saline seeps in southwestern North Dakota. In 'Regional Saline Seep Control Symposium Proceedings'. Bull. No. 1132, pp. 303–16. (Montana State University: Bozeman, Mont.)

Ferguson, H. (1976). The salt status of saline seep area soils. In 'Regional Saline Seep Control Symposium Proceedings'. Bull. No. 1132, p. 86.

(Montana State University: Bozeman, Mont.)

Freeze, R. A. and Witherspoon, P. A. (1967). Theoretical analysis of regional groundwater flow. 2. Effect of water table configuration and subsurface permeability variation. Water Resour. Res. 3, 632–4.

Freeze, R. A. and Witherspoon, P. A. (1968). Theoretical analysis of regional groundwater flow. 3. Quantitative interpretations. Water Resour. Res. 4, 581–90.

Fried, J. J. and Combarnous, M. A. (1971). Dispersion in porous media. Adv. Hydrosci. 7, 169–282.

Gardner, W. R. (1965). Movement of nitrogen in soil. In 'Soil Nitrogen'. Agron. No. 10, pp. 550–72. (Am. Soc. Agron.: Madison, Wis.)

Greenlee, G. M., Pawluk, S. and Bowser, W. E. (1968). Occurrence of soil salinity in the drylands of southwestern Alberta. Can. J. Soil Sci. 48, 65–75.

Halvorson, A. D. (1973). Saline seeps in eastern Montana and what a wet season does to saline seeps. In 'Governor's Saline Seep Emergency Meeting Proceedings'. pp. 31–2. (Montana State University: Bozeman, Mont.)

Halvorson, A. D. and Black, A. L. (1974). Saline seep development in dryland soils of northeastern Montana. J. Soil Water Conserv. 29, 77–81.

Halvorson, A. D. and Reule, C. A. (1976). Controlling saline seeps by intensive cropping of recharge areas. In 'Regional Saline Seep Control Symposium Proceedings'. Bull. No. 1132, pp. 115–24. (Montana State University: Bozeman, Mont.)

Halvorson, A. D. and Rhoades, J. D. (1976). Field mapping soil conductivity to delineate dryland saline seeps with four-electrode technique. Proc. Soil Sci. Soc. Am. 40, 571–5.

Hartley, R. (1976). Focus on the Kangaroo Island salinity problem. J. Agric. S. Aust. 79, 34–8.

Hillel, D. (1971). 'Soil and Water: Physical Principles and Processes.' (Academic Press: New York.)

Hingston, F. J. and Galaitis, V. (1976). The geographic variation of salt precipitated over Western Australia. Aust. J. Soil Res. 14, 319–35.

Holmes, J. W. and Colville, J. S. (1968). On the water balance of grassland and forest. Trans. 9th Int. Congr. Soil Sci., Adelaide, Vol. 1., pp. 39–46.

Jacobsen, T. and Adams, R. M. (1958). Salt and silt in ancient Mesopotmian agriculture. Science (New York) 128, 1251–8.

Kovda, V. A., van den Berg, C. and Hagan, R. M. (1973). 'Irrigation, Drainage and Salinity.' (UNESCO/FAO: Paris).

Krogman, K. (1973). Cropping to control ground-water. In 'Alberta Dryland Salinity Workshop Proceedings'. pp. 61–6. (Alberta Department of Agriculture: Edmonton, Alberta.)

Lightfoot, L. C., Smith, S. T. and Malcolm, C. V. (1964). Salt land survey 1962: report of a survey of soil salinity. J. Agric. West. Aust. 5, 396–410.

Logan, J. M. (1958). Erosion problems on salt affected soils. J. Soil Conserv. Serv. N. S. W. 14, 220–42.

Loh, I. C. and Hewer, R. A. (1977). Salinity and flow simulation of a catchment reservoir system. Inst. Eng. Aust., Natn. Conf. Publ. No. 77/5, pp. 192–3.

Lyles, L. and Allen, R. R. (1966). Landforming for leaching of saline soils in a non-irrigated area. J. Soil Water Conserv. 21, 57–60.

Malcolm, C. V. and Stoneman, T. C. (1976). Salt encroachment – the 1974 saltland survey. J. Agric. West. Aust. 17, 42–9.

Matheson, W. E. (1968). When salt takes over. J. Agric. South Aust. 71, 266–72.

McCoy, D. (1977). South Warner drainage project. In 'Dryland Salinity and Seepage in Alberta'. (Alberta Dryland Salinity Committee: Lethbridge, Alberta.)

McCracken, L. J. (1973). The extent of the problem. In 'Alberta Dryland Salinity Workshop Proceedings'. pp. 3–15. (Alberta Department of Agriculture: Edmonton, Alberta.)

Miller, M. R. (1971). Hydrogeology of saline-seep spots in dryland farm areas – a preliminary evaluation. In 'Saline Seep-Fallow Workshop Proceedings'. (Highwood Alkali Control Assn: Highwood, Mont.)

Miller, M. R. (1973). Saline-seep development in Montana and adjacent areas. Hydrogeological aspects. In 'Governor's Saline Seep Emergency Meeting Proceedings'. pp. 23–8. (Montana State University: Bozeman, Mont.)

Miller, M. R., Vander Pluym, H., Holm, H. M., Vasey, E. H., Adams, E. P. and Bahls, L. L. (1976). An overview of saline-seep programs in the States and Provinces of the Great Plains. In 'Regional Saline Seep Control Symposium Proceedings'. Bull. No. 1132, pp. 4–17. (Montana State University: Bozeman, Mont.)

Nielsen, G. L. (1973). Geology and macro-ground water movement, saline lands, Alberta. In 'Alberta Dryland Salinity Workshop Proceedings'. pp. 16–43. (Alberta Department of Agriculture: Edmonton, Alberta.)

Northcote, K. H. and Skene, J. K. M. (1972). Australian soils with saline and sodic properties. CSIRO Aust. Soil Publ. No. 27.

Oosterveld, M. (1977). Increasing evaporation to reduce saline areas. In 'Dryland Salinity and Seepage in Alberta'. (Alberta Dryland Salinity Committee: Lethbridge, Alberta.)

Pauli, H. W. (1972). Relationships between land use and water pollution. In 'Water Pollution'. Rep No. 38. pp. 5.1–5.13. (Water Research Foundation of Australia: Sydney.)

Peck, A. J. (1976). Interactions between vegetation and water quality in Australia. In 'Proceedings of the US/Australia Workshop on Management of Range and Forest Lands', pp. 149–55. (Utah State University Water Research Laboratory: Logan, Utah.)

Peck, A. J. (1977). Development and reclamation of secondary salinity. In 'Soil Factors in Crop Production in a Semi-Arid Environment'. pp. 301–19. (University of Queensland Press: St. Lucia.)

Peck, A. J. (1978). Note on the role of an unconfined water table in dryland salinity. Aust. J. Soil Res. 16, 237–40.

Peck, A. J. and Hurle, D. H. (1973). Chloride balance of some farmed and forested catchments in south-western Australia. Water Resour. Res. 9, 648–57.

Peck, A. J., Williamson, D. R., Bettenay, E., and Dimmock, G. M. (1973). Salt and water balances of some catchments in the South-West Coast drainage division. Inst. Eng., Aust., Nat. Conf. Publ. No. 73/3, pp. 1–4.

Pennefather, R. R. (1950). Report on salinity problems in W.A (Mimeo.) (CSIRO Aust.: Canberra.)

Philip, J. R. (1970). Flow in porous media. Annu. Rev. Fluid Mech. 2, 177–204.

Raats, P. A. C. and Gardner, W. R. (1974). Movement of water in the unsaturated zone near a water table. In 'Drainage for Agriculture'. Agron. No. 17, pp. 311–57. (Am. Soc. Agron.: Madison, Wisc.)

Rhoades, J. D. and Halvorson, A. D. (1977). Electrical conductivity methods for detecting and delineating saline seeps and measuring salinity in northern great plains soils. U.S. Dep. Agric. Res. Serv. Rep. ARS–W–42.

Rowan, J. N. (1971). Salting on dryland farms in North-Western Victoria. (Soil Conservation Authority of Victoria: Melbourne, Vic.)

Selim, H. M., Selim, M. S. and Kirkham, D. (1975). Mathematical analysis of steady saturated flow through a multilayered soil with a sloping surface. Proc. Soil Sci. Soc. Am. 39, 445–53.

Smith, C. M. (1973). Soil and water management for dryland crop production related to saline seep control. *In* 'Governor's Saline Seep Emergency Meeting Proceedings'. pp. 33–6. (Montana State University: Bozeman, Mont.)

Smith, D. M. (1975). Salty soils and saline seep. Definitions–identification. Circ. No. 1166. (Montana State University: Bozeman, Mont.)

Smith, S. T. (1961). Soil salinity in Western Australia. *J. Agric. West. Aust.* 2, 757–60.

Smith, S. T. (1962). Some aspects of soil salinity in Western Australia. M.Sc. (Agric.) Thesis, Univ. of Western Australia, Perth.

Smith, S. T. (1966). The relationship of flooding and saline water tables. *J. Agric. West. Aust.* 7, 334–40.

Sommerfeldt, T. G (1976). Mole drains for saline-seep control. *In* 'Regional Saline Seep Control Workshop Proceedings'. Bull. No. 1132, pp. 296–302. (Montana State University: Bozeman, Mont.)

Sommerfeldt, T. G. (1977). Warner drainage project. *In* 'Dryland Salinity and Seepage in Alberta'. (Alberta Dryland Salinity Committee: Lethbridge, Alberta.)

Stretch, D., and Vulcan, D. A. (1977). Champion soil salinity project report. *In* 'Dryland Salinity and Seepage in Alberta'. (Alberta Dryland Salinity Committee: Lethbridge, Alberta).

Talsma, T. (1963). The control of saline ground-water. *Meded. Landbouwhogesch. Wageningen,* 63(10), 1–68.

Teakle, L. J. H. and Burvill, G. H. (1938). The movement of soluble salts in soils under light rainfall conditions. *J. Agric. West. Aust.* 15, 218–45.

Toth, J. (1971). Groundwater discharge: a common generator of diverse geologic and morphologic phenomena. *Bull. Int. Assoc. Sci. Hydrol.* 16, 7–24.

United States Environmental Protection Agency (1972). Managing irrigated agriculture to improve water quality. (Graphics Management Corporation: Washington.)

United States Salinity Laboratory Staff (1954). Diagnosis and improvement of saline and alkali soils. U.S. Dep. Agric. Handb. No. 60.

Van Schiffgaarde, J. (1974). (Ed.) 'Drainage for Agriculture.' Agron. No. 17. (Am. Soc. Agron.: Madison, Wisc.)

Venables, J. R. C. (1970). Factors affecting salinity in Western District streams. *Aqua,* Sept. 20–1.

Wagner, R. (1957). Salt damage on soils of the Southern Tablelands. *J. Soil Cons. Serv. N. S. W.* 13, 33–9.

Wood, W. E. (1924). Increase of salt in soil and streams following the destruction of the native vegetation. *J. R. Soc. West. Aust.* 10, 35–47.

9

LAND USE AND SOIL EROSION IN PREHISTORIC AND HISTORICAL GREECE*

T. H. Van Andel, E. Zangger and A. Demitrack

Introduction

Soil erosion has been regarded as the inevitable outcome of human land exploitation (Brown 1981) ever since such disasters as the North American "dust bowl" of the 1930s (Borchert 1971) drew public attention to the devastating effect of soil erosion on agricultural productivity and the environment. Wolman (1967) has elegantly illustrated its consequences in his study of a woodland area in Maryland which, in its original state, lost 0.2cm of soil per 1000 years. The spread of farming in the 19th century raised this fifty-fold but when, in the early and middle 1900s, part of the land was returned to forest, the rate dropped back to 5 cm/1000 yrs.

Clearing land for farming, farming itself, deforestation for timber and by grazing, and man-made fires are the most important causes of accelerated anthropogenic soil erosion (Butzer 1982: 123–145), but there are many others (Park 1981). The resultant loss of soil in the uplands and catastrophic sedimentation in valleys and coastal plains are obvious today in many parts of the world (Butzer 1974). Judson (1968) has estimated that the total sediment load of the rivers of the world has increased nearly threefold since the arrival of human beings on earth.

In the Mediterranean, Forbes and Koster (1976) have pointed out the consequences of farming and overgrazing, Hughes (1983) and Thirgood (1981) those of timber cutting for shipbuilding, and Wertime (1983) the effect of firewood exploitation and industrial charcoal-making. It is therefore not surprising that the barren character of much of the Mediterranean landscape has been widely regarded as the result of human carelessness, thus dating the erosion to the Holocene.

To examine these issues further, the senior author in 1979 began a series of studies of prehistoric and historical soil erosion in Greece. The first of these, a part of the Argolid Exploration Project of Stanford University (figure 9.1), yielded a model that related soil erosion and alluviation primarily to human land use (Pope and van Andel 1984). Additional data were subse-

* Originally published in *Journal of Field Archaeology*, 1990, vol. 17, pp. 379–96.

Figure 9.1 [*orig. figure 1*] Study areas and other geographic locations in Greece referred to in the text. SA: Southern Argolid; AP: Argive plain: LB: Larissa basin, Thessaly.

quently obtained in the different settings of the Argive plain (Finke 1988) and the Thessalian Larissa basin (Demitrack 1986) to test and refine this model. Below we present, after a brief synopsis of Vita-Finzi's (1969) scheme for the late Quaternary history of Mediterranean alluviation, a synthesis of the Stanford project and compare it with data on Greek soil erosion culled from the literature. We emphasize that ours is a geoarchaeological, not an archaeological, perspective and that, in contrast to the work of, for example, Osborne (1987) or Alcott (1989), we touch only in passing on matters of land use practices and rural economics.

Previous Concepts of Mediterranean Alluviation

In 1969, Vita-Finzi, using a large body of evidence from across the entire Mediterranean, presented a simple history of late Quaternary stream deposition that has provoked much debate and has had considerable influence on archaeological thinking. In his stimulating book he defined two major phases of alluviation, the Older and the Younger Fill, each silting up stream channels, valley floors, and coastal plains

that had been incised during a preceding erosional phase. Because the Older Fill tends toward red tones, the Younger one to browns and greys, these units can be recognized even from afar. Renewed incision, continuing today in most valleys, terminated the Younger Fill. Vita-Finzi placed the Older Fill in the late Pleistocene (*ca.*50,000–10,000 BP), while archaeological data suggested to him that the Younger Fill had been deposited between late Roman (*ca.*AC 400) and early modern times. Both events were attributed to climatic factors.

Vita-Finzi's model has been applied in Greek archaeology by Bintliff (1976a, 1976b, 1977). Believing that the Older Fill required much higher rainfall than occurs at present, he correlated it with a presumed pluvial phase of the early or middle part of the last glacial. Like Vita-Finzi he attributed the Younger Fill to climate changes thought to have taken place between the middle of the first millennium AC and late Medieval times.

The Mediterranean is large and diverse in terms of human history, bedrock, tectonic state, climate, and vegetation. This renders such a simple model suspect from the start; criticism soon emerged (Butzer 1969), and alluviation events were described that differed regionally in age and character and indicated the existence of more than two units (Davidson 1971, 1980; Eisma 1964, 1978; Kraft, Rapp, and Aschenbrenner 1975; Kraft, Aschenbrenner, and Rapp 1977; Raphael 1968, 1973, 1978). Such diversity also suggests causes other than climatic change, which is likely to have a more uniform regional effect. Consequently, Wagstaff (1981), after a comprehensive analysis of the evidence for the complex history and local variability of late Holocene alluviation, concluded that anthropogenic rather than climatic factors might have been responsible for the Younger Fill.

It thus seems that the Vita-Finzi model as it has been applied to Greek prehistoric and historic land use is too simple or perhaps even erroneous. The three Stanford field studies summarized below were designed to examine this issue in more detail. At the same time, other archaeological surveys in Greece (e.g., Aetolia [Bommeljè 1987];

Boeotia [Bintliff and Snodgrass 1985]; Melos [Davidson and Tasker 1982], Nemea [Cherry et al. 1988]) and elsewhere in the Mediterranean (e.g., Barker et al. 1986; Brückner 1983, 1986; Delano Smith 1979, 1981; Gilman and Thornes 1985) have begun to include geological studies as well. Clearly, Vita-Finzi's ideas, although now obsolete, have raised much interest in the history of Mediterranean soil erosion and alluviation.

Three Case Histories

The three areas of study that form the core of this paper (figure 9.1) have quite distinct characteristics. The Southern Argolid (van Andel 1987) is a small peninsula until recently isolated from the rest of Greece except by sea. Its climate is semiarid Mediterranean, the individual drainages are small, and the short, steep streams flow perhaps once every 10–15 years. The coastal plains are narrow. Relative to Greece as a whole farming was introduced early here (Jacobsen 1976), but until the latest Neolithic and Early Bronze Age it remained limited to the vicinity of a single site (Jameson, Runnels, and van Andel in press; Runnels and van Andel 1987).

The second region, the Argive plain, is the heartland of Greek prehistory and has maintained an important position throughout historical times as well. Extensive human exploitation began earlier here than in the Southern Argolid (Finke 1988; Theocharis 1973: 33–110) and continued on a larger scale throughout the following millennia (Dickinson 1982; Hope, Simpson and Dickinson 1979; Kilian 1984; Pullen 1985). Its climate, vegetation and geological history are similar to the Southern Argolid, but its rivers are larger and its well-integrated drainage system ends in a wide coastal plain (Finke 1988). Thus, despite their proximity, the Argive plain contrasts with the Southern Argolid in geomorphology, settlement patterns, and history.

The Larissa basin in eastern Thessaly (Demitrack 1986), a large inland plain traversed by the Peneios River and remote from the influence of the sea and changing sea levels, is the third study area. Although the climate is Mediterranean with almost all precipitation in the winter months, the surrounding mountains are much wetter than in the other two regions (Philippson 1948; Furlan 1977), and the main rivers flow all year although with highly variable discharges. Most importantly for our purpose, extensive farming began much earlier in the Neolithic here than in the Peloponnese (Halstead 1984).

The Southern Argolid

This small, rugged peninsula lacks extensive lowlands or coastal plains. The northern half is traversed by steep limestone ridges with large exposures of bedrock, fringed here and there by remnants of red, semi-consolidated Pleistocene fans (figure 9.2). The intervening valleys are filled with alluvium. The softer Pliocene sediments of the southern half are deeply dissected, but the upland marls locally retain remnants of once-extensive, deep woodland soils. An integrated archaeological and geological survey was carried out in this region between 1979 and 1985; Jameson, Runnels, and van Andel (in press), Pope and van Andel (1984), Runnels and van Andel (1987), van Andel, Runnels and Pope (1986), and van Andel and Runnels (1987) provide the documentation for the following summary.

The Late Quaternary sediment sequence in the area (figure 9.3) comprises seven depositional units separated by soil horizons. Each unit represents an episode of erosion of the headwaters and slopes that resulted in alluviation on the valley floors and small coastal plains. Each unit ends with a loam and a soil profile that indicates a long period of slope stability during which the slow process of soil formation took place. At the same time the stream channels became incised and sedimentation, except for the intermittent deposition of overbank loams in the lower courses of the streams, virtually ceased. The semiarid soils have a thin, rarely preserved upper (A) and a distinct lower (B) horizon. As the soil matures, the B horizon turns darker red in color, acquires a higher clay content and a blocky structure, and develops a lower calcareous horizon

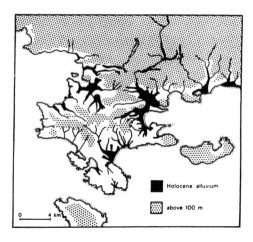

Figure 9.2 [*orig. figure* 2] Relief and drainage systems of the Southern Argolid. Area above 100 m above sea level stippled; Holocene alluvium black. After van Andel, Runnels, and Pope (1986: figs 1 and 3).

(Bca) which evolves from carbonate flecks and stringers by way of well-developed carbonate nodules to a thick, hard calcareous bank. A fully mature soil profile of this type takes many thousands of years to form.

The age-related characteristics of the soils (Birkeland 1984: 203–225; Harden 1982) make it possible to correlate depositional units from one valley to another (Kraus and Bown 1986), and to construct a composite stratigraphic section (figure 9.3). In the Southern Argolid, the units of this section have been dated with the aid of prehistoric and historical sites resting on or buried under them, supplemented by the use of imbedded artifacts and ^{14}C and uranium/thorium disequilibrium dates (Pope and van Andel 1984: table 3).

In the small valleys of this region, where the sediments are laid down close to their sources, three sediment types can be recognized (figure 9.3): (1) chaotic, ill-sorted gravels in which the fine fraction supports the coarse components, a feature typical of debris flows (Innes 1983); (2) stratified, well-sorted sands and gravels laid down by streams; and (3) sandy loams formed by overbank flooding. Debris flows occur mainly in the upper reaches of a drainage and are evidence of catastrophic sheet erosion of slopes when decreasing precipitation or human activity reduces the protective plant cover. Streamflood deposits form when gully cutting is enhanced by increased runoff or as a result of damage by livestock or humans. They dominate in the middle course. Overbank loams are the result of floods and are most common in the lower valleys and on the small coastal plains.

Three alluvial units date to the Pleistocene, at *ca.*272,000, 52,000 and 33, 000 BP (Pope, Runnels, and Ku 1984), whereas four mark the last 5000 years of the Holocene. The voluminous Pleistocene units combine all three sediment types, but the thin Holocene ones consist either of debris flows (Pikrodhafni and Upper Flamboura) or of streamflood deposits (Lower Flamboura and Kranidhi), always topped with loam (figure 9.3).

The Holocene alluvia, although thinner and more restricted in extent than the Pleistocene alluvia, are much more closely spaced in time, but all except the youngest one have soil profiles indicative of prolonged slope stability between erosion events. By comparison, most erosion events were brief. The Lower Flamboura event is bracketed by dates that allow a duration of at most a few centuries, and inspection shows that several meters of the youngest (Kranidhi) unit have accumulated since products made of plastic were introduced in the region (Pope and van Andel 1984).

During the Final Neolithic and Early Bronze Age (mid-4th to mid-3rd millennium BC), the deep woodland soils of the hills and some valley bottoms of the Southern Argolid were widely settled by farmers (Runnels and van Andel 1987). Evidence for soil erosion, however, is lacking until the end of the 3rd millennium when Pikrodhafni debris flows covered the valleys of those drainages that were occupied by settlers. They are broadly bracketed to between *ca.*2300 and 1600 BC by the enclosed late Early Helladic II sherds and superimposed Late Helladic sites. This date eliminates initial woodland clearing as the

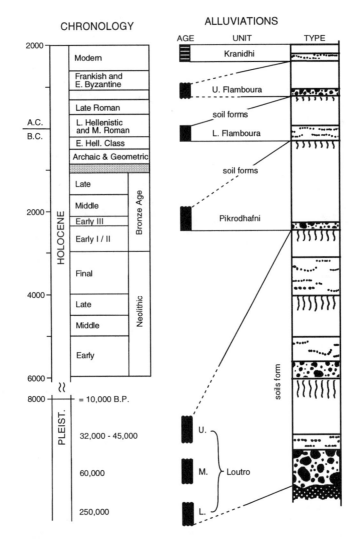

CHRONOLOGY

ALLUVIATIONS

Figure 9.3 [*orig. figure 3*] Chronology and stratigraphy of Late Quaternary alluvia and soils in the Southern Argolid. Archaeological timescale after Runnels and van Andel (1987: table 3); shading indicates "dark ages." Date and duration of each alluviation event shown in column labeled *Age* (broken bar: intermittent deposition; wavy terminations: uncertain age boundaries). Note change in age scale between 6000 and 8000 BC. Black cobbles: debris flows; strings of pebbles: streamflood deposits; blank: loam. Length of wavy vertical lines proportional to soil maturity. Unit heights roughly proportional to thickness. After van Andel, Runnels, and Pope (1986: fig. 4).

cause of the implied catastrophic sheet erosion of slopes, and van Andel, Runnels, and Pope (1986) have attributed it instead to gradual intensification of land use with shorter fallow, expansion onto steeper, less stable slopes, and the introduction of the plow (ard).

During the Late Bronze Age (Mycenaean), which brought more extensive use of the same soils after a sharp decrease in site density in the early 2nd millennium, soil erosion appears to have been kept in check. This development may be attributed to the introduction of soil conservation by means of terracing and gully checkdams, although no securely-dated terrace walls of this period are known. No soil erosion occurred during the post–Mycenaean "dark age" of the 11th–10th centuries BC either, probably because the natural vegetation is capable of rapid recuperation in the absence of tillage or grazing (e.g., Naveh and Dan 1973; Rackham 1982, 1983). Recolonization began in the 8th and culminated in the late 5th to early 3rd centuries BC in a major increase in site numbers. The site pattern resembles that of a classic market-oriented central-place distribution (Runnels and van Andel 1987), and the dry, stony fans and alluvia apparently utilized for the first time are well-suited to olive culture (and are so used today) but not for cereal and pulse farming.

Subsequently, during the last few centuries BC, extensive, well-sorted and stratified streamflood deposits (Lower Flamboura) were laid down in the valleys, simultaneously with a sharp decrease in the number of sites, abandonment of the city of Halieis, and decay of the city of Hermion (Runnels and van Andel 1987). A similar decline has been observed elsewhere in Greece at this time (Alcock 1989), and there is historical evidence for it as well. Alcock (1989) has analyzed this evidence on a much broader base and shows that it can be interpreted either as a rural economic depression or in terms of a different exploitation system not based on single-family farms. Provided this latter mode of land use was to be of reduced intensity, it too could account for the evidence of serious soil loss.

In response to economic stress modern Greek farmers withdraw to their best soils, turning over more distant or poorer fields to pasturage (van Andel, Runnels, and Pope 1986). Without an incentive to repair damage caused by livestock, it takes but a few decades for terrace walls to tumble and gully erosion to strip the stored soil and lay it down in the valley bottoms as streamflood deposits. The evidence for an alternate cause, a climatically-induced increase in runoff, is weak.

Widespread settlement on all usable lands reappeared in Late Roman times (3rd through 6th century AC), apparently with good soil conservation practices, because erosion and alluviation did not occur. The landscape continued to remain stable, presumably because of rapid recolonization by the Mediterranean shrub vegetation called *maquis*, during the next period of depopulation which began in the 7th century. Then, possibly as early as the 9th century AC, upland and headwater areas away from the sea were resettled, while extensive deposition of debris flows took place in the valleys below the new settlements (Upper Flamboura), but stability returned eventually. The final alluviation episode (Kranidhi) began in early modern times. It is localized in extent, happened at different times in different parts of the region, and continues today in several places. Its relation to local economic conditions is clear: land speculation related to a booming tourist industry is more rewarding than olive groves, and the valued crops, such as citrus and vegetables, can only be grown on the best soils. Hence terraces are allowed to decay and land is carelessly cleared with bulldozers (Pope and van Andel 1984; Sutton 1987).

Thus we see a strong case for attributing the frequent but quantitatively minor alluviations of the middle and late Holocene to human activity. Once the Greek landscape had been controlled by soil conservation measures, its equilibrium became precarious, the price of maintaining the equilibrium was high, and economic perturbations were only too likely to disturb it.

The Argive Plain

The Argive plain (Lehmann 1937) was chosen to test, in a different setting, concepts developed in the Southern Argolid. The investigation, undertaken between 1984 and 1987, was similar in approach to that of the Southern Argolid except that the existing archaeological database was used instead of an archaeological survey. Finke (1988) furnishes the documentation for the

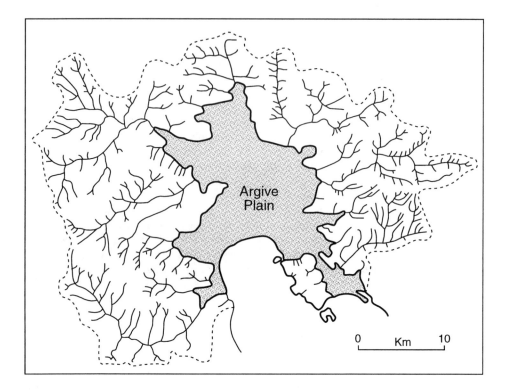

Figure 9.4 [*orig. figure 4*]   The Argive plain and its drainage system. Within the plain, stream courses are not shown because they tend to be ill-defined or have been canalized. After Finke (1988: fig. 9).

following summary. We note that, because in this lowland area river loams and fine-grained coastal sediments predominate, the distinction between debris flows and streamflood deposits, so useful in the Southern Argolid, is not applicable.

The Argive plain, 243 sq km in area, occupies a subsiding coastal basin bordered by the steep slopes of 400–700 m-high mountain ranges and open to the Gulf of Argos. Instead of the small drainages and short streams typical of the Southern Argolid, the region has an integrated drainage system of 1167 sq km (figure 9.4). Sediments eroded from this large region are transported mainly by the seasonal Inachos River which skirts the western margin of the plain and deposited mainly near the present coast, at times causing rapid seaward progradation of the shore (Kraft, Aschenbrenner, and Rapp 1977).

Many large alluvial fans of Middle to Late

Pleistocene age (Koutsouveli-Nomikou 1980) fringe the central plain, their mature soils indicate a long-lasting surface stability. In the plain itself, these deep red paleosols are buried under a Holocene alluvium that reaches a maximum thickness of 8 m at the present coast.

The early and middle Holocene land surface consisted of Pleistocene strata formed during the last glacial when sea level was *ca.*100 m lower than today. This surface, marked by a dark-brown organic soil with many roots and locally abundant charcoal and pottery sherds indicating Neolithic or later occupation (Finke 1988), is only preserved where it is buried under younger alluvium.

In the coastal zone, an early Holocene alluvium, consisting of coarse, poorly sorted sediments like those of the Pleistocene fans, rests on this land surface (figure 9.5), but its extent inland is not known. Since it has

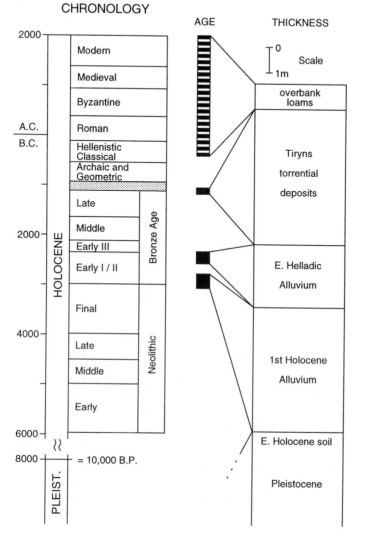

Figure 9.5 [orig. figure 5]   Stratigraphy and chronology of Late Quaternary alluviations of the Argive Plain. Archaeological time scale simplified after Runnels and van Andel (1987: table 3); shading indicates "dark age." Date and duration of alluviation events shown in column labeled *Age* (broken bar: intermittent deposition; wavy terminations: uncertain boundary age). Third column: name and thickness of the alluvial units. After Finke (1988).

buried a Middle Neolithic site, it must be later than *ca.*5000–4000 BC but probably predates 3000 BC. After this alluviation event the landscape stabilized and a soil formed on the deposits. At the same time, the still-rising postglacial sea continued to push the coastline landward until it reached its northernmost position around 2500 BC. Coastal out-building then began, a process that has continued intermittently to the present day (van Andel and Lianos 1983, 1984).

The most pervasive environmental changes in the Argive plain came late in the

3rd millennium BC (Early Helladic II). Floodplain deposits, the equivalent of the overbank loams of the Southern Argolid and 1–3 m thick, spread across the early Holocene plain where today they form most of its surface (figure 9.6). This Early helladic alluvium, easily identified by its reddish brown color, good consolidation, and ubiquitous Early Helladic (Early Bronze Age) pottery, is most extensive on the inner plain and along its streams, but thickest in the coastal zone. Slope stability then returned and lasted until nearly the end of the Late Bronze Age (Late Helladic IIIB), long enough for a soil to form on the Early Helladic alluvium.

This Early Helladic alluviation phase, the largest in the area during the Holocene, resulted from a major soil erosion event that stripped the Pliocene marls and Pleistocene fans of the foothills along the eastern, northern, and NW margins of the plain of most of their brown woodland soils. Bintliff (1976a, 1977) has claimed that those marls were the only ones exploited (and exploitable!) during the Bronze Age. There is no doubt that these soils, as in the Southern Argolid and the Nemea basin (Cherry et al. 1988), were preferentially used and later seriously eroded; only remnants are found today. That erosion, however, took place well before the beginning of the Middle Bronze Age, and its alluvial deposits in the plain would by Mycenaean (Late Bronze Age) times have been much like the old woodland soils themselves in quality or better. Moreover, the swampiness of the Argive plain regarded by Bintliff (1976a, 1977) as an insuperable obstacle to its use as a cropland, was in reality very limited in area (figure 9.6; Finke 1988). The ever increasing number of Late Bronze Age, Classical, Hellenistic, and Roman sites found in the plain confirms that the alluvium there was indeed inhabitable and extensively used from at least Mycenaean times on.

Since the late 3rd millennium BC, no alluviation events have affected the Argive plain in its entirety, although major changes have taken place in the coastal zone. At the end of the Early Bronze Age (Early Helladic III),

immediately after the peak of the marine transgression and approximately simultaneous with the Early Helladic alluvium, the rate of sediment supply to the coast increased and a rapid progradation of the shore began. The out-building of the coast, although slower during the subsequent period of stream incision, has continued intermittently ever since, its focus episodically shifting from the eastern to the western segment of the coast and back.

This should not be taken to mean, however, that there has been no erosion and sedimentation inland on the Argive plain for 4000 years. Late in the Bronze Age (Late Helladic IIIB), torrential flooding, possibly associated with anthropogenic soil erosion in the large inland valleys on the east side of the plain, buried parts of the lower town of Tiryns under several meters of alluvium (Finke 1988; Kilian 1978). The problem was solved by the construction of a large dam and a diversion channel upstream (Balcer 1974), but this merely displaced the deposition area farther to the SE.

Otherwise, the landscape has remained fairly stable. The remains of Classical, Hellenistic, and later settlements and isolated buildings generally lie less than 1 m below the surface, demonstrating that for the last few millennia sedimentation in the Argive plain has been less than in the Southern Argolid and much less than the thick "Younger Fill," as assumed by Bintliff (1977). The only exceptions are black, unsorted deposits of Classical age near Argos which may be the result of landslides after forest or brush fires.

The present appearance of the Argive plain has thus been shaped by three regional soil erosion and alluviation events, which occurred in the Pleistocene, in the later Neolithic, and in Early Helladic II respectively. Except for episodic progradation of the coast and the intermittent deposition of overbank loams along the Inachos River and its tributaries (figure 9.6), landscape changes since about 2000 BC have been of minor extent, although some had a significant local effect.

The two regional alluviation events between 5000 and 2000 BC are approxi-

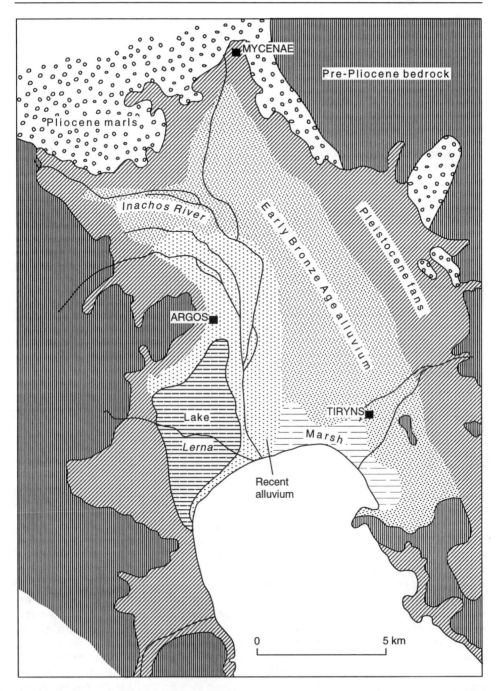

Figure 9.6 [*orig. figure 6*]   Late Quaternary deposits of the Argive plain. The Early Bronze Age alluvium, derived from the Pliocene marls and Pleistocene alluvium of the surrounding hills, was deposited late in the 3rd millennium BC. A large lake existed since the Bronze Age in the plain south of Argos but is now reduced to a small swamp. Flood deposits of the Inachos River, its tributaries, and various small ephemeral streams have been laid down intermittently over the past few thousand years. After Finke (1988: fig. 18).

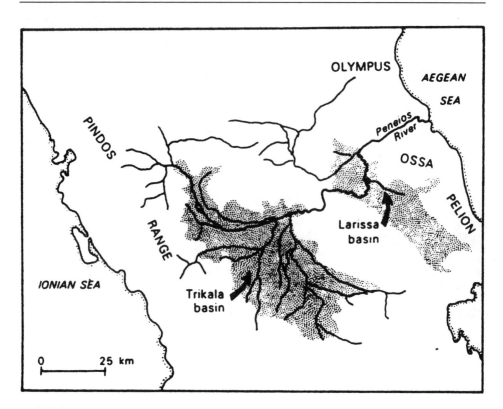

Figure 9.7 [*orig. figure 7*]   The Peneios drainage system in Thessaly and the Trikala and Larissa basins (shaded).

mately contemporaneous with the maximum invasion of the sea. Of course, the transgression itself did not increase the rates of erosion and sediment supply, its impact being limited to the coastal zone. The increasing Late Bronze Age population of the Argive plain (Dickinson 1982; Kilian 1984), on the other hand, must have required land clearance or intensified agriculture for subsistence. This could not fail to produce slopes seasonally unprotected by vegetation, soil erosion, alluviation of the plain itself, and an increased sediment supply to the coastal zone as well.

This version of the Holocene history of the Argive plain is at variance with Bintliff's reconstruction (1977) which rests on his interpretation of Vita-Finzi's Younger and Older Fill scheme. Bintliff saw the Argive plain as a swamp of little economic value until *ca.*1500 years ago when the Younger Fill, his only Holocene phase of soil erosion,

began to bury the wet lowlands. By his estimate, many meters of alluvium were deposited on the swampy plains during historical time. Only when this phase ended about 200 years ago did the soil exist that now supports the thriving agriculture of the region.

In reality, the main erosional event occurred much earlier, and only a single meter of sediments has been deposited in the last 3000 years or more. Furthermore, except for Lake Lerna and some small, spring-fed ponds and bogs that existed locally prior to the overexploitation of groundwater since the middle of this century, the plain was never a swamp (Finke 1988). Therefore, the soils that are the source of the current prosperity of the Argive plain were available to prehistoric and early historical farmers as well and exploited by them since the Bronze Age.

The Larissa Basin in Thessaly

The Peneios River, rising far to the NW in the high Pindos range, crosses the Thessalian plain before it finds its way, joined by several tributaries, through narrow gorges across the Pelion-Ossa-Olympus coastal massif to the Aegean Sea (figure 9.7). The plain itself, one of the largest in Greece, is divided into an eastern (Larissa) and a western (Trikala) basin by a low NW–SE trending ridge of Pliocene marls (Schneider 1968). Today trees are rare in this region, but before major deforestation took place during the last few millennia, the plain and surrounding hills were covered with an open woodland characterized by *Ostrya* and *Carpinus* and dominated by oaks (Bottema 1979; van Zeist and Bottema 1982).

The Larissa basin was occupied in the

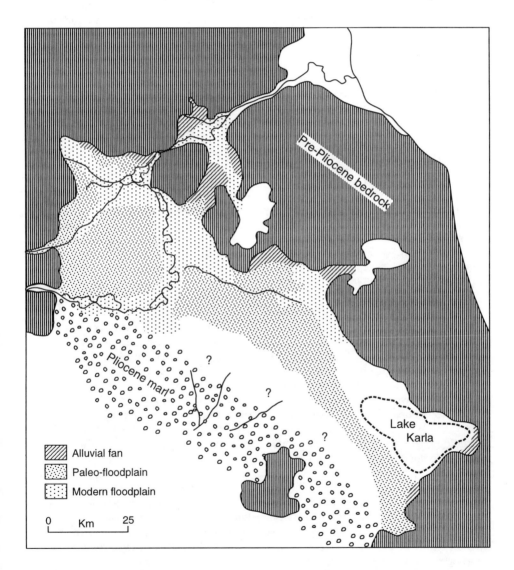

Figure 9.8 [*orig. figure 8*]   Late Quaternary floodplain and fan deposits of the Larissa Basin, Thessaly. The Agia Sophia, Mikrolithos, and Girtoni floodplains are shown together as Niederterrasse. After Demitrack (1986: fig. 6).

Middle and early Upper Paleolithic, but appears to have been (mostly?) deserted during the later Upper Paleolithic and Mesolithic (Runnels 1988). Settled again early in the Neolithic, the population expanded slowly throughout the Neolithic and Bronze Age (Halstead 1977, 1981).

Beginning in 1983, we undertook a detailed study of this basin with methods similar to those employed in the Southern Argolid, using the archaeological background compiled by Halstead (1984). For documentation of the following summary see Demitrack (1986). We note that, as in the Argive plain, we deal here with a lowland river plain where the distinction between debris flows and streamflood deposits used in the Southern Argolid is not applicable.

Alluvial fans fringe the tectonically-active northern rim of the Larissa basin, but the basin itself is covered with river deposits (figure 9.8). As in the Southern Argolid and Argive plain, deposition has been episodic throughout the late Quaternary, each unit ending with a paleosol indicative of a long period of slope stability during which the streams incised their valleys.

Eight fan units, separated by paleosols and ending around 54,000 BP, constitute the earliest dated Pleistocene sequence (Old Red fans: figure 9.9). After a period of tectonic activity and stream incision, fan building (New Red fans) resumed during the last glacial maximum until stream incision took over once again around 14,000 BP. The minor Rodia fan unit formed a few millennia later. In the Holocene fan building was reactivated twice, between 5000 and 4000 BC and in historical times (Old and New Deleria fans).

The floodplain deposits form two groups now at different elevations: 1) an older, Late Pleistocene to Middle Holocene set called the Niederterrasse (Schneider 1968), now well above the river; and 2) a lower, historical pair (figures 9.8, 9.9) that forms the present floodplain. The earliest and most extensive unit of the Niederterrasse (the Agia Sophia alluvium) dates to the middle of the last glacial (ca.40,000–27,000 BP), and is topped by a mature paleosol (Agia Sophia soil). Deposition resumed between 14,000 and 10,000 BP with the Mikrolithos alluvium on which the Noncalcareous Brown soil formed during an early Holocene period of landscape stability. The construction of the higher floodplain was completed in the middle Holocene with the deposition of the Girtoni alluvium, topped by the Girtoni soil.

The present floodplain, built in two stages that could be archaeologically dated, lies 5–15 m below the Niederterrasse. The first episode (Early Peneios alluvium) seems to have Roman structures on it, and has an immature (Deleria) soil, but without further work its precise age cannot be established. The Late Peneios alluvium is from the last few centuries and too young to have a well-developed soil.

Numerous Neolithic and Bronze Age settlement mounds rest on the old floodplain surfaces of the Larissa basin, of which a subset was correlated with the various alluvial units (Demitrack 1986: 33–39, table 5). At least some of the earliest Neolithic settlements were built ca.6000 BC on the late Pleistocene Agia Sophia soil, which had by then been eroded down to its calcareous lower B horizon. There is, on the other hand, no evidence that the Mikrolithos surface, formed between 12,000 and 8000 BC (Demitrack 1986: table 3), was occupied or exploited until the Middle Neolithic (5000–4500 BC).

In the Bronze Age many sites were established on top of the Girtoni alluvium, thus dating its deposition to ca.4500–4000 BC, about 1000 years after the high Thessalian floodplain began to be farmed (the Thessalian Bronze Age [Halstead 1984] begins 4000 BC, being partly synchronous with the Final Neolithic of figures 9.3 and 9.5). At least one Late Neolithic site also occurs on the Girtoni surface. Its edges are covered by more than a meter of Girtoni alluvium, with a soil profile suggesting that it was subject to intermittent slow sedimentation during spring floods.

The depositional history of fan and floodplain sediments in the Larissa basin is a function of distant events in the high Pindos and at the Peneios River mouth, as well as of intrabasin climatic, tectonic, and anthropogenic factors. The coincident beginning of

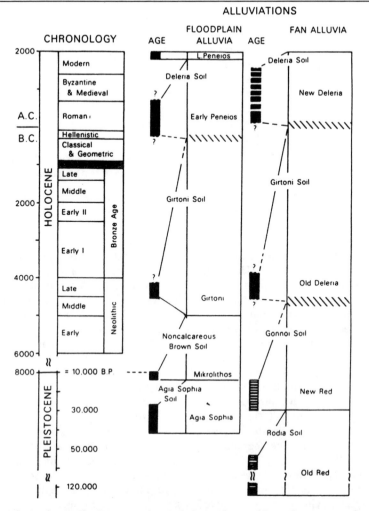

Figure 9.9 [*orig. figure 9*] Stratigraphy and chronology of Late Quaternary floodplain and fan alluvia and their soils in the Larissa basin, Thessaly. Archaeological chronology after Halstead (1984: section 4.2). Date and duration of alluviation events shown in column labeled *Age* (broken bars: multiple units; wavy terminations; uncertain boundary age). Note change in age scale between 6000 and 8000 BC. After Demitrack (1986).

the dry (Bottema 1979; van Zeist and Bottema 1982) glacial maximum and cessation of Agia Sophia aggradation in the floodplain, and the renewal of floodplain deposition (Mikrolithos alluvium) during the shift from dry late glacial to more humid post-glacial conditions (van Zeist and Bottema 1982) imply climatic control. Fan building, on the other hand, did occur also during the dry late glacial, responding to intermittent tectonic activity rather than to climate alone.

Here the human factor is of greatest interest. The Girtoni alluviation, coming about 1000 years after the Larissa basic was first settled, points to a causal relationship between land use and the resumption of soil erosion and floodplain aggradation in the middle Holocene. During those 1000 years the number of sites, and presumably the population, increased steadily (Halstead 1977, 1984) without serious loss of soil. Therefore, as in the Argive plain and Southern Argolid, the initial land clearance cannot be held responsible for the erosion.

Eventually, however, soil erosion did take place. Halstead (1981, 1987, 1989) has argued that, given the low population density of Neolithic Thessaly and the good fertility of its virgin soil, the continuous use of a small area of cultivation adjacent to each settlement would have sufficed to provide the cereals and pulses needed for the slowly growing population. He visualized tiny, self-sufficient villages, isolated in woodland clearings, that exploited rain-fed, animal-fertilized fields (Halstead 1987, 1989). At the same time, an increasing proportion (and number?) of goats and cattle relative to sheep were being grazed in the woodlands between the settlements, producing progressive woodland degradation (Halstead 1981). He did not, on the other hand, see evidence that the slopes of the Larissa basin were exploited to any great degree.

The pollen record for Thessaly, admittedly difficult to interpret in terms of human interference (Bottema 1982), fails to provide solid evidence for extensive forest clearing during the Thessalian Neolithic (Bottema 1979); the trend towards a more open and drier woodland seen after *ca*.8000 BP can be explained adequately by climatic change, but could indicate progressive degradation by grazing as well.

Halstead's concept of little villages fails to account for the record of Neolithic erosion. First, the overlap of Girtoni alluvium onto the feet of several mounds indicates that some sites were located in, and their inhabitants presumably exploited, lands that flooded when the river was highest in the spring, with the great advantage that the renewable fertility of such fields lessened dependence on rain as well as on animal fertilization. Second, the considerable soil erosion producing the deposits of the Girtoni alluvium appears to have stripped the Agia Sophia and Mikrolithos (Noncalcareous Brown) surfaces of the Niederterrasse, as truncated soil profiles under the earliest Neolithic settlements on the Agia Sophia surface attest (Demitrack 1986). Because the Niederterrasse has a low relief, the erosion producing the Girtoni deposits must have

been extensive and not likely to come mainly from degraded woodland, no matter how degraded this may have been. Detailed study of soil types and patterns around settlement mounds, in the manner of the archaeological map of the Netherlands (van Es, Sarfatij, and Woltering 1988: 101–108), would help to address these issues.

Discussion: Land Use and Soil Erosion in Ancient Greece

Our three studies cast little light on Pleistocene slope erosion and stream aggradation, because detailed stratigraphic sections and a chronology of sufficient reliability and resolution are not available. The Thessalian sequence shows how complex the dependence of late Quaternary alluviation had been on changes in neotectonic activity, climate, sea level, runoff, slope stability, and vegetation: a conclusion not unexpected for this time of major tectonism and climatic change. Neither the Argive nor the Thessalian sequence correlates with global glacial-interglacial or stadial-interstadial climatic changes, suggesting that a NW European sense of the geomorphological effect of the Pleistocene upon the landscape needs to be adjusted when we deal with the eastern Mediterranean.

This is regrettable, because there is the suggestion that these were the times that, in essence, shaped the present Greek landscape and that it was stripped by nature much more than by man, as Hutchinson (1969) has suggested for Epiros and Rohdenburg and Sabelberg (1973) for the western Mediterranean.

The Holocene alluviation history (figure 9.10) presents an altogether different case. The last two decades have seen a considerable deepening of our understanding of the depositional and postdepositional processes in continental settings, and the resolving power of our chronological and environmental methods has much increased. The stratigraphic use of paleosols has added a new dimension to correlation and chronology, and integration with archaeological surveys also has proved to be a powerful approach.

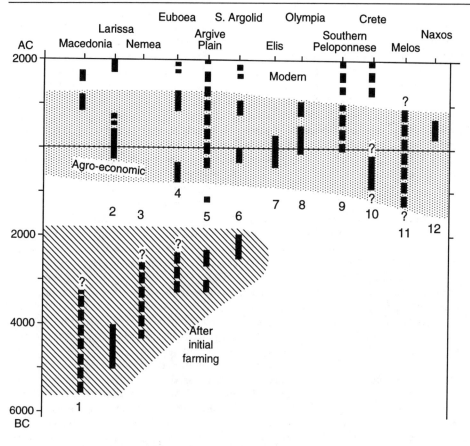

Figure 9.10 [*orig. figure 10*]   Chronology of Holocene alluviation events in Greece and the Aegean. Broken bars are dated uncertainly or represent intermittent deposition. Dates taken from the original publications: (1) L. Faugères in Delibrias (1978); (2) Demitrack (1986); (3) Cherry et al. (1988); (4) Genre (1988) and Rust (1978); (5) Finke (1988); (6) Pope and van Andel (1984); (7) Raphael (1968, 1973, 1978); (8) Dufaure (1967) and Büdel (1965); (9) Hempel (1982, 1984); (10) Hempel (1982, 1984); (11) Davidson and Tasker (1982); (12) Renault-Miskowsky (1983).

The Holocene histories of soil erosion and valley deposition of the Southern Argolid, the Argive plain, and the Larissa basin agree only in part, even if we allow for large uncertainties in dating (figure 9.10). Somewhat surprisingly, in view of the major climatic change involved, which includes the postglacial warming and greatly increased precipitation (van Zeist and Bottema 1982), the landscapes of all three regions seem to have remained stable from the late Pleistocene through the early Holocene. Major slope destabilization and alluviation did not come until much later, about 1000 years after the first spread of settlement and farming. In the Argive plain and Thessaly, these early alluviations were also the most extensive and voluminous ones; their regional extent and influence on the landscape exceed that of all later episodes. In the Southern Argolid the early phase, although quite marked, is smaller than, and mostly buried under, the deposits of the later Hellenistic–Early Roman erosional event.

Subsequently, the histories of the individual areas diverge. The torrential floods of the Late Bronze Age that affected the important but small area of the Argive plain near Tiryns are absent in the Southern Argolid. There, however, very extensive alluviations took place in the late first millennium BC, and again in early medieval times. Other erosion/alluviation events, smaller and more dispersed in space and time, have intermittently troubled various Argolid valleys since the 17th century. These early medieval and more recent alluviation episodes of the Southern Argolid probably correspond to those of the low Peneios floodplain in Thessaly, although the dating of the latter is uncertain. The Argive plain, on the other hand, has remained essentially stable since the Bronze Age.

Some alluvia are dated closely, for example by the common occurrence of datable pottery fragments in Argive plain sediments or by the accurate positioning of sites of known age on or under alluvial units in the Southern Argolid, to show short durations and rapid sedimentation. Near Tiryns, 4.8 m of Late Bronze Age (Late Helladic IIIB2) sediments were deposited in as little as 50 years, and the Late Hellenistic–Early Roman event in the Southern Argolid (Lower Flamboura) did not last more than two or three centuries. The age of many alluvia is not as well constrained, but many of them also have been quite brief. In contrast, the degree of maturity of the late Pleistocene and all but the latest Holocene soils shows that the episodes of stability and soil formation have lasted thousands to tens of thousands of years. Overall, stability appears to have been the prevailing state of the Greek landscape during the last 100,000 years or more (see Thornes and Gilman [1983] for a similar conclusion concerning the Iberian peninsula), destabilization being a rare and often brief event.

The same spatial and temporal diversity of the Holocene alluviation history can be inferred for other parts of Greece from the literature (figure 9.10). Unfortunately, few data are available for the Neolithic and Early Bronze Age, surely in part because the earliest Holocene alluvia resemble the "Older Fill" in color, and are easily but wrongly assigned to the Pleistocene. A loam

deposit postdating 5500 BC and widespread in Macedonia (L. Faugères in Delibrias 1978) may be a case of soil erosion associated with Neolithic land use, although early settlement was less dense there than in Thessaly (Jarman, Bailey, and Jarman 1982: 146–154), and major deforestation came much later (Wijmstra 1969). On Euboea, Genre (1988) has noted alluviation well before 2000 BC, and in the Nemea basin brown woodland soils like those preferred in the Southern Argolid were extensively occupied in the Middle Neolithic (Cherry et al. 1988). The soils have been washed away, but the time of erosion has not yet been determined. The same woodland soils appear to have been preferred elsewhere, too (e.g., Delano Smith 1972).

There are few reports of alluviation in the later Bronze Age (2nd millennium BC). Other than our own data for Tiryns and the suggestion by Davidson and Tasker (1982) that soil erosion began on Melos in the late 2nd millennium BC, no reasonably documented cases are known to us. Apparently, soil erosion and alluviation were not troublesome in Mycenaean times, except locally; given the considerable density of settlement, effective soil management, such as the use of terracing, seems the only plausible explanation.

Turning to historical times, one finds a good deal more, although the data are sometimes poorly constrained in time or geologically weak. It is also difficult to decide whether the relative wealth of information derives from a greater interest in the period or is evidence for a real increase in soil erosion and sedimentation. The alluviation of later Hellenistic and Early Roman times in the Southern Argolid seems to have been quite widespread. Besides its occurrence in Thessaly, it has been noted in Elis between ca.350 BC and AC 300 (Raphael 1968, 1978). Dufaure (1976) has placed an extensive alluviation of the Alpheios valley above Olympia in the 2nd–6th centuries AC, but the dating of this earliest of his two alluviation phases is not robust. Anecdotal references to the burial under alluvium of Classical, Hellenistic, and Roman structures abound (Vita-Finzi 1969), but once again the dating tends to be less than exact.

Rapid coastal progradation, such as occurred after 3000 BC near Pylos in the Bay of Navarino (Kraft, Rapp, and Aschenbrenner 1980), may indicate a period of enhanced soil erosion inland. On the Aegean coast of Turkey (figure 9.1), Eisma (1964, 1978) found a major phase of fill dating 500–100 BC in the Küçük Menderes valley near Ephesus, and Aksu, Piper, and Konuk (1987) placed the main delta advance there between 900 BC and AC 100. In the next valley to the south, the main depositional activity of the Büyük Menderes came slightly later, ca.100–300 BC (Eisma 1978) or, more broadly, between 500 BC and AC 500 (Aksu, Pipe, and Konuk 1987). Caution is advisable, however, when deducing soil erosion from coastal accretion, because long-shore drift bringing sediment from elsewhere, and small changes in the rise and fall of sea level can be easily mistaken for changes in sediment supply and hence in inland erosion rates (Curray 1964).

Even leaving this class of data aside, there seems to be sufficient reason to suggest that the last few centuries BC and the first few of our era were a time of widespread but by no means ubiquitous destabilization of the Greek landscape (figure 9.10). This does not mean, however, that the preceding Archaic and Classical periods enjoyed complete freedom from soil erosion problems. Genre (1988) and Rust (1978) document two brief but troublesome events of alluviation in central Euboea between 720 and 680 BC, and another at the end of the 5th and beginning of the 4th century BC. They hold deforestation in the upper parts of the drainage basin responsible for both. As at Tiryns in the Argive plain, the remedy was found in sophisticated engineering works. Localized soil erosion also took place between 700 and 200 BC in the southern Peloponnese and on Crete (Hempel 1982, 1984). On Melos a historical phase of erosion and alluviation may have begun in the first quarter of the 1st millennium BC, reaching its peak around AC 500 (Davidson and Tasker 1982).

Alluvial deposits of medieval to modern times are common. Büdel (1965) and later Dufaure (1976) have described major alluvi-

ation at Olympia between the 7th and 14th centuries AC. Renault-Miskovsky (1983) says that alluvial loams were emplaced on Naxos between the 3rd and 7th centuries AC. Major stream aggradation in the 9th–12th centuries AC, comparable to and roughly synchronous with the Upper Flamboura of the Southern Argolid, has been described by Genre (1988) for central and northern Euboea. L. Faugères (in Delibrias 1978) mentions alluviation in Macedonia from the 9th century AC onward. Many other cases, ranging from the 9th century (Middle Byzantine) through the Turkish period, have been compiled by Wagstaff (1981) in his analysis of Vita-Finzi's "Younger Fill." From this he concluded (as we do) that, over the last 1500 years, stream aggradation was episodic and localized in the eastern Mediterranean, with intensities and dates that varied from place to place. This implicates human interference with slope equilibrium as the principle cause, rather than climatic changes such as the Little Ice Age (15th through mid-19th century AC).

This view contrasts with that of Vita-Finzi (1969) and Bintliff (1976a, 1977), who attributed both the Older and the Younger Fill to climatic changes. Hassan (1985), in his review of stream aggradation in semiarid and arid regions, also stressed climatic factors. Genre (1988) and Brückner (1986), on the other hand, concluded after a thorough consideration of other potential causes that neither climate nor changes in the relative levels of land and sea could have induced the observed alluviation events. Instead, they too opted for the human factor.

This is not to say that climatic and sea level changes of tectonic or eustatic origin are not potentially important factors in Holocene geomorphology (Hassan 1985; Nir 1983; Thornes 1987), and they should not be casually dismissed. For the time being, however, we regard these natural factors as of minor importance in the Holocene alluviation history of Greece. First, changes of sea level and/or stream base level have been small (a few meters) during the period considered here, except for a few areas where neotectonic uplift and subsidence are well documented, as in the Gulf of Corinth. Climatic changes, on the other hand, are usually invoked merely by reference to the Holocene climatic history of NW Europe, an inappropriate analogy, and hard local evidence for their existence has not yet been presented. In our view, the burden of proof rests for the time being on those who propose climatic, tectonic, or sea level changes as causes of landscape destabilization in the Aegean.

The complex nature of the processes that have produced the alternating stability and destabilization of the Greek landscape is somewhat bewildering and encourages over-simplification. It is, for example, widely believed that valley aggradation and coastal accretion tend to be unrelated and are seldom concurrent. This is rarely true. The response of a stream system to a change in conditions is complex and a single cause, e.g., increased slope erosion, can set off a chain of down-valley consequences that evokes different responses in different parts of the system (Schumm 1977, 1981; Schumm, Harvey, and Watson 1984; Patton and Schumm 1983). The aggradation accompanying slope destabilization in the upper reaches of a drainage often, although not always, begins at the shore and proceeds up-valley with time, accompanied by changes in sediment type (Patton and Schumm 1983; Pope and van Andel 1984; Schumm 1981). Nondeposition, erosion, and the formation of various different sediment types may thus simultaneously take place in the same drainage.

Conclusions

Alternations between stability and destabilization analogous to those cited above have occurred elsewhere in the Mediterranean as well and appear to support our conclusions as, for example, the work of Delano Smith (1979, 1981) shows. Also in southern Italy, the earliest valley fill, assigned by Brückner (1983) to the late Pleistocene or early Holocene, is weakly dated and might well be Neolithic in age. A subsequent stable period lasted for much of the last two millennia BC; it was followed by extensive deposition

between the 5th and the 3rd century BC, attributed to land clearing and land use during the Greek colonization (Brückner 1983, 1986). A more poorly-dated early medieval erosion phase caused by resettlement and by cultivation of hill lands ended in the 11th–12th, or alternatively the 14–15th century. Finally, large-scale 19th to early 20th century alluviation was caused by the extensive deforestation (Brückner 1986) that accompanied the opening-up of wooded hinterlands by roads and railways.

Overall, this and other Mediterranean depositional histories compiled by Brückner (1986: fig. 7) demonstrate the same local and temporal variability that is found in Greece and the Aegean.

Thus the evidence that can be brought to bear on the problem of natural versus human-induced landscape destabilization in the Aegean, although still limited, appears to us to point first and foremost to the dominant role of human activity. The chronology remains less certain than one might wish, and this is also true for our understanding of how the processes of soil erosion and alluviation related to land use practices and rural economics (see Alcott 1989). A few inferences regarding the role of cultivation techniques come from the small drainages of the Southern Argolid (van Andel, Runnels, and Pope 1986), but much more research is needed.

We suspect that the soil preferences of the early Greek farmers were not strong and played only a secondary role in land use patterns. The brown woodland soils that had formed in late and postglacial times on Pliocene marls and shales and late Quaternary alluvial loams seem to have been preferred for cereals and pulses, as they still are to some extent. The coarser deposits of alluvial fans and slopes were not widely exploited until the introduction of large-scale olive culture, probably not before the Late Bronze Age (Runnels and Hansen 1986). Beyond this broad generalization, and while we recognize that the properties of a soil measured today cannot tell us very much about their quality when they were farmed in the distant past, we see little that points to a large influence of soil quality on past land use. On the whole, water more than soil appears to have determined land use and settlement patterns in Greece from the early Neolithic to the 19th century, as far as the current data show.

Many problems remain, such as our lack of knowledge of the time of introduction of terrace agriculture or the important question of how "catastrophic" (Brückner 1986) the consequences of anthropogenic soil erosion in the Holocene have really been. A reasonably secure estimate for the Southern Argolid (Jameson, Runnels, and van Andel in press) suggests that on average less than 40 cm were stripped from that area. This is insignificant in geomorphic terms and confirms the view expressed by Rohdenburg and Sabelberg (1973) and Thornes and Gilman (1983:75) that the present Mediterranean landscape was shaped mainly during the Pleistocene. These solitary examples, of course, need to be augmented before we can put a finger on the dominant cause of the barren state of so many Greek slopes.

Acknowledgments

Two of the regional studies summarized here, the Southern Argolid and Thessaly, were supported by the National Science Foundation, the Argive plain investigation by the Deutsches Archäologisches Institut in Athens, and all three by donations from private Stanford supporters who shared our interests. We thank these sponsors for their confidence and the Director and staff members of the Institute of Geology and Mineral Exploration (GME) in Athens for permission to do the fieldwork and for other valuable assistance. Others too numerous to acknowledge except as a group have been essential each in their own way, but we do owe special thanks to Michael Jameson, Klaus Kilian, and Curtis Runnels for their encouragement and support. The senior author has been inspired by Karl W. Butzer in ways best illustrated by the latter's small, thoughtful paper on the dating and correlation of Holocene alluvial sequences (1980).

References

Aksu, A. E., David J. N. Piper, and T. Konuk. 1987. "Quaternary Growth Patterns of Büyük Menderes and Küçük Menderes Deltas, Western Turkey," *Sedimentary Geology* 52: 227–250.

Alcock, Susan. 1989. "Roman Imperialism in the Greek Landscape," *Journal of Roman Archaeology* 2: 5–34.

Balcer, Jack M. 1974. "The Mycenaean Dam at Tiryns," *American Journal of Archaeology* 78: 141–149.

Barker, Graeme, S. Coccia, D. Jones, and J. Sitzia. 1986. "The Montarrentic Survey, 1985: Problems in Integrating Archaeological, Environmental and Historical Data," *Archeologia Medievale* 13: 291–320.

Bintliff, John L. 1976a. "Sediments and Settlement in Southern Greece," In D. A. Davidson and M. L. Shackley, eds., *Geoarchaeology*. Duckworth: London, 267–275.

―― 1976b "The Plain of Western Macedonia and the Neolithic Site of Nea Nikomedia," *Proceedings of the Prehistoric Society* 42: 241–262.

―― 1977 *Natural Environment and Human Settlement in Prehistoric Greece. BAR Supplementary Series* 28 (i,ii). Oxford: B.A.R.

Bintliff, John L., and Anthony M. Snodgrass. 1985. "The Cambridge/Bradford Boeotian Expedition: The First Four Years," *Journal of Field Archaeology* 12: 125–161.

Birkeland, Peter W. 1984. *Soils and Geomorphology.* Oxford: Oxford University Press.

Bommeljè, Sebastiaan. 1987. *Aetolia and the Aetolians: Towards the Interdisciplinary Study of a Greek Region. Studia Aetolica.* 1. Utrecht: Parnassus Press.

Borchert, J. R. 1971. "The Dust Bowl in the 1930's," *Annals of the Association of American Geographers* 61: 1–22.

Bottema, Sytze. 1979. "Pollenanalytical Investigations in Thessaly, Greece," *Palaeohistoria* 21: 19–40.

―― 1982. "Palynological Investigations in Greece with Special Reference to Pollen as an Indicator of Human Activity," *Palaeohistoria* 24: 257–289.

Brown, L. R. 1981. "World Population Growth, Soil Erosion, and Food Security," *Science* 214: 995–1002.

Brückner, Helmut. 1983. "Holozäne Bodenbildungen in den Alluvionen Süditalienischer Flüsse," *Zeitschrift für Geomorphologie, Supplement Band* 48: 99–116.

―― 1986. "Man's Impact on the Evolution of the Physical Environment in the Mediterranean Region in Historical Times," *Geo-Journal* 13: 7–17.

Büdel, Julius. 1965. "Aufbau und Verschüttung Olympias: Mediterrane Flusztätigkeit seit der Frühantike," *Deutsche Geographische Tagung, Heidelberg 1963, Tagungsberichte und*

Wissenschaftliche Abhandlungen: 179–183. Heidelberg.

Butzer, Karl. W. 1969. "Changes in the Land" (review of C. Vita-Finzi: *The Meidterranean Valleys: Geological Changes in Historic Times), Science* 165: 52–53.

—— 1974. "Accelerated Soil Erosion: A Problem of Man-Land Relationship," in I. R. Manners and M. W. Mikesell, eds., *Perspectives on the Environment.* Washington, D.C.: Association of American Geographers, 57–78.

—— 1980. "Holocene Alluvial Sequences: Problems of Dating and Correlation," in R. A. Cullingford, D. A. Davidson, and J. Lewin, eds., *Timescales in Geomorphology.* New York; John Wiley, 131–142.

—— 1982. *Archaeology as Human Ecology.* Cambridge: Cambridge University Press.

Cherry, John F., Jack L. Davis, Anne Demitrack, Eleni Mantzourani, Thomas F. Strasser, and Lauren Talalay. 1988. "Archaeological Survey of an Artifact-rich Landscape: A Middle Neolithic Example from Nemea, Greece," *American Journal of Archaeology* 92: 159–176.

Curray, Joseph R. 1964. "Transgressions and Regressions," in R. L. Miller, ed., *Papers in Marine Geology: Francis P. Shepard Commemorative Volume.* New York: MacMillan, 175–203.

Davidson, Donald A. 1971. "Geomorphology and Prehistoric Settlement of the Plain of Drama," *Revue de Géomorphologie Dynamique* 20: 22–26.

—— 1980. "Erosion in Greece during the First and Second Millennia B. C., " in R. A. Cullingford and D. A. Davidson, eds., *Timescales in Geomorphology.* New York: John Wiley, 143–158.

Davidson, Donald A., and Catriona Tasker. 1982. "Geomorphological Evolution during the Late Holocene", in C. Renfrew and J. M. Wagstaff, eds., *An Island Polity – The Archaeology of Exploitation in Melos.* Cambridge; Cambridge University Press, 82–94.

Delano Smith, Catherine. 1972. "Late Neolithic Settlement, Land Use, and *Garrique* in the Montpellier Region, France," *Man* 7: 397–407.

—— 1979. *Western Mediterranean Europe: A Historical Geography of Italy, Spain and Southern France since the Neolithic.* New York: Academic Press.

—— 1981. "Valley Changes: Some Observations from Recent Field and Archive Work in Italy," in G. Barker and R. Hodges, eds., *Archaeology and Italian Society: Prehistoric, Roman and Medieval Studies.* BAR International Series 102. Oxford: B.A.R., 239–257.

Delibrias, G., ed. 1978. *Evolution des Paysages sur les Rives Nord-Méditerranéennes au Course du Postglaciaire.* Orsay: 6" R.A.S.T.

Demitrack, Anne. 1986. *The Late Quaternary Geologic History of the Larissa Plain, Thessaly, Greece: Tectonic, Climatic and Human Impact on the Landscape.* Ph.D. dissertation, Stanford University, CA; Ann Arbor, Michigan: University Microfilms.

Dickinson, Oliver T. P. K. 1982. "Parallels and Contrasts in the Bronze Age of the Peloponnese," *Oxford Journal of Archaeology* 1: 125–138.

Dufaure, Jean-Jacques. 1976. "La Terrasse Holocène d'Olympie et ses Équivalents Méditerranéens," *Bulletin de l'Association Géographique Française* 433: 85–94.

Eisma, Doeke. 1964. "Stream Deposition in the Mediterranean Area in Historical Times," *Nature* 203: 1061.

—— 1978. "Stream Deposition and Erosion by the Eastern Shore of the Aegean," in W. C. Brice, ed., *The Environmental History of the Near and Middle East since the Last Ice Age.* London: Academic Press, 87–81.

Finke, Eberhard A. W. 1988. *Landscape Evolution of the Argive Plain, Greece: Paleoecology, Holocene Depositional History and Coastline Changes.* Ph.D. Dissertation, Stanford University, CA Ann Arbor: University Microfilms.

Forbes, Hamish A., and Harold A. Koster. 1976. "Fire, Ax, and Plow: Human Influence on Local Plant Communities in the Southern Argolid," *Annals of the New York Academy of Science* 268: 109–126.

Furlan, Dietrich. 1977. "The Climate of Southeast Europe," in C. C. Wallen, ed., *World Survey of Climatology 6: Climates of Central and Southern Europe.* Amsterdam: Elsevier, 185–235.

Genre, C. 1988. "Les Alluvionnements Historiques en Eubée, Grèce: Caractères Principaux, Chronologie, Signification," *Actes de la Table Ronde "Géomorphologie et Dynamique des Bassins Versants Élémentaires en Régions Méditerranéennes."* Poitiers: Etudes Méditerranéennes 12: 229–258.

Gilman, Anthony, and J. B. Thornes. 1985. *Land-Use and Prehistory in Southeast Spain.* London: Allen and Unwin.

Halstead, Paul L. J. 1977. "Prehistoric Thessaly: The Submergence of Civilisation," in J. L. Bintliff, ed., *Mycenaean Geography.* Cambridge: British Association for Mycenaean Studies, 23–29.

—— 1981. "Counting Sheep in Neolithic and Bronze Age Greece," in I. Hodder, G. Isaac, and N. Hammond, eds., *Pattern of the Past: Studies in Honour of David Clarke.* Cambridge:

Cambridge University Press, 307–339.

—— 1984. *Strategies for Survival: An Ecological Approach to Social and Economic Change in the Early Farming Communities of Thessaly, Northern Greece*. Unpublished Ph.D. dissertation, Cambridge University, Cambridge.

—— 1987. "Traditional and Ancient Rural Economy in Mediterranean Europe: Plus Ça Change?" *Journal of Hellenic Studies* 107: 77–87.

—— 1989. "The Economy Has a Normal Surplus: Economic Stability and Economic Change among Early Farming Communities in Thessaly," in Paul L. J. Halstead and John O'Shea, eds., *Bad Year Economics: Cultural Responses to Risk and Uncertainty*. Cambridge: Cambridge University Press, 68–80.

Harden, Jennifer W. 1982. "A Quantitative Index of Soil Development from Field Descriptions: Examples from a Chronosequence in Central California," *Geoderma* 28: 1–28.

Hassan, Fekri A. 1985. "Fluvial Systems and Geoarchaeology in Arid Lands: With Examples from North Africa, the Near East, and the American Southwest," in J. K. Stein and W. R. Farrand, eds. *Archaeological Sediments in Context*. Orono: University of Maine, 53–68.

Hempel, L. 1982. "Jungquartäre Formungs-prozesse in Südgriechenland und auf Kreta," *Forschungsbericht des Landes Nordrhein-Westphalen* 3114: 1–80.

—— 1984 "Geoökodynamik im Mittelmeerraum während des Jüngquartärs: Beobachtungen zur Frage 'Mensch und/oder Klima?' in Südgriechenland und auf Kreta," *Geoökodynami* 5: 99–104.

Hope Simpson, Richard, and Oliver T. P. K. Dickinson. 1979. *A Gazetteer of Aegean Civilisation in the Bronze Age: 1. The Mainland and the Islands. Studies in Mediterranean Archaeology 52*. Göteborg.

Hughes, J. Donald. 1983. "How the Ancients Viewed Deforestation," *Journal of Field Archaeology* 10: 437–445.

Hutchinson, J. 1969. "Erosion and Land-use: The Influence of Agriculture on the Epirus Region of Greece," *Agricultural History Review* 17: 85–90.

Innes, John L. 1983. "Debris Flows," *Progress in Physical Geography* 7: 469–501.

Jacobsen, Thomas W. 1976. "17,000 Years of Greek Prehistory," *Scientific American* 234: 76–87.

Jameson, Michael H., Curtis N. Runnels, and Tjeerd H. van Andel in press. *A Greek Countryside: The Southern Argolid from Prehistory to the Present Day*. Stanford, CA: Stanford University Press.

Jarman, Michael R., Geoffrey B. Bailey, and Heather N. Jarman. 1982. *Early European Agriculture: Its Foundation and Development*. Cambridge: Cambridge University Press.

Judson, Sheldon. 1968 "Erosion of the Land – or What's Happening to our Continents?" *American Scientist* 56: 356–374.

Kilian, Klaus. 1978. "Ausgrabungen in tiryns 1976," *Archäologischer Anzeiger* 4: 449–470.

—— 1984. "Die Verwaltungspolitische Organ-isation in Messenien und in der Argolis während der Mykenischen Epoche," *Praktiki 2; Topikou Synergiou Messeniakon Spoudon*: 55–68. Athens.

Koutsouveli-Nomikou, A. 1980. *Contribution à l'Étude Quaternaire de la Région d'Argos (Nord-est du Péloponnèse, Grèce)*. Thèse de Doctorat, Université d'Aix, Marseille.

Kraft, John C., Stanley E. Aschenbrenner, and George Rapp, Jr. 1977. "Paleogeographic Reconstructions of Coastal Aegean Archaeo-logical Sites," *Science* 195: 941–947.

Kraft, John C., George Rapp, Jr., and Stanley E. Aschenbrenner. 1975. "Holocene Paleogeography of the Coastal Plain of the Gulf of Messenia, Greece, and its Relationship to Archaeological Setting and Coastal Change," *Bulletin of the Geological Society of America* 86: 1191–1208.

—— 1980. "Late Holocene Palaeogeomorphic Reconstructions in the Area of the Bay of Navarino: Sandy Pylos," *Journal of Archaeo-logical Science* 7: 187–210.

Kraus, M. J., and T. M. Bown. 1986. "Paleosols and their Resolution in Alluvial Stratigraphy," in V. P. Wright, ed., *Paleosols – Their Recognition and Interpretation*. Oxford: Blackwell, 180–207.

Lehmann, Herbert. 1937. *Argolis – Landeskunde der Ebene von Argos und ihrer Randgebiete*. Athens: Deutsches Archäologisches Institut.

Naveh, Z., and J. Dan. 1973. "The Human Degradation of Mediterranean Landscapes in Israel," in F. di Castri and H. A. Mooney, eds., *Mediterranean Type Ecosystems: Origin and Structure*. New York: Springer Verlag, 373–390.

Nir, Dov. 1983. *Man, a Geomorphological Agent*. Dordrecht: Reidel Publishers.

Osborne, Robin. 1987. *Classical Landscapes with Figures*. London: George Philip.

Park, C. C. 1981. "Man, River Systems and Environmental Impacts," *Progress in Physical Geography* 5: 1–31.

Patton, Peter C., and Stanley A. Schumm. 1983. "Ephemeral Stream Processes: Implications for Studies of Quaternary Valley Fills,"

Quaternary Research 15: 24–43.

Philippson, A. 1948. *Das Klima Griechenlands.* Bonn: Dümmler Verlag.

Pope, Kevin O., Curtis N. Runnels, and Teh-Lung Ku. 1984. "Dating Middle Palaeolithic Red Beds in Southern Greece," *Nature* 312: 264–266.

Pope, Kevin O., and Tjeerd H. van Andel. 1984. "Late Quaternary Alluviation and Soil Formation in the Southern Argolid: Its History, Causes and Archaeological Implications," *Journal of Archaeological Science* 11: 281–306.

Pullen, Daniel J. 1985. *Social Organization in Early Bronze Age Greece: A Multidimensional Approach.* Ph.D. dissertation, Indiana University, Bloomington. Ann Arbor: University Microfilms

Rackham, Oliver. 1982. "Land Use and the Native Vegetations of Greece," in M. Bell and S. Limbrey, eds., *Archaeological Aspects of Woodland Ecology. BAR International Series* 146. Oxford B.A.R., 177–198.

—— 1983. "Observations on the Historical Ecology of Boeotia," *The Annual of the British School at Athens* 78: 291–351.

Raphael, Constantine N. 1968. *Geomorphology and Archaeology of the Northwest Peloponnese.* Ph.D. dissertation, Lousiana State University, Baton Rouge. Ann Arbor: University Microfilms.

—— 1973. "Late Quaternary Changes in Coastal Elis, Greece," *Geographical Reveiw* 63: 73–89.

—— 1978. "The Erosional History of the Plain of Elis in the Peloponnese," in W. C. Brice, ed., *The Environmental History of the Near and Middle East since the Last Ice Age.* New York: Acadmeic Press, 51–66.

Renault-Miskowsky, J. 1983. "Les Connaissances Actuelles sur les Végétations et les Climats Quaternaires de la Grèce, Grace aux Donhées de l'Analyse Pollinique," *Actes de la Table Ronde: Les Cyclades Antiques.* Paris: C.N.R.S., 99–109.

Renfrew, Colin. 1972. *The Emergence of Civilisation: The Cyclades and the Aegean in the Third Millennium B.C.* London: Methuen.

Rohdenburg, H., and U. Sabelberg. 1973. "Quartäre Klimazyklen im westlichen Mediterrangebiet un ihre Auswirkungen auf die Relief und die Boden-Entwicklung," *Catena* 1: 71–180.

Runnels, Curtis N. 1988. "A Prehistoric Survey of Thessaly: New Light on the Greek Middle Paleolithic," *Journal of Field Archaeology* 15: 277–290.

Runnels, Curtis N., and Julie Hansen. 1986. "The Olive in the Prehistoric Aegean: The Evidence

for Domestication in the Early Bronze Age," *Oxford Journal of Archaeology* 5: 299–308.

Runnels, Curtis N., and Tjeerd H. van Andel. 1987. "The Evolution of Settlement in the Southern Argolid, Greece," *Hesperia* 56: 303–334.

Rust, Uwe. 1978. "Die Reaktion der fluvialen Morphodynamik auf die anthropogene Entwaldung östliches Chalkis (Insel Euboea, Griechenland)," *Zeitschrift für Geomorphologie, Supplement Band* 30: 183–203.

Schneider, Horst E. 1968. "Zur Quartärgeologischen Entwicklungsgeschichte Thessaliens (Griechenland)," *Beiträge zur Ur- und Frühgeschichte des Mittelmeerischen Kulturraumes* 6: 1–127.

Schumm, Stanley A. 1977. *The Fluvial System.* New York: John Wiley and Sons.

—— 1981. "Evolution and Response of the Fluvial System: Sedimentological Implications," *Society of Economic Paleontologists and Mineralogists, Special Publication* 31: 19–29.

Schumm, Stanley A., M. D. Harvey, and C. C. Watson. 1984. *Incised Channels: Morphology, Dynamics and Control.* Littleton, CO. : Water Resources Publications.

Sherratt, Andrew G. 1981. "Plough and Pastoralism: Aspects of the Secondary Products Revolution," in I. Hodder, G. Isaac, and N. Hammond, eds., *Pattern of the Past: Studies in Honour of David G. Clarke.* Cambridge: Cambridge University Press, 261–305.

Sutton, Susan B. 1987. "Landscape and People of the Franchthi Region, Part II. The People," in T. W. Jacobsen, ed., *Excavations at Franchthi Cave, Greece* 2. Bloomington: Indiana University Press, 63–79.

Theocharis, Demetrios R. 1973. "The Neolithic Civilisation: A Brief Survey," in D. R. Theocharis, ed., *Neolithic Greece.* Athens: National Bank of Greece, 17–128.

Thirgood, J. V. 1981. *Man and the Mediterranean Forest.* New York: Academic Press.

Thornes, John B. 1987. "The Palaeoecology of Erosion," in J. W. Wagstaff, ed., *Landscape and Culture.* Oxford: Blackwell, 37–55.

Thornes, John B., and Anthony Gilman. 1983. "Potential and Actual Erosion around Archaeological Sites in South-east Spain," *Catena Supplement* 4: 91–113.

van Andel, Tjeerd H. 1987. "Landscape and People of the Franchthi Region, Part I. The Landscape," in T. W Jacobsen, ed., *Excavations at Franchthi Cave, Greece* 2. Bloomington: Indiana University Press; 1–62.

van Andel, Tjeerd H., and Nikolaos Lianos. 1983.

"Prehistoric and Historic Shorelines of the Southern Argolid Peninsula: A Subbottom Profiler Study," *International Journal of Nautical Archaeology and Underwater Exploration* 12: 303–324.

—— 1984. "High Resolution Seismic Reflection Profiles for the Reconstruction of Postglacial Transgressive Shorelines: An Example from Greece," *Quaternary Research* 22: 31–45.

van Andel, Tjeerd H., and Curtis N. Runnels. 1987. *Beyond the Acropolis: A Rural Greek Past.* Stanford, CA: Stanford University Press.

van Andel, Tjeerd H., Curtis N. Runnels, and Kevin O. Pope. 1986. "Five Thousand Years of Land Use and Abuse in the Southern Argolid, Greece," *Hesperia* 55: 103–128.

van Es, Willem A., H. Sarfatij, and P. J. Woltering. 1988. *Archeologie in Nederland.* Amsterdam: Meulenhoff.

van Zeist, Willem, and Sytze Bottema. 1982. "Vegetational History of the Eastern Mediterranean and the Near East during the Last 20,000 Years," in J. L. Bintliff and W. van Zeist, eds., *Palaeoclimates, Palaeoenvironments and Human Communities in the Eastern Mediterranean Region in Later Prehistory. BAR International Series* 133. Oxford: B.A.R., 277–321.

Vita-Finzi, Claudio. 1969. *The Mediterranean Valleys: Geological Changes in Historical Times.* Cambridge: Cambridge University Press.

Wagstaff, John M. 1981. "Buried Assumptions: Some Problems in the Interpretation of the 'Younger Fill' Raised by Recent Data from Greece," *Journal of Archaeological Science* 8: 247–264.

Wertime, Theodore A. 1983. "The Furnace Versus the Goat: The Pyrotechnologic Industries and Mediterranean Deforestation in Antiquity," *Journal of Field Archaeology* 10: 445–452.

Wijmstra, T. A. 1969. "Palynology of the First 30 metres of a 120 m Deep Section in Northern Greece," *Acta Botanica Neerlandica* 18: 511–527.

Wolman, Martin G. 1967. "A Cycle of Sedimentation and Erosion in Urban River Channels," *Geografiska Annaler* 49A: 385–395.

10

SOIL EROSION IN BRITAIN: A REVIEW*

J. Boardman and R. Evans

1 Introduction

The potential for widespread soil damage in Britain was first noted in 1971, with the emphasis placed on water erosion of upland peat and mineral soils, and wind erosion on lowland arable fields (Evans, 1971). Eight years later Reed (1979) listed soils at risk of erosion, mainly from publications of the Soil Survey of England and Wales. He also demonstrated that in the West Midlands, erosion by water occurred frequently and caused spectacular gullying. A survey of erosion in Britain by Morgan (1980) pointed out that many cases had occurred in the past, and were recorded in local data sources such as engineer's reports and newspapers. These data sources have been largely ignored. Speirs and Frost (1985) review the reasons for a recent increase in the incidence of erosion in Scotland.

In 1982 an erosion monitoring scheme was set up by the Soil Survey of England and Wales and the Ministry of Agriculture, Fisheries and Food. Seventeen localities in England and Wales were monitored each year between 1982 and 1986, using aerial photography, ground checking and measurement (Evans, 1988). The localities were selected on the basis of known or suspected erosion and the possibility of land-use change from pasture to arable. Some of the study areas were assumed to be at low risk of erosion. In each locality aerial photographs were taken annually at a scale of 1:10,000 or 1:15,000, covering an area of 15 to 30km in length, and 2 to 3km wide. This survey forms the principal data source of any assessment of the extent, frequency and magnitude of erosion on agricultural land in England and Wales. Preliminary findings are discussed by Evans and Cook (1986) and results for the first three years of the five year monitoring period are presented in Evans (1988; 1990a). Since 1989, a scaled-down version of the scheme has operated but no results have been published.

Data from other sources can supplement those from the SSEW/MAFF monitoring scheme. The National Soil Map (1:250,000) shows soil associations that are at risk of erosion by water and wind (Mackney et al., 1983). The map was compiled from data available in 1981 but there are omissions. For example, Andover 1 Association soils on chalk have subsequently proved to be at risk

* Originally published in R. J. Rickson (ed.) *Conserving Soil Resources: European Perspectives* (Wallingford: CAB International, 1994).

of erosion. A more up-to-date listing of soils at risk is in Evans (1990b). Morgan's (1985a) assessment of erosion risk is based on the National Soil Map data, allied to a rainfall erosion index (KE>10) taken from Morgan (1980).

Suspended sediment yields from rivers and volumes of sediment trapped in reservoirs are frequently used to estimate rates of erosion in catchments. The main problem with this approach is that it takes no account of sediment storage between the site of erosion and the sampling point. A delivery ratio may be used but there are many theoretical and practical difficulties (Walling, 1983). The average suspended sediment yield from 56 catchments in Britain ranging in size from 0.1 to 6850km² is 0.63t ha⁻¹yr⁻¹. The longer term average from reservoir sedimentation studies is 0.3t ha⁻¹yr⁻¹ (Walling and Webb, 1981). Erosion rates from catchments of mixed land use and varying size are therefore in the order of 0.5t ha⁻¹yr⁻¹ (Boardman, 1986). This, however, is not a very useful figure in an agricultural context since fields within a catchment may experience high rates of erosion with little immediate impact on streams due to storage of eroded soil within the field or in flood plain sites (Trimble, 1983).

Other data sources include single storm measurements (e.g. Morgan 1985b); estimates based on caesium-137 (Quine and Walling, 1991); regional studies (e.g. Reed, 1979; Colborne and Staines, 1985; and Boardman, 1990a), and data from several studies using experimental plots (Morgan, 1985b; Fullen and Reed, 1986; Robinson and Boardman, 1988). However, there are many difficulties in integrating data from these different sources to the extent that apparently simple questions such as 'how serious is erosion on agricultural land in Britain?' are not easy to answer!

2   Areas at Risk and Rates of Erosion

Areas of lowland arable farming most at risk of erosion are reasonably well known as a result of studies of individual erosion events, the SSEW/MAFF monitoring scheme and information presented on the National Soil Map. They have been identified as:

1   Lower Greensand soils of southern England including the Isle of Wight;
2   Sandy and loamy soils in Nottinghamshire and the West Midlands;
3   Sandy and loamy soils in Somerset and Dorset;
4   Sandy and loamy soils in parts of East Anglia;
5   Chalk soils of the South Downs, Cambridgeshire, Yorkshire and Lincolnshire Wolds, Hampshire and Wiltshire;
6   Sandy soils in South Devon;
7   Loamy soils in eastern Scotland.

In most of these areas the main agent of erosion is runoff, but wind erosion occurs locally on sandy and peaty soils in the Vale of York, the Breckland of East Anglia and on peat mosses in Lancashire (Mackney et al., 1983; ADAS, 1985; Evans, 1990b). Sandy soils in the West Midlands and Nottinghamshire are at risk of both wind and water erosion. Evans (1990b) lists 144 soil associations as being at moderate to very high risk of water erosion and 20 at similar risk of wind erosion.

Fourteen soil associations are at moderate to very high risk of erosion in the uplands of England and Wales (Evans, 1990b). The limited amount of information on rates of upland erosion is reviewed by Evans (1989; 1990c). Soils at risk include deep peats, where gullying occurs, and mineral soils.

The SSEW/MAFF monitoring scheme quantified rates of erosion by measurement of volumes of soil lost from rills and gullies, or volumes of soil deposited in fans (Evans, 1988). The data for erosion rates is very left skewed, with many low and few high values. For this reason the median rather than the mean is a better reflection of central tendency. Median values for the years 1982–84 are generally below 1m³ha⁻¹ with higher values in a few localities (table 10.1). Silty soils in Somerset generally have the highest median values, but the highest was in Kent in 1984 when all the eroded fields were sown to irrigated market garden crops

(Boardman and Hazelden, 1986). Areas of sandy soils in the Isle of Wight, Nottinghamshire, Shropshire and Staffordshire also have high median rates of erosion. The maximum values of erosion show a wider range than the median values.

In an area of about 35km² of agricultural land on the South Downs Boardman (1990a and unpublished) monitored erosion from 1982–92. Median annual erosion rates ranged from 0.5 to 5.0m³ha⁻¹ measured on the basis of contributing area rather than field area. For one year (1987–88) the median rates were 3.3m³ha⁻¹ on a field basis and 5.0m³ha⁻¹ on a contributing area basis (Boardman, 1988).

Rates of erosion estimated using caesium-137 for 13 sites in England and Wales are presented by Quine and Walling (1991) (table 10.2). The locations were in areas where erosion is known to be a problem and covered a range of soil textures. Each sampling site is one field, which may not be representative of the wider area. Erosion rates are average annual rates for approximately the last 30 years. The ability to estimate erosion retrospectively makes the technique potentially very useful. It is however based on several assumptions and a calibration procedure. The net erosion rate represents the amount of soil which leaves the field. The gross rate is calculated by dividing the total mass of eroded sediment by the area of the field (Quine and Walling, 1991). The gross figure is therefore comparable to the figures quoted by Evans (1988 and table 10.1), and Boardman (1990a). The gross erosion rates in table 10.2 are higher than the median figures in table 10.1, sometimes by a considerable amount. This may be because Quine and Walling sampled sites with unusually high rates of erosion, whereas Evans used the median value of all eroding fields in a locality. At the Lewes site (table 10.2), a gross rate of erosion of 4.3t ha⁻¹yr⁻¹ compares with a rate estimated

Table 10.1 [*orig. table 1*]   Erosion rates per field in 17 monitored localities (data from Evans, 1988).

	1982			1983			1984		
	n	m	M	n	m	M	n	m	M
Cambridge/ Bedfordshire	21	0.32	2.33	14	0.25	2.36	2	0.20	0.25
Cumbria	3	0.31	3.62	–	–	–	1	0.01	0.01
Devonshire	1	1.29	–	6	1.67	5.94	–	–	–
Dorset	12	0.58	7.14	35	1.45	22.20	5	0.74	1.41
Gwent	19	0.95	4.90	26	0.66	15.62	4	0.70	1.15
Hampshire	3	0.81	4.90	7	0.66	21.22	6	0.61	1.82
Hereford	47	0.87	9.44	5	0.62	7.64	1	0.63	2.07
Isle of Wight	13	1.40	4.79	31	2.0	20.45	2	1.00	3.69
Kent	8	0.61	3.46	12	1.13	7.09	7	7.92	12.76
Norfolk East	3	0.90	1.46	48	0.80	6.75	9	1.47	4.77
Norfolk West	5	0.05	1.20	42	0.58	8.54	10	0.14	3.11
Nottingham	9	0.33	3.89	85	1.81	47.25	16	0.24	1.50
Shropshire	29	0.99	28.84	69	1.35	35.24	21	0.40	8.57
Somerset	19	4.30	39.74	37	2.75	29.52	8	3.42	7.85
Stafford	30	0.90	77.34	85	1.31	55.24	22	0.75	10.77
Sussex East	–	–	–	10	0.56	6.72	3	0.13	0.48
Sussex West	–	–	–	30	0.43	7.15	6	0.21	0.41

n = number of fields.
m = median volume of eroded soil (m³ha⁻¹).
M = maximum volume of eroded soil in each locality (m³ha⁻¹).

Table 10.2 [orig. table 2]   Caesium-137 based soil erosion rates from Quine and Walling (1991) (t ha^{-1}yr^{-1}).

Site	Soil texture*	Net	Gross
Yendacott, Devon	Loamy	1.9	5.3
Mountfield, Somerset	Silty	2.2	4.6
Higher, Dorset	Clayey	3.1	5.2
Fishpool, Gwent	Silty	1.9	5.1
Wootton, Herefordshire	Silty	2.8	6.4
Dalicott, Shropshire	Sandy	6.5	10.2
Rufford, Nottingham	Sandy	10.5	12.2
Brook End, Bedford	Clayey	1.2	3.6
Keysoe Park, Bedford	Clayey	0.6	2.2
Manor House, Norfolk	Coarse loamy	2.4	6.3
Hole Farm, Norfolk	Sandy	3.0	6.3
West Street, Kent	Fine silty	4.3	7.7
Lewes, Sussex	Silty	1.4	4.3

*From description of soil association (Mackney et al., 1983).

by measurement of rills and gullies over eight years of approximately 1t ha^{-1}yr^{-1} (Boardman, unpublished). Walling and Quine (1991) recognise this problem but regard the values as showing reasonable agreement, considering the differences between the two approaches. It seems unlikely that the discrepancy is due to unmeasured sheet and wind erosion since these processes appear to be insignificant on the South Downs at present. It also seems unlikely that rates of erosion were higher before 1982 when Boardman began monitoring. All available evidence suggests the reverse. There is therefore an unexplained contrast in rates obtained by the two methods which would repay further investigation.

In many localities winter cereals are most at risk of erosion because of the extensive area under this crop. However, rates of erosion are generally higher on sugar beet, potatoes, market garden crops and soft fruits (Evans, 1988).

3   Causes of Erosion

There has been considerable increase in erosion on agricultural land in Britain in the last fifteen years (Evans and Cook, 1986; Boardman, 1990a). This is the result of the intensification of agriculture (Morgan, 1980; Speirs and Frost, 1985; Bullock, 1987; Boardman, 1990b), with heavier and more powerful vehicles leading to compaction problems (Fullen, 1985), and cultivation of steeper slopes. Contour working is not possible for reasons of safety. Fields have been enlarged by the removal of hedges, banks, walls and ditches, so that larger machinery is easier to use. There is some evidence that a decrease in the organic matter content of soils has made them more susceptible to degradation and erosion (Morgan, 1985c). Farmers have been encouraged to produce fine seedbeds by the use of the power harrow (Frost and Speirs, 1984). On some soils fields drilled to cereals, oil seed rape or grass are rolled after drilling to produce smooth surfaces with low micro-topography on which runoff readily occurs.

The association of erosion with winter cereal crops is very clear (Boardman, 1984; Colborne and Staines, 1985; Evans and Cook, 1986; Speirs and Frost, 1985). In England and Wales between 1969 and 1983 there was a threefold increase in the area of winter cereals (Evans and Cook, 1986) and in Scotland a fourfold increase between 1967 and 1986 (Evans, 1989). Winter cereals are susceptible to erosion following drilling until the fields have about 30% crop cover (Robinson and Boardman, 1988). This period is frequently the wettest of the year with rilling initiated on bare smooth surfaces (Boardman, 1992).

Furthermore, under winter cereals farmers are often unable to complete drilling operations on all their land before early November due to periods of wet weather or late harvesting of some previous crops (e.g. potatoes and sugar beet). This means that fields drilled late will not attain sufficient crop cover to inhibit erosion before the spring. Fields will also be drilled when soils are damp and compaction is more likely to occur. Large areas of a single crop increase the likelihood of adjacent fields being bare at the same time. In some parts of the South Downs winter cereals occupy 60% of the landscape and runoff is able to travel uninhibited from one bare field to the next. Under these conditions runoff and eroded soil have moved up to 5km along normally dry valleys and into a housing estate (Boardman and Evans, 1991).

There is no evidence that the recent increase in erosion can be explained by a change in the amount or character of rainfall. Wetter seasons undoubtedly occur, such as the autumns of 1982 and 1987 on the South Downs. However, in each autumn current farming practices and the area of land under arable crops means that the landscape is primed for erosion. It is a question of 'How much?' and 'Where?' rather than 'If?' and 'When?' In areas such as the West Midlands and Nottinghamshire, the intensification of agriculture has meant that much erosion occurs on sugar beet and potatoes as a result of spring and summer convective rain (Evans, 1988; Fullen, 1985; Fullen and Reed, 1986).

Data on the extent and rate of wind erosion on agricultural land in Britain is lacking. Soils at risk are known, but the magnitude and frequency of erosion events is unknown. Wilson and Cooke (1980) give a range of soil loss between 20 and 44t ha^{-1}yr^{-1}. Hepworth (1987) recorded widespread damage and significant costs in a study area south east of York in April 1986. Wind erosion is difficult to measure because much soil is redistributed within the field and fine components may be moved long distances. However, estimates based on material trapped in ditches and behind field boundaries could be made and are needed.

Erosion in the uplands is a result of water, wind, frost, fire, animal and human agencies (such as footpaths, ski impacts, and afforestation). In the peat areas of the southern Pennines a phase of gullying seems to have been initiated by pollution at the beginning of the Industrial Revolution which killed the protective cover of sphagnum moss. The gullying has continued to the present day (Burt and Labadz, 1990). In many upland areas a large increase in sheep numbers since the Second World War has caused overgrazing problems.

4   Implications of Erosion in Britain

Erosion affects crop yields due to the decrease in soil depth and loss of nutrients. The full economic impact may, however, be delayed for hundreds of years, depending on the rate of erosion, the depth of the soil and the economics of farming. Erosion damages current crops which may have to be redrilled or abandoned. Arden-Clarke and Evans (in press) review the impact of erosion on yields in Britain (table 10.3). On thick soils the reduction in cereal yields with rates of erosion typical of Britain (e.g. 3t ha^{-1}yr^{-1}) is significant. On thin soils similar rates of erosion give more cause for concern (table 10.4). Where topsoils are thin and subsoils are rapidly exposed by erosion, or are incorporated into the topsoil, a less workable seedbed results and yield losses from erosion quickly become evident. On eroding valley sides in Cambridgeshire, where chalky boulder clay had been brought to the surface, winter wheat yields were only 59% of those on the deeper soils of the valley floor (Evans and Catt, 1987).

Off-farm costs associated with erosion incidents are far higher than on-farm costs (Crosson, 1984). In Britain many costs are unquantified. Most attention has focused on the flooding of properties by soil-laden water from farmers' fields, mostly in autumn and winter from cereal fields. Recently, there appears to have been a substantial increase in flooding, although the

Table 10.3 [*orig. table 3*]   Estimated percentage reduction in yield of cereals over 100 years at various rates of erosion.

| Rates of erosion (t ha⁻¹yr⁻¹) | Reduction in yield (%) | | | |
| | Spring Barley | | Winter Wheat | |
	Hempstead	Maxey	Hempstead	Mean
1.0	0.3	0.5	1.4	0.7
3.0	1.0	1.8	3.6	2.1
12.0	5.3	7.9	13.1	8.8

Initial soil depth at Hempstead = 0.9m.
Initial soil depth at Maxey = 1.2m.
After Evans (1981).

Table 10.4 [*orig. table 4*]   Calculated life-span of soils.

Soil Type	Sandy loam			Clay Loam		
Bulk density (g.cm⁻³)	1.4			1.3		
Minimum soil depth (m)	0.2			0.15		
Available soil depths (m)	0.50	0.30	0.25	0.30	0.25	0.20
Life span (yr) at mean annual erosion rates of (t ha⁻¹)						
2	7800	1400	700	1950	1300	650
5	700	350	175	488	325	163
10	311	156	78	217	144	72
20	147	74	37	103	68	34
50	57	29	14	40	27	13
100	28	14	7	20	13	7

After Morgan (1987).

phenomenon is not new (e.g. at Shepton Beauchamp in Somerset, June 1966 and July 1968, in Morgan, 1980). On the South Downs prior to the 1980s there are few records of erosion and associated flooding – even in the wet autumn of 1976 (Boardman, 1990a). In the 1980s there have been many incidents. In 1982, Lewes District Council spent about £12,000 in emergency and protective measures at Highdown (Stammers and Boardman, 1984). In 1987, total off-farm costs attributable to flooding at four major sites around Brighton were in excess of £660,000 (Robinson and Blackman, 1990).

The impact of erosion on rivers in Britain is reviewed by Evans (1989). This includes the impact on fish stocks and the reasons for increased turbidity in many chalk streams of southern England. Discolouration of water by inwashed peat or soil has caused severe difficulties with water filtration at the Cray Reservoir in South Wales (Stretton, 1984), near Huddersfield, South Yorkshire (Austin and Brown, 1982), and in Nidderdale, North Yorkshire (Edwards, 1986). The annual loss of water storage capacity due to sediment deposition is reviewed by Butcher (1992) for six reservoirs in the southern Pennines. The loss of capacity per century ranges from 3 to 75% for individual reservoirs, varying with erosion rate and reservoir size.

5   Conclusion

The data from the SSEW/MAFF monitoring scheme have been related to soil texture, topography, climate and farming practices.

The causes of erosion under current conditions are now reasonably well understood (Evans, 1988, 1990a). Unfortunately the scheme was abandoned prior to 1987 when severe erosion occurred in parts of southern England (e.g. Boardman, 1988). Boardman et al. (1990) show that considerable increases in erosion can be expected as a result of changes in future climate and land use.

By world standards rates of erosion in Britain are probably low, although little reliable data exists for most countries. In Britain, locally high rates of erosion on loamy, silty and sandy soils under intensive arable cultivation cause concern. Caesium-137 measurements suggest that certain fields have been subject to surprisingly high rates of erosion for the last thirty years. Most attention has been focused on the issue of flooding by runoff from fields of winter cereals and there have been local attempts to deal with the problem by engineering approaches and land use change (e.g. Robinson and Blackman, 1990; Boardman and Evans, 1991) but no coherent policy exists. Government response to the threat of soil erosion, flooding and associated pollution problems has been disappointing. Reform of the EC Common Agricultural Policy, with less incentive for farmers to grow arable crops or graze sheep on unsuitable land, is generally agreed to be necessary for environmental and economic reasons including the control of erosion.

References

Adas 1985. Soil erosion by wind. Ministry of Agriculture Fisheries and Food Leaflet 891.

Arden-Clarke, C. and Evans, R. In press. Soil erosion and conservation in the UK. In: Pimental, D. (ed.). *World soil erosion and conservation*. Cambridge University Press.

Austin, R. and Brown, D. 1982. Solids contamination resulting from drainage works in an upland catchment, and its removal by flotation. *Journal of the Institution of Water Engineers and Scientists* 36:281–288.

Boardman, J. 1984. Erosion on the South Downs. *Soil and Water* 12:19–21.

Boardman, J. 1986. The context of soil erosion. *Seesoil* 3:2–13.

Boardman, J. 1988. Severe erosion on agricultural land in East Sussex, UK, October 1987. *Soil Technology* 1:333–348.

Boardman, J. 1990a. Soil erosion on the South Downs: a review. In: Boardman, J., Foster, I. D. L. and Dearing, J. A. (eds). *Soil erosion on agricultural land*. John Wiley, Chichester. pp. 87–105.

Boardman, J. 1990b. Soil erosion in Britain: costs, attitudes and policies. Social Audit Paper No. 1. Education Network for Environment and Development. University of Sussex, Brighton.

Boardman, J. 1992. The sensitivity of Downland arable land to erosion by water. In: Thomas, D. S. G. and Allison, R. J. (eds). *Environmental sensitivity*. John Wiley, Chichester.

Boardman, J. and Evans, R. 1991. Flooding at Steepdown. Report for Adur District Council.

Boardman, J., Evans, R., Favis-Mortlock, D. T. and Harris, T. M. 1990. Climate change and soil erosion in England and Wales. *Land Degradation and Rehabilitation* 2:95–106.

Boardman, J. and Hazelden, J. 1986. Examples of erosion on brickearth soils in east Kent. *Soil Use and Management* 2:105–108.

Bullock, P. 1987. Soil erosion in the UK – an appraisal. *Journal Royal Agricultural Society of England* 148:144–157.

Burt, T. and Labadz, J. 1990. Blanket peat erosion in the Southern Pennines. *Geography Review* 3(4):31–35.

Butcher, D. P. 1992. The Southern Pennines: peat erosion and reservoir sedimentation. In: Boardman, J. (ed.). *Post Congress Tour Guide*. European Society for Soil Conservation, Silsoe.

Colborne, G. J. N. and Staines, S. J. 1985. Soil erosion in south Somerset. *Journal of Agricultural Science, Cambridge*. 104:107–112.

Crosson, P. 1984. New perspectives on soil conservation policy. *Journal of Soil and Water Conservation* 39:222–225.

Edwards, A. M. C. 1986. Land use and reservoir gathering grounds. In: Solbe, J. F. de L. G. (ed.). *Effects of land use on fresh waters*. Ellis Horwood, Chichester. pp. 534–537.

Evans, R. 1971. The need for soil conservation. *Area 3*. pp. 20-23.

Evans, R. 1981. Assessments of soil erosion and peat wastage for parts of East Anglia, England. A field visit. In: Morgan, R. P. C. (ed.). *Soil conservation: problems and prospects*. John Wiley, Chichester. pp. 521–530.

Evans, R. 1988. *Water erosion in England and Wales 1982–1984*. Soil Survey and Land Research Centre, Silsoe.

Evans, R. 1989. Soil erosion – the nature of the problem. Paper given to Symposium, Scottish Geographical Society.

Evans, R. 1990a. Water erosion in British farmers' fields – some causes, impacts, predictions. *Progress in Physical Geography* 14:199–219.

Evans, R. 1990b. Soils at risk of accelerated erosion in England and Wales. *Soil Use and Management* 6:125–131.

Evans, R. 1990c. Erosion studies in the Dark Peak. *Northern England Soils Discussion Group Proceedings* 24:39–61.

Evans, R. and Catt, J. A. 1987. Causes of crop patterns in eastern England. *Journal of Soil Science* 38:309–324.

Evans, R. and Cook, S. 1986. Soil erosion in Britain. *Seesoil* 3:25–58.

Frost, C. A. and Speirs, R. B. 1984. Water erosion of soils in south-east Scotland – a case study. *Research and Development in Agriculture* 1:145–152.

Fullen, M. A. 1985. Compaction, hydrological processes and soil erosion on loamy sands in east Shropshire, England. *Soil and Tillage Research* 6:17–29.

Fullen, M. A. and Reed, A. H. 1986. Rainfall, runoff and erosion on bare arable soils in east Shropshire, England. *Earth Surface Processes and Landforms* 11:413–425.

Hepworth, P. 1987. Soil erosion by wind in the Vale of York. BA Dissertation, Brighton Polytechnic.

Mackney, D., Hodgson, J. M., Hollis, J. M. and Staines, S. J. 1983. Legend for the 1:250,000 soil map of England and Wales. *Soil Survey of England and Wales*. Harpenden.

Morgan, R. P. C. 1980. Soil erosion and conservation in Britain. *Progress in Physical Geography* 4:24–47.

Morgan, R. P. C. 1985a. Assessment of soil erosion risk in England and Wales. *Soil Use and Management* 1:127–130.

Morgan, R. P. C. 1985b. Soil erosion measurement and soil conservation research in cultivated areas of the UK. *The Geographical Journal* 151:11–20.

Morgan, R. P. C. 1985c. Soil degradation and erosion as a result of agricultural practice. In: Richards, K. S., Arnett, R. R. and Ellis, S. (eds). *Geomorphology and soils*. George Allen & Unwin. pp. 379–395.

Morgan, R. P. C. 1987. Sensitivity of European soils to ultimate physical degradation. In: Barth, H. and L'Hermite, P. (eds). *Scientific basis for soil protection in the European Community*. Proceedings of Commission for the European Communities Symposium October 1986. Berlin. Elsevier, London.

Quine, T. A. and Walling, D. E. 1991. Rates of erosion on arable fields in Britain: quantitative data from Caesium-137 measurements. *Soil Use and Management* 7:169–176.

Reed, A. H. 1979. Accelerated erosion of arable soils in the United Kingdom by rainfall and runoff. *Outlook on Agriculture* 10:41–48.

Robinson, D. A. and Blackman, J. D. 1990. Some costs and consequences of soil erosion and flooding around Brighton and Hove, autumn 1987. In: Boardman, J., Foster, I. D. L. and Dearing, J. A. (eds). *Soil erosion on agricultural land*. John Wiley, Chichester. pp. 369–382.

Robinson, D. A. and Boardman, J. 1988. Cultivation practice, sowing season and soil erosion on the South Downs, England: a preliminary study. *Journal of Agricultural Science, Cambridge*. 110:169–177.

Speirs, R. B. and Frost, C. A. 1985. The increasing incidence of accelerated soil water erosion on arable land in the east of Scotland. *Research and Development in Agriculture* 2:161–167.

Stammers, R. and Boardman, J. 1984. Soil erosion and flooding on downland areas. *The Surveyor* 164:8–11.

Stretton, C. 1984. Water supply and forestry – a conflict of interests: Cray Reservoir, a case study. *Journal of the Institution of Water Engineers and Scientists* 38:232–330.

Trimble, S. W. 1983. A sediment budget for Coon Creek Basin in the Driftless area, Wisconsin, 1853–1977. *American Journal of Science* 283:545–574.

Walling, D. E. 1983. The sediment delivery problem. *Journal of Hydrology* 65:129–144.

Walling, D. E. and Quine, T. A. 1991. Use of ^{137}Cs measurements to investigate soil erosion on arable fields in the UK: potential applications and limitations. *Journal of Soil Science* 42:147–165.

Walling, D. E. and Webb, B. W. 1981. Water quality. In: Lewin, J. (ed.). *British rivers*. Allen & Unwin. pp. 126–169.

Wilson, S. J. and Cooke, R. U. 1980. Wind erosion. In: Kirkby, M. J. and Morgan, R. P. C. (eds). *Soil erosion*. John Wiley, Chichester. pp. 217–249.

PART III

Water Impacts

INTRODUCTION

In a recent review of the world's freshwater resources, Gleick (1993, p. 1) summed up the importance of water in a few trenchant sentences:

Fresh water is a fundamental resource, integral to all environmental and societal responses. Water is a crucial component of ecological cycles. Aquatic ecosystems harbour diverse species and offer many valuable services. Human beings require water to run industries, to provide energy, and to grow food.

Because water is so important to human affairs, humans have sought to control water resources in a whole variety of ways, and because water is such an important component of so many natural and human systems, its quantity and quality have undergone major changes as a consequence of human activities. Again one can quote Gleick (1993, p. 3):

As we approach the 21st century we must now acknowledge that many of our efforts to harness water have been inadequate or misdirected. Rivers, lakes, and groundwater aquifers are increasingly contaminated with biological and chemical wastes. Vast numbers of people lack clean drinking water and rudimentary sanitation services. Millions of people die every year from water related diseases such as malaria, typhoid, and cholera. Massive water developments have destroyed many of the world's most productive wetlands and other aquatic habitats.

In part III we look at some of the ways in which the quantity and quality of water has been modified in some of the world's freshwater systems – rivers, groundwater and lakes.

Chapter 11 deals with what has been described as the greatest environmental catastrophe of the former Soviet Union: the desiccation of the Aral Sea. The Soviet government carried out major schemes for wholesale reorganization of drainage and for massive inter-basin water transfers. Fears were expressed in the 1980s that major river diversions could modify the salinity and freezing characteristics of the Arctic Ocean, which in turn might modify regional climates (Semtner, 1984). Such fears were essentially speculative: the desiccation of the Aral is in contrast, a reality, and Philip Micklin brought this clearly to the attention of the world scientific community in the chapter reproduced here.

The Aral disaster has generated a large scientific literature, including a big volume in French (Létoll and Mainguet, 1993), a special issue of *Post-Soviet Geography* (Micklin, 1992), and an analysis of the political roots of the crisis (Glantz, Rubinstein and Zonn, 1993).

Dam construction, inter-basin water transfers, the construction of embankments, and channelization are among the deliberate methods that have been used by humans to modify the hydrosphere. As the Aral example demonstrates, however, the consequences of such deliberate actions are not always beneficial. No less significant changes in the hydrological cycle are also achieved non-deliberately and indirectly as a result of land-use changes. In chapter 8 we noted that deforestation can, for example, lead to a rise in water-tables and the development of saline seeps. In chapter 12 Bari and Schofield, also working in Australia, demonstrate that the reverse can occur, with reforestation causing water-tables to be lowered.

Later chapters in part III turn to the equally important question of the modification of water quality. Chapter 13 derives from a very important series of studies that were undertaken in the Hubbard Brook Experimental Forest catchments in New Hampshire, USA. It demonstrates the value of paired catchment studies in which a deliberately disturbed catchment is compared with an undisturbed catchment. In this case the disturbed catchment was clear cut and the effects on nutrient loss were dramatic. It needs to be remembered, however, that elegant though this study was, it was simulating extremely severe land-use change of a type that cannot be considered as normal. Regrowth was deliberately suppressed by the application of herbicide!

Another major form of land-use change that has clear implications for water quality is urbanization (chapter 14). The classic paper on this, and one that has been much emulated, was prepared by two distinguished geomorphologists, M. G. Wolman and A. P. Schick, who worked in the Baltimore and Washington D.C. areas of the United States. The chapter is rich in data on comparative rates of sediment yield under different land-use types and also looks at some of the issues raised by high rates of sediment yield from construction sites. Likewise, as chapter 15 demonstrates, cities generate very special hydrological conditions in terms of runoff.

The intensification of agriculture as a result of the use of synthetic fertilizers and the stall-feeding of domestic stock has also had numerous implications for water quality, and has been implicated in the widespread process of cultural eutrophication. Chapter 16, by Burt and Haycock, looks at the nitrogen cycle: how it is disturbed, at some of the implications, and, very importantly, at some of the potential solutions.

One of the great areas of research in recent decades has been in the study of acid precipitation, acid rain or acid deposition (see chapter 27). However, the acidification of surface waters, including lakes, could result from a whole series of causes, of which acid deposition is but one. An elegant way of establishing that acidification of lake waters has occurred and also of determining when the acidification was initiated, has been to extract, analyse and date sediment cores from lakes. In particular, the record of diatoms in the core sediment gives a picture of changing pH (acidity) levels through time. This is a technique that has been used extensively and effectively by a team led by Professor R. W. Battarbee at University College, London (chapter 17). They review the competing hypothesis to explain acidification and, on balance, favour the view that acid precipitation is to blame.

However, emission controls and changes in economic activity have in certain parts of the world caused a reduction in the amount of acidity being deposited. In Britain, for example, sulphur dioxide emissions, one of the key components of acid rain, have declined. The question therefore arises as to whether or not there has been a concomitant reversal in the acidification of surface waters. In chapter 18, Professor Battarbee's team used their diatom technique to show that the lakes of Galloway in Scotland have responded quickly to a decrease in acid deposition in recent years.

References

Glantz, M. H., Rubinstein, A. Z., and Zonn, I. 1993, Tragedy in the Aral Sea basin. Looking back to plan ahead? *Global Environmental Change* 3:174–98.

Gleick, P. (ed.) 1993, *Water in Crisis: A Guide to the World's Fresh Water Resources*. New York: Oxford University Press.

Létoll, R., and Mainguet, M. 1993, *Aral*, Paris: Springer Verlag.

Micklin, P. P. 1992, The Aral crisis. *Post-Soviet Geography* 33(5).

Semtner, A. J. 1984, The climatic response of the Arctic Ocean to Soviet river diversion. *Climatic Change* 6:109–30.

11

Desiccation of the Aral Sea: A Water Management Disaster in the Soviet Union*

P. P. Micklin

The Aral Sea is a huge, shallow, saline body of water located in the deserts of the south-central Soviet Union (figures 11.1 and 11.2). A terminal lake (having no outflow), its secular level is determined by the balance between river and ground-water inflow and precipitation on its surface on the one hand and evaporation from the sea on the other.

The Aral depression has repeatedly been flooded and desiccated since the Pliocene (1; 2, pp. 277–297). The most recent filling began in the late Pleistocene, around 140,000 years ago, when the Syr Dar'ya flowed into the lowest part of the hollow. The lake did not attain great size until the beginning of the Holocene (Recent) Epoch when inflow was increased some threefold by capture of the Amu Dar'ya. Marine fossils, relict shore terraces, archeological sites, and historical records point to repeated major recessions and advances of the sea during the past 10,000 years. Until the present century, fluctuations in its surface level were at least 20 m and possibly more than 40 m (1, 3). Significant cyclical variations of sea level during this period resulted from major

changes in river discharge into it caused by climatic alteration, by natural diversions of the Amu Dar'ya away from the Aral, and during the past 3000 years by man. Human impacts included sizeable withdrawals for irrigation from the Amu Dar'ya and diversions of this river westward into lower lying channels and hollows because of the destruction of dikes, dams, and irrigation systems during wars (1, 4).

From the middle 18th century until 1960, sea level varied 4 to 4.5 m (1, 5). Beginning in 1910, when accurate and regular level observations began, to 1960, the lake was in a "high" phase with level changes of less than 1 m (6). However, during the past 28 years the sea's surface has dropped precipitously. In 1960, sea level was 53.4 m, area 68,000 km², volume 1090 km³, average depth 16 m, and average salinity near 10 g/liter (7, 8). The Aral was the world's fourth largest lake in area, behind the Caspian Sea, Lake Superior, and Lake Victoria. By the beginning of 1987, sea level had fallen 12.9 m, area decreased by 40%, volume diminished by 66%, average depth dropped to 9 m, and

* Originally published in *Science*, 1988, vol. 241, pp. 1170–5.

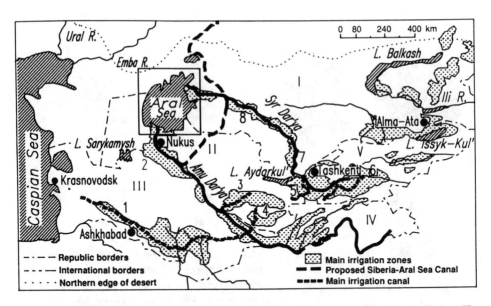

Figure 11.1 [*orig. figure 1*] The Aral Sea, located in the driest part of the Soviet Union. The population of this region is nearly 40 million, predominantly Muslim, and growing rapidly. Agriculture with extensive irrigation is the mainstay of the local economy. Major irrigation complexes are shown: 1, Karakum Canal; 2, Amu Dar'ya Delta; 3, Amu-Bukhara Canal; 4, Karshi Steppe; 5, Golodnaya Steppe; 6, Fergana Valley; 7, Kzyl-Kum Canal; and 8, Kzyl-Orda Canal. Roman numerals: I, Kazakh Republic; II, Uzbek Republic; III, Turkmen Republic; IV, Tadzhik Republic; and V, Kiryiz Republic. Area in the inset is enlarged in figure 11.2.

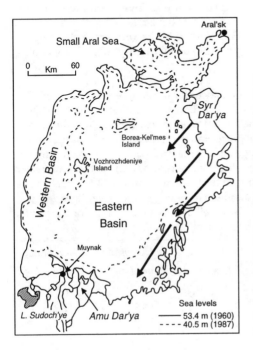

Figure 11.2 [*orig. figure 2*] The Aral Sea, fourth largest lake in the world by area in 1960. The Aral Sea has shrunk significantly because of a nearly total cutoff of river inflow from the Amu Dar'ya and Syr Dar'ya as a result of heavy withdrawals for irrigation. The deeper Western Basin has been less affected than the shallow Eastern Basin. The large area of exposed former bottom along the eastern shore is a source of major dust and salt storms (the black arrows indicate the source and direction of major storms) that are causing significant ecological and agricultural damage for hundreds of kilometers inland. The former ports of Aral'sk and Muynak are now tens of kilometers from the sea. Compiled from (22, figure 4) and satellite imagery (61).

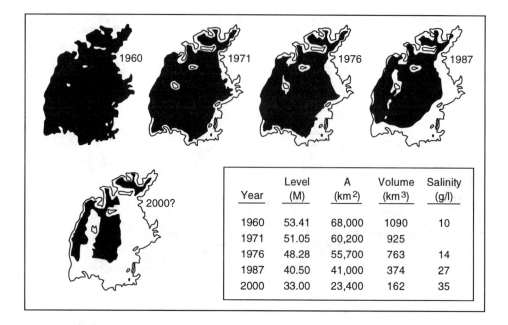

Figure 11.3 [*orig. figure 3*]   The changing profile of the Aral Sea, 1960 to 2000 (*7, 61, 62*).

average salinity risen to 27 g/liter (figure 11.3). The sea had dropped to sixth in area among the world's lakes.

The recent recession has been the most rapid and pronounced in 1300 years (*1*). Human actions have been the primary cause. Desiccation continues at a rapid pace and if unchecked will shrink the sea to a briny remnant in the next century. Severe and widespread ecological, economic, and social consequences that are progressively worsening have resulted from the Aral's recession. The scale of impacts on this large body of water for such a short period is unprecedented. Soviet commentators in recent years have referred to the Aral situation as "one of the very greatest ecological problems of our century" (*9*), an "impending

Table 11.1 [*orig. table 1*]   Average annual water balances for the Aral Sea, 1926 to 1985. The computational form of the annual water balance equation for the Aral Sea is $Qr + Qu + (PF)/10^6 = (EF)/10^6 \pm (\delta hF)/10^6$, where Qr is the annual river inflow (km³), Qu is the annual net ground-water inflow (km³), P is the annual precipitation on the sea (mm), E is the annual evaporation from the sea (mm), F is the average annual sea area (km²), δh is the net annual sea level change (mm), and 10^6 is a proportionality constant (mm/km). Net ground-water inflow is small (around 3 to 4 km³) and is ignored in the table calculations. The water balance was in essential equilibrium from 1926 to 1960 and sea level was stable. River flow declined substantially between 1960 and 1970, and sea level fell at a moderate rate. For the period 1970 to 1985, river flow fell drastically and sea level declined rapidly (*7*).

Period	Average area (km²)	Gain (km³)			Evaporation loss (km³)	Net volume change (km³)
		River flow	Precipitation	Total		
1926–60	65,780	55.2	8.2	63.4	64.1	0.7
1960–70	64,470	42.8	8.4	51.2	63.3	-12.2
1970–85	53,660	16.3	6.6	22.9	56.2	-33.3

disaster" (*10*), and as "a dangerous experiment with nature" (*11*).

Water Balance Changes

As in the past, the cause of the modern recession of the Aral is a marked diminution of inflow from the Syr Dar'ya and Amu Dar'ya, the sea's sole sources of surface water inflow, that has increasingly shifted the water balance toward the negative side (table 11.1). The trend of river discharge has been steadily downward since 1960 (figure 11.4).

A shrinking body of water is dominantly a negative feedback mechanism, that is, one that resists change and promotes stability. Evaporative losses significantly diminish as area decreases, pushing the water balance system toward equilibrium. Hence, in the future, assuming some level of surface- and ground-water inflow, the Aral should stabilize. However, this is not likely to occur for decades. The primary determinant of level change, the difference between inflow and net evaporation, is currently large and negative. It will only decrease slowly as the sea shrinks to a much smaller size.

The causes of reduced inflow since 1960 are both climatic and anthropogenic. A series of dry years in the 1970s, particularly 1974–75, lowered discharge from the zones of flow formation of the Amu Dar'ya and Syr Dar'ya around 30 km³ per year (27%) compared to the average during the preceding 45 years (*8; 12*, p. 227). The 1982 to 1986 period has also suffered low flows (*12–14*). Nevertheless, the most important factor reducing river flow has been large consumptive withdrawals (that is, water withdrawn from rivers that is not returned to them), overwhelmingly for irrigation. Average annual river flow in the zones of formation of these rivers (high mountains to

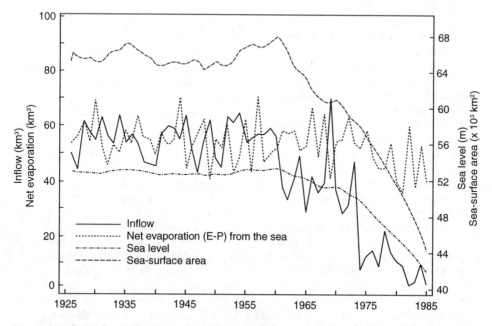

Figure 11.4 [*orig. figure 4*] Annual hydrologic parameters of the Aral Sea, 1926 to 1985. For the period 1926 to 1960, there was little difference between inflow and net evaporation (sea-surface evaporation minus sea-surface precipitation). Consequently, sea level and area varied insignificantly. Since 1960, with the exception of the unusually heavy flow year of 1969, discharge to the Aral Sea has trended steadily downward. Net evaporation has diminished somewhat as the sea's area contracted but the gap between it and river flow has grown steadily larger, accelerating the rate of level drop and area decrease (*7*).

the southeast of the Aral Sea) averaged 111 km³ from 1926 to 1970 (12). Under "natural" conditions only about half of this would reach the Aral because of losses to evaporation, transpiration, and filtration as these rivers cross the deserts and flow through their deltas (12).

Irrigation has been practiced in the lower reaches of the Amu Dar'ya and Syr Dar'ya for several millennia (4). In 1900 more than 3 million hectares were under irrigation in the Aral Sea basin, growing to 5 million by 1960 when consumptive withdrawals for it reached an estimated 40 km³ (1, 15). However, irrigation withdrawals before the 1960s did not measurably reduce inflow to the Aral. These artificial losses were compensated by correspondingly large reductions of natural evaporation, transpiration, and filtration, particularly in the deltas of the Syr Dar'ya and Amu Dar'ya where truncated spring floods diminished floodplain inundation, the area of deltaic lakes, and the expanse of phreatophytes (12, 15, 16). Also, the installation of drainage networks increased irrigation return flows to these rivers.

By 1980, the irrigated area in the Aral Sea basin had grown to nearly 6.5 million hectares (17; 18, pp. 226–230). Withdrawals from the Amu Dar'ya and Syr Dar'ya for all purposes were 132 km³ with consumptive use, including evaporation from reservoirs, of 85 km³ (18, pp. 212–215). Irrigation accounted for 120 km³ of withdrawals (91%) and for 80 km³ of consumptive use (94%). Extrapolation, from data on area and rates of growth of irrigation for administrative units in the Aral Sea basin for the period 1980 to 1984 and 1980 to 1986, indicates that in 1987 about 7.6 million hectares were irrigated (17). Between 1980 and 1987, there was a major improvement in irrigation efficiency in the Aral Sea basin which lowered average withdrawals from 18,500 to 13,700 m³/ha (19). Thus a 17% larger area was irrigated with considerably less water (104 km³). Information on consumptive use in 1987 is not available but it probably remained near the 1980 figure because of the efficiency gains (that is, a higher percentage of withdrawn water was used by crops and

a lower percentage was return flows).

Factors that compensated the earlier growth of consumptive withdrawals reached their limits in the 1960s (2, 12, 15, 16). Hence, as irrigation expanded during the past three decades, the increase in water use has not been balanced by commensurate reductions in natural losses. Furthermore, the irrigation of huge new areas such as the Golodnaya (Hungary) Steppe along the Syr Dar'ya consumed huge volumes of water to fill soil pore spaces (20), newly created giant reservoirs required filling and heightened evaporative losses, increased flushing of soils to counteract secondary salinization raised water use, and new irrigation systems discharged their drainage water into the desert or large hollows where it evaporated.

The Karakum Canal has been the single most important factor contributing to the diminution of inflow to the Aral in recent decades. The largest and longest irrigation canal in the Soviet Union, it stretches 1300 km westward from the Amu Dar'ya into the Kara-Kum Desert (figure 11.1). Between 1956 and 1986, 225 km³ were diverted into it as annual withdrawals rose from less than 1 km³ to more than 14 km³ (21). All of the water sent along the Karakum Canal is lost to the Aral.

Environmental Impacts

During planning for a major expansion of irrigation in the Aral Sea basin, conducted in the 1950s and 1960s, it was predicted that this would reduce inflow to the sea and substantially reduce its size. At the time, a number of experts saw this as a worthwhile tradeoff: a cubic meter of river water used for irrigation would bring far more value than the same cubic meter delivered to the Aral Sea (6, 22–25). They based this calculation on a simple comparison of economic gains from irrigated agriculture against tangible economic benefits from the sea. Indeed, the ultimate shrinkage of the Aral to a residual brine lake as all its inflow was devoted to agriculture and other economic needs was viewed as both desirable and inevitable.

These experts largely dismissed the possibility of significant adverse environmental consequences accompanying recession. For example, some scientists claimed the sea had little or no impact on the climate of adjacent territory and, therefore, its shrinkage would not perceptibly alter meteorological conditions beyond the immediate shore zone (6). They also foresaw little threat of large quantities of salt blowing from the dried bottom and damaging agriculture in adjacent areas (22). This theory rested, in the first place, on the assumption that during the initial phases of the Aral's drying only calcium carbonate and calcium sulfate would be deposited on the former bottom. Although friable and subject to deflation, these salts have low plant toxicity. Second, it was assumed that the more harmful compounds, chiefly sodium sulfate and sodium chloride, which would be deposited as the sea continued to shrink and salinize, would not be blown off because of the formation of a durable crust of sodium chloride. Some optimists even suggested the dried bottom would be suitable for farming (22).

Although a small number of scientists warned of serious negative effects from the sea's desiccation, they were not heeded (14, 24). Time has proved the more cautious scientists not only correct but conservative in their predictions. A brief discussion of the most pronounced impacts follows.

Bottom exposure and salt and dust storms. The Aral contained an estimated 10 billion metric tons of salt in 1960, with sodium chloride (56%), magnesium sulfate (26%), and calcium sulfate (15%) the dominant compounds (22). As the sea shrank, enormous quantities of salts accumulated on its former bottom. This results from capillary uplift and subsequent evaporation of heavily mineralized ground water along the shore, seasonal level variations that promote evaporative deposition, and winter storms that throw precipitated sulfates on the beaches (25–27).

Much of the 27,000 km^2 of bottom exposed between 1960 and 1987 is salt-covered. In contrast to earlier predictions that were based on a faulty understanding of the geochemistry of a shrinking and salinizing Aral, not only have calcium sulfate and calcium carbonate deposited but sodium chloride, sodium sulfate, and magnesium chloride have as well (24). Because of the concentration of toxic salts in the upper layer, a friable and mobile surface, and lack of nutrients and fresh water, the former bottom is proving extremely resistant to natural and artificial revegetation (26, 28).

However, the most serious problem is the blowing of salt and dust from the dried bottom. There is as yet no evidence of the formation of a sodium chloride crust that would retard or prevent deflation (24). The largest plumes arise from the up to 100-km-wide dried stripe along the sea's northeastern and eastern coast and extend for 500 km (figure 11.2) (11, 25). Recent reports state traces of Aral salt have been found 1000 km to the southeast of the sea in the fertile Fergana Valley, in Georgia on the Black Sea coast, and even along the arctic shore of the Soviet Union (29, 30).

Soviet scientists report major storms as beginning in 1975 when they were first detected on satellite imagery. Between 1975 and 1981, scientists confirmed 29 large storms from analysis of *Meteor* (a high-resolution weather satellite) images (11). During this period, up to ten major storms occurred in 1 year. Recent observations by Soviet cosmonauts indicate the frequency and magnitude of the storms are growing as the Aral recedes (31). Sixty percent of the observed storms moved in a southwest direction which carried them over the delta of the Amu Dar'ya, a region with major ecological and agricultural importance (11). Twenty-five percent traveled westward and passed over the Ust-Yurt plateau, which is used for livestock pasturing.

An estimated 43 million metric tons of salt annually are carried from the sea's dried bottom into adjacent areas and deposited as aerosols by rain and dew over 150,000 to 200,000 km^2 (11, 13, 32). The dominant compound in the plumes is calcium sulfate but they also contain significant amounts of sodium chloride, sodium sulfate, magnesium sulfate, and calcium bicarbonate (33). Sodium chloride and sodium sulfate are especially toxic to plants, particularly during flowering. In spite of the expected increase in the area of former bottom, salt export is predicted to diminish slightly to 39 million metric tons per year by the year 2000 as a result of the exhaustion of deflatable materials, the leaching of salt into deeper layers, and through the process of diagenesis of the older surface (32).

Loss of biological productivity. As the sea has shallowed, shrunk, and salinized, biological productivity has steeply declined. By the early 1980s, 20 of 24 native fish species disappeared and the commercial catch (48,000 metric tons in 1957) fell to zero (2, pp. 507–524; 13, 26). Major fish canneries at Aral'sk and Muynak, formerly ports but now some distance from the shore, have slashed their work forces and barely survive on the processing of high cost fish brought from as far away as the Atlantic, Pacific, and Arctic oceans (29, 30, 34, 35). Both plants in 1988 will be switched to *khozraschet* (economic principles of management) and may be forced to close (30). Residual commercial fishing continues in the two largest irrigation drainage water lakes that have formed (Sarykamysh and Aydarkul'). However, levels of pesticides and herbicides, from cotton field runoff, in fish taken from Sarykamysh and Aydarkul' are dangerously high, prompting a halt to commercial fishing in the former in 1987 (14, 29).

Employment directly and indirectly related to the Aral fishery, reportedly 60,000 in the 1950s, has disappeared (36). The demise of commercial fishing and other adverse consequences of the sea's drying has led to an exodus from Aral'sk and Muynak whereas many former fishing villages have been completely abandoned (30, 34). During recent years, more than 40,000 have left the districts of Kzyl-Orda Oblast that abut the Aral on the east and northeast (30).

Deterioration of deltaic ecosystems. The shrinking of the Aral along with the greatly diminished flow of the Syr Dar'ya and Amu Dar'ya has had particularly devastating effects on these rivers' deltas (11, 13, 14, 26, 37). Prior to 1960, these oases surrounded by desert not only possessed great ecological value because of the richness of their flora and fauna but provided a natural feed base for livestock, spawning grounds for commercial fish, reeds harvested for industry, and opportunities for commercial hunting and trapping. Deltaic environments deteriorated as river flow diminished and sea level fell, leading to the drying or entrenchment of distributary and even main channels, the cessation of spring inundation of floodplains, and the shrinking or disappearance of lakes. Between 1960 and 1974, the area of natural lakes in the Syr Dar'ya Delta decreased from 500 km^2 to several tens of square kilometers, whereas in the Amu Dar'ya Delta from the 1960s until 1980, 11 of the 25 largest lakes disappeared and all but 4 of the remainder significantly receded (38, 39).

Native plant communities have degraded and disappeared. *Tugay* forests, composed of dense stands of phreatophytes mixed with shrubs and tall grasses fringing delta arms and channels to a depth of several kilometers, have particularly suffered. The expanse of *Tugay* in the Amu Dar'ya Delta, estimated at 13,000 km^2 in the 1950s, had been halved by 1980 (37). The major cause of deltaic vegetation impoverishment has been the 3 to 8 m drop of ground water along with the end of floodplain inundation.

Degradation of vegetational complexes and drops in the water table have initiated desertification in both deltas. Satellite imagery and photography from manned spacecraft indicate that desert is spreading rapidly (11). Livestock raising has also suffered considerable damage because of a decline in yields and a reduction of suitable areas. In the Amu Dar'ya Delta between 1960 and 1980 the area of hayfields and pastures decreased by 81% and yields fell by more than 50% (26).

Habitat deterioration has harmed delta fauna, which once included muskrat, wild boar, deer, jackal, many kinds of birds, and even a few tigers. At one time 173 animal species lived around the Aral, mainly in the deltas; 38 have survived (26, 30). Commercial hunting and trapping have largely disappeared. The harvest of muskrat skins in the Amu Dar'ya Delta has fallen to 2,500 per year from 650,000 in 1960 (14).

Climate changes. Earlier claims to the contrary notwithstanding, research over the past two decades has established that the Aral affects temperature and moisture conditions in an adjacent stripe estimated to be 50 to 80 km wide on its north, east, and west shores and 200 to 300 km wide to the south and southwest (*13, 26, 40*). With contraction, the sea's influence on climate has substantially diminished. Summers have become warmer, winters cooler, spring frosts later, and fall frosts earlier, the growing season has shortened, humidity has lowered, and there has been an overall trend toward greater continentality. The most noticeable changes have occurred in the Amu Dar'ya Delta. At Kungrad, now located about 100 km south of the Aral, comparison of the period 1935 to 1960 with that of 1960 to 1981 indicates that relative humidity diminished substantially, the average May temperature rose 3 to 3.2 degrees Celsius, and the average October temperature decreased 0.7 to 1.5 degrees Celsius (*13*). The growing season in the northern Amu Dar'ya Delta has been reduced an average of 10 days, forcing cotton plantations to switch to rice growing (*14, 26*).

Ground-water depression. The drop in the level of the Aral has been accompanied by a reduction of the pressure and flow of artesian wells and a decline of the water table all around the sea (*13*). Soviet scientists have estimated that a 15-m sea level drop, likely by the early 1990s, could reduce ground-water levels by 7 to 12 m in the coastal zone and affect the water table 80 to 170 km inland (*41*). The sinking water table has had significant adverse impacts outside the Amu Dar'ya and Syr Dar'ya deltas, drying wells and springs and degrading natural plant communities, pastures, and hayfields.

Water supply and health concerns. The reduction of river flow, salinization and pollution of what is left, and lowering of ground-water levels has caused drinking water supply problems for communities around the sea. Problems are especially acute in the more heavily populated deltas (*13, 26*). To provide a reliable, safe water

supply to Nukus (1987 population of 152,000) in the Amu Dar'ya Delta, a 200-km pipeline costing 200 million rubles (officially a ruble is about $1.60) is under construction from the upstream Tyuyamuyun Reservoir. The declining quality of drinking water is cited as the main factor increasing intestinal illnesses, particularly among children, and throat cancer incidence in the lower reaches of the Amu Dar'ya and Syr Dar'ya (*26, 34, 35*). There is fear of epidemics because of the deterioration of the quality of the water supply and the increasing rodent population (*8, 35*). Desert animals who use the Aral Sea as a drinking source have died from its greatly increased mineral content (*26*).

Economic losses. There are no accurate figures on damages associated with the Aral's recession. Soviet scientists and economists have attempted to estimate the costs of the more tangible consequences. A 1979 study concluded that aggregate damages within the Uzbek Republic, which has suffered the greatest harm, totaled 5.4 to 5.7 billion rubles (*42*). A 1983 evaluation concluded that annual damages in the lower course of the Amu Dar'ya were 92.6 million rubles with the following distribution: agriculture, 42%; fisheries, 31%; hunting and trapping, 13%; river and sea transport, 8%; and living and working conditions, 6% (*26*). A recent popular article listed, without elaborating, a figure of 1.5 to 2 billion rubles as the annual losses for the entire Aral Sea region (*14*).

The Fate of the Aral

What does the future hold for the Aral Sea? If surface inflow remains at the low levels of recent years, it averaged only 5.2 km³/year between 1981 and 1985, and was reportedly near zero in 1986 (*7, 14, 43*), shrinkage will continue into the next century. By the year 2000, the sea could consist of a main body in the south with the salinity of the open ocean and several small brine lakes in the north (figure 11.3). Subsequently, assuming a residual inflow of irrigation drainage water and ground water totaling around 10 km³, the southern sea will separate into two parts with an aggregate area around 12,000 km²,

8% of the Aral's size in 1960 (44). Salinity would rise to 140 g/liter.

This scenario is not inevitable. The sea's recession could be halted if considerably more water reached it. Water balance calculations indicate that to maintain the 1987 size (41,000 km^2) would require river inflow around 30 km^3/year (27, table 2). This discharge is possible if consumptive irrigation withdrawals from the Amu Dar'ya and Syr Dar'ya were to be markedly reduced. However, irrigation is the economic mainstay of the Aral Sea basin where over 90% of the harvest comes from irrigated lands (45). Although plans for irrigation expansion in the Aral Sea basin have been somewhat scaled back under the Gorbachev regime in light of the region's ecological problems and strained water balance, many water management experts see continued growth of this sector a necessity (45, 46).

There is a national campaign to improve irrigation water use. Reclamation agencies are implementing, among other measures, reconstruction of old irrigation systems, automation and remote control of water allocation and delivery systems for entire river basins, use of more efficient water application techniques (for example, sprinklers, drip and subsurface), and "programming" of harvests, involving the use of simulation-optimization models to minimize inputs and maximize outputs given a set of production objectives and constraints (45, 47).

The average efficiency of irrigation systems (ratio of water used productively at the fields to headworks withdrawals) was around 60% in the Aral Sea basin in the early 1980s, the lowest of any region in the Soviet Union (48). On the basis of 1980 irrigation withdrawals of 120 km^3, raising average system efficiency from 60% to between 74 and 80%, the goal (49, 50) would allow irrigation of the same area with 23 to 30 km^3 lower annual withdrawals. However, the net addition to river flow would be less because of the diminution of return flows from irrigated areas associated with the increase in efficiency. Furthermore, a water use limitation program, introduced for the region in 1982 because of the increasingly

dire water supply situation, mandated lower crop application rates and may already have raised average efficiencies to near 70% (19). Using 1987 withdrawals (104 km^3) and assuming an efficiency of 70%, the improvement to 74 to 80% would only save 6 to 13 km^3/year. The most knowledgeable Soviet experts estimate realistic future water savings from renovation of irrigation systems in the Aral Sea basin at 10 to 22 km^3/year (43, 46, 49, 50).

Modernization of irrigation in the Aral basin is necessary not only to save water but to improve yields, prevent secondary salinization, and cope with waterlogging. Nevertheless, it is an expensive and time-consuming process. Cost of a comprehensive program could reach 95 billion rubles (51). Furthermore, most of the "freed" water will be needed to irrigate new lands to provide more food for the region's rapidly expanding population, growing around 2.7% annually, as well as to meet increasing municipal and industrial water needs (46, 47, 50).

Ground water could make a larger contribution to regional water supplies. Subsurface storage is huge but little used (47). However, much of the reserve lies at great depth or is heavily mineralized. Up to 17 km^3/year of ground water could be consumed in the Aral Sea basin without adversely affecting river flow (18, pp. 182–183).

Another means of supplementing the Aral's water balance would be to channel irrigation drainage water to it. Soviet experts estimate that 34 km^3 of drainage were generated annually in the Aral basin in the early 1980s (12). Approximately 21 km^3 returned to rivers, leaving 13 km^3 to evaporate from the desert or accumulate in depressions (12). The lakes formed in the latter hold around 40 km^3 (50). Perhaps 10 to 12 km^3 of drainage water annually could be sent to the Aral by collectors running parallel to the Amu Dar'ya and Syr Dar'ya (9). However, drainage water is saline, frequently above 3 g/liter, and is pesticide- and herbicide-laden; drainage should be purified and demineralized before discharge to the sea (27, 29, 52). Indeed, the need to keep this flow out of the two rivers stimulates interest in such a scheme as much as the need to provide more water to the Aral. Work on an enormous project to collect drainage water along 1500 km of the right bank of the Amu Dar'ya for delivery to the Aral has started (53). At the same time, the program to improve irrigation efficiency will significantly reduce the amount of drainage water available for delivery to the Aral.

Channeling irrigation drainage water to the sea will dry the two largest lakes supported from this source, Aydarkul' and Sarykamysh, with areas in excess of 2000 km^2 each (figure 11.1). Since their origins in the 1960s, each has developed considerable wildlife, fishery, and recreation importance (47).

Schemes to Preserve the Aral

Delivery of 12 km^3 of irrigation drainage water plus 4 km^3 of net ground-water inflow to the Aral would support a sea of only 20,000 km^2 whose salinity would be high (40 to 50 g/liter) and ecological value and economic uses minimal (27, 43). Hence, additional measures will be necessary if the Aral is to be preserved as a greatly shrunken but viable body of water and to reduce the adverse impacts of its recession. One approach, first suggested in the 1970s, is to partition the sea with dikes to preserve low salinity conditions in a portion of it while allowing the remainder to dry or become a residual brine lake receiving outflow from the freshened part (5, 27, 41). Most of the designs are obsolete since they would require considerably more surface inflow (25 to 30 km^3/year) than realistically will be available. A scheme put forward in 1986 to preserve a 12,000-km^2 sea with a salinity of 8 g/liter in the Eastern Basin (figure 11.2) shows some promise as it needs inflow of only 8 to 9 km^3/year (44).

A recent proposal, which assumes meager future inflow to the sea, focuses on restoring and preserving the deltas of the Amu Dar'ya and Syr Dar'ya because of their great ecological and economic value (13). The plan for the former would involve constructing a 225-km dike in front of the delta to create a system of polders with a surface elevation 8 m above current sea level but 5 m below that of 1960. This would raise ground- and surface-water levels in the delta. Low earth dams and regulating reservoirs would be built in the delta to provide further water control.

A mixture of fresh river water and saline irrigation drainage water would be delivered to the polders. The dried seabed in front of the delta would be stabilized to prevent the encroachment of sand dunes and the blowing of salt and dust. Additional efforts would be undertaken to restore plant and animal communities as well as improve irrigation, livestock raising, fisheries, and trapping. The scheme would require drainage water and fresh flow totaling 8 to 9 km^3/year. The estimated cost is 406 million rubles. A similar plan for the Syr Dar'ya Delta would require 7 km^3/year.

Regardless of what, if any, scheme is implemented to preserve a residual Aral Sea, it is essential to stabilize the exposed bottom to reduce the blowing of salt and dust. There has been some success in establishing salt-tolerant xerophytic shrubs (for example, black saksaul – *Haloxylon aphullum*). But this program is so far limited to relatively small areas with the most favorable conditions and the survival rate is low (52, 54). Scientists are also investigating the feasibility of using mechanical and chemical means of binding the loose surface (13, 28).

The Aral's water balance could also be improved by importing water from more humid regions. Such a project was formulated in the 1970s and early 1980s by the National Water Management Design Institute (*Soyuzgiprovodkhoz*), a subagency of the Ministry of Reclamation and Water Management (55). Providing more water for irrigation was the plan's main purpose but it would have helped the Aral as well. Central Asian party and government officials enthusiastically supported the scheme. Part of the flow from the arctic draining Ob' and Irtysh rivers, situated to the north in Western Siberia, would be transferred southward. Water would be sent 2500 km to the Amu Dar'ya by a system of low dams, pumping stations, and a huge canal (popularly named

"Sibaral," Siberian to the Aral Sea Canal) (figure 11.1). The project's first stage (27 km³/year) was undergoing final engineering design in 1985 and was scheduled for implementation by the late 1980s or early 1990s.

Following Gorbachev's ascension to Soviet leadership in March 1985, the fortunes of the Siberian scheme, as well as a companion project for the European part of the country, waned. He and his advisers see north–south water transfer projects as a poor investment of scarce resources and believe less costly, more effective local means of solving water supply problems in the arid regions of the Soviet Union are available. The diversion schemes had been periodically attacked during the 1970s and early 1980s by some scientists and a group of Russian national writers who foresaw severe ecological, economic, and cultural damage occurring in northern regions of water export. In a dramatic policy reversal, the Communist Party and Soviet government, in August 1986, ordered a cessation of construction and design work on these projects (56). However, the decree directed that research on the scientific problems associated with water diversions, stressing ecological and economic concerns, continue.

In spite of the suspension of work on water transfers, critics have remained on the offensive. They have bitterly denounced in the popular Soviet media those directly or indirectly involved with project planning or evaluation (57). Evidently, they fear that the projects could be revived. The most vociferous opponents have engaged in personal attacks as well as exaggeration and misrepresentation (59).

Conclusions

The modern recession of the Aral Sea, the most severe in 1300 years, has resulted from excessive consumptive use of river inflow to the sea. Processes of potential ecological change were not carefully evaluated nor clearly understood when the water management decisions leading to the drop in the sea's level were made. Water management planners ignored warnings of dire conse-

quences from some scientists. The future is not bright. River inflow by the mid-1980s was near zero, and the sea continues to rapidly shrink and salinize. The Aral could become several residual, lifeless, brine lakes early in the next century. Already substantial ecological damages and economic losses will grow worse.

Scientific study of the "Aral problem" and its amelioration has been a national effort since 1976 under the aegis of the State Committee on Science and Technology (25, 52). The August 1986 decree ordering the cessation of work on water diversion projects directed that Soviet scientific and planning agencies devise a comprehensive program for the development of Central Asia to the year 2010, considering the demographic, water management, and agricultural situation (56). Because of worsening conditions, a special government commission was appointed in December 1986 to study ecological problems around the Aral (46). Its 1987 report recommended several measures to improve drinking water supplies and health conditions for people living near the sea. The commission also supported a plan to preserve the delta of the Amu Dar'ya.

In spite of all the studies and recommendations, other than starting construction on a water collector to carry irrigation drainage from the Amu Dar'ya basin to the sea, a project that will take years to complete, the government has taken no concrete measures to improve the condition of the Aral. Help may come too late: some say that the sea may be beyond rescue (60).

However, local inhabitants are far from accepting this grim fate. Although party and government officials from the two republics adjacent to the Aral (Uzbekistan and Kazakhstan) have been silent for the last several years, scientists, writers, and journalists from the region continue to plead angrily and sometimes eloquently, in the regional as well as national press, for action to save the Aral (9, 10, 29, 30, 34, 35).

As the situation worsens, those living around the sea will put great pressure on the national government to resurrect the Siberian diversion plan in order to provide minimum inflow to the Aral while maintaining irrigation. The campaign has already begun. In March 1988, the president of the Uzbek Academy of Sciences along with a well-known expert on the Aral Sea problem publicly stated that the ecological and social and economic difficulties of the Aral region could not be solved without diversion of water from Siberian rivers (46). The Moscow correspondent of the *Manchester Guardian* reported that Gorbachev, during his April 1988 visit to Tashkent, capital of the Uzbek Republic, after pleas from local officials, agreed to a new feasibility study of the project (59).

References and Notes

1   A. S. Kes', *Izv. Akad. Nauk SSSR Ser. Geogr.* 4, 8 (1978).
2   L. A. Zenkevich, *Biologiya Morey SSSR* (Akademiya Nauk SSSR, Moscow, 1963).
3   O. V. Zuyeva, *Probl. Osvoyeniya Pustyn'* 3, 40 (1987).
4   M. S. Lunezheva, A. K. Kiyatkin, V. P. Polishchuk, *Gidrotekh, Melior.* 10, 65 (1987).
5   M. I. L'vovich and I. D. Tsigel'naya, *Izv. Akad. Nauk SSSR Ser. Geogr.* 1, 42 (1978).
6   V. N. Kunin, *Priroda* 1, 36 (1967).
7   Annual data on the Aral Sea water balance, 1926 to 1985, were compiled by A. Asarin and V. Bortnik and provided by the Institute of Water Problems, USSR Academy of Sciences, and the Hydro Facilities Design Institute (*Gidroproyekt*), December 1987.
8   D. Ya. Ratkovich, V. I. Kuksa, L. V. Ivanova, *Vodn. Resur.* 6, 42 (1987).
9   E. Yusupov, *Pravda Vostoka*, 10 September 1987, p. 3.
10   T. Kaipbergenov, *Literaturnaya Gazeta*, 6 May 1987, p. 6.
11   Al. A. Grigor'yev, *Probl. Osvoyeniya Pustyn'* 1, 2 (1987).
12   I. A. Shiklomanov, *Anthropogenyy Izmeneniya Vodnosti Rek* (Gidrometeoizdat, Leningrad, 1979), pp. 225–240.
13   V. A. Dukhovnyy, P. M. Razakov, I. B. Ruziyev, K. A. Kosnazarov, *Probl. Osvoyeniya Pustyn'* 6, 3 (1984).
14   V. Kovalev, *Zvezda Vostoka*, 12, 3 (1986).
15   Goskomgidromet and Gosudarstvennyy Gidrologicheskiy Institut, *Mezhzonal'noye pereraspredeleniye vodnykh resursov*, A. A. Sokolov and I. A. Shiklomanov, Eds.

(Gidrometeoizdat, Leningrad, 1980), pp. 312–322.
16   I. B. Vol'ftsun, *Chelovek i stikhiya '84* (Gidrometeoizdat, Leningrad, 1983), pp. 96–98.
17   *Narodnoye khozyaystvo SSSR za 70 let* (Finansy i statistika, Moscow, 1987), p. 245; *Narodnoye khozyaystvo Kazakhstana v 1984g.* (Kazakhstan, Alma-Ata, 1985), pp. 97–98.
18   G. V. Voropayev et al., *Ekonomiko-Geograficheskiye Aspekty Formirovaniya Territorial'nykh Edinits v Vodnom Khozyaystve Strany* (Nauka, Moscow, 1987).
19   K. Lapin, O. Lebedev, V. Dukhovnyy, *Pravda Vostoka*, 29 June 1987, p. 8.
20   N. T. Kuznetsov and T. P. Gryaznova, *Probl. Osvoyeniya Pustyn'* 1, 10 (1987).
21   I. Bogdanov, *Soviet Life* (September 1987), pp. 35–36; B. T. Kirsta, *Probl. Osvoyeniya Pustyn'* 1, 19 (1988); also reported at a meeting with officials of "Karakumstroy", Ashkhabad, Soviet Union, 19 October 1987.
22   Institut geografii, *Problema Aral'skogo Morya*, S. Yu. Geller, Ed. (Nauka, Moscow, 1969), pp. 5–25.
23   B. Fedorovich, *Deserts Given Water* (Foreign Languages Publishing House, Moscow, 1958), pp. 85–96.
24   V. M. Borovskiy, *Izv. Akad. Nauk SSSR, Ser. Geogr.* 5, 35 (1978).
25   V. M. Kotlyakov et al., *Obshchestv. Nauki Uzb.* 10, 23 (1987).
26   I. P. Gerasimov et al., *Probl. Osvoyeniya Pustyn'* 6, 22 (1983).
27   I. M. Chernenko, *ibid.* 1, 3 (1986).
28   S. K. Kabulov and Kh. Sheripov, *ibid.* 2, 21 (1983); S. K. Kabulov, *ibid.* 3, 16 (1984); M. Sh. Ishankulov, *Geogr. Prirodnyye Resur.* 2, 45 (1985).
29   A. Kudryashov, *Pravda Vostoka*, 10 January 1988, p. 2.
30   A. Lapin, *Sobesednik* 13, 6 (March 1988).
31   I. Lein, *Pravda Vostoka*, 17 November 1987, p. 4.
32   I. V. Rubanov and N. M. Bogdanova, *Probl. Osvoyeniya Pustyn'* 3, 9 (1987).
33   M. Ye. Bel'gibayev, *ibid.*, *Probl. Osvoyeniya Pustyn'* 1, 72 (1984).
34   A. Nurpeisov, *Ogonyok* 1, 23 (1988).
35   S. Azimov, *Literaturnaya Gazeta*, 26 November 1986, p. 11.
36   D. Ratkovich. A. Frolov, V. Kuksa, Institute of Water Problems, USSR Academy of Sciences, meeting, 19 September 1987.
37   A. B. Bakhnev, N. M. Novikova, M. Ye. Shenkareva, *Vodn. Resur.* 2, 167 (1987)
38   V. A. Bondarev, *Tr. Sredneaziat. Reg.*

Nauchno-Issledovatel'skogo Inst. 96 (no. 177), 12 (1983).

39 I. A. Klyukanova and Ye. N. Minayeva, *Izvestiya Akad. Nauk SSSR Ser. Geogr.* 1, 50 (1986).

40 S. K. Kabulov, *ibid.* 2, 95 (1985).

41 I. M. Chernenko, *Probl. Osvoyeniya Pustyn'* 3, 18 (1983).

42 K. I. Lapkin and E. D. Rakhimov, *ibid.* 2, 84 (1979).

43 G. V. Voropayev et al., Institute of Water Problems, USSR Academy of Sciences, meeting, 16 September 1987.

44 I. M. Chernenko, *Probl. Osvoyeniya Pustyn'* 4, 53 (1987).

45 N. F. Vasilyev, *Gidrotekh. Melior.* 11, 6 (1987).

46 P. Khabibullayev and V. Dukhovnyy, *Pravda Vostoka*, 3 March 1988, p. 3.

47 P. Micklin, *Soviet Geography Studies in Our Time*, L. Holzner and J. M. Knapp, Eds. (Univ. of Wisconsin Press, Milwaukee, 1987), pp. 229–261.

48 V. Kotlyakov, *Pravda*, 14 April 1988, p. 3.

49 N. R. Khamrayev, *Probl. Osvoyeniya Pustyn'* 1, 11 (1988).

50 A. A. Rafikov, *Geogr. Priordny. Resur.* 4, 44 (1986).

51 N. Reymers, *Trud*, 1 December 1987, p. 4.

52 A. S. Kes' and Yu. N. Kulikov, *Izv. Akad. Nauk SSSR Ser. Geogr.* 4, 143 (1987).

53 G. Dimov, *Izvestiya*, 20 September 1987, p. 2.

54 L. Perlov, *Lesnaya Promyshlennost'*, 9 April 1988, p. 4; L. Levin, *Pravda Vostoka*, 11 May 1988, p. 2.

55 P. Micklin, *Soviet Geogr.* 5, 287 (1986); *Cent. Asian Sur.* 2, 67 (1987).

56 "V Tsentral'nom Komitete KPSS i Sovete Ministrov SSSR," *Pravda*, 20 August 1986, p. 1.

57 S. Zaligin, *Kommunist* 13, 63 (1985); *Nash Sovrem.* 1, 113 (1987); S. P. Zaligin, *Povorot* (Mysl', Moscow, 1987); V. Leybovskiy, *Ogonyok* 40, 24 (1987).

58 "Interv'yu po pros'be chitateley," *Gidrotekh. Melior.* 5, 66 (1987); *Nov. Mir* 7, 181 (1987); V. Korzun, *Zvezda Vostoka* 9, 122 (1987).

59 M. Walker, *Manchester Guardian Weekly*, 24 April 1988, p. 8.

60 During a fall 1987 visit to the Soviet Union, P. P. M. spoke with several Soviet water management experts who held this opinion.

61 NOAA-9 AVHRR image tape (GIL 1 226:N:46 08606 2 92AN9 LVS, 14 August 1986).

62 S. L. Vendrov, *Problemy Preobrazovaniya Rechnykh Sistem SSSR* (Gidrometeoizdat, Moscow, 1979), p. 56, figure 4.

63 Supported by the National Council for Soviet and East European Research, the National Academy of Sciences, and the Lucia Harrison Fund of the Department of Geography, Western Michigan University. I thank the Institute of Water Problems, Institute of Geography, and Desert Institute of the Soviet Academy of Sciences for providing data and other information that aided this research and B. Fogle for assistance in the preparation of the figures.

12

LOWERING OF A SHALLOW, SALINE WATER TABLE BY EXTENSIVE EUCALYPT REFORESTATION*

M. A. Bari and N. J. Schofield

Introduction

Stream and land salinisation is a major environmental and resource problem in the south west of Western Australia (Schofield et al., 1988; Schofield and Ruprecht, 1989). At present, only 48% of the divertible surface water resources remain fresh (less than 500 mg l⁻¹ total soluble salts (TSS)) (Western Australian Water Resources Council, 1986), and some 440 000 ha of once productive farmland (Australian Bureau of Statistics, personal communication, 1989) has become salinised. Man-induced salinity also affects other parts of Australia (Peck et al., 1983), North America (Brown et al., 1983; McKell et al., 1986) and several other semi-arid or arid regions of the world (Dudal and Purnell, 1986).

In Western Australia increasing stream and land salinity is a by-product of the agricultural development of the State. Replacement of the deep-rooted, perennial native vegetation (forests and woodlands) with shallow-rooted annual crops and pasture has disturbed the hydrologic balance. The resulting reduction in evapotranspiration has led to increased groundwater recharge and rising groundwater levels (Peck and Williamson, 1987; Ruprecht and Schofield, 1989). The salts (principally NaCl) 'stored' in the unsaturated zone are dissolved and ultimately discharged at the land surface or into the streams (Williamson, 1986; Williamson et al., 1987; Ruprecht and Schofield, 1992).

One approach to reclaiming the salinised streams and land is to reverse the process by reforesting farmland. Partial reforestation of the cleared land was considered promising in south west Western Australia because annual potential evaporation substantially exceeds annual rainfall across most of the region. Consequently, in theory at least, only a part of the farmland would require reforesting to control rising groundwater and hence land and stream salinity (Schofield, 1990).

During the late 1970s a number of experimental sites were established with various reforestation strategies embracing different layouts and densities of trees (Schofield et

* Originally published in *Journal of Hydrology*, 1992, vol. 133, pp. 273–91.

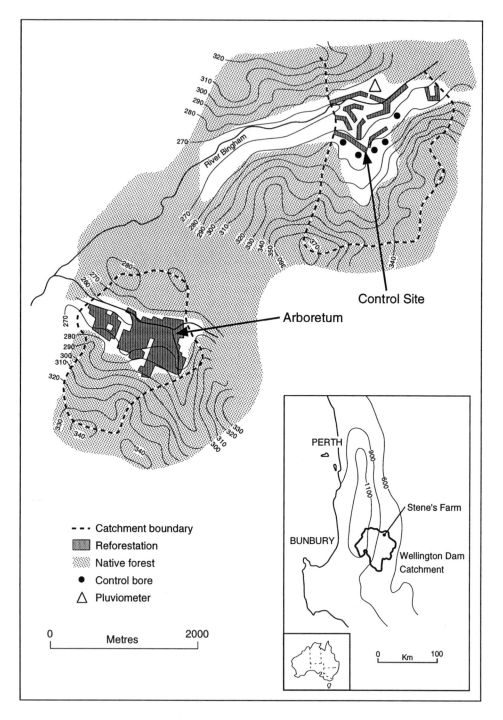

Figure 12.1 [*orig. figure 1*]   Location of the study area.

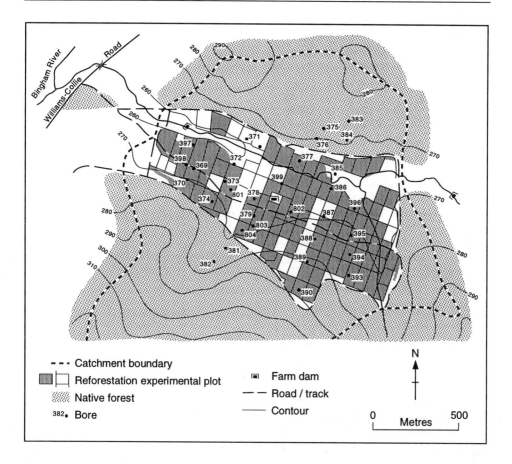

Figure 12.2 [orig. figure 2]   Reforestation layout and hydrometric network.

al., 1989). One of these sites is the Stene's Farm arboretum where near-total reforestation of the cleared land was attempted. This paper presents results from this site as a demonstration of the effects of extensive eucalypt reforestation on groundwater level, groundwater salinity and soil salinity.

Site Description

Location and climate

Stene's Farm arboretum site is located in the Wellington Dam catchment some 150 km south of Perth and 40 km north of Collie (figure 12.1). The area has a Mediterranean climate with cool, wet winters and dry, hot summers. About 80% of the annual rainfall occurs in the winter

months (May–October). The long-term (1926–1988) annual rainfall at the site is 713 mm year^{-1}. The annual average pan evaporation is about 1600 mm year^{-1} (Luke et al., 1988). Temperatures range from a maximum in excess of 40°C, which occurs in January–February, to a minimum of less than 0°C, occurring in June or July.

Site history and layout

In the 1950s, Stene's Farm was partially cleared for pasture development. The site was later purchased by the State Government (in the 1970s) as part of a reforestation programme to control the increasing inflow salinity of the Wellington Reservoir (Loh, 1988), the highest-yielding reservoir in the south of the State.

The arboretum site has a catchment area

of 284 ha. The valley floor and lower to middle slopes of the catchment were cleared in the 1950s. In 1979, 63 Eucalyptus and 2 Pinus species were established, each species covering a minimum 0.5 hectare block. A few small areas were not planted and some species failed (figure 12.2). In total, 70% of the cleared farmland was successfully planted.

To account for the effects of climatic variation, a nearby pasture control site (figure 12.1) was also monitored. The control site lies approximately 3.5 km north east of the arboretum site. This site had a clearing pattern similar to that of Stene's arboretum. In 1976–1980, 14% of the cleared area was reforested in strips (figure 12.1). The control bores are located in pasture upslope of these strips.

Topography, soil and geology

The topography of the arboretum site is illustrated in figure 12.2. The upslope forested portion of the catchment is slightly steeper (6.5%) than the reforested zone, which has an average 4% slope. The soil profiles of a valley cross-section, determined from the core samples, are shown in figure 12.3. The surface soils consist of silty sand of high permeability, typical of the Darling Range (Bettenay et al., 1980). The rainfall intensity very rarely exceeds the infiltration capacity of the soil (Sharma et al., 1987). The topsoil is underlain by a sandy and gravelly silt to sandy and gravelly clay horizon to 2 m depth. This is underlain by a deep (10–20 m) gravelly clay to clay horizon.

The physiography of the pasture control site is similar to that of the arboretum site. The mean slope of the control catchment is 3.8%.

Vegetation

The native forest of the arboretum and control sites consists of jarrah (Eucalyptus marginata), marri (E. calophylla), Western Australian flooded gum (E. rudis) and wandoo (E. wandoo). After clearing, the pasture was dominated by rye grasses (Lolium spp.), barley grasses (Hordium spp.) and other grasses with a legume subclover (Trifolium subterraneum). The reforestation consists of 63 species of Eucalyptus and two species of Pinus.

Experimental Methods

Plantation establishment and management

In 1979, 63 Eucalyptus and two Pinus species were planted at the arboretum site, with an initial stem density of 625 stems per hectare (sph). The cleared area was divided into 123 plots of about 0.5 ha size and each plot was planted with a single species. On the plots with the most salt and waterlogging conditions tree survival was poor, and in 1988 the tree densities varied from nil to 600 sph with an overall average of 340 sph. Trees were not culled or pruned. Crown cover was measured in December 1987 using a Crownometer similar to the one described by Montana and Ezcurra (1980).

Rainfall and groundwater monitoring

Rainfall data were continuously recorded with a pluviometer situated about 3.5 km northeast of the arboretum site (figure 12.1). For the period of missing record (4.5% by time during 1980–1989), data were estimated using a regression between this pluviometer and its nearest neighbour (9 km away).

The groundwater bore networks at the control and the arboretum sites are shown in figures 12.1 and 12.2 respectively. The control bores were installed to bedrock at the mid-slope and had an average slotted screen length of 11.5 m. At the arboretum site 31 bores were installed, and included a transect across the valley. The bores were approximately evenly spaced over the reforested area (figure 12.2). All but one bore had a screen length of 3 m.

All bores were sampled on a monthly basis for groundwater level. Salinity was determined at installation and again in May 1989. At the latter measurement all bores were pumped by greater than two standing volumes before sample collection.

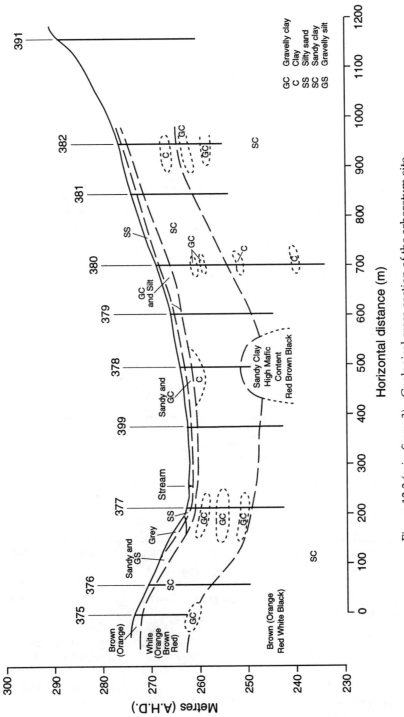

Figure 12.3 (*orig. figure 3*) Geological cross-section of the arboretum site.

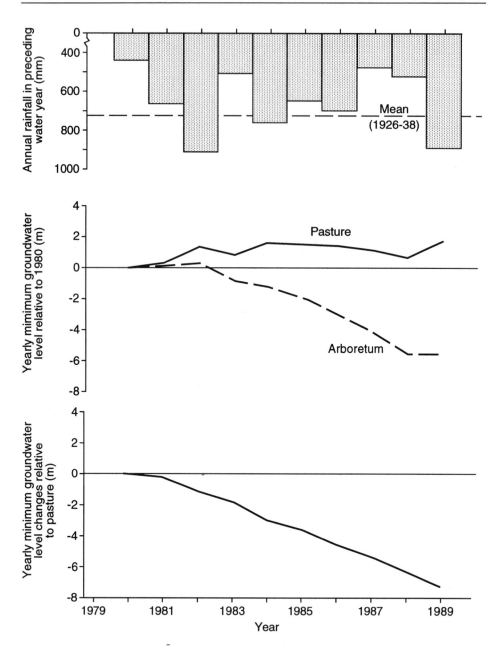

Figure 12.4 [orig. figure 4]   Annual changes in rainfall and minimum groundwater levels beneath pasture and reforestation.

Results

Rainfall

During the 1980–1989 period, annual rainfall varied from 473 to 910 mm (figure 12.4). The average for the period, 655 mm, was 8% below the long-term (1926–1988) average of 713 mm. Three years (1981, 1983 and 1988) had rainfall higher than long-term average. Most of the rainfall (about 80%) occurred during May–October.

Plantation cover

In December 1987, crown cover of the arboretum site was measured and reported by Bell et al. (1988). Ignoring blocks with no, or very few trees (those shown blank in figure 12.2), the crown cover of individual blocks ranged from 14 to 66% and averaged 38%.

Initial groundwater conditions

At the arboretum site, the depth of minimum groundwater level in 1980 varied from 1.0 m (bore 399) to 14.93 m (bores 389 and 393), with a mean of 6.77 m. The variation of maximum groundwater level was from 0.0 m (bore 393) to 14.0 m (bore 389), with an average of 4.71 m. The groundwater salinity was highly spatially variable. In 1980, bore salinities ranged from 338 mg l^{-1} TSS (bore 373) to 12 227 mg l^{-1} TSS (bore 393), with a mean of 5012 mg l^{-1} TSS. Saline groundwater discharge in the valley floor was apparent.

At the control site, bores are located at the mid-slopes. In 1980, the average depths of minimum and maximum groundwater level were 9.54 m and 8.08 m, respectively. Bore salinities varied from 229 (bore 8) to 4915 mg l^{-1} TSS (bore 44), with an average of 2404 mg l^{-1} TSS. In 1980, saline seeps were evident along the river flat.

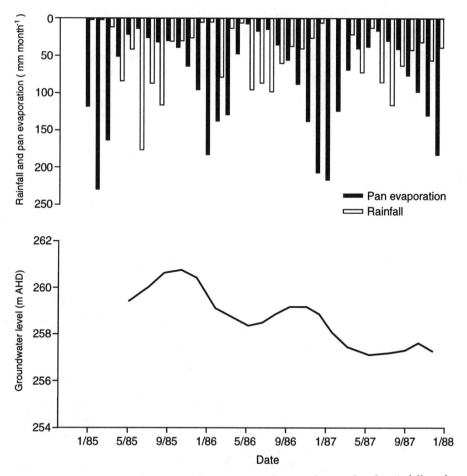

Figure 12.5 [*orig. figure 5*]   Seasonal variations of groundwater level, rainfall and pan evaporation.

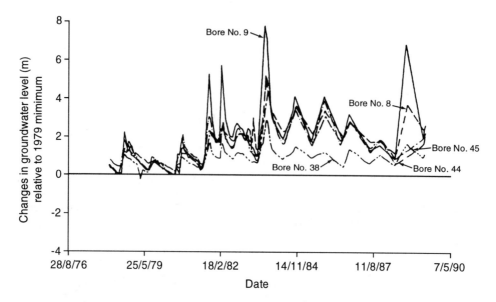

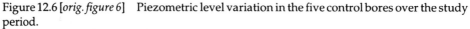

Figure 12.6 [*orig. figure 6*]   Piezometric level variation in the five control bores over the study period.

Seasonal water balance

Monthly pan evaporation, rainfall and groundwater level during 3 years are shown in figure 12.5. After the onset of the rainy season (April/May) the groundwater level begins to rise (May/June). This trend continues to October/November. While rainfall exceeds pan evaporation it can be assumed that groundwater recharge occurs. The annual groundwater level peak lags rainfall by about 4 months. From October to April pan evaporation exceeds rainfall and the water table declines. Again, the annual minimum groundwater level lags peak evaporation by about 4 months.

Groundwater levels beneath pasture control

Hydrographs of the control bores show large seasonal variations and a rise of the annual minima over the study period (figure 12.6). The variation of the annual maxima is more than that of the annual minima, suggesting that the minima are a more stable indicator of groundwater level trends. Most of the bores in the control set have reasonably similar hydrographs.

The variation of the yearly minimum groundwater level relative to 1980, averaged for the five bores, is shown in figure 12.4. Groundwater levels increased in 1982, 1984 and 1989, as a result of above average rainfall in the preceding 'water years' (1 April–31 March). Comparing 1989 with 1980, there was a rise of 1.8 m. Similar to the minimum, the maximum groundwater level showed rises in 1982, 1984 and 1989, and falls in other years. During 1980–1989, there was a net rise of 2.0 m in maximum groundwater level.

Groundwater levels beneath arboretum

The groundwater levels beneath the arboretum had strong declining trends. The decline varied across the site as shown in figure 12.7. The annual minimum groundwater level changes relative to 1980, averaged for 18 bores, are shown in absolute (relative to the ground surface) and relative (relative to the pasture control) terms in figure 12.4. In absolute terms, groundwater levels did not start to decline until 3–4 years after planting. However, relative to the pasture control, the decline began somewhat earlier. The rate of decline in absolute terms

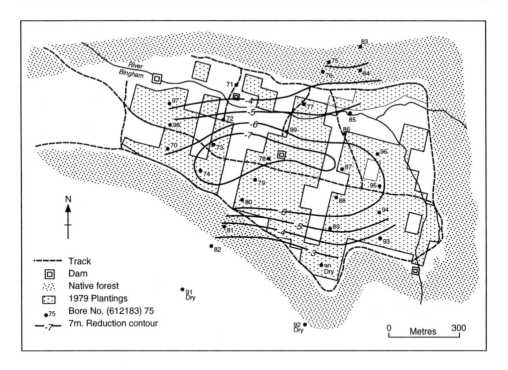

Figure 12.7 [*orig. figure 7*]   Contours of the reductions of minimum groundwater levels between 1980 and 1989.

has generally shown little response to annual rainfall, except in the wettest years (e.g. 1988). Since 1981 the rate of decline of the annual minimum relative to the control has been near-uniform at about 800 mm year^{-1} (figure 12.4). During 1980–1989, the absolute reduction of minimum groundwater level was 5.5 m and the relative reduction was 7.3 m.

The behaviour of annual maximum groundwater level was generally similar to that of the annual minimum. It was responsive to annual rainfall variation in the early years of the plantation (to 1983) but less so in subsequent years. By 1989, the absolute reduction of annual maximum groundwater level was 5.8 m and the relative reduction was 7.8 m.

The annual minimum potentiometric surfaces along the bore transect 375–391 of the arboretum site are shown in figure 12.8 for the years 1980, 1985 and 1989. Over time, the groundwater level across the valley has been lowered substantially by reforestation.

This result indicates that such extensive reforestation is an effective strategy for lowering shallow, saline groundwater tables in this region.

Modelling of reforestation area and groundwater level reduction

To improve the hydrological interpretation of the results, a simple water balance model (Schofield, 1990) was applied. The conceptual model (figure 12.9) can be applied to this site, noting that in this case partial reforestation of the 'seep area' occurred. For a particular reforestation area A_{r7} the corresponding rate of groundwater level reduction (Z) across the site is

$$Z = \frac{A_r(E_r - E_c) + E_c(1 - A_f) - A_s(E_c - E_s) - E_f(1 - A_f)}{\theta}$$

where E_f is annual evaporation from native forest, E_s is annual evaporation from saline seep, E_c is annual evaporation from pasture, E_r is annual evaporation from reforestation, A_f is area of remnant forest, A_s is area of

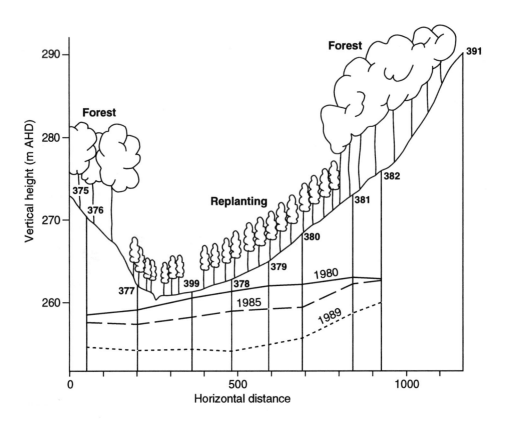

Figure 12.8 [*orig. figure 8*] Comparison of the valley cross-section potentiometric surfaces of 1980, 1985 and 1989.

saline seep, A_c is area of pasture, A_r is area of reforestation and θ is change in soilwater content after desaturation.

The rationale for determining the values of the above variables for a 750 mm year^{-1} rainfall zone in Western Australia is described by Schofield (1990). For Stene's arboretum site, the evaluated or assigned values are given in table 12.1. The model predicts a decline of groundwater level at 770 mm year^{-1}.

This prediction is comparable with

Table 12.1 [*orig. table 1*]   Data for the water balance model.

Variable	Reforestation	Pasture Control	Units
E_f	650	650	mm year^{-1}
E_c	390	390	mm year^{-1}
E_r	1600	1600	mm year^{-1}
E_s	150	150	mm year^{-1}
A_r	0.245	0.044	m^2 m^{-2}
A_f	0.65	0.69	m^2 m^{-2}
A_c	0.055	0.216	m^2 m^{-2}
A_s	0.05	0.05	m^2 m^{-2}
θ	0.25	0.25	m^3 m^{-3}

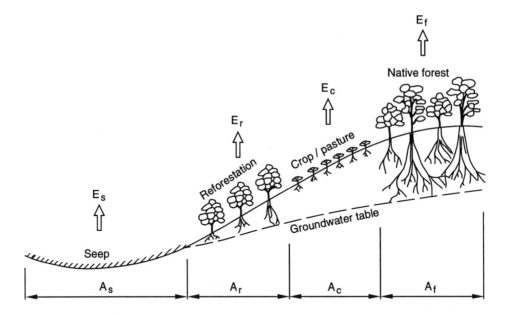

Figure 12.9 [*orig. figure 9*]   Conceptual representation of the water balance model.

the observed mean rate of decline of 610 mm year^{-1}. It also predicts a rise in groundwater table of 160 mm year^{-1} under the pasture control, whereas the actual rise was 200 mm year^{-1}.

Groundwater flow characteristics

Groundwater level contours across the site indicate that, at the commencement of reforestation, a northwesterly flow prevailed. This feature was still partially evident in May 1989, although at that time a near-flat groundwater table was apparent below the valley floor and lower slopes.

Groundwater salinity

Comparing groundwater salinities in May 1980 and May 1989 (table 12.2) shows significant variation in salinity change over time amongst the individual bores. On average, groundwater salinity was reduced by 41% and 11% beneath the pasture control and the arboretum site respectively, over the 1980–1989 period.

Soil salinity

The soil salinity profiles of the bore transect 375–391 (figure 12.2) measured in 1979 and

1989 are illustrated in figure 12.10. With the exception of bore 380, there was relatively little change between the salt profiles in 1980 and 1989. In the case of bore 380, there has been a substantial reduction in salt content through the profile, except in the 4–6 m depth range. The salinity of the water measured in the bore has also decreased substantially. Bore 378 shows an accumulation of salt in the 0.5–7 m depth range.

Discussion

Rainfall and groundwater level response

The average annual rainfall for the study period was 8% below the 1926–1988 average. If the longer-term average rainfall had occurred, the magnitude of the groundwater level reduction would probably have been smaller. However, it is difficult to determine an appropriate mean annual rainfall as the region may be subject to 'drying' as a consequence of the greenhouse effect (Pittock, 1988).

Table 12.2 [*orig. table 2*]   Comparison of groundwater salinities at arboretum and pasture control sites between 1980 and 1989.

Bore	6 May 1980	31 May 1989	% Change
	Salinity (mg l⁻¹)	Salinity (mg l⁻¹)	Individual
Pasture control			
8	229	210	−8.3
9	397	203	−48.9
38	3093	1776	−42.6
44	4915	2828	−42.5
45	3386	2130	−37.1
Average	2404	1429	−40.5
Arboretum site			
369	5146	893	−82.7
370	5603	4979	−11.1
372	1669	436	−73.9
373	338	4112	1116.6
374	6834	9138	33.7
377	2267	1359	−40.1
378	5030	4602	−8.5
379	4309	3268	−24.2
380	1825	505	−72.3
386	1008	208	−79.4
388	7015	5565	−20.7
389	9961	10179	2.2
393	12227	13571	11.0
394	11929	10578	−11.3
396	3292	216	−93.4
397	4739	4495	−5.2
398	1375	856	−37.8
399	5654	5260	−7.0
Average	5012	4457	−11.1

Water balance model

The comparability between the observed and predicted rates of groundwater level decline partly validates the model and the water balance concept it utilizes. However, the model prediction is sensitive to the variation of parameter values (Schofield, 1990), and some of parameter values (e.g. E_r, E_s and E_c) have only moderate accuracy.

Suitability of pasture control

The pasture control site has 14% strip reforestation of the cleared land, mainly at the valley floor and on lower slopes. This reforestation may have some time-varying influence on the groundwater control bores located on the mid-slope. However, as a slight rise in the groundwater level beneath pasture has occurred over the study period, the control bores are considered fairly representative of agricultural pasture in which

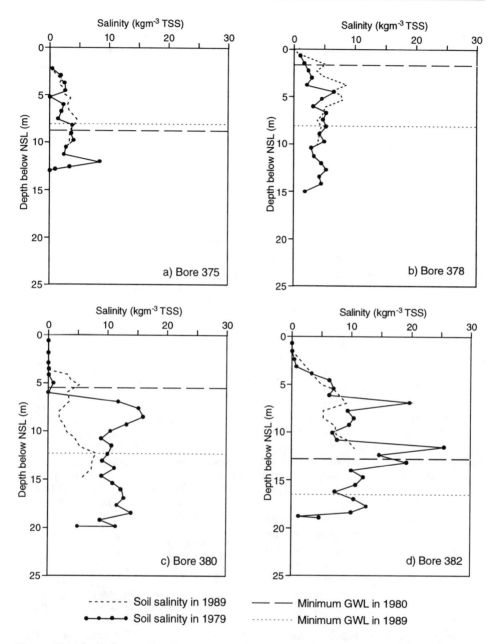

Figure 12.10 [*orig. figure 10*]   Changes in soil salinity profiles between 1980 and 1989.

groundwater has approached equilibrium (Ruprecht and Schofield, 1989).

Groundwater and soil salinity

The decline in groundwater salinity beneath the arboretum (11%) was lower than for the control (41%). However, the main significance of the result is that the average groundwater salinity has not increased under reforestation as a result of evaporative concentration as was considered likely by a number of authors (Conacher, 1982; Morris and Thomson, 1983; Williamson, 1986). From the limited data available, the cause of salinity reduction under reforestation is uncertain. One possibility is the accumulation of salt in the unsaturated zone when the water table falls. This is seen to occur for bore 378 (figure 12.10). However, bore 380 shows a substantial decline in bore-water salinity and soil profile salinity, which may be indicative of lateral leaching. The variation that does occur from samplings close to bores 375, 378 and 382 could be partly attributed to the high spatial variability of salt content, even at a small scale (Johnston, 1987). Measurement errors may also be significant. In the case of bores 375 and 382, which have remained undisturbed under native forest, little change in salinity profile is expected.

Use of the reforestation strategy in salinity control

The results clearly demonstrate that the extensive, dense reforestation strategy used at the arboretum site is successful in lowering the saline groundwater table substantially across the valley floor, thus eliminating saline groundwater discharge to the stream in a short period of time (approximately 10 years). The absolute rate of decline of the minimum groundwater level was the highest (610 mm year^{-1}) among a number of different reforestation strategies tested in the area (Bell et al., 1990).

Clearly, though, this extensive reforestation strategy would have major implications for agriculture in the region and consequently it is likely to be applied in specific salinity management priority areas.

Conclusions

Seventy per cent reforestation has lowered the yearly minimum groundwater level of a partly salinised agricultural site by 5.5 m relative to the ground surface and a 7.3 m relative to a pasture control in 10 years. The groundwater level reduction, relative to the control, was near-uniform with time (about 800 mm year^{-1}) and had a continuous downward trend.

The reforestation also lowered the annual

maximum groundwater level by a similar magnitude.

The groundwater salinity beneath the arboretum was reduced by 11%, whereas the groundwater salinity under pasture control decreased by 41% over the 1980–1989 period.

The soil salinity profiles beneath the reforestation changed substantially for only one of the bores sampled. The changes were difficult to interpret.

The extensive reforestation strategy has been successful in salinity control but its application will be limited to specific purposes because of its implications for agriculture.

References

Bell, R. W., Anson, B. and Loh, I. C., 1988. Groundwater response to reforestation in the Darling Range of Western Australia. Surface Water Branch, Water Auth. WA, Perth, Rep. WS 24, 89 pp.

Bell, R. W., Schofield, N. J., Loh, I. C. and Bari, M. A., 1990. Groundwater response to reforestation in the Darling Range of Western Australia. J. Hydrol., 119: 179–200.

Bettenay, E., Russell, W. G. R., Hudson, D. R., Gilkes, R. J. and Edmiston, R., 1980. A description of experimental catchments in the Collie area, Western Australia. CSIRO Div. Land Resour. Manage., Perth, Tech. Pap. 7, 36 pp.

Brown, P. L., Halvorson, A. D., Siddoway, F. H., Mayland, H. F. and Millar, M. R., 1983. Saline seep diagnosis, control and reclamation. US Dep. Agric. Conserv. Res. Rep. 30, 20 pp.

Conacher, A., 1982. Dryland agriculture and secondary salinity. In: W. Hanley and M. Cooper (Editors), Man and the Australian Environment. McGraw-Hill, Sydney, pp. 113–125.

Dudal, R. and Purnell, M. F., 1986. Land resources: salt affected soils. Reclam. Reveg. Res., 5: 1–9.

Johnston, C. D., 1987. Distribution of environmental chloride in relation to subsurface hydrology. J. Hydrol., 94: 76–88.

Loh, I. C., 1988. The history of catchment and reservoir management on Wellington Reservoir catchment, Western Australia. Surface Water Branch, Water Auth. WA, Perth, Rep. WS 35, 39 pp.

Luke, G. J., Burke, K. L. and O'Brien, T. M., 1988. Evaporation data for Western Australia. West Australian Dep. Agric. Div. Resour. Manage.,

Perth, Tech. Rep. 65, 29 pp.

McKell, C. M., Goodin, J. R. and Jeffries, R. L., 1986. Saline land of United States and Canada. *Reclam. Reveg. Res.*, 5: 159–165.

Montana, C. and Ezcurra, E., 1980. A simple instrument for quick measurement of crown projections. *J. For.*, 78: 699.

Morris, J. D. and Thomson, L. A. J., 1983. The role of trees in dryland salinity control. *Proc. R. Soc. Vic.*, 95: 123–131.

Peck, A. J. and Williamson, D. R., 1987. Effects of forest clearing on groundwater. *J. Hydrol.*, 94: 47–65.

Peck, A. J., Thomas, J. F. and Williamson, D. R., 1983. Salinity issues: Effects of man on salinity in Australia. Water 2000 Consultants Rep. 8, Aust. Gov. Publ. Serv., Canberra, 78 pp.

Pittock, A. B., 1988. Actual and anticipated changes in Australia's climate. In: G. I. Pearman (Editor), *Greenhouse – Planning for Climate Change*. CSIRO, Melbourne, pp. 35–51.

Ruprecht, J. K. and Schofield, N. J., 1989. Analysis of steamflow generation following deforestation in south-west Western Australia. *J. Hydrol.*, 105: 1–18.

Ruprecht, J. K. and Schofield, N. J., 1992. Effects of partial deforestation on hydrology and salinity in high salt storage landscapes. I. Extensive block clearing. *J. Hydrol.*, (in press).

Schofield, N. J., 1990. Determining reforestation area and distribution for salinity control. *J. Hydrol. Sci.*, 35: 1–19.

Schofield, N. J. and Ruprecht, J. K., 1989. Regional analysis of stream salinisation in south-west Western Australia. *J. Hydrol.*, 112: 18–38.

Schofield, N. J., Ruprecht, J. K. and Loh, I. C., 1988. The impact of agricultural development on the salinity of surface water resources of south-west Western Australia. Water Auth. WA, Perth, Rep. WS 27, 69 pp.

Schofield, N. J., Loh, I. C., Scott, P. R., Bartle, J. R., Ritson, P., Bell, R. W., Borg, H., Anson, B. and Moore, R. (1989). Vegetation strategies to reduce stream salinities of water resource catchments in south-west Western Australia. Water Auth. WA, Perth, Rep. WS 33, 98 pp.

Sharma, M. L., Barron, R. J. W. and Fernie, M S., 1987. Areal distribution of infiltration parameters and some soil physical properties in lateritic catchments. *J. Hydrol.*, 94: 109–127.

Western Australian Water Resources Council, 1986. Water Resources Perspectives Western Australia: Rep. 2. Water Resources and Water Use, Summary of Data for the 1985 National Survey. Publ. WRC 7/86, Perth, 164 pp.

Williamson, D. R., 1986. The hydrology of salt affected soils in Australia. *Reclam. Reveg. Res.*, 5: 181–196.

Williamson, D. R., Stokes, R. A. and Ruprecht, J. K., 1987. Response of input and output of water and chloride to clearing for agriculture. *J. Hydrol.*, 94: 1–28.

13

Nutrient Loss Accelerated by Clear-cutting of a Forest Ecosystem[*]

F. H. Bormann, G. E. Likens, D. W. Fisher and R. S. Pierce

One-third of the land surface of the United States supports forest, and much of it is occasionally harvested. Yet, apart from nutrient losses calculated on the basis of extracted timber products, we have little quantitative data on the effects of harvesting on either the nutrient status of forest eco-systems or the chemistry of stream water draining from them – which is closely related to the increasingly important problem of eutrophication of stream and river water (1). This paucity of infor-mation partly reflects the difficulties of measuring characteristics of massive forest ecosystems, and the fact that nutrient cycles are closely tied to hard-to-measure hydro-logic parameters (2).

The input and output of chemicals can be measured and nutrient budgets constructed for forest ecosystems by use of the small-watershed approach (2). For several years we have measured these parameters on six small undisturbed watersheds in the Hubbard Brook Experimental Forest in cen-tral New Hampshire (3); here the bedrock is practically impermeable (4), and all liquid

water leaves the watersheds by way of first-or second-order streams; the runoff pattern is typical of northern regions having deep snow packs (5). Additional information on topography, geology, climate, and biology is given by Likens et al. (4).

Chemical relations for these undisturbed forest ecosystems (watersheds) are being established by weekly measurements of dissolved cations and anions entering the ecosystem in all forms of precipitation and leaving the system in stream water. These data, combined with measurements of precipitation and stream flow, enable computation of the input and output of these various elements, as well as annual budgets (2). These results have been reported (4, 6, 7).

In 1965 the forest of one watershed (W-2) was clear-cut in an experiment designed to: (i) determine the effect of clear-cutting on stream flow, (ii) examine some of the funda-mental chemical relations of the ecosystem, and (iii) evaluate the effects of forest manipulation on nutrient relations and on eutrophication of stream water. This is a preliminary report of chemical effects

[*] Originally published in *Science*, 1968, vol. 158, pp. 882–4.

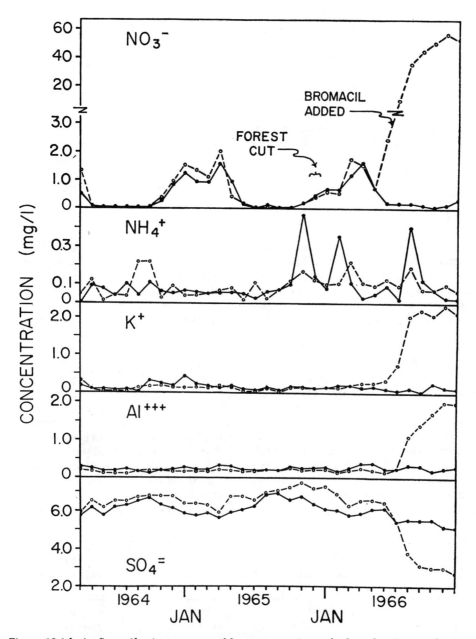

Figure 13.1 [*orig. figure 1*] Average monthly concentrations of selected cations and anions in stream water draining from forest ecosystems undisturbed (solid lines) and clear-cut during the winter of 1965–66 (dashed lines).

observed during the subsequent year.

The experiment began during the winter of 1965–66 when the beech-maple-birch forest (15.6 hectares) was leveled by the U.S. Forest Service. All trees, saplings, and shrubs were cut, dropped in place, and limbed so that no slash was more than 1.5 m above ground. No products were removed from the forest, and great care was taken to minimize erosion of the surface. On 23 June 1966, regrowth of vegetation was inhibited by application of the herbicide Bromacil ($C_9H_{13}BrN_2O_2$) at 28 kg/hectare; approximately 80 percent of the mixture applied was Bromacil; 20 percent was largely inert carrier (8).

Samples of stream water were collected and analyzed weekly, as they had been for 2 years before the cutting; the loss of ions was calculated in terms of kilograms per hectare. Similar measurements on adjacent undisturbed watersheds provided comparative information.

The cutting had a pronounced effect on runoff, which began to increase in May 1966; the cumulative runoff value for 1966 exceeded the expected value by 40 percent. The greatest difference occurred during June through September, when runoff values were 418 percent greater than expected. The difference is directly attributable to the removal of transpiring surface and probably reflects wetter conditions within the soil profile.

The striking loss of nitrate nitrogen in stream water (figure 13.1) suggests that alteration of normal patterns of nitrogen flow played a major role in loss of nutrients from the cutover ecosystem. This loss is best understood by consideration of nitrogen patterns in the undisturbed ecosystem. Runoff data from such systems (7) (figure 13.1) indicate a strong and reproducible seasonal cycle of concentration of nitrate in stream water. High concentrations are associated with the winter period from November through April, while low concentrations persist from April through November.

The decline of nitrate concentrations during May and the low concentrations throughout the summer correlate with heavy nutrient demands by the vegetation and increased heterotrophic activity associated with warming of the soil. The winter concentration pattern of NO_3^- may be explained in strictly physical terms, since the input of nitrate in precipitation from November through May largely accounts for nitrate lost in stream water during this period. Evaporation from the snow pack may account for some increase in concentration of NO_3^- in the stream water in the spring. Also, since early input of nitrate in precipitation exceeds losses in stream water (table 13.1), concentrations in stream water provide little conclusive evidence of the occurrence of nitrification in these undisturbed acid soils.

Results from the cut watershed demon-

Table 13.1 [*orig. table 1*]   Partial nitrogen budgets for watersheds 6 (13.2 hectares) and 2 (15.6 hectares): all data are in kilograms of elemental nitrogen per hectare. Gains by biological fixation and losses by volatilization are not included. Watershed 2 was cut in the winter of 1965–1966.

Year	Input in precipitation		Output in stream water		Net gains (+) losses (−)	
	NH_4^-N	NO_3^-N	NH_4^-N	NO_3^-N	NH_4^-N	NO_3^-N
	Watershed 6 (undisturbed)					
1965	2.1	2.8	0.6	1.0	+1.5	+1.8
1966	2.0	4.5	0.5	1.5	+1.5	+3.0
	Watershed 2 (cutover)					
1965	2.1	2.8	0.4	1.3	+1.7	+1.5
1966	2.0	4.5	1.2	58.1	+0.8	−53.6

strate nitrogen relations of such an ecosystem and indirectly those of the undisturbed ecosystem. Comparison of nitrate concentrations in stream water from watersheds W-6 (undisturbed) and W-2 (cutover) indicates a similar pattern of concentrations throughout 1964 and 1965, prior to cutting, and through May of 1966 (figure 13.1). Beginning on 7 June 1966, 16 days before application of the herbicide, nitrate concentrations in W-2 show a precipitous rise, while the undisturbed ecosystem shows the normal late-spring decline. Allison (9) has documented similar losses of nitrate from uncropped fields or fields carrying poorly established crops. The increase in nitrate concentrations is a clear indication of the occurrence of nitrification in the cutover ecosystem. Since an NH_4^+ substrate is required, the occurrence of nitrification also indicates that soil C : N ratios were favorable for the production of NH_4^+, in excess of heterotrophic needs, sometime before 7 June.

Some of these conclusions must hold for the undisturbed ecosystem; that is to say, sometime prior to 7 June C : N ratios were favorable for the flow of ammonium either to higher plants or to the nitrification process. The low levels of NH_4^+ and NO_3^- in the drainage water of the undisturbed ecosystem (W-6) may attest to the efficiency of the oxidation of NH_4^+ to NO_3^-, and to the efficiency of the vegetation in utilizing NO_3^-. However, Nye and Greenland (10) state that growing, acidifying vegetation represses nitrification; thus the vegetation may draw directly on the NH_4^+ pool, and little nitrate may be produced within the undisturbed ecosystem. In this case, one must assume that cutting drastically altered conditions controlling the nitrification process.

The action of the herbicide in the cutover watershed seems to be one of reinforcing the already well-established trend, of loss of NO_3^-, induced by cutting alone. This action is probably effected through herbicidal destruction of the remaining vegetation – herbaceous plants and root sprouts. In the event of rapid transformation of all nitrogen in the Bromacil, this source could at best contribute only 5 percent of the nitrogen lost as nitrate.

During 1966 the cutover area showed a net loss of N of 52.8 kg/hectare, compared to a net gain of 4.5 kg/hectare for the undisturbed system (table 13.1). If one assumes that the cutover system would have normally gained 4.5 kg/hectare, the adjusted net loss from the cutover system is about 57 kg/hectare. The annual turnover of nitrogen in undisturbed systems is approximately 60 kg/hectare on the basis of an equilibrium system in which annual leaf fall is about 3200 kg/hectare (11) and annual losses of roots are about 800 kg/hectare. Consequently an amount of elemental nitrogen, equivalent to the annual turnover, was lost during the first year following cutting.

Nitrogen losses from W-2 do not take into account volatilization, which accounted for about 12 percent of the total losses from 106 uncropped soils (9). Moreover, denitrification, an anaerobic process, requires a nitrate substrate generated aerobically (12); consequently, for substantial denitrification to occur in fields, aerobic and anaerobic conditions must exist in close proximity. The large increases in subsurface flow of water from the cutover watershed suggests that such conditions may have been more common throughout the watershed.

A high level of nitrate ion in the soil solution implies a corresponding concentration of cations and ready leaching (10); precisely this situation prevailed in W-2. Simultaneously with the rise of nitrate, concentrations of Ca^{++}, Mg^{++}, Na^+, and K^+ rose ultimately severalfold. These increases, in combination with the increase in drainage water, led to net losses 9, 8, 3, and 20 times greater, respectively, than similar losses from five undisturbed ecosystems between June 1966 and June 1967. Concentrations of Al^{+++} rose about 1 month later than the initial rise in nitrate, while sulfate showed a sharp drop in concentration, coincident with the rise in nitrate (figure 13.1).

These results indicate that this ecosystem has limited capacity to retain nutrients when the bulk of the vegetation is removed. The accelerated rate of loss of nutrients is related to the cessation of uptake of nutrients by plants and to the larger quantities of drainage water passing through the system. Accelerated losses may also relate to increased mineralization resulting from changes in the physical environment, such as change in temperature or increase in available substrate.

However, the effect of the vegetation on the process of nitrification cannot be overlooked. In the cutover ecosystem the increased loss of cations correlates with the increased loss of nitrate; consequently, if the intact vegetation inhibits the process of nitrification (13) and if removal of the vegetation promotes nitrification, release from inhibition may account for major losses of nutrients from the cutover ecosystem.

These results suggest several conclusions important for environmental management:

(1) Clear-cutting tends to deplete the nutrients of a forest ecosystem by (i) reducing transpiration and so increasing the amount of water passing through the system; (ii) simultaneously reducing root surfaces able to remove nutrients from the leaching waters; (iii) removal of nutrients in forest products; (iv) adding to the organic substrate available for immediate mineralization; and (v), in some instances, producing a microclimate more favorable to rapid mineralization. These effects may be important to other types of forest harvesting, depending on the proportion of the forest cut and removed. Loss of nutrients may be greatly accelerated in cutover forests where the soil microbiology leads to an increase of dissolved nitrate in leaching waters (10).

(2) Management of forest ecosystems can significantly contribute to eutrophication of stream water. Nitrate concentrations in the small stream from the cutover ecosystem have exceeded established pollution levels (10 parts per million) (14) for more than 1 year, and algal blooms have appeared during the summer.

Acknowledgments

Supported by NSF grants GB 1144, GB 4169, GB 6757, and GB 6742. We thank Noye Johnson, David Pramer, William Smith, and Garth Voigt for suggestions and for reviewing the manuscript. Contribution 4 of the Hubbard Brook Ecosystem Study; a contribution to the U.S. program for the International Hydrological Decade, and to the International Biological Program.

References and Notes

1   W. H. Carmean, *Science* 153, 695 (1966); E. S. Deevey, *ibid.* 154, 68 (1966).
2   F. H. Bormann and G. E. Likens, *ibid.* 155, 424 (1967).
3   Hubbard Brook Experimental Forest is part of the Northeastern Forest Experiment Station.
4   G. E. Likens, F. H. Bormann, N. M. Johnson, R. S. Pierce, *Ecology* 48, 772 (1967).
5   W. E. Sopper and H. W. Lull, *Water Resources Res.* 1, 115 (1965).
6   F. H. T. Juang and N. M. Johnson, *J. Geophys. Res.* 72, 5641 (1967).
7   D. Fisher, A. Gambell, G. E. Likens, F. H. Bormann, in preparation.
8   H. J. Thome, personal communication, 1967.
9   F. E. Allison, *Advan. Agron.* 3, 213 (1955).
10   R. H. Nye and D. J. Greenland, *Tech. Commun. 51* (Commonwealth Bureau of Soils, Harpenden, England, 1960).
11   G. Hart, R. E. Leonard, R. S. Pierce, *Forest Res. Note 131* (Northeast Forest Exp. Sta., Upper Darby, Pa., 1962).
12   S. L. Jansson, *Kgl. Lantbrukshogskol. Ann.* 24, 101 (1958).
13   E. L. Rice, *Ecology* 45, 824 (1964).
14   F. H. Rainwater and L. L. Thatcher, *U.S. Geol. Surv. Water-Supply Paper 1454* (U.S. Government Printing Office, Washington, D.C., 1960).

14

EFFECTS OF CONSTRUCTION ON FLUVIAL SEDIMENT, URBAN AND SUBURBAN AREAS OF MARYLAND*

M. G. Wolman and A. P. Schick

Sediment Yield, Transport, and Deposition

The type area

The principal area of study included metropolitan Baltimore and an area considered part of metropolitan Washington in the State of Maryland. Both cities are rapidly expanding urban centers of about 2,000,000 inhabitants each in their metropolitan areas (figure 14.1). Physiographically the areas under construction are Coastal Plain or Piedmont, with slopes generally 1 to 10%, but sometimes 20% or more. Soil is very deep nearly everywhere, and bedrock is encountered in the construction process in only a few localities.

Mean annual precipitation in the type area is about 42 inches (1100 mm). This amount is nearly evenly distributed throughout the twelve months of the year. Intensities are, however, much higher in the summer. A rainfall of 2 inches per hour recurs once in about ten years. Some of the winter

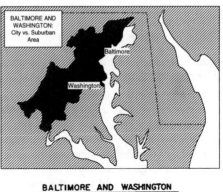

BALTIMORE AND WASHINGTON:
City vs. Suburban Area

Baltimore

Washington

BALTIMORE AND WASHINGTON
City vs. Suburban Area

Figure 14.1 [*orig. figure 1*] The Metropolitan areas of Baltimore and Washington.

* Originally published in *Water Resources Research*, 1967, vol. 3, pp. 451–64.

Table 14.1 [*orig. table 1*]   Sediment yield from rural drainage areas.

Stream and Location	Drainage Area, sq mi	Sediment Yield, tons/mi²/yr	Condition
Helton Branch nr. Somerset, Ky.	0.85	15	wooded [Collier et al., 1962]
Watts Branch, Rockville, Md.	3.7	516	rural [Wark and Keller, 1963]
Northwest Branch Anacostia River nr Colesville, Md.	21.3	470	rural [Wark and Keller, 1963]
Georges Creek at Franklin, Md.	72.4	207	rural, wooded
Gunpowder Falls, Prettyboy Reservoir, Hereford, Md.	80	913	rural; 1933–1943 [Holeman, 1965]
Gunpowder Falls, Prettyboy Reservoir, Hereford, Md.	80	500	rural; 1943–1961 [Holeman, 1965]
Seneca Creek, Dawsonville, Md.	101	320	rural [Wark and Keller, 1963]
Gunpowder Falls, Loch Raven Dam, Towson, Md.	300	808	rural; 1914–1943 [Holeman, 1965]
Same	300	233	rural; 1943–1961 [Holeman, 1965]
Monocacy River, Frederick, Md.	817	327	rural [Wark and Keller, 1963]

precipitation is snow, but snow cover rarely persists for more than a fortnight.

Sediment production and yield from areas under construction

Nearly always there is a variable 'background' quantity of sediment provided to all or to virtually all streams. Therefore, in attacking the problem of sediment pollution, one must first establish the level of this background at any given time and place if the possible problems posed by sedimentation and their amelioration are to be considered in a realistic framework.

A representative selection of large and small drainage basins in Maryland or in comparable areas which are either rural or wooded (table 14.1) indicates that the average sediment yield is on the order of 200 to 500 t/mi²/yr. The data in table 14.1 are derived from sampling at stream gaging stations along the rivers and are not estimates of sediment yield based on the distribution of land use in the watershed. Figures as low as 15 t/mi²/yr were recorded for a small stream in the Appalachians in Kentucky, in a heavily wooded area comparable to Western Maryland. Piedmont areas that are being farmed produce sediment yields on the order of 500 t/mi²/yr, as illustrated by Watts Branch at Rockville, Maryland.

Although all figures in table 14.1 are for rural or wooded watersheds, the figures by periods for Gunpowder Falls at Loch Raven Dam and at Prettyboy Reservoir are particularly significant. For the period 1914–1943, the sediment yield to Loch Raven was 800 t/mi²/yr. A later survey, however, indicated that from 1943 to 1961 the yield had dropped to one quarter of its previous value, to 233 t/mi²/yr. Similarly, during 1933–1943 the rate of sedimentation in Prettyboy Reservoir was 913 t/mi²/yr, whereas between 1943 and 1961 the rate had decreased to 500 t/mi²/yr.

The earlier sedimentation rates, particularly in Loch Raven from 1914 to 1943, appear to reflect the intensive farming activity in the state, particularly in the Piedmont area, during the period 1880 to approximately 1910. Between 1880 and 1900, the acreage in farms in Maryland reached its maximum. Since then there has been a

Table 14.2 [*orig. table 2*]   Sediment yield from selected drainage basins: Maryland and other areas.

Reference Number	Stream and Location	Drainage Area, sq.mi.	Sediment Yield, tons/mi³/yr	Condition
1	Johns Hopkins University Baltimore, Md.	0.0025	140,000	construction site
2	Tributary, Minebank Run Towson, Md.	0.031	80,000	commercial
3	Tributary, Kensington, Md.	0.091	24,000	housing subdivision [Guy, 1963]
4	Tributary, Gwynne Falls, Md.	0.094	11,300	housing (yield computed from small stilling basin, probably low trap efficiency)
5	Oregon Branch, Cockeysville, Md.	0.236	72,000	industrial park
6	Cane Branch, near Somerset, Ky.	0.67	1,147	strip mine [Collier et al., 1962]
7	Greenbelt Reservoir, Greenbelt, Md.	0.83	5,600	housing [Guy and Ferguson, 1962]
8	Little Falls Branch, Bethesda, Md.	4.1	2,320	urban & development (includes urban area as well as area undergoing development). [Wark and Keller, 1963]
9	Lake Barcroft, near Fairfax, Va.	9.5	32,500	housing subdivision (for maximum year). [Holeman and Geiger, 1959]
10	Northwest Branch Anacostia River near Hyattsville, Md.	49.4	1,850	urban & development (includes urban area as well as area undergoing development). [Wark and Keller, 1963; Keller, 1962]
11	Rock Creek, Sherrill Drive Washington, D.C.	62.2	1,600	urban & development (includes urban area as well as area undergoing development). [Wark and Keller, 1963; Keller, 1962]
12	Northeast Branch Anacostia near Riverdale, Md.	72.8	1,060	urban & development (includes urban area as well as area undergoing development). [Wark and Keller, 1963; Keller, 1962]

steady decline in the total area in farms in the state. The decline in cultivated area on farms is particularly rapid in a representative area peripheral to a metropolitan center such as Baltimore County where, after 1920, the area of cultivated land declined continuously at a rate of about 1200 acres per year.

From the sediment yield data and from

the analysis of land area in farms, it is clear that throughout much of the state sediment yield from agricultural and wooded lands is relatively low. In addition, the difference between the present conditions and those in 1900 is perhaps most striking in areas adjacent to metropolitan regions. Here a good deal of land is no longer farmed and is instead growing up in brush and woodland until it is bought for subdivision and housing development. Hence, the sediment yield from such areas is perhaps at its lowest point during recent historic time.

Against this background of sediment yields ranging from 200 to 500 tons per square mile per annum, we may compare yields from drainage basins undergoing development through subdivision and highway construction. The figures in table 14.2 are based upon measurements in stream channels and, like those for the rural area in table 14.1, are not estimates of yield based on rainfall, soil, and topographic characteristics. Data are quite limited, and, as of the date of compilation, table 14.2 provided the largest summary of available information known to the authors.

Data for locations 1, 2 and 5 in table 14.2

are derived from the present study. Locations 1 and 5 represent periodic samplings at particular construction sites. The data for the tributary of Minebank Run are based upon a survey of deposition in a large alluvial fan downstream from a construction site.

For the small streams sampled in this study it was necessary to develop flow frequency curves by correlation with nearby gaging stations or from rainfall records and, from a relation between flow and measured sediment concentrations, to compute the annual sediment yield. This was done at Oregon Branch and at The Johns Hopkins University site.

A glance at table 14.2 shows that sediment yields from areas undergoing construction range from one thousand to roughly 100,000 t/mi²/yr. The highest figures are derived from the smallest unit areas, and in some cases the actual area under construction is exceedingly small. Thus, the yields extrapolated to a unit square mile may appear unusually large. In examples 1 and 2 (table 14.2) the contributing area is the total drainage area given. In all of the others, the area under construction is considerably less

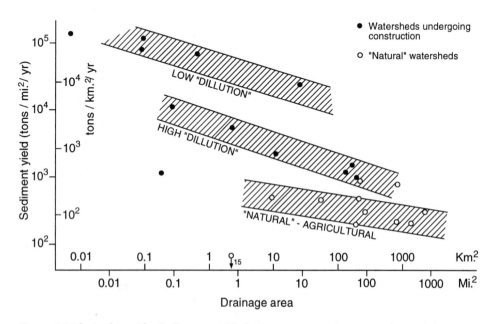

Figure 14.2 [*orig. figure 2*] Sediment yield, drainage area, and construction activity.

than the drainage area sampled. At Kensington (item 3, table 14.2), for example, Guy and Ferguson [1962] note that the disturbed area at any one time was roughly 2 to 10 acres of the total drainage area of 58 acres (0.091 mi^2). The figures in table 14.2 suggest that the sediment yield expressed in weight per unit area per year declines with increasing drainage area. It is important to note, however, that for the larger drainage area the entire area is not at any one time undergoing construction, whereas for the smaller areas most of the land is in fact under construction. Thus the declining unit yield is a dilution factor in the case of the larger drainage areas in table 14.2.

The relationship between sediment yield, drainage area, and construction activity is expressed in figure 14.2, which incorporates the data of tables 14.1 and 14.2. The stratification according to 'dilution' is a rough separation based on the percentage of the area under construction. Figure 14.2 suggests that the ratio of sediment yielded by watersheds undergoing construction to that contributed by 'natural' agricultural watersheds may increase with a decrease in drainage area.

In contrast to runoff from rural areas, individual measurements of sediment concentration in streams often show concentrations as high as 60,000 ppm, whereas those for rural areas rarely, even in extreme events, have concentrations greater than one thousand ppm. Comparison of concentrations measured at tributary junctions where streams draining source areas under construction join streams from undisturbed watersheds confirms both the source and an expected 10- to 20-fold difference in concentration, as illustrated by Towson Run north of Baltimore, where concentration in the main channel upstream from the junction was 1500 ppm, whereas downstream concentration was 16,000 ppm as a consequence of tributary inflow with a concentration of 20,000 ppm. In Oregon Branch near Cockeysville, Maryland (table 14.2, reported in detail in Wolman [1964]), concentrations below a 150-acre industrial park reached 30,000 ppm, whereas those upstream did not exceed 1500 ppm. Curves

relating sediment concentration and discharge indicate that on the average at high flows concentrations downstream are about five times those measured above the construction site. Keller [1962, p. C130] reports for the Anacostia River in the vicinity of Hyattsville, Maryland, roughly a 6-fold increase in sediment yield for the drainage areas undergoing construction on the Northwest Branch of the Anacostia River between Colesville and Hyattsville, Maryland.

The data on annual yields for small areas are subject to sampling errors and to errors resulting from the shortness of the record. It is important to recognize, however, that, because of the very high sediment yields measured, even a large percentage reduction in the reported values in no way reduces either the significance of the results or the conclusions one draws from them. Thus it is clear from a comparison of tables 14.1 and 14.2 that the quantity of sediment derived from areas undergoing construction is from 2 to 200 times as large as that derived from comparable areas in a rural or wooded condition.

Although the sediment yields given in table 14.2 are for areas subject to both subdivision and highway construction, some data are available on the yields from highway slopes or roadcuts alone. Diseker and Richardson [1962] found that soil loss, expressed in tons per acre per year, was a function of inclination of the slope, rainfall, and direction of exposure. During the winter months, in one instance at least, south-facing slopes subject to more frequent freeze and thaw provided nearly three times as much sediment as the north-facing slopes. Soil loss per square mile from the roadside cuts in the Piedmont of Georgia, i.e., for roughly the same rock type as in Maryland but for slightly higher rainfall, is of the same order of magnitude as the soil loss reported in table 14.2 for the small catchments under construction. The quantity of sediment derived from roadside areas was on the order of 50,000 to 150,000 t/mi^2/yr.

Comparable figures for sediment yield are suggested by an analysis of the volume of material eroded from road cuts in the form

Table 14.3 [orig. table 3]   Bank and slope erosion in diverse areas.

Location	Site	Region and Soil Type	Years of Observation	Rate of Erosion, in./yr	Angle of of Slope, degrees	Source of Data
Cartersville, Ga.	Roadcut	Piedmont, Cecil clay	3	0.96	32	Diseker and Richardson, 1962
				0.88	37	
				0 to 1.15	17	
Oxford, Miss.	Gully headcuts	Coastal Plain silts, sands	5	7.3	90	Miller et al., 1962
Perth, Amboy, N.J.	Clay & sand fill	Coastal Plain	1*	0.92	43	Schumm, 1956
Bethany, Mo.	Fallow plot	Great Plains Shelby Loam	10	0.48	5	Smith et al., 1945
Urim, Israel	Badlands	Loess-like sand and silt	1	5.0	8	Aghassy, 1957
				5.7	20	
				6.3	34	
				5.5	48	

* Actual period 5 weeks; estimated total erosion might be as much as 5 inches.

of rills and sheet erosion. Measurements were made of the frequency of rills and the dimensions of rills and alluvial fans at the break in slope at the base of 35 road cuts in the Baltimore area. The volume of sediment in the rills represents a rough estimate of the volume of sediment removed, assuming that the divides adjacent to the rills have not been eroded. The volume applies to a strip 15 feet wide along the face of the slope. In every case the volume represented by the rills is less than the volume of material deposited in the fan at the base of the slope, indicating that sheet or surficial erosion on the divide areas also contributes to the volume of material in the fan. This result was confirmed by detailed study at several sites. The depth of erosion as shown by fans and rills was on the order of 0.1 to 0.2 feet over time intervals of generally less than one year. These rates are comparable to those observed in the road cuts in the Piedmont of Georgia and are of the same order of magnitude as those observed in several other studies of erosion on steep slopes (table 14.3).

Soil loss from roadside slopes may also be expressed in terms of sediment yield per linear mile of highway construction. The exposed area per one linear mile of a divided highway ranges from 13 acres per mile on the Eastern Shore to 26 acres per mile in Central Maryland. For two-lane highways, the range is from 9 on the Eastern Shore to 16 acres per mile in Western Maryland (table 14.4). The lower unit areas are applicable to the low relief of the Coastal Plain of the Eastern Shore, whereas the higher figures apply to the rolling topography of the Piedmont and the high relief of the Appalachian Valley and Ridge region. Richardson and Diseker [1961] imply a somewhat larger area, about 30 acres per linear mile for major interstate highways. Where the right-of-way on major highways has been designed to accommodate additional lanes in each direction, the cleared area per mile for a dual-lane highway in the Piedmont of Maryland would be increased by 12 to 15% to a value of about 30 acres per mile. For the major highway, these figures may be somewhat low, inasmuch as they do not include adjacent areas used for maneuvering equipment and for stockpiling of soil.

A computed estimate of the total sediment yield that might be expected from a mile of dual-lane highway construction in the Piedmont of Maryland based on measured sediment yields from roadcuts in the Piedmont of Georgia [Diseker and Richardson, 1962], considering separately the areas exposed in the flatter slopes in the

Table 14.4 [*orig. table 4*] Examples of areas exposed during highway construction (data provided by Maryland State Roads Commission).

Central Maryland: Frederick County	
4-Lane Divided or Dual Highway	
Area occupied by surfacing and shoulders	9.47 acres/mile
Area occupied by median, slopes and ditches	15.70 acres/mile
Area occupied by concrete gutters	0.33 acres/mile
Area exposed during construction—	25.50 acres/mile
23% of exposed area in cut	
77% of exposed area in fill	
2-Lane Highway	
Area occupied by surfacing and shoulders	5.66 acres/mile
Area occupied by slopes and ditches	7.33 acres/mile
Area exposed during construction—	12.99 acres/mile
47% of exposed area in cut	
53% of exposed area in fill	
Eastern Shore: Worcester County	
4-Lane Divided or Dual Highway	
Area occupied by surfacing and shoulders	7.63 acres/mile
Area occupied by median, slopes and ditches	5.60 acres/mile
Area exposed during construction—	13.23 acres/mile
10% of exposed area in cut	
90% of exposed area in fill	
2-Lane Highway	
Area occupied by surfacing and shoulders	4.90 acres/mile
Area occupied by slopes and ditches	3.92 acres/mile
Area exposed during construction—	8.82 acres/mile
20% of exposed area in cut	
80% of exposed area in fill	
Western Maryland: Washington County	
4-Lane Divided or Dual Highway	
Area occupied by surfacing and shoulders	6.66 acres/mile
Area occupied by median, slopes and ditches	14.00 acres/mile
Area exposed during construction—	20.66 acres/mile
55% of exposed area in cut	
45% of exposed area in fill	
2-Lane Highway	
Area occupied by surfacing and shoulders	6.18 acres/mile
Area occupied by slopes and ditches	9.68 acres/mile
Area exposed during construction—	15.86 acres/mile
36% of exposed area in cut	
64% of exposed area in fill	

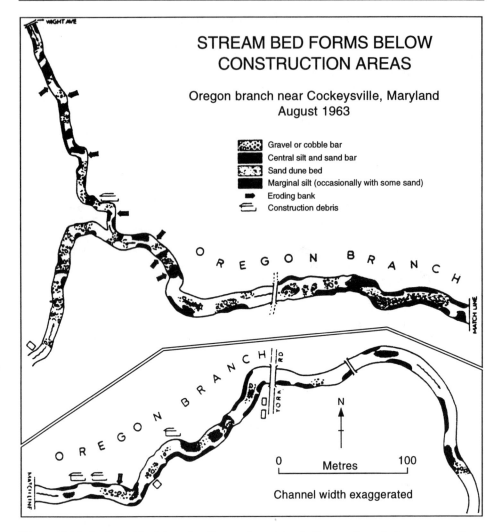

Figure 14.3 [*orig. figure 3*] Map of sediment deposition in Oregon Branch downstream from construction area.

median, surface, and shoulder of the road, and in the steeper cut and fill areas, suggests a yield of about 3000 tons per linear mile.

Expressed on an areal basis, it appears that the sediment contributed to drainage channels from areas undergoing construction either in subdivisions or in highways is on the order of from 2 times to several hundred times as great per unit area as it is from rural lands or woodland.

Sediment delivery

Not all sediment eroded from exposed surfaces actually makes its way to river courses. Roehl [1962], for example, has shown that for the southeastern United States the sediment delivery ratio ranges from 0.45 at a drainage area of 1/10 square mile to 0.1 at a drainage area of 100 square miles. At some stages in the development of subdivisions, the topography is greatly modified both by earth moving equipment and by the initial construction of cellars, sidewalks, driveways, and streets. Sediment derived from the irregular and hummocky topography associated with areas under construction may be carried short distances to the nearest depressions. These depressions may be provided behind curbs, associated with cellar walls, or adjacent to unfinished terrace surfaces. Observations on one construction site revealed that the noncontributing area, that is, basins or essentially flat surfaces, constituted approximately 40% of the exposed area. Elsewhere deposition takes place in alluvial fans where the gradient flattens abruptly; often, of course, it is aided by vegetation. In our observations, the distance from the apex of the fan to the apron or limit of deposition will be approximately equal to the height of the exposed surface.

Although there are some moderating influences on the delivery of sediment from areas exposed during construction, some of which can be useful in ameliorating sediment supplied to streams, it must be emphasized that the data provided in table 14.2 and on figure 14.3 describe conditions prevailing within streams in integrated drainage systems. Lag either in the arrival time of such sediments at downstream points in the drainage basin, or in storage within the stream channels, constitutes a change in the regimen of transport and deposition in some reaches of the channel. There will be instances in which the exposed area will not contribute directly to the network of stream channels. On the other hand, if the drainage network is integrated from the construction site to the downstream reaches, it appears that at present the unit yield figures will apply to the area as a whole. Like the estimation of the effect of sediment delivery, reduction of total yields due to limited time of exposure and to estimated coincidence of rainfall periods and construction also continues to require an educated guess.

Physical and Biological Effects of Sediment on Streams

Imposition of large quantities of sediment on streams previously carrying relatively small quantities of primarily suspended materials produces a variety of changes in the physical and biological characteristics of a stream channel. These changes include deposition of channel bars, coarsening of suspended sediments in the channel, erosion of channel banks as a result of deposition within the channel, obstruction of flow and increased flooding, shifting configurations of the channel bottom, blanketing of bottom dwelling flora and fauna, alteration of the flora and fauna as a result of changes of light transmission and abrasive effects of sediment, and alteration of the species of fish as a result of changes produced in the flora and fauna upon which the fish depend. Many of these effects were observed in the Maryland area. Oregon Branch provides an illustration.

A map of the distribution of sediment derived from construction of an industrial park on Oregon Branch shows the location of five kinds of sediments in the stream channel (figure 14.3). Covering a distance of one mile downstream from the industrial park, including for comparative purposes a tributary reach of Beaverdam Run that has been free of construction activity in the recent past, the map shows both the persistent influence of the sediment to the mouth of Oregon Branch at Loch Raven Reservoir near Cockeysville, and the marked contrast between the sediment-laden Oregon Branch and the upstream portion of Beaverdam Run unaffected by construction. This upper reach of Beaverdam Run consists of a series of cobble riffles and pools. Such fine sediments as there are constitute a fine humic zone on the inside or convex portions of the normal river bends. In contrast, virtually no sections of the reach of Oregon Branch and Beaverdam Run below the industrial park are without traces of sand and silt deposits. In addition, a considerable stretch of the stream contains deposits of sand up to depths of two feet in some places (mapped as sand dune bed). In these reaches the character of the stream has been completely altered. In contrast to the pool and riffle sequence, the sand reaches resemble more the typical sand-bed channels of many rivers in the western United States. The sand moves along the bottom in dunes and completely blankets the fauna and flora previously associated with the cobble bottom.

A rough estimate of the volume of sediment in the channel from the industrial park to the backwater section of Loch Raven indicates that approximately ⅓ of the material removed from the industrial park is at present in the stream channel. The fact that the upper portion of Beaverdam Run, traversed by construction of a major interstate highway seven years ago, is now nearly free of sediment derived from construction, suggests that a channel of this slope and flow may be cleared of sediment in a period of seven years or less.

Construction derived sediment was recognized in Oregon Branch from its source to its confluence with Loch Raven Reservoir. Observations on a number of other streams in the area demonstrate, as one would expect, that 'the distance of influence' of the construction site varies with the nature, quantity, and caliber of sediment load derived from the exposed area as well as with the flow and topographic characteristics of the drainage basin. No simple correlations could be derived, nor has a 'limiting' distance been observed in this study, as sediment reached tidewater or reservoirs in every case in distances of less than 2 miles.

It has been recognized that large amounts of sediment can alter the reproductive and growth rates of fish, severely reducing the population and, in many instances, the number and composition of fish species [Cordone and Kelley, 1961; Tarzwell and Gaufin, 1953; Allanson, 1961]. Although it is likely that the transformation of a stream bed from pools and riffles to dune sand might be accompanied by a change in fish species from game fish to scavengers, as suggested by analogy with Van Deusen's [1954] stream classification for Maryland, satisfactory data are not available with which to verify such alterations in Maryland streams. Older residents report such changes, however, and observations elsewhere [U.S. Senate, 1963, p. 827] indicate that highway and railroad construction have

virtually eliminated trout from 78 miles of stream channel in central Montana, owing to sedimentation associated with channel straightening, land clearing, and construction. Fifty-four miles, 45% of the total length of the original stream channel of the Little Big Horn River, is considered lost to trout fishing as a result of construction and channel alteration. On a tributary of Clark Fork of the Columbia River, successive studies of the trout population in a reach of river altered by highway construction show a 94% reduction in both numbers and weight of large size game fish in a period of two years.

Erosion of sediment from construction sites produces high sediment concentrations as well as channel and reservoir deposits. Although deposits may alter the conveyance of stream channels as well as the flora and fauna of both channels and bodies of standing water, neither the extent nor the storage time of sedimentary deposits is predictable at present.

Magnitude of the Problem Posed by Sediment Derived from Construction Activities

Duration and amount of land exposed during construction

To determine exposure time during construction, as well as a measure of the rate at which subdivision development takes place, an analysis was made of a random sample of one hundred building permits issued in Baltimore County in the 10-year period 1952–1961. The following data were obtained from the record: date of granting of the permit, date of completion of the structure, lot size, type of permit (house, commercial, etc.), and value. Field checks indicated that construction usually began shortly after issuance of permits.

Approximately 50% of the sites were open for eight months, and 60% were open for nine. Read in reverse, the sample indicated that 40% of the sites were open for more than nine months, and 25% were open for more than one year. Reports from builders on 49 units constructed in the past five years, comprising about 1000 acres, showed a median completion time of ten months, 75% completed in less than twelve months, and two units of less than seven acres each completed in four months. However, field observations also demonstrated that many commercial, school, and industrial sites are open for periods of one, two, or more years.

In addition, data from the random sample of building permits showed that the total land area exposed to construction varied little throughout the year, a condition due primarily to the initiation of new construction in the late fall and even winter months. Although land is continuously exposed, intense summer rainfall appears to be the primary cause of rapid erosion.

From records of highway, subdivision, and utility construction, it was estimated that in any given year the total area cleared for construction in each of the four metropolitan counties in Maryland bordering Baltimore and Washington was from 1 to 2.5 square miles [Wolman, 1964]. More recent studies of the U.S. Geological Survey (J. Wark, personal communication) indicate that land exposed in the Washington area may exceed these estimates by 1.5 to 2 times.

Metropolitan Baltimore County is growing at the rate of about 13,000 people per year. Assuming a sediment yield of 10,000 tons per square mile exposed to construction per year, a low estimate in terms of data in table 14.2, sediment production is on the order of 1800 tons per 1000 increase in population. For Prince Georges County near Washington, with a somewhat larger population increase and less exposed land, the yield would be 700 tons per 1000 increase in population.

Social evaluation of problems posed by sediment

Because sediment derived from construction activities has been shown to have significant physical and biological effects on streams, reservoirs, and estuaries, an attempt was made to evaluate the social and economic significance attached to these effects. Evaluations were based upon three approaches: (1) responses to questionnaires sent to builders and public officials; (2)

reports of the Water Pollution Control Commission; and (3) estimates of costs associated with sediment removal, accumulation, or damage. The questionnaires were designed not only to determine whether or not public officials and builders deemed a problem to exist but to elicit as well some evaluation of the need for legal regulation to abate the yield of sediment from construction activities.

Among the 18 public engineers responding from a total of 23 counties, 7 indicated that a problem existed and agreed in general terms that some kind of regulation might be warranted. Others suggested that, without regulation, considerable control could be effected if information on control measures and manuals designed to fit local conditions were made available. Responses to a letter of inquiry clearly indicated that the problem was severe in the major metropolitan regions but not ubiquitous.

Of 75 builders to whom questionnaires were sent, only 14 responded, and only one indicated that erosion or sedimentation posed a severe problem for a downstream property owner. Most indicated that the problems derived from sediment were slight either to adjacent property owners, to the community, or to themselves. For the builders, problems were confined to gullying of graded land and sedimentation in drains or foundations. It is interesting to note, however, that a number of engineering firms that provide designs and supervise construction for large developments indicated an awareness of the problems posed by sediment, and that some firms regularly recommended that sediment control measures be instituted on specific projects. As a rule, these measures have been recommended to forestall complaints, perhaps suits, from downstream property owners.

Among 17 reports collected by the staff of the Maryland Water Pollution Control Commission in the period from February to December 1963, six kinds of damages were reported to result from erosion and deposition of sediment: (1) stream deposition and consequent overflow; (2) turbid water unsuited for municipal use; (3) turbid water unsuited for industrial use; (4) failure of pumping equipment; (5) clogging of drains; (6) despoiling of recreation areas. Damage to municipal and industrial water supplies was reported in several areas. Where water is used for such industrial operations as vegetable processing or cloth manufacture, even small amounts of sediment may pose considerable problems. In municipal use, where storage facilities are lacking and water is pumped directly from a river, slugs of sediment associated with periods of rainfall may pose severe problems, inasmuch as the intake of water cannot be discontinued for more than very short periods of time. Highway officials have recognized this damage, and in some cases municipalities have been compensated for the construction of additional storage facilities and for changing the location of water intakes.

The public, regulatory agencies, engineers, and the courts have recognized problems posed by sediment derived from construction. However, in choosing alternative courses of action and in evaluating the magnitude of the problem, economic measures of the damages and costs associ-

Table 14.5 [*orig. table 5*]   Alternative estimates of economics of sediment damage.

Location of Sediment	Unit Cost or Value	Method of Estimation
Reservoir	$100/acre-ft	Cost of storage of water: range $60 to $145 per acre-ft
	$0.03/yr/acre-ft	Annual cost assuming rate of depletion of storage, Liberty Reservoir, Patapsco R., 0.03%/yr sedimentation at prevailing rates.
	$1/yr/acre-ft	Annual cost assuming rate of depletion of storage, Liberty Reservoir, 1%/yr, urbanization: extremely high sediment yield 80,000 t/sq mi^2
Reservoir	$4000/acre-ft	Present value of storage per acre-ft (Liberty Reservoir)
Reservoir	$22,000 to $78,000/yr	Loss of reservoir use: alternative sources and emergency pumping, Worcester, Mass.
Reservoir	$2/yd^3	Dredging of small lake (Lake Barcroft, Va.)
	$1.25/yd^3	Dredging of small reservoir (Tollgate, Md.)
Reservoir	Estimated	Recreation: dependent upon % loss capacity, turbidity, etc., at $1.00/visitor/day
Estuary	$0.60 to $1.25/yd^3	Dredging: Baltimore Harbor ⎫ – much dependent
	$0.19	Dredging: Anacostia Area ⎭ on disposal
Channel	$0.80/yd^3	Removal of sediment: spoil placed adjacent to channel
	$1.20/yd^3	Removal of sediment: spoil removed
Channel	Value unknown	Increase in flood damage due to channel obstruction
Channel	Value unknown	Deposition of sediment during floods
Channel	Value unknown	Fish kill, substitution of less desirable species, or recreation time lost due to poor fishing
Channel	Variable	Increased costs of water treatment, $23,400/yr in treatment of 180 mgd (Washington D.C.)
Riparian Lands	Damage equivalent	Legal award for damages equal to cost of restoration (if less than diminution in value) plus value of loss of use

ated with sediment would be exceedingly useful. Because we have been unable to locate or to construct a valid relationship between incremental concentrations of sediment in streams or of sediment accumulations in reservoirs and economic measures such as the loss in income from recreation facilities or the incremental costs of water treatment, we have compiled in table 14.5 a number of estimates of the economic damages associated with erosion and deposition of sediment. It is perhaps a reflection of the state of the art that most of the figures in table 14.5 are commonly recognized costs and are not necessarily an appropriate measure of value. For these reasons only a few are commented upon here.

Although there is a good deal of disagreement as to the most appropriate economic measures to be used in evaluating damages, a present cost or an assumed annual cost for the value of storage, or a cost per unit weight for dredging, appear to represent the simplest and most easily recognized dollar amounts (costs not values) attributable to damage from sediment deposition. Assuming various rates of loss of storage due to sediment accumulation, annual costs can be computed based either on the cost of providing equivalent storage, the cost of a particular reservoir, or the present value of an existing reservoir and its appurtenances. Several of these alternatives are given in table 14.5. In addition, customary dredging costs appropriate to specific localities are listed for comparison.

Where reservoirs are located adjacent to areas undergoing rapid urban development, however, dollar values of sediment damages due to loss of storage capacity, to increased costs of water treatment, and on occasion to temporary loss of the use of the reservoir itself may be large. Thus, at Worcester, Massachusetts, loss of the use of a reservoir because of sediment from nearby airport construction resulted in expenditures of $22,000 and $78,000 in two successive years for supply and pumping from alternative sources (table 14.5, personal communication, C. B. Hardy, March 20, 1964).

Data are also available (table 14.5) that show that the cost of water treatment for both municipal and industrial uses can increase when the source becomes highly turbid. Jackson (personal communication, 1963, see Guy et al. [1963]) has calculated that for the water supply of Washington D.C., reducing turbidities in the Potomac River could produce annual savings of approximately $25,000. With increasing populations served, annual savings might increase. Simple extrapolation of the figures is difficult, however, inasmuch as the unit treatment costs may vary with scale. For the same reason, and because the turbidities of natural streams are highly variable, treatment costs and savings cannot be safely transposed from one area to another [Garin and Forster, 1940, pp. 13–14].

Lastly, lost recreational opportunities may represent another measure of the economic significance of sediment. In all the environments of deposition listed in table 14.5, sediment poses problems to the recreational use of the resource. Not only may high concentrations of sediment produce poor fishing conditions, but deposition of sediment changes the ecology of channels and of standing water bodies. Brown [1942, p. 79] estimated, for example, that 49,090 person-days were lost to recreation in the Meramac River watershed near Saint Louis, Missouri, in 1940 as a result of above normal but below floodstage flows of high turbidity. At the much disputed figure of $1.00 per recreation-day, this would represent an annual loss of $49,090. The computation, however, involves many assumptions, including the reduction in recreation due to floods and to 'above normal' flow, the value of the fishing-day, and the allocation of losses specifically attributable to sediment.

Because economic data and sophisticated analyses are limited, it is relatively easy to minimize their significance. In evaluating the problems posed by sediment, then, it is unwise to measure the degree of public interest and therefore the values associated with such problems solely on the basis of the apparent economic data. Public officials, householders, fishermen, engineers, contractors, and courts have all recognized that sediment derived from construction does create problems. Recent legal controls adopted in Montgomery County, Maryland, as well as regulations of the Maryland State Department of Water Resources, suggest that citizens not only recognize the problem but place significant value on amelioration.

Conclusion

The physical and biological effects of sediment erosion and deposition are clearly apparent even to the casual observer in the field. No one who has seen the streams, ponds, and reservoirs that have been affected by large concentrations and large quantities of sediment can have any doubt as to either the source of such material or as to its possible effects. The damages have been made most apparent where plaintiffs in court have successfully recovered costs and reimbursement for damages sustained. Less likely to be able to seek a legal remedy is the public, who will ultimately bear the cost of the destruction of fishing, of removal of material from channels and estuaries, and of loss of esthetic values in its environment.

The areas most affected by sediment derived from construction are obviously those undergoing most rapid development. It is also true that it is in these same areas that people will live most closely together, and the activities of one group of society will most closely affect the interests of all the others. It is in these regions of high population density that the maximum use and value can be derived from the adjacent water resources. The fact that these values cannot always be expressed in simple economic terms does not lessen their significance. The physical effects on the environment of large quantities of sediment have been demonstrated. Relatively simple measures can be instituted that will reduce or moderate the quantities of sediment contributed to the natural environment in metropolitan regions. These include reduction of the time of exposure, vegetative traps, and diverse types of detention structures well documented in the agricultural conservation literature [Guy et al., 1963]. Should public

policy call for some action to ameliorate the quantity of sediment derived from construction activities, as we believe it should, it should also be recognized that we must deal in the realm of the reasonable and possible and not with extremes. It is unrealistic to expect that all of the sediment derived from construction can at reasonable costs be prevented from reaching the stream channels. It is equally unreasonable to demand that an effort be made to achieve such an objective. It is not unreasonable, however, to suggest that modest controls can be effective in moderating the quantities of sediment now being contributed to water bodies in areas undergoing major urban development.

Acknowledgments
The authors are indebted to the Maryland Water Pollution Control Commission, Annapolis, Maryland, to whom an original report [Wolman, 1964] upon which this paper is based was presented. G. B. Anderson of the Soil Conservation Service, Fairfax County, Virginia, and Henry Silbermann of the Water Pollution Control Commission provided continuing help and information during the study. Essential data were compiled and made available by the Maryland State Roads Commission. Both the Homebuilders Association of Maryland and the Associated Builders and Contractors provided valuable assistance. Our debt is particularly great to the 14 builders who took time to reply to an elaborate questionnaire. The U.S. Geological Survey provided basic data on sediment quantities in streams. E. R. Keil, State Conservationist, Soil Conservation Service, and his staff provided a review of problem areas in Maryland. Professor R. L. Green of the University of Maryland visited the field with the senior author. Thanks are due to the firm of Matz, Childs and Associates for providing data and technical assistance. R. C. Zimmerman, at the time a graduate student at The Isaiah Bowman Department of Geography, The Johns Hopkins University, Baltimore, Maryland, served as field assistant in summer 1963. Lastly, it is a pleasure to thank Mrs. Roger Stenerson for collecting invaluable field data.

References
Aghassy, Y., The morphology of badlands in Palestine, M. S. thesis, Dept. of Geography, The Hebrew University of Jerusalem, 1957.
Allanson, B. R., Investigations into the ecology of

polluted inland waters in the Transvaal, *Hydrobiologia*, *18*, 94 pp., 1961.

Brown, C. B., Floods and fishing, *Land Quarterly*, 78–79, 1942.

Collier, C. R., et al., Influence of strip mining on the hydrologic environment of parts of Beaver Creek basin, Kentucky, 1955–1959, *U.S. Geol. Surv.*, open-file report, 276 pp., 1962.

Cordone, A. J., and D. W. Kelley, The influence of inorganic sediment on the aquatic life of streams, *Calif. Fish and Game*, *47*, 189–228, 1961.

Diseker, E. G., and E. C. Richardson, Roadside sediment production and control, *Am. Soc. Agr. Engrs.*, *4*, 62–68, 1961.

Diseker, E. G., and E. C. Richardson, Erosion rates and control measures on highway cuts, *Am. Soc. Agr. Engrs.*, *5*, 153–155, 1962.

Garin, A. N., and G. W. Forster, Effect of soil erosion on the costs of public water supply in the North Carolina Piedmont, *U.S. Dept. Agr.*, SCS-EC-1, 106 pp., 1940.

Guy, H. P., Residential construction and sedimentation at Kensington, Maryland, paper presented at Federal Inter-Agency Sedimentation Conference, Jackson, Miss., Jan., 1963, 16 pp., 1963.

Guy, H. P., and G. E. Ferguson, Sediment in small reservoirs due to urbanization, *Am. Soc. Civ. Engrs., Proc., J. Hydraul. Div.*, *88*, 27–37, 1962.

Guy, H. P., N. E. Jackson, K. Jarvis, C. J. Johnson, C. R. Miller, and W. W. Steiner, A program for sediment control in the Washington Metropolitan region, *Interstate Comm. Potomac River Basin, Tech. Bull. 1963–1*, 48 pp., 1963.

Holeman, J. N., and A. F. Geiger, Sedimentation of Lake Barcroft, Fairfax County, Va., *U.S. Dept. Agr. SCS-TP-136*, 12 pp., 1959.

Holeman, J. N., and A. F. Geiger, Sedimentation of Loch Raven and Prettyboy Reservoirs, Baltimore County, Md., *U.S. Dept. Agr. SCS-TP-145*, 17 pp., 1965.

Keller, F. J., Effect of urban growth on sediment discharge, Northwest Branch Anacostia River basin, Maryland, *U.S. Geol. Surv. Prof. Paper 450-C*, pp. C129–131, 1962.

Miller, C. R., R. Woodburn, and H. R. Turner, Upland gully sediment production, *Int. Assoc. Scient. Hydrol. Publ. 59*, Symposium of Bari, pp. 83–104, 1962.

Richardson, E. C., and E. G. Diseker, Roadside mulches, *Crops and Soils*, *13*(5), 1 pp., 1961.

Roehl, J. W., Sediment source areas, delivery ratios, and influencing morphological factors, *Int. Assoc. Scient. Hydrol., Publ. 59*, Symposium of Bari, pp. 202–213, 1962.

Schumm, S. A., Evolution of drainage systems and slopes in badlands at Perth Amboy, N.J., *Geol. Soc. Am. Bull.*, *3*, 600–615, 1956.

Smith, D. D., D. M. Whitt, A. W. Zingg, A. G. McCall, and F. G. Bell, Investigations in erosion control and reclamation of eroded Shelby and related soils at the conservation experiment station, Bethany, Mo., 1930–1942, *U.S. Dept. Agr. Tech. Bull.*, *883*, 175 pp., 1945.

State of Maryland, State Roads Commission, Specifications for materials, highways, bridges, and incidental structures, 507 pp., 1962.

Tarzwell, C. M., and A. R. Gaufin, Some important biological effects of pollution often disregarded in stream surveys, *Purdue Univ. Eng. Bull., Proc. 8th Industrial Waste Conf.*, 33 pp., 1953.

U.S. Senate, 88th Congress, *Congressional Record*, statement by E. B. Welch, Destruction of natural fish habitat is ruining Montana's fishing streams, entered in record by Sen. Metcalf (Mont.), pp. 826–841, 1963.

Van Deusen, R. D., Maryland freshwater stream classification by watershed, *Chesapeake Biol. Lab. Contr. 106*, 30 pp., 1954.

Wark, J. W., and F. J. Keller, Preliminary study of sediment sources and transport in the Potomac River basin, *Interstate Comm., Potomac River Basin, Tech. Bull. 1963–11*, 28 pp., 1963.

Wilson, J. N., Effects of turbidity and silt on aquatic life, *U.S. Public Health Serv. Seminar*, 235–239, 1962.

Wolman, M. G., Problems posed by sediment derived from construction activities in Maryland, Rept. to Md. Water Pollution Control Comm. (now Dept. of Water Resources), 125 pp., 1964.

Supplementary References

Diseker, E. G., and E. C. Richardson, Roadside sediment production and control, *Am. Soc. Agr. Engrs.*, *4*, 62–68, 1961.

State of Maryland, State Roads Commission, *Specifications for Materials, Highways, Bridges, and Incidental Structures*, 507 pp., 1962.

Wilson, J. N., Effects of turbidity and silt on aquatic life *U. S. Public Health Serv. Seminar*, 235–239, 1962.

15

RAIN, ROADS, ROOFS AND RUNOFF: HYDROLOGY IN CITIES*

G. E. Hollis

Urban Hydrology

Defined by UNESCO as the interdisciplinary science of water and its interrelationship with urban people, urban hydrology is a broad canvas (Berthelot and Lindh, 1979). The field spans the modelling of storm sewer systems (Standing Technical Committee on Sewers and Water Mains, 1981) to the erosion of river channels by floods enlarged by urbanisation (Park, 1977); ecological approaches to urban systems (Whyte, 1985) to the pollution of groundwater by waste tips and spillages (Mather and Parker, 1979); and the increased frequency of thunder rainfall over cities (Yperlaan, 1977) to "societal development . . . and water utilization" (Lindh, 1983).

Urbanisation has a profound effect on the functioning of the natural hydrological cycle where it interrupts and rearranges the storages and pathways of water. Urbanisation introduces new transfers of water around the urban area, alters water storages well outside the city limits and can involve efforts to offset some of the adverse and inadvertent effects of the land use change.

From the perspective of scientific hydrology, Hall (1984) has provided a clear system of linkages between urbanisation

and the three key problems of water resources, pollution control and flooding (figure 15.1). In this paper the wide field of urban hydrology converges rapidly onto the hydrological effects of a change to urban land use and the resulting environmental impacts. From this the focus narrows to the modelling of runoff from urban impermeable surfaces. A research project measuring the proportion of rain running off individual roofs and pieces of road provides a very specific, but surprising, set of results. The paper concludes with some low budget ideas in response to the question, "Can we do any project work on these subjects within school?"

The Hydrology of Urbanisation

The imposition of urban impermeable surfaces and their related activities onto a landscape has a series of separable elements each of which has consequences for the functioning of the hydrological cycle.

First, *the replacement of vegetated soils with impermeable surfaces* typically covers 20 per cent of post-war urban areas, up to 90 per cent in sub-catchments in city centres, but under 5 per cent with suburban detached housing. This replacement:

* Originally published in *Geography*, 1988, vol. 73, pp. 9–18.

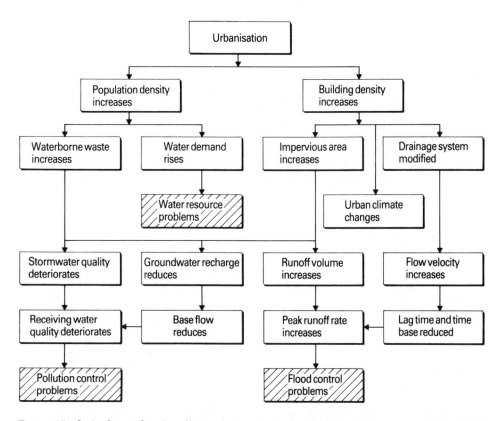

Figure 15.1 [*orig. figure 1*]   The effects of urbanisation on hydrological processes (after Hall, 1984).

(i)   reduces storage on the surface and in the soil and so increases the percentage of rain that runs off;

(ii)   increases the velocity of overland flow;

(iii)   decreases evapotranspiration because urban surfaces are usually dry; and

(iv)   reduces percolation to groundwater because the surface is impermeable.

Second, *augmentation of the channel network by storm water sewers* occurs once every downpipe and road gully is served by a drain where a limited channel network used to operate. This increase in drainage density:

(i)   reduces the distance that overland flow has to travel before reaching a channel;

(ii)   increases the velocity of flow because sewers are smoother than natural channels; and

(iii)   reduces storage in the channel system because sewers are designed to drain completely as soon as possible.

Third, *building activity* during the construction of houses, roads and bridges:

(i)   initially clears vegetation which exposes the soil to the elements and also facilitates overland flow;

(ii)   disturbs and churns the soil which increases its erodibility;

(iii)   regrades slopes within the urban landscape and usually leaves these slopes with a disturbed soil profile and limited vegetation cover; and

(iv)   finally, protects the soil with an armour coating of concrete, asphalt or tiles etc. which effectively stops any further erosion, except for the weathering and transport of the urban materials themselves.

Fourth, *encroachment on the river channel* by embankments, reclamation, and river-side roads:

(i) usually reduces the channel width which inevitably increases the height of floods in the restricted channel; and

(ii) bridges etc. in the river can restrict the free discharge of floods and so increase the inundation level upstream.

Fifth, *the rainfall climatology of urban areas* is affected by the greater aerodynamic roughness of buildings compared to rural landscapes, the promotion of convection by the urban heat island and the burning of fossil fuels, and the profusion of condensation nuclei in and down wind of the urban area. These changes often cause:

(i) more rainfall especially in the summer; and

(ii) heavier and more frequent convective rainstorms and thunder.

The combined effects of these changes is that the flow regime, the flood hydrology, the sediment balance and the pollution load of streams are radically altered. The following examples illustrate these alterations.

A 530 per cent increase in paved area on Long Island, New York has led to a 270 per cent increase in total runoff (Seaburn, 1969). Annual runoff from the city centre of Kursk in the USSR is 109mm and 45mm drains from suburban catchments. In the surrounding ploughlands the runoff varies between 7 and 14mm depending upon the cropping whilst surface runoff from virgin steppelands is 0mm (Lvovich and Chernishov, 1977). At Harlow in Essex, the paving of 15 per cent of the clay catchment increased total runoff by almost 60mm and made it 130 per cent of that of surrounding rural areas (Hollis, 1974).

No study has unequivocally related a fall in groundwater levels to reduced percolation through urban impermeable surfaces because of the presence of groundwater abstraction through wells and enhanced recharge of groundwater through soakaway drainage. Indeed, Thomson and Foster (1986) showed that for the limestone aquifer of Bermuda, recharge of groundwater is highest (740mm out of precipitation of 1460mm) in high density residential areas with compulsory soakaways. Grassland and woodland have a recharge of only 365mm because evapotranspiration returns 75 per cent of the rainfall to the atmosphere.

Urbanisation has been shown to increase the peak of the mean annual flood in virtually all the studies published; for example, a 243 per cent increase resulted from the construction of Stevenage and an 85 per cent rise followed the building of Skelmersdale (Knight, 1979). Likewise, the peak of the unit hydrograph (typically the hydrograph resulting from 25mm of effective rainfall in one hour) grew threefold and the lag time declined by 40 per cent following the paving of an extra 6 per cent of the basin of the Silk Stream in north London (Hall, 1977). However, Hall found virtually no change in the flood hydrographs of the neighbouring Dollis Brook despite considerable building because the new developments were largely infilling or redevelopment and the runoff was partially throttled by the existing storm sewer network.

However it is fallacious to believe that urbanisation increases the magnitude and frequency of all floods. It has been argued that after continued heavy rainfall there is no hydrological difference between asphalt and saturated soil, that urban sewer systems serve to throttle flows in excess of their surcharged capacity and that surface flooding is more extensive and deeper in the 'rough and irregular' urban morphology of kerbs, walls and hollows than equivalent rural fields. Consequently, it is now widely accepted that beyond a certain threshold, which is hard to determine, the land use has little effect on flood magnitudes (figure 15.2; Packman et al., 1976).

Climatologists have been increasingly successful in demonstrating the urban effect on precipitation since Ashworth (1929) found that, during the early part of this century, Rochdale had significantly less rainfall on Sundays when the mills and factories were not producing smoke (condensation nuclei). The vast resources devoted to the Metropolitan Meteorological Experiment (METROMEX) at St. Louis in the

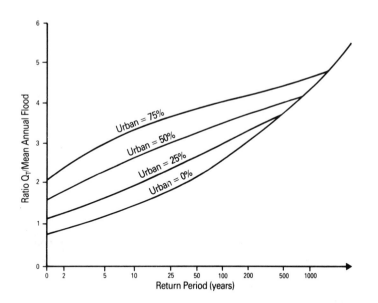

Figure 15.2 [*orig. figure 2*]   The effect of urbanisation on floods of increasing magnitude. The vertical axis represents the ratio of the flood to the rural mean annual flood (after Packman et al., 1976).

USA have advanced all aspects of urban climatology (Lee, 1984). An urban-induced rainfall anomaly of 30–35 per cent was observed each summer some kilometres downwind of the city. The enhanced precipitation came mainly from cold fronts and squall lines; radar analyses of convective cells showed enlargement and coalescence especially over the urban area. The city appeared to have its major impacts between 2 and 5pm when solar heating was at a maximum and between 9pm and midnight when the urban heat island was well developed (Changnon, 1978). Similarly, the localised thunderstorm that brought 170mm of precipitation to Hampstead, London, in August 1975 was affected to a significant extent by the urban heat island (Atkinson, 1977).

A 5 to 10 fold increase in sediment load and a 30 per cent increase in solute runoff during building activity in 25 per cent of a suburban basin in Exeter has been reported (Walling, 1979). The interlinkage of human activity, sediment load and stream channel stability was clearly demonstrated by Wolman's (1967) cycle of erosion and sedimentation in urban river channels (figure 15.3). The enlargement of river channels downstream of urban areas as a result of enlarged floods and reduced sediment discharges after the completion of the building work have been demonstrated in the UK below Stevenage, Skelmersdale and Woodbury, Devon (Knight, 1979; Park, 1977).

Urban storm water passes directly to open watercourses via the storm sewers. Foul water passes to the sewage works via separate foul sewers or old fashioned combined sewers. However, the storm water that washes off the roads and roofs of urban areas is not clean and unpolluted. During the last ten years it has been appreciated that, at least during the 'first flush' of urban runoff, the quality of the water can be worse than foul sewage and it can carry seriously polluting discharges of heavy metals, volatile solids and organic chemicals. For example, Ellis (1976) has presented records of the pollutant discharges associated with flood flows in the Silk Stream. Hall and Ellis

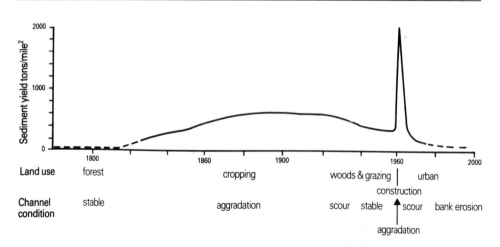

Figure 15.3 [*orig. figure 3*]   Land use, sediment yield and channel response in the Piedmont region near Washington DC (after Wolman, 1967).

(1985) have synthesised the results of a wide range of studies to show that 20–40 per cent of stormwater sediments are organic in origin and most are biodegradable. They found that highway runoff had 5 to 6 times the concentration of heavy metals of roof runoff and Ellis (1985) has shown that the Silk Stream is severely polluted with faecal coliform bacteria. A Transport and Road Research Laboratory study (Colwill et al., 1984) found that the annual runoff from 1 km of a single carriageway of the M1 included 1.5 tons of suspended sediment, 4kg of lead, 126kg of oil and 18g of hazardous polynuclear aromatic hydrocarbons.

Modelling the Process of Urban Runoff

There has been a progressive transformation of models of urban runoff. They have developed from simple, whole catchment, whole storm models to complex representations of individual roads, roofs and subcatchments with time steps of around 1 minute. A recent meeting of engineers and scientists even discussed the development of a further model for the real-time, continuous simulation of runoff quantity and quality for stormwater drains, combined sewers and the receiving watercourse.

A historic breakthrough in modelling urban runoff occurred when Watkins (1962)

published the computer-based Transport and Road Research Laboratory (TRRL) method. This simulated the above-ground hydrological process of overland flow and the below-ground hydraulic process of pipeflow. It set a trend continued by virtually all subsequent models, that each section of pipe and each impermeable surface should be represented as realistically as possible.

The method was calibrated against a series of instrumented catchments where the outflow from a surface water sewer system was monitored. The procedure gave good results but subsequent studies have shown that the simulation was not in accord with the actual processes operating. The model assumed 100 per cent runoff from paved surfaces, 0 per cent runoff from permeable surfaces draining to the sewer system and it assumed that all of the rainfall on the paved surfaces entered the sewer system 2 minutes after reaching the ground. There was, therefore, no above-ground attenuation of the peak flows and all of the attenuation that was observed in the outflow hydrographs was ascribed to below-ground, pipe-flow effects.

The next breakthrough was brought about by a series of measurements of the runoff from urban surfaces (Kidd, 1978). These studies measured runoff as it entered the sewer system. They showed that there was considerable attenuation of peaks of intense rainfall during overland flow over roads and that runoff was very rarely 100 per cent of rainfall. The British studies of urban surface runoff were combined with the work of the, then, Hydraulics Research Station into the Wallingford Storm Sewer Procedure (WSSP) for the design, evaluation and optimisation of storm sewer systems (Standing Technical Committee on Sewers and Water Mains, 1981). Equations control the estimation of depression storage, percentage runoff and the parameters of the non-linear reservoir but the minimum percentage runoff that is permitted from paved surfaces in the model is 70 per cent. The whole sewer system is simulated and each inlet point to the system is considered. The variation of flow rate and flow depth in the sewer system is simulated and the surcharging of man-holes is incorporated in the model. This simulation package is now undergoing further development at Hydraulics Research Ltd at Wallingford particularly in relation to the simulation of the proportion of rainfall that runs off to the sewers (runoff coefficient). In relation to these developments in design practice, there have been a number of detailed studies of hydrological processes on very small areas of roof and road. Such a study by the author is described below.

The Runoff Coefficient in Urban Areas

Although widely assumed, it is not true that all the rainfall on urban 'impermeable' surfaces runs off. It is a comfortable assumption for those concerned with land use change in large catchments and it has proved to be a safe, but expensive, assumption when designing storm drainage systems. On all urban 'impermeable' surfaces there is initial wetting of the surface, perhaps some absorption of water by the surfaces, certainly the filling of depressions and irregularities, and throughout and after virtually all rainstorms there must be evaporation from the surface into the atmosphere. Roofs are free from the infiltration of water but when the capacity of their gutters is exceeded in a severe storm

there is an overflow to the ground. Finally, roads and paved surfaces at ground level have some infiltration through gaps in their construction, cracks that develop from uneven loading and frost activity, and the inherent permeability of some materials.

An Urban Catchment Study and a Road Irrigation Experiment

In order to understand rather better the precise processes operating on roads and roofs, the Natural Environment Research Council funded a project to examine the hydrology of roofs and roads in some detail.

Rainfall and runoff were monitored simultaneously for one year from a residential road, a car park, nine sections of road draining to individual gullies, two house roofs, two garage roofs and three types of factory roof. The sites, which included an automatic weather station, were in Redbourn, Hertfordshire, on permeable soils overlying chalk. The 2906 quality controlled 'station-storms' represented 193 rain storms and involved 57.2 per cent of the annual rainfall; 1174 storms had more than 1.4mm of rain, whilst 77 had over 10mm.

The per cent runoff averaged 11.4 per cent for roads and 56.9 per cent for roofs (28.3 per cent and 90.4 per cent for rainfalls > 5mm). Per cent runoff from the roads was cyclic with a peak during the summer months but there was a huge variation in monthly per cent runoff within and between sites. A regression analysis was undertaken to attempt to explain per cent runoff in terms of rainfall, catchment characteristics and antecedent conditions. The same analysis was undertaken, in turn, for sites, roads, and roofs. At each stage the analysis considered all storms; > 1 per cent runoff events; > 5mm rainfalls; and events with > = 4mm rain and > = 5 per cent runoff. The per cent runoff could not be explained satisfactorily with statistical methods. The most important explanatory variables for roads were short term rainfall intensity and rainfall amount; the former was the most important for roofs. 'Seasonal' variables had a positive relationship for roads which shows that the per cent runoff from roads is higher in summer than winter. The antecedent variables showed that percentage runoff from roads and roofs is increased by antecedent rainfall. Seasonal factors and evaporation were unimportant for the per cent runoff from roofs. This surprising result from the 'field' study was checked with a tightly controlled experiment.

An irrigation experiment was conducted to check the results of monitoring actual storms and to confirm that there were no instrumental errors in the hydrometric network. Six sections of the residential road (75mm bituminous macadam with granite chippings over 200mm lean mix concrete) that drained to the already instrumented individual gully pots were irrigated along the kerb and then over the whole road approximately monthly for a year. The aim was the determination of terminal infiltration losses, initial losses, percentage runoff and infiltration curves for the kerb and road surface (figure 15.4).

The results were not as expected from the literature. There was a cycle of infiltration losses at the kerb with a winter peak caused by frost action that was 3.2 times greater than the terminal loss rate at kerbs in summer. The terminal loss rate for an 'average catchment' was 6.425 litres per minute from the road surface and 14.25 l min^{-1} in summer and 46.25 l min^{-1} in winter at the kerb. Evaporation was usually more than an order of magnitude less

significant than infiltration.

The wide variation in initial losses before runoff commenced (depression storage) was inexplicable. Two sections of road behaved in the classic manner with initial losses averaging 0.8mm, two other catchments had highly variable initial losses in the range 1.2 to 8.8mm, and the last two pieces of road were even more erratic. The percentage runoff was normally under 10 per cent and never more than 60 per cent. There were no significant simple or multiple regression relationships between percentage runoff from the kerb or the whole road and irrigation amount, slope, an antecedent rainfall index and soil moisture deficit.

Infiltration curves, for kerb and road irrigation (figure 15.5), were so diverse that they do not represent the "simple impervious surfaces" envisaged at the start of the experiment and described in the literature.

Urban Hydrology in Schools

The conduct of projects and studies in urban hydrology in schools is feasible despite their limited stores of equipment, lack of sophisticated data logging devices and budgetary constraints. Indeed, as residential field classes become more expensive, the local environment and the school itself can provide for fieldwork in hydrology. Enthusiasm, ingenuity, some inexpensive equipment and a few ideas can surmount

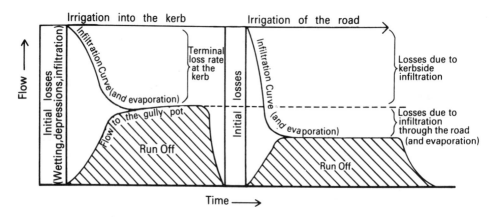

Figure 15.4 [*orig. figure 4*]   The methodology of the irrigation experiment.

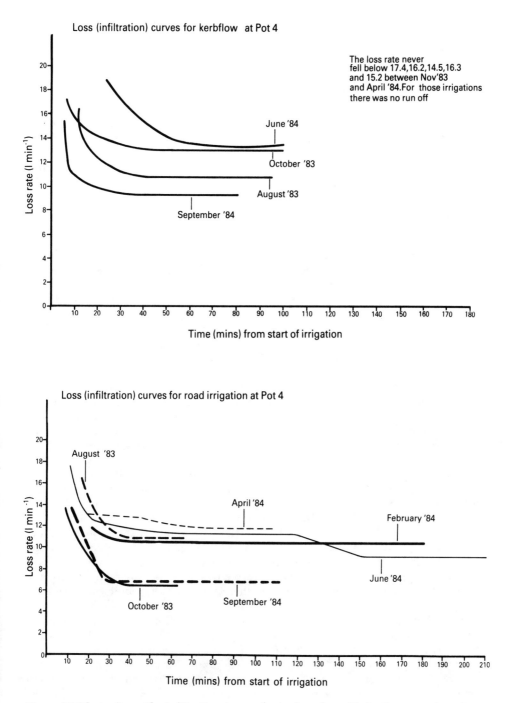

Figure 15.5 [*orig. figure 5*]   Infiltration curves for kerb and road irrigation experiments.

most problems but vandalism afflicts both school projects and research sites alike. The ensuing ideas have mostly developed from low-cost project work undertaken by undergraduates at University College London.

Runoff from a modest sized house roof or a garage can be monitored by leading the flow into a graduated water butt. The graduation can be done with a container of known volume and the butt needs to be emptied after each storm. Rainfall can be measured by using one of the published designs employing a cut-down 2 litre plastic bottle. Infiltration tests can be done on road surfaces with a bottomless oil drum sealed to the road with flexible mastic designed for sealing the edges of baths. Uunk and van de Ven (1984) have used this method for professional engineering studies in Lelystad in the Netherlands. Irrigation experiments can be done with a hose, a watch, two buckets and ingenuity. The flow from the hose can be assessed by repeated measurements of the time taken to fill a bucket and the inflow to the gully can be assessed by the time taken to fill a bucket suspended in the gully and fed with water funnelled from the edge of the grating. Where a simple current meter and level are available, it is possible to determine the frictional roughness of urban (and rural) river channels using Manning's equation:

$$Q = A V = A \frac{R^{\frac{2}{3}} S^{\frac{1}{2}}}{n}$$

where: Q is discharge, A is cross-sectional area of the wetted channel, V is velocity of flow, R is hydraulic radius found from the division of A by the wetted perimeter of the channel, and S is the slope of the water surface (often approximated by the slope of the bed). When Manning's n, the roughness coefficient, has been determined it can be used to calculate theoretical bankfull flow or perhaps its changing value can be monitored as aquatic plants grow in a channel during the summer. A geomorphological study of the effect of urbanisation on channel capacity could plot the variations of the width, depth and cross-sectional area of a stream against the catchment area draining to the measured section. The growth of channel capacity with drainage area often shows a sharp rise when sections downstream of an urban area are plotted (Park, 1977). Documentary records of flooding in the archives of local newspapers or libraries can provide a different perspective on urban hydrology. Gloucester library, for instance, has newspaper reports of floods in the city from the 1700s and London newspapers have recorded the River Brent's battle with the North Circular Road on the occasion of each flood.

The examination of water quality often demands instrumentation but liaison with colleagues in Biology or Chemistry may help. A UCL student produced a fascinating map showing poorer quality water on major roads and in the centre of London after a one week dry spell. She used an inexpensive portable dissolved oxygen meter and examined a stratified sample of six gullies near to the majority of London Underground stations. Another student examined the quantities of particulates, their grain size and chemistry on a variety of road surfaces. The sampling device was a domestic vacuum cleaner and a large supply of bags. This project followed Ellis' (1979) demonstration that street cleaning machines tend to redistribute rather than remove particulates. A third project examined the suspended load in urban rivers during floods with the aid of a bucket, rope, filter apparatus and 'Pooh sticks' amongst other things. There are a series of packs available for the assessment of pollution loadings by the determination of the invertebrate assemblages of streams in springtime. Surveys above and below sewage work outlets, and above and below storm water sewer inflows, have demonstrated the relative effects of these two sources of pollution. Students taking geography and chemistry, who may be more numerous with the advent of AS levels, could examine the chemistry of cores of sediment taken from an urban lake and a farm pond. Those with some training in geology could examine the most recent Holocene strata in the geological column that are currently being deposited by our urbanised rivers. Douglas (1985) quotes new terms and a new dating methodology in urban sedimentology; "the lower dustbinian" has no plastic interbedded in it, but "the upper dustbinian" consists of "ill-sorted quartz sand particles, plate shaped pebbles of brick, fragments of tar, and remnants of plastic bags and polythene sheets". Finally, it would be fair to assume that urban rivers are warmer than their rural counterparts and that they become warmer downstream. Such an idea was tested in a project with a simple thermometer but the design of the experiment was carefully planned with a 'rural' control catchment and some allowance for the meteorological factors affecting stream temperature.

Conclusion

The conclusion must be that urban hydrology provides a physical geography context for studies of relationships between society and the environment. It furnishes a vehicle for environmental impact inquiries, it exemplifies the use of statistical and computer methods and opportunities for field studies in urban hydrology abound. Finally and significantly, it illustrates the bonds between social, historical, planning and environmental aspects of urban areas.

References

Ashworth, J. R. (1929) "The influence of smoke and hot gases from factory chimneys on rainfall", *Quarterly Journal Royal Meteorological Society*, 55, pp. 334–350.

Atkinson, B. W. (1977) *Urban Effects on Precipitation: an Investigation of London's Influence on the Severe Storm in August 1975*, London: Queen Mary College, Department of Geography Occasional Paper 8.

Berthelot, R. M. and Lindh, G. (1979) *Socio-economic Aspects of Urban Hydrology*, Paris: UNESCO Studies and Reports in Hydrology 27.

Changnon, S. A. (1978) "Urban effects on severe local storms at St. Louis", *Journal of Applied Meteorology*, 17, pp. 578–586.

Colwill, D. M., Peters, C. J. and Perry, R. (1984) *Water Quality of Motorway Runoff*, Crowthorne: Transport and Road Research Laboratory Supplementary Report 823.

Douglas, I. (1985) "Urban sedimentology", *Progress in Physical Geography*, 9(2), pp. 255–280.

Ellis, J. B. (1976) "Sediments and water quality of urban stormwater", *Water Services*, 80, pp. 730–734.

Ellis, J. B. (1979) "The nature and sources of urban sediments and their relationship to water quality: a case study from north-west London", in Hollis, G. E. (ed.) *Man's Impact on the Hydrological Cycle in the UK*, Norwich: Geo Books, pp. 199–216.

Ellis, J. B. (1985) "Water and sediment microbiology of urban rivers and their public health implications", *Public Health Engineer*, 13(2), pp. 95–98.

Hall, M. J. (1977) "The effect of storm runoff from two catchment areas in North London", in *Effects of Urbanization and Industrialization on the Hydrological Regime and on Water Quality; Proceedings of the Amsterdam Symposium*, Paris: UNESCO, pp. 144–152.

Hall, M. J. (1984) *Urban Hydrology*, London: Elsevier.

Hall, M. J. and Ellis, J. B. (1985) "Water quality problems in urban areas", *GeoJournal*, 11(3), pp. 265–275.

Hollis, G. E. (1974) "The effects of urbanisation on floods in the Canon's Brook, Harlow, Essex", in Gregory, K. J. and Walling, D. E. (eds.) *Fluvial Processes in Instrumented Watersheds*, London: Institute of British Geographers Special Publication 6, pp. 123–139.

Kidd, C. H. R. (1978) *Rainfall-runoff Processes over Urban Surfaces: Proceedings of an International Workshop*, Wallingford: Institute of Hydrology Report 53.

Knight, C. (1979) "Urbanization and natural stream channel morphology: the case of two English new towns", in Hollis, G. E. (ed.) *Man's Impact on the Hydrological Cycle in the UK*, Norwich: Geo Books, pp. 181–198.

Lee, D. O. (1984) "Urban climates", *Progress in Physical Geography*, 8(1), pp. 1–31.

Lindh, G. (1983) *Water and the City*, Paris: UNESCO.

Lvovich, M. I. and Chernishov, E. P. (1977) "Experimental studies of changes in the water balance of an urban area", in *Effects of Urbanization and Industrialization on the Hydrological Regime and on Water Quality; Proceedings of the Amsterdam Symposium*, Paris: UNESCO, pp. 63–67.

Mather, J. D. and Parker, A. (1979) "The disposal of domestic and hazardous waste and its effect on groundwater quality", in Hollis, G. E. (ed.) *Man's Impact on the Hydrological Cycle in the UK*, Norwich: Geo Books, pp. 217–228.

Packman, J. C., Lynn, P. P., Beran, M. A., Lowing, M. J. and Kidd, C. H. R. (1976) *The Effect of Urbanisation on Flood Estimates*, Wallingford:

Institute of Hydrology Urban Drainage Research Note to National Water Council (WP-HDSS-76/17) and CIRIA (SG/244/5).

Park, C. C. (1977) "Man-induced changes in stream channel capacity", in Gregory, K. J. (ed.) *River Channel Changes*, London: Wiley, pp. 121–144.

Seaburn, G. E. (1969) *Effects of Urban Development on Direct Runoff to East Meadow Brook, Nassau County, Long Island, New York*, Washington DC: US Geological Survey Professional Paper 627-B.

Standing Technical Committee on Sewers and Water Mains (1981) *Design and Analysis of Urban Storm Drainage: The Wallingford Procedure. Volume 1 Principles, Methods and Practice*, London: National Water Council.

Thomson, J. A. M. and Foster, S. S. D. (1986) "Effect of urbanization on groundwater of limestone islands: an analysis of the Bermuda case", *Journal of the Institution of Water Engineers and Scientists*, 50(6), pp. 527–540.

Uunk, J. B. and van de Ven, F. H. M. (1984) "Water budgets for the town of Lelystad", in Balmer, P., Malmquist, P-A., and Sjoberg, A. (eds.) *Proceedings of the Third International Conference on Urban Storm Drainage*, Göteborg, pp. 1221–1230.

Walling, D. (1979) "The hydrological impact of building activity: a study near Exeter", in Hollis, G. E. (ed.) *Man's Impact on the Hydrological Cycle in the UK*, Norwich: Geo Books, pp. 135–152.

Watkins, L. H. (1962) *The Design of Urban Sewer Systems*, London: HMSO, Road Research Technical Paper 55.

Whyte, A. (1985) "Ecological approaches to urban systems: a retrospective and prospective look", *Nature and Resources*, 21(1), pp. 13–19.

Wolman, M. G. (1967) "A cycle of sedimentation and erosion in urban river channels", *Geografiska Annaler*, 49A, pp. 385–395.

Yperlaan, G. J. (1977) "Statistical evidence of the influence of urbanization on precipitation in the Rijnmond area", *Effects of Urbanization and Industrialization on the Hydrological Regime and on Water Quality; Proceedings of the Amsterdam Symposium*, Paris: UNESCO, pp. 20–30.

16

FARMING AND NITRATE POLLUTION*

T. P. Burt and N. E. Haycock

Modern agriculture is now recognised by many farmers and environmentalists as a significant source of water pollution. Sediment from eroded soil, pesticides, fertilisers and farmyard wastes can adversely affect the quality of surface and ground waters. These 'non-point' pollutants enter rivers over wide areas of a drainage basin (in contrast to 'point' sources such as the effluent from a sewage works). The environmental effects of agricultural pollutants may be of little consequence to the farmer, but this may not be so for other members of the community. In Europe, most concern has been expressed about increases in the nitrate concentration of rivers and aquifers over recent years.

Figure 16.1 shows an example of the nitrogen cycle; the nitrate ion (NO_3^-) is only one part of this complex cycle. In studying the nitrogen cycle we are concerned with both storage of nitrogen in different forms and its movement between various parts of the cycle. Given its variety of forms, we often express the concentration of nitrogen present, rather than the concentration of the compound (eg. nitrate) so that comparisons may be made. Thus, a concentration of 50 mg l^{-1} nitrate (NO_3^-) is identical to a concentration of 11.3 mg l^{-1} nitrate-nitrogen ($NO_3^-N^-$) because 22.58 per cent of the nitrate molecule is nitrogen. Nitrate is frequently expressed using either of these units so some care is needed when interpreting published figures.

Nitrogen exists in various forms, both organic and inorganic, and may be present as solid matter, in solution or as a gas. The major forms found in freshwater are nitrate, ammonium and particulate organic matter. Most of the nitrogen in soil is organic, but our interest is mainly in the inorganic fraction which comprises only a few per cent of the total soil nitrogen. Biological transformations of nitrogen comprise the most important processes within the nitrogen cycle:

(i)   fixation of nitrogen gas to ammonium (NH_4^+) and organic nitrogen by microorganisms;

(ii) assimilation of inorganic forms (nitrate and ammonium) by plants and microorganisms to form organic matter;

(iii) heterotrophic conversion of organic nitrogen from one organism to another;

(iv) mineralisation (or ammonification) of organic matter by bacteria to produce ammonium during the decomposition of organic matter;

* Originally published in *Geography*, 1990, vol. 76, pp. 60–3.

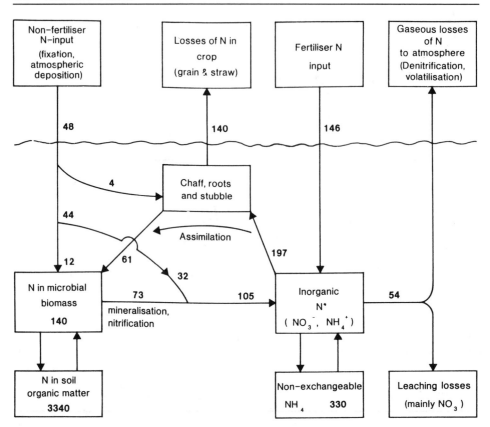

Figure 16.1 [*orig. figure 1*] The nitrogen cycle for a plot under continuous winter wheat receiving approximately 144 kg ha^{-1} N every year (adapted from Powlson et al., *Journal of Agricultural Science*, Cambridge, 1986). Figures in kg N ha^{-1} in boxes; in Kg N ha^{-1} a^{-1} other-wise*. Since the level of inorganic nitrogen present is highly variable, no figure is given; as a rough guide, 30–60 kg N ha^{-1} would be typical for this soil in autumn falling to 2–16 kg N ha^{-1} in spring before fertiliser is applied.

(v) nitrification (or oxidation) of ammonium to nitrite (NO_2^-) and then to nitrate (NO_3^-); and

(vi) denitrification (or reduction) of nitrate to form nitrous oxide (N_2O) and nitrogen (N_2) gases under anaerobic conditions.

Nitrate is formed naturally even in unfertilised situations: nitrogen assimilated into organic matter may be mineralised by soil bacteria into ammonium (NH_4^+); this may be used directly as a nutrient by some plants. Ammonium may then be nitrified to nitrite and then nitrate. Ammonium is a positively charged ion (cation) and is easily held in the soil on the surfaces of clay particles which are negatively charged. However, nitrate is a negatively charged ion (anion) and remains in the soil solution from where it is easily washed out, or leached. Nitrate may be used as a plant nutrient, leached from the soil, or may be subject to gaseous losses by denitrification or, if converted to ammonium, volatilisation (loss of ammonia, NH_3).

There are three reasons why high concentrations of nitrogen compounds in rivers are a cause for concern. Firstly, loss of fertiliser is an economic loss to the farmer. Thus, good farming practices which maintain productivity but reduce losses of nitrate make good sense. Secondly, and most contentiously, high nitrate concentrations in drinking water may affect human health. Recent epidemiological research provides no evidence that nitrate induces cancer in humans. Nor is there evidence that infantile methaemoglobinaemia (the 'blue-baby' syndrome) is caused by bacteriologically sound water supplies containing nitrate concentrations up to 100 mg l^{-1} NO_3. Nevertheless, both the World Health Organisation (WHO) and the European Community (EC) have decided that the maximum acceptable concentration for nitrate in drinking water should be fixed at 50 mg l^{-1} NO_3 (11.3 mg l^{-1} NO_3^-N). In the UK farmers, water companies and the National Rivers Authority will have to learn to live with these limits. Thirdly, nitrogen compounds can cause undesirable effects in aquatic ecosystems, particularly excessive growth of macrophytes and microphytes (eg. algae). Most interest has focused on the role of nitrate in the eutrophication, or nutrient enrichment, of fresh and marine waters. In lakes, the role of phosphate may be more crucial than that of nitrate, but in rivers the situation is less certain. In estuarine and marine waters, nitrate is the main cause of eutrophication.

The popular misconception that the nitrate problem is caused by farmers applying too much nitrate fertiliser to crops, so that the surplus left after harvest is leached away in the following winter, is too simplistic. Nevertheless, there is now little doubt that the high concentrations of nitrate in fresh waters noted in recent years have resulted mainly from drainage from agricultural lands and that the progressive intensification of agricultural practices, with increasing reliance on the use of nitrogen fertiliser, has contributed significantly to this problem. In other words, where the amount of fertiliser added is large, the loss of nitrate can be *equivalent* to a significant proportion of the fertiliser added. However, since a complex transfer of the applied nitrogen through various compartments of the nitrogen cycle is involved, it is not necessarily the same molecules of nitrogen applied as fertiliser which are lost by leaching. Figure 16.1 shows the nitrogen cycle for a plot on which winter wheat has been grown continuously for many years. Inputs of nitrogen in fertiliser, rainfall and from biological fixation are balanced by crop uptake, denitrification, and leaching. Fertiliser is applied in the spring and leaching losses are usually low; nevertheless, only about 70 per cent of the total input of nitrogen is recovered in the crop. Note that most of the soil nitrogen is immobilised in soil organic matter. For a nearby field where some fertiliser (50 kg ha^{-1}) was added in the autumn, losses increased considerably with 40–80 per cent lost, mainly by leaching.

Even when no fertiliser is applied in autumn, arable soils often contain much inorganic nitrogen. Some of this will be fertiliser not used by the previous crop, but most will come from mineralisation of organic matter generated from autumn ploughing and the lack of crop cover. The more mineral nitrogen remaining the greater will be the potential for leaching. Such losses tend to increase if the previous crop used nitrogen inefficiently or if the soil contains much organic matter. Ploughing is most important, particularly of grassland: this releases vast quantities of nitrogen by mineralisation, but unless the new crop is planted quickly, much can be lost by leaching. A final influence is climate. There is normally much mineralisation in the autumn when warm soils begin to wet up; grass can absorb the nitrate produced as it is still growing, but, with bare soil, this nitrate is prone to leaching. This problem is especially severe where a wet autumn follows a dry summer: much soil organic matter may be mineralised and leached at such times.

Since 1945, agriculture in the UK has become much more intensive. Considerably more fertiliser is used; fields are ploughed more frequently; more land is devoted to arable crops; and grassland is also farmed more intensively. All these factors combine to amplify the nitrogen cycle, one effect of which is to increase nitrate leaching losses. Though sewage and air pollution may be partly to blame, it is generally acknowledged that the long-term increase in nitrate

Table 16.1 [*orig. table 1*]   Annual mean nitrate (NO_3-N) concentrations (mg l^{-1}) for three English rivers, for the water years 1972–1988.

Year	Slapton Wood (Devon) (Area = 1 km²)	Windrush (Gloucs.) (Area = 300 km²)	Thames at Oxford (Area = 3,400 km²)
1972	5.9	n.a.	6.5
1973	5.6	5.1	5.9
1974	5.9	6.0	6.7
1975	6.4	6.1	6.4
1976	6.9	4.5	5.4
1977	8.3	7.7	10.7
1978	7.7	7.6	7.8
1979	6.8	7.7	8.8
1980	6.9	7.3	7.4
1981	6.7	7.6	7.9
1982	7.8	7.2	7.8
1983	6.8	7.7	8.6
1984	6.9	7.3	7.9
1985	8.3	9.1	9.4
1986	9.2	8.5	7.7
1987	9.2	8.5	7.7
1988	8.8	9.0	8.0

Data sources: Slapton Ley Field Centre; P. J. Johnes; Thames Water Authority.

concentration noted for many British and European rivers is mainly the result of modern agricultural practices. Table 16.1 shows annual mean nitrate concentrations for three rivers of different catchment size from 1972 to 1988. A clear upward trend can be detected. Climate does have some influence in that nitrate concentrations tend to be higher in wetter years: the early 1970s were dry whereas the late 1970s and early 1980s were wetter. Also, as noted above, nitrate concentrations may be very high in the year following a drought. Nevertheless, statistical analysis shows that the main effect is a steady increase in nitrate levels over time which is independent of climate. If trends continue, the mean nitrate concentration of many rivers in the UK will soon be above the EC limit (11.3 mg l^{-1} NO_3-N); in many cases this limit is already exceeded during the winter when nitrate concentrations reach their maximum level. In catchments where groundwater discharge is important, this long-term upward trend may be particularly prolonged since it may take years for nitrate to percolate down to the saturated zone. In

such basins, nitrate pollution may remain a problem for decades to come.

What can be done? Various options are available to the water companies. In terms of water treatment, the removal of nitrate from water is costly. The cheapest method of treatment is to blend high and low nitrate waters, though this may still be very expensive in areas like East Anglia where low nitrate water is scarce. Storage of water in reservoirs to allow denitrification is also helpful, but again is not always an available option. It is clear that good farming practices can help to reduce nitrate losses: for example farmers should avoid autumn applications of fertiliser, should sow crops in autumn to take up any available soil nitrate, and should avoid spreading fertiliser if heavy rain is forecast. However, since voluntary schemes may not be sufficient, the Government has recently introduced the 'Nitrate Sensitive Areas' scheme. Farmers close to boreholes may be obliged to restrict their operations beyond the degree which can be regarded as good practice, but will be compensated for doing so. Given the time taken for groundwater to respond, it may be some years before we can judge whether this scheme has been a success. Our own research is examining the role of floodplains as buffer zones between farmland and rivers. Preliminary results from the 1989/90 winter suggest that groundwater moving through floodplain soils may lose much of its nitrate through a combination of assimilation and denitrification. The use of undrained flood-plains as barriers to nitrate movement could allow modern farming and water supply to co-exist in the same basin. Such buffer zones would also trap eroded soil (and associated pollutants) and additionally could provide habitats of high conservation value. This suggests that strategically placed 'set aside' land might solve both the difficulties of over-production and the pollution problems associated with modern agriculture.

Further Reading

Royal Society (1983) *The Nitrogen Cycle of the United Kingdom*, London: Royal Society.

Trudgill, S. T. (1989) "The nitrate issue", *Geography Review*, 2(5), pp. 28–31.

17

DIATOM AND CHEMICAL EVIDENCE FOR REVERSIBILITY OF ACIDIFICATION OF SCOTTISH LOCHS*

R. W. Battarbee, R. J. Flower, A. C. Stevenson, V. J. Jones, R. Harriman and P. G. Appleby

The cause–effect relationship between acid deposition and lake acidification is well established. Attention now focuses on the response of acidified lakes to reductions in SO_2 emissions.[1] In the United Kingdom there has been a roughly 40% decline in national SO_2 emissions since the maximum in 1970 (ref. 2), and reductions in both the concentration of non-marine sulphate in precipitation[2,3] and the wet deposition of sulphate[2] have been recorded in Scotland since the mid-1970s. The trend is especially clear at Eskdalemuir, an international monitoring site[2] (figure 17.1), and a site close to acidified lakes in Galloway, southwestern Scotland.[4,5] Two of these lakes have been resampled. Diatom analysis of sediment cores taken between 1981 and 1986 show a trend towards progressively decreasing acidity in the uppermost sediments for both

sites. Water chemistry records show that there have been significant decreases in both proton and sulphate concentrations since 1978.

We present data for Loch Enoch and the Round Loch of Glenhead. We have chemical records for both sites from repeated surveys at 2–3 month intervals in 1978–79 and 1984–86 and we have diatom data for two cores for Loch Enoch (sampled in 1982 and 1986) and for three cores for the Round Loch of Glenhead (sampled in 1981, 1984 and 1986). Site details are published elsewhere.[6]

Sampling and analytical methods used for water chemistry are described in ref. 7. The most abundant cations were determined by atomic absorption spectroscopy. Before 1986 anions were determined by an ion exchange method,[8] and afterwards by ion chromatography. Both methods were used for April

* Originally published in *Nature*, 1988, vol. 332, pp. 530–2.

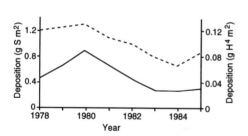

Figure 17.1 [*orig. figure 1*] Annual wet deposition of non-marine sulphate (broken line) and hydrogen ion from 1978–1985 for Eskdalemuir (from ref. 2).

1986 samples and there was no significant difference between the results from the two methods. Total monomeric and labile forms of aluminium were measured by the pyrocatechol violet technique.[9,10]

Sediments were sampled using Mackereth and Kajak corers and cores were sectioned at 0.5 cm intervals. [210]Pb dates were derived for all cores except for the 1984 Round Loch core, which was dated by biostratigraphic correlation with the 1981 core.[11] Diatom analysis used standard procedures[12] and correspondence analysis used the program CANOCO. [13]

A comparison of the water chemistry of Loch Enoch between 1978–79 and 1984–86 (table 17.1) indicates a rise in the mean pH of 0.22 pH units (H+ concentration 16 µeq l⁻¹, from 4.40 ± 0.12 (95% confidence limits) to 4.62 ± 0.04). The sample means for H+ were significantly different between the two periods ($P = 0.05$). Mean excess sulphate values were also significantly different ($P = 0.05$), falling from 77 ± 10 to 52 ± 3 µeq l⁻¹ (table 17.1).

The Round Loch of Glenhead (table 17.1) showed no statistically significant difference ($P = 0.05$) in mean pH and H+ between the two dates, although the mean pH increased from 4.65 ± 0.16 to 4.77 ± 0.05, a decline in proton concentration of 5 µeq l⁻¹. Mean excess sulphate concentrations were significantly different, falling from 94 ± 12µeq l⁻¹ to 67 ± 14 µeq l⁻¹.

Diatom diagrams for the upper 15 cm of

the two cores from Loch Enoch are shown in figure 17.3. A general comparison between the assemblages of the two cores shows a good agreement, although the 1986 core has a peak of *Asterionella ralfsii* between 8 and 10 cm which is absent from the 1982 core. The 1986 core has a somewhat faster accumulation rate. [210]Pb dates show that the 1970 level for the 1982 and 1986 cores occurs at ~ 2.5 and 4.5 cm respectively, and [134]Cs measurements show the presence of post-Chernobyl radioactive fallout in the 1986 core. The 1982 core shows that the uppermost centimetre, representing sediment accumulation from about 1977–1982, contains lower proportions of acidobiontic species, whereas in the 1986 core the uppermost 1.5 cm (1981–86) has lower proportions of these species.

Although diatom changes in sediment cores are often used to reconstruct pH trends,[14,15] the changes described here (0.1–0.2 pH units) are less than the errors in currently available models for pH recon-

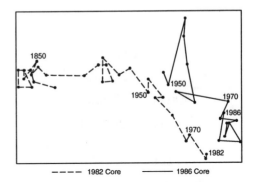

Figure 17.2 [*orig. figure 3*] Comparison of diatom assemblages from a 1982 (dashed line) and 1986 (continuous line) core from Loch Enoch using Correspondence Analysis.[13] Only axes 1 and 2 are shown. Successive stratigraphic levels are linked together to indicate the time-track of the individual core. Pre-1850 samples for the 1982 core show a high degree of similarity. Post-1850 samples show a directional change along axis 1. The 1986 core begins at about 1950 and shows a similar time-track, but the uppermost samples show a high degree of similarity and then track in the reverse direction.

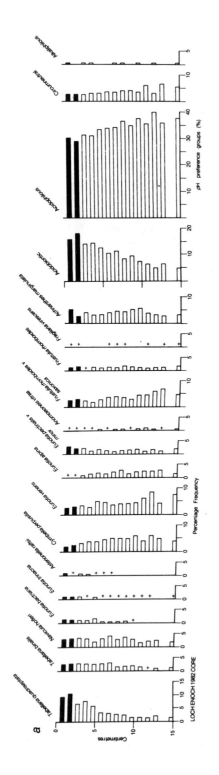
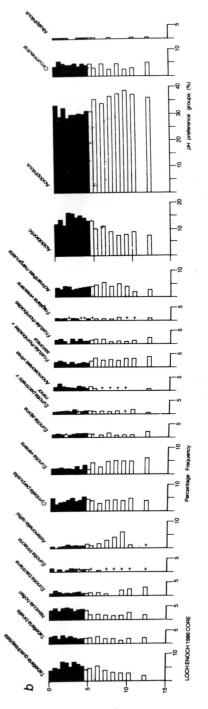

Figure 17.3 [*orig. figure* 2]　Summary of diatom diagrams for two cores from Loch Enoch: *a*, 1982 and *b*, 1986. Post-1970 levels are shaded.

Table 17.1 [*orig. table 1*]   Physical and chemical characteristics for Loch Enoch and Round Loch of Glenhead (1978–79 and 1984–86)

	Loch Enoch		Round Loch of Glenhead	
Altitude (m)	493		299	
Lake area (ha)	50		12.6	
Catchment area (ha)	186		95	
Catchment: lake	3.7		7.5	
Max. depth (m)	36		13.5	
	1978–79 ($n = 6$)	1984–86 ($n = 6$)	1978–79 ($n = 5$)	1984–86 ($n = 6$)
pH	4.40 (4.54)	4.62	4.65 (4.67)	4.77
H^+	40	24	22	17
Conductivity ($\mu s \ cm^{-1}$)	38	40	36	41
Na^+	101	168	122	191
K^+	8	8	7	9
Ca^{2+}	15	27	27	41
Mg^{2+}	36	45	42	51
Cl^-	139	205	140	224
SO_4^{2-}	91 (89)	72	108 (112)	89
NO_3^-	10	15	6	9
Alkali	0	0	0	0
Absorbance (250 nm)	–	0.047	–	0.107
Total organic carbon (mg l^{-1})	– (1.1)	1.0	– (2.5)	2.2
Al (*l*) ($\mu g \ l^{-1}$)	–	73	–	61
Al (*nl*) ($\mu q \ l^{-1}$)	–	15	–	36
Al (total soluble $\mu g \ l^{-1}$)	114 (150)		139 (165)	

Unless otherwise indicated values are expressed as microequivalents per liter ($\mu eq \ l^{-1}$). Catchment area excludes lake area. Aluminium values are shown for labile (*l*) and non-labile (*nl*) fractions. Values for 1979 from ref. 19 and indicated in parentheses.

struction. The extent to which a change in trend has occurred can be evaluated more sensitively using ordination techniques that allow the degree of similarity between samples in a core to be shown. If the samples are linked in stratigraphic order, the series forms a time-track. In this case, correspondence analysis, an ordination method most suited to biological data, has been used. In figure 17.2 it can be seen that except for the deviation on axis 2 caused by the *A. ralfsii* peak, the two cores follow similar time-tracks until most recent times. In the 1986 core there is a clear floristic reversal and the surface sample (~ 1982–86) has an almost identical composition to samples at 4–5 cm (1968–1972).

The diatom data for the Round Loch of Glenhead have been treated similarly. Correspondence analysis time-tracks for the three cores are shown in figure 17.4. There is no evidence for reversal at the top of the 1981 core, but both the 1984 and 1986 cores show clear floristic reversals. The 1984 core has an accumulation rate more than twice that of the 1986 core, allowing the recent history of the lake to be recorded with finer resolution. The data for this core show that little change occurred in the 1970s, indicating no further acidification, and that the improvement

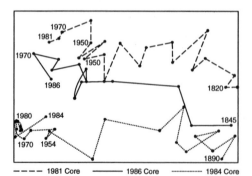

----- 1981 Core   ——— 1986 Core   ·········· 1984 Core

Figure 17.4 [orig. figure 4] Comparison of diatom assemblages from three cores from the Round Loch of Glenhead using Correspondence Analysis.[13] The 1981 core (dashed line) shows no indication of reversibility although post-1970 samples are quite similar. But the 1984 (dotted line) and 1986 (continuous line) cores show reversed time-tracks in the uppermost levels.

dates from about 1980. The uppermost samples of the 1984 and 1986 cores from Round Loch have a diatom composition similar to that of the 1950s.

Loch Enoch and Round Loch of Glenhead are still highly acidified lakes, but the data presented here show that no further acidification has taken place at either site since the mid-1970s, and that a small improvement has taken place since 1980. We cannot say whether concentrations of aluminium or total organic carbon have changed between the sampling periods as these were not measured in the 1978 survey.

Current diatom assemblages are not dissimilar from those recorded in the sediment between 1950 and 1970. As the period of rapid acidification at these sites occurred between 1930 and 1970 (ref. 6), it might be argued that a further significant reduction in acid deposition in Scotland could lead to a relatively larger response in water quality with the re-emergence of circum-neutral diatom taxa, such as *Anomoeoneis vitrea* and *Achnanthes microcephala*, in the diatom flora.

This trend towards improved conditions has also been observed at nearby Loch Valley (1978–9, pH 4.47, $n = 6$; 1984–6,

pH 4.75, $n = 6$) (ref. 16), and cores taken in 1985 at several sites with moorland catchments in other parts of Scotland show diatom reversals in the uppermost sediment (unpublished observations; also ref. 17).

These data suggest that there is little delay (~10 yr) in the response of Galloway lakes to a decrease in acid deposition. Lakes have not continued to acidify as suggested by some models,[18] and some have improved. Indeed ordination of the core data suggests that the trajectory of the improvement is not only in the reverse direction but, in the case of diatoms, occurs without hysteresis.

We thank the Forestry Commission, John Thornicroft and Central Television, Stuart Phethean, Simon Patrick, Annette Kreiser and Sarah Wells for help with field work, and John Birks for discussion. This work was supported by the Royal Society (SWAP Project), The Natural Environmental Research Council, the Central Electricity Generating Board, and the Department of Environment (UK).

References

1   Dillon, P. J., Reid, R. A. and de Grosbois, E. *Nature* 329, 45–48 (1987).

2   Barrett, C. F. et al. *Acid deposition in the United Kingdom, 1981–1985* (Warren Spring Laboratory, Stevenage, 1987).

3   Harriman, R. and Wells, D. E. *J. Wat. Pollut. Cont.* 84, 215–222 (1985).

4   Flower, R. J. and Battarbee, R. W. *Nature* 305, 130–133 (1983).

5   Battarbee, R. W., Flower, R. J., Stevenson, A. C. and Rippey, B. *Nature* 314, 350–352 (1985).

6   Flower, R. J., Battarbee, R. W. and Appleby, P. G. *J. Ecol.* 75, 797–824 (1987).

7   Harriman, R., Morrison, B. R. S., Caines, L. A., Collen, P. and Watt, A. W. *Wat. Air and Soil Pollut.* 32, 89–112 (1987).

8   Mackereth, F. J. H., Heron, J. and Talling, J. F. *Freshwater Biol. Assn* 36 (1978).

9   Seip, H. M., Muller, L. and Naas, A. *Wat. Air and Soil Pollut.* 23, 81–95 (1984).

10   Driscoll, C. J. Jr, Baker, J. P., Bisogni, J. J. Jr and Schofield, C. *Nature* 284, 161–164 (1980).

11   Jones, V. J. thesis, Univ. London (1987).

12   Battarbee, R. W. in *Handbook of Holocene Palaeoecology and Palaeohydrology* (ed. Berglund, B. E.) 527–570 (Wiley, Chichester, 1986).

13   Ter Braak, C. S. F. *Ecology* 67, 1167–1179 (1986).

14   Battarbee, R. W. *Phil. Trans. R. Soc.* 305, 451–477 (1984).

15   Flower, R. J. *Hydrobiologia* 143, 93–103 (1987).

16   Harriman, R. and Wells, D. E. in *Acid Rain: Scientific and Technical Advances* (eds Perry, R., Harrison, R. M., Bell, J. N. B. and Lester, J. N.) 287–292 (Selper, London, 1987).

17   Flower, R. J., Darley, J., Rippey, B., Battarbee, R. W. and Appleby, P. G. *J. Appl. Ecol.* (in the press).

18   Cosby, B. J., Wright, R. F., Hornberger, G. M. and Galloway, J. N. *Wat. Resourc. Res.* 21, 1591–1601 (1985).

19   Wright, R. F. and Henriksen, A. *Regional Survey of Lakes and Streams in Southwestern Scotland* Internal Report 72/70 SNSF Project, Oslo, Norway.

18

LAKE ACIDIFICATION IN GALLOWAY: A PALAEOECOLOGICAL TEST OF COMPETING HYPOTHESES*

R. W. Battarbee, R. J. Flower, A. C. Stevenson and B. Rippey

Possible causes of lake acidification in Britain include acid precipitation, heathland regeneration, afforestation and post-glacial natural acidification (long-term change). In Galloway, south-west Scotland, several lakes with non-afforested catchments have been acidified by approximately 1 pH unit since about 1840,[1,2] which clearly precludes long-term change and afforestation as causes of the present acidity, but does not discriminate between the possible effects of heathland regeneration and acid precipitation; it could be argued that a decline in upland farming in this area led to an increase in acid heathland communities capable of promoting lake acidification during this period of time. Here we use pollen analysis to show that there is no evidence for an increase in *Calluna vulgaris*, the most important heathland species, over the past 200 yr or so, and we substantiate the acid precipitation hypothesis by demonstrating substantial increases of the heavy metals Pb,

Cu and Zn in the sediment since about AD 1800.

Surface water acidification has been variously ascribed to the effects of acid precipitation,[3] land-use change,[4] afforestation[5,6] and long-term change,[7] the importance of which can be assessed by palaeoecological techniques. We consider the last two of these processes first.

Long-term acidification is well known, having been recognized in Scandinavia,[8-11] Britain[12-15] and North America (ref. 16 and J. Ford, unpublished results). In Britain it has been suggested that this process is the cause of contemporary acidity in upland tarns of the English Lake District and that certain lakes were so acidic by 1800 that no further change can have occurred since then.[7] Although this supposition has not yet been examined in the context of the Lake District, it is clear that the present acidity of lakes with granitic catchments in Galloway cannot be explained in this way; diatom

* Originally published in *Nature*, 1985, vol. 314, pp. 350–2.

Table 18.1 [*orig. table 1*] Physical and chemical characteristics for Loch Enoch

Altitude (m)	493	
Lake area (ha)	50	
Catchment area (ha)	186	
Catchment: lake area	3.7	
Maximum depth (m)	~36	
Temperature (°C)	7.0 (4.4)	
Conductivity (μS_{18} cm^{-1})	30.0 (8.0)	37*
pH (range)	4.4–4.7	4.5*
H^+ (μeql^{-1})	28.9 (6.8)	28
Na^+ (μeql^{-1})	58.3 (24.4)	126
K^+ (μeql^{-1})	6.7 (4.6)	6
Ca^{2+} (μeql^{-1})	10.5 (3.5)	21
Mg^{2+} (μeql^{-1})	16.5 (7.2)	37
Al (μeql^{-1})	10.0 (4.4)	16
Cl^- (μeql^{-1})	100.9 (18.0)	146
SO_4^{2-} (μeql^{-1})	44.5 (12.9)	89

Numbers in parentheses indicate standard deviations for the mean values calculated from eight measurements made between November 1981 and November 1982. The catchment area excludes lake area.

* Data in the last column are from ref. 33.

analysis indicates that rapid acidification of these lakes has occurred since about 1840, after long periods of stability.[1,2]

Recent afforestation, though responsible for accelerated soil erosion[17] and other undesirable ecological changes, is also of little significance as lakes with non-afforested catchments were acidified, and those with afforested catchments began to be acidified, in the late nineteenth and early twentieth centuries, well before the beginning of afforestation.[1]

A hypothesis involving change in land use has been proposed to explain recent lake acidification in Norway.[4,18] It maintains that acidity generated by soil processes is substantially greater than that received from atmospheric deposition and that an increase in the formation of acid humus in soils as a result of a decline in upland agriculture is likely to be of more significance than an increase in the acidity of precipitation.

In Galloway it may be expected that a similar decline in grazing would lead to the greater dominance of C vulgaris in particular, and to a regeneration of heathland communities in general.[19] This hypothesis would be rejected if it could be shown that

such a land-use change was absent or was inadequate to explain the observed acidification of the lakes concerned, and we would then need to evaluate the acid precipitation hypothesis.

To carry out these tests we have used diatom analysis, pollen analysis and trace metal analysis of a dated lake sediment core from Loch Enoch, a 50-hectare, 36-m deep multi-basin lake situated at an altitude of 493 m on the Merrick in the Rhinns of Kells region of Galloway. The catchment is non-afforested and the vegetation is dominated by *Molinia caerulea* and *C. vulgaris*.

We obtained a sediment core from the centre of the south-east basin of the lake and determined a linear sediment accumulation rate of ~1 mm yr^{-1} using ^{210}Pb analysis.[17] Diatom analysis shows the onset of recent acidification at ~30 cm, dated to about 1840 (figure 18.1). The index B pH reconstruction[11] gives a *c.*1840 pH of 5.2 and a present pH of 4.3, slightly more acidic than the measured range of pH values (see table 18.1). The strong increase in acidobiontic diatoms registered in the diagram (figure 18.1) include *Tabellaria binalis* (Ehr.) Grun. and *Tabellaria quadriseptata* Knudson, species that

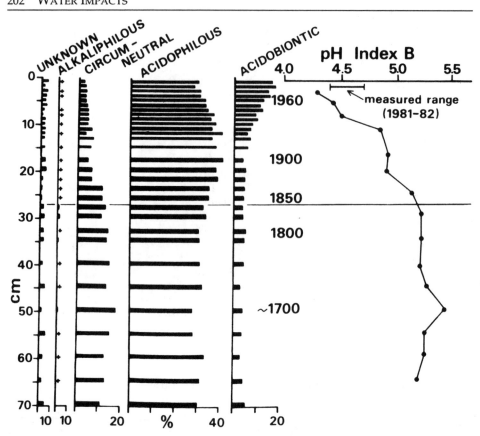

Figure 18.1 [*orig. figure 1*]   Summary diatom diagram from Loch Enoch showing changes in the proportions of the various pH groups through time. Dominant taxa are as follows: (1) circumneutral, *Anomoeoneis vitrea* (Grun.) Ross; (2) acidophilous, *Eunotia veneris* (Kütz.) O. Müll., *E. alpina* (Naeg.) Hust., *Tabellaria flocculosa* (Roth.) Kütz., *Frustulia rhomboides* v. *saxonica* (Rabh.) de Toni; (3) acidobiontic, *T. quadriseptata* Knudson, *T. binalis* (Ehr.) Grun., *Navicula höfleri* Cholnoky. Dates are based on ²¹⁰Pb measurements,[17] and pH is reconstructed using index B.[11] Horizontal line signifies point of change.

typically increase in a similar way at other sites in the region.[1,20]

If acidification was caused by a change in land use in the catchment of the lake, in accordance with the above hypothesis, we would expect historical records to show a decline in grazing and consequent regeneration and expansion of heathland on approximately the same timescale. As records of sheep numbers over the past 150 yr are available only for large administrative regions,[21] it is more appropriate to test this prediction using pollen analysis of the lake sediment core itself. This has been performed using standard procedures[22] on samples selected at 5-cm intervals from the

top of the core. The full pollen diagram shows very little change over the past 200–300 yr, except in the relative proportions of *C. vulgaris* and Gramineae, the taxa of greatest interest (figure 18.2). However, rather than an increase in *C. vulgaris*, as would be expected were the land-use hypothesis valid at these sites, there is a gradual decrease in this species and a reciprocal expansion of Gramineae; similar results have been obtained from the analysis of cores from other lakes in the area both in terms of percentage and flux change (unpublished results). These results are not unexpected as they agree with contemporary observations that all the non-afforested

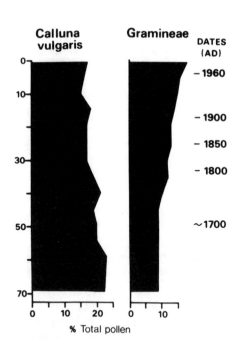

Figure 18.2 [*orig. figure* 2] Proportions of C. vulgaris and Gramineae pollen from the Loch Enoch sediment core.

lake catchments in this area are grazed and repeatedly burnt over, a process that in the west of Scotland favours M. caerulea and keeps C. vulgaris young and palatable for both cattle and sheep.[23]

Having rejected three hypotheses for the recent acidification of Loch Enoch, we retain only the acid precipitation hypothesis. This hypothesis would be questionable if we could not show that Loch Enoch and other similar Galloway lakes have received atmospheric pollutants that may be associated with industrial emissions. We therefore analysed the trace metal content of the sediments (figure 18.3).

Total sedimentary Zn, Cu and Pb concentrations were determined by flame atomic absorption after $HF/HNO_3/HClO_4$ digestion, the precision of the results are, respectively, 4.7, 0.9 and 3.8 $\mu g\ g^{-1}$ ($n = 14$). The background concentrations (table 18.2) show that there are no geochemical anomalies in the catchment. The slightly lower Cu and Zn and somewhat higher Pb concentrations, compared with the mean for freshwater sediments,[24] are a result of the granites in the catchment.[25] The increased

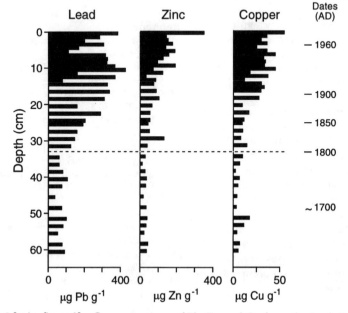

Figure 18.3 [*orig. figure 3*] Concentrations of Pb, Zn and Cu from the Loch Enoch sediment core. Dotted line signifies point of change.

Table 18.2 [*orig. table* 2]   Background sedimentary concentrations, fluxes and recent fluxes of Zn, Cu and Pb in a sediment core from Loch Enoch.

	Zn	Cu	Pb
Background concentration, $\mu g\ g^{-1}$	25.4	6.2	74.6
Standard deviation (less than 32 cm depth, $n = 12$)	7.4	4.6	23.3
Background sedimentary flux, $mg\ m^{-2}\ yr^{-1}$ (average value for 32–61 cm)	6.1	1.5	17.8
Recent sedimentary flux, $mg\ m^{-2}\ yr^{-1}$ (average value for 0–10 cm)	35.5	8.3	67.0
Anthropogenic sedimentary flux, $mg\ m^{-2}\ yr^{-1}$ (recent, background flux)	29.4 (83)	6.8 (82)	49.2 (73)

Calculations were made using trace metal concentrations (figure 18.3) and sediment wet density, dry weight and accumulation rate (ref. 21). Numbers in parentheses indicate % anthropogenic flux of total metal influx.

concentrations (figure 18.3) and sedimentary fluxes (table 18.1) can be accounted for by increased deposition from the atmosphere,[26] a situation common in remote and rural lakes.[27-29] It has been shown that atmospheric contamination of lake waters by trace metals is correlated with acidification,[30,31] and that contamination of sediments is correlated with sedimentary evidence of acidification.[26,27,32] This latter relationship holds true for Loch Enoch and all these relationships are consistent with expectations derived from the acid precipitation hypothesis.

We cannot prove that an increase in acid deposition was responsible for the acidification of Loch Enoch and similar lakes in Galloway, but we have shown that alternative hypotheses, as presently formulated, are inadequate. Land-use change may play a part in other areas, but at present it is clearly not applicable to the Galloway sites. For this hypothesis to be tested by palaeoecological means, examination of an acidified site in an area of low acid deposition, where heathland regeneration has occurred concurrently, is required.

Acknowledgments

We thank the Central Electricity Generating Board for funding this work, CERL, Leatherhead, for carrying out chemical analyses of Loch Enoch water samples, Stuart Phethean for help with fieldwork, Kevin Odell for radiometric data, Peter Appleby for ^{210}Pb computations, and Richard Skeffington and Gwyneth Howells for comments on the manuscript.

References

1   Flower, R. J. and Battarbee, R. W. *Nature* 305, 130–133 (1983).
2   Battarbee, R. W. *Phil. Trans. R. Soc.* B305, 451–477 (1984).
3   Jensen, K. W. and Snekvik, E. *Ambio* 1, 223–225 (1972).
4   Rosenquist, I. T. *Sci. tot. Envir.* 10, 39–49 (1978).
5   Harriman, R. and Morrison, B. R. S. *Hydrobiologia* 88, 251–263 (1984).
6   Nilsson, S. I., Miller, H. G. and Miller, J. D. *Oikos* 39, 40–49 (1982).
7   Pennington, W. *A. Rep. Freshwat. Biol. Ass.* 28–46 (1984).
8   Quennerstedt, N. *Acta phytogeogr. suec.* 36, 1–208 (1955).
9   Digerfeldt, G. *Folia limnol. scand.* 16, 1–104 (1972).
10   Renberg, I. *Early Norrland* 9, 113–160 (1976).
11   Renberg, I. and Hellberg, T. *Ambio* 11, 30–33 (1982).
12   Round, F. E. *New Phytol.* 56, 98–126 (1957).
13   Mackereth, F. J. H. *Proc. R. Soc.* B161, 295–309 (1965).

14 Pennington, W., Haworth, E. Y., Bonny, A. P. and Lishman, J. P. *Phil. Trans. R. Soc.* B264, 191–294 (1972).

15 Evans, G. H. and Walker, R. *New Phytol.* 78, 221–236 (1977).

16 Engstrom, D. R. and Wright, H. E. in *Lake Sediments and Environmental History* (eds Haworth, E. Y. and Lund, J. W. G.) (Leicester University Press, 1984).

17 Battarbee, R. W., Appleby, P. G., Odell, K. and Flower, R. J. *Earth Surf. Processes* (in the press).

18 Krug, E. E. and Frink, C. R. *Science* 221, 520–525 (1983).

19 Gimmingham, C. H. *J. Ecol.* 48, 455–483 (1960).

20 Flower, R. J. and Battarbee, R. W. *Br. phycol. Bull.* 21, (in the press).

21 Flower, R. J. and Battarbee, R. W. *Working Pap.* No. 6 (Palaeoecology Research Unit, University College London, 1983).

22 Moore, P. D. and Webb, J. A. *An Illustrated Guide to Pollen Analysis* (Hodder & Stoughton, Dunton Green, 1978).

23 McVean, D. N. and Lockie, J. D. *Ecology and Land Use in Upland Scotland* (Edinburgh University Press, 1969).

24 Forstner, U. *Arch. Hydrobiol.* 50, 172–191 (1977).

25 Mason, B. and Moore, C. B. *Principles of Geochemistry* (Wiley, New York, 1982).

26 Galloway, J. N., Thornton, J. D., Norton, S. A., Volchok, H. L. and McLean, R. A. N. *Atmos. Envir.* 16, 1677–1700 (1982).

27 Davis, R. B., Norton, S. A., Brakke, D. F., Berge, F. and Hess, C. T. in *Ecological Impact of Acid Precipitation* (eds Drablos, D. and Tollan, A.) (SNSF project, Oslo, 1980).

28 Rippey, B., Murphy, R. J. and Kyle, S. W. *Envir. Sci. Technol.* 16, 23–30 (1982).

29 Wong, H. K. T., Nragu, J. O. and Coker, R. D. *Chem. Geol.* 44, 187–201 (1984).

30 Henricksen, A. and Wright, R. F. *Wat. Res.* 12, 101–112 (1978).

31 Wright, R. F. and Henricksen, A. *Limnol. Oceanogr.* 23, 487–498 (1978).

32 Holdren, G. R., Brunelle, T. M., Matisoff, G. and Wahlen, M. *Nature* 311, 245–248 (1984).

33 Wright, R. F. and Henricksen, A. *Regional Surv. Lakes Streams Southwestern Scotland* (SNSF project IR 72/80, 1980).

PART IV

Climatic and Atmospheric Impacts

INTRODUCTION

Until about the 1960s studies of the human impact on the environment were very much concerned with local or regional effects. From that time on, however, there was an increasing realization that humans might be able to cause changes at a global or planetary scale by modifying the composition of the atmosphere. At the same time, because of the availability of a whole array of new tools – satellite images, powerful electronic computers, deep sediment cores from the oceans, lakes and ice cores – climatologists were gaining a clearer impression of how the global atmosphere works now and has worked in the past. Since the mid-1960s, therefore, global atmospheric and climatic changes, both natural and anthropogenic, have become a vibrant, controversial and rapidly evolving area for scientific analysis. Themes such as global warming and ozone depletion have also emerged into the political arena.

The literature on such topics is now enormous, and the table opposite provides a guide to some of the more recent general textbooks on global atmospheric and climatic changes.

Chapter 19 is a classic review by one of the great figures in the field of human modification of climates. It dates from 1970 and thus provides a good picture of how this issue was perceived at the start of 'the global change revolution'. Landsberg, it will be noted, is very cautious about the potential significance of global warming caused by increasing levels of atmospheric carbon dioxide. He is equally cautious about the role of anthropogenically derived dust in the atmosphere and suggests that overall 'the

evidence for man's effect on global climate is flimsy at best'.

Chapter 20 appeared over ten years later than the first and takes a much more robust view on the climatic impact of increasing atmospheric carbon dioxide. It predicted, correctly, that there was a high probability of warming in the 1980s, and asserted: 'the global warming projected for the next century is of almost unprecedented magnitude' and that 'the climate change induced by anthropogenic release of CO_2 is likely to be the most fascinating global geophysical experiment that man will ever conduct'.

However, there are greenhouse enhancing gases other than carbon dioxide. These include chlorofluorocarbons (CFCs), nitrous oxide and methane. In combination they add considerably to the greenhouse gas loadings of the atmosphere, for although they only occur in very modest quantities they are, molecule for molecule, extremely effective at trapping outgoing solar radiation. Evidence of air bubbles from dated polar ice cores enables us to establish changes in the levels of certain atmospheric gases over long time spans. This is the case with respect to methane (chapter 21). Its concentration in the atmosphere has increased exponentially over the last few centuries because of a range of agricultural and industrial activities.

Greenhouse gases are not, however, the only important substances that are being emitted into the atmosphere by human activities. In recent years it appears that sulphate aerosols, which show particularly high levels over the industrial areas of the Northern Hemisphere, may not only

Selected recent text books on global climatic and atmospheric changes

Author	Title	Date	Publisher
Eisma, D. (ed.)	Climate Change: Impact on Coastal Habitation	1995	Lewis Publishers
Graedel, T. E., and Crutzen, P. J.	Atmospheric Change: An Earth System Perspective	1993	Freeman
Graedel, T. E., and Crutzen, P. J.	Atmosphere, Climate, and Change	1995	Scientific American Library
I. D. Whyte	Climatic Change and Human Society	1995	Arnold
J. D. Houghton et al. (eds)	Climate Change 1994	1995	Cambridge University Press
Kemp, D. D.	Global Environmental Issues: A climatological Approach (2nd edition)	1994	Routledge
Williams, M. A. J., and Balling, R. C.	Interactions of Desertification and Climate	1996	Arnold

contribute to acid rain (see chapters 17 and 18), but also cause some cooling by reflecting solar radiation back into space (chapter 22).

Changes in land use may be important at the regional scale. A substantial literature has, for example, emerged, concerning the possible effects of tropical deforestation on surface albedo, surface roughness, evapotranspiration rates, temperatures and precipitation. This is well reviewed in a special issue of *Climatic Change* volume 19 (1991), edited by Norman Myers. From that collection comes chapter 23. The effects are discussed at a whole variety of spatial scales.

Other chapters in part IV move from a consideration of anthropogenic climatic changes to a consideration of changes in atmospheric composition. The most important of these changes is the reduction in stratospheric ozone levels and the production of an 'ozone hole' over Antarctica. Work from British scientists at the British Antarctic Survey was instrumental in identifying the hole in the mid-1980s (chapter 24), and the cause of the hole became the subject of detailed discussion (chapter 25). Since that time ozone depletion has been recognized as being important in other areas, including high latitudes in the Northern Hemisphere.

The role of chlorofluorocarbons (CFCs) in the depletion process led, through the Montreal Protocol, to international measures to control emissions of CFCs from aerosol cans, fridges and other sources.

Longer used technologies are also capable of causing changes in atmospheric composition and no major technology has longer antecedents than the use of fire. It was probably first deliberately used by our forbears over a million years ago, and it remains one of the most powerful means at our disposal for environmental transformation. The burning of biomass to convert land from forest or woodland for agricultural or pastoral use is a widespread phenomenon, especially in the tropics, and most particularly in the savannas of Africa. As Crutzen and Andreae show (chapter 26) it impacts not only upon atmospheric chemistry but also upon biogeochemical cycles on the ground.

Acid rain became a major transnational pollution issue in the 1970s. It was found that, especially since the Second World War, rain and surface waters were becoming acidified (see also chapters 17 and 18) as a result of the emission of gases like sulphur dioxide and nitrogen oxides from the burning of fossil fuels (coal, oil and gas).

Phenomena like forest decline, the disappearance of fish from lakes over large swathes of Scandinavia and North America, and the accelerating decay of buildings of great cultural and architectural significance are regarded as being some of the undesirable consequences of acid rain. Chapter 27, based in part like chapter 13 from studies at the Hubbard Brook Experimental Forest in New Hampshire, USA, was an early review of the phenomenon.

As recent studies of acidification have shown (see chapter 18), when emissions of particular pollutants are reduced, whether because of environmental legislation and regulation, or because of technological and industrial changes, systems may adjust with some rapidity. This is also true with respect to the long-distance atmospheric transport of metal pollutants such as lead, cadmium and zinc. Chapter 28 studies the marked reductions that have occurred in these pollutants in the snow of Greenland over recent years.

19

MAN-MADE CLIMATIC CHANGES*

H. E. Landsberg

Climate, the totality of weather conditions over a given area, is variable. Although it is not as fickle as weather, it fluctuates globally as well as locally in irregular pulsations. In recent years some people have voiced the suspicion that human activities have altered the global climate, in addition to having demonstrated effects on local microclimates. There have also been a number of proposals advocating various schemes for deliberately changing global climate, and a number of actual small-scale experiments have been carried out. For most of the larger proposals, aside from considerations of feasibility and cost, one can raise the objection that a beneficial effect in one part of the earth could well be accompanied by deterioration elsewhere, aside from the inevitable disturbances of the delicate ecological balances.

But the question "Has man inadvertently changed the global climate, or is he about to do so?" is quite legitimate. It has been widely discussed publicly – unfortunately with more zeal than insight. Like so many technical questions fought out in the forum of popular magazines and the daily press, the debate has been characterized by misunderstandings, exaggerations, and distortions. There have been dire predictions of imminent catastrophe by heat death,

by another ice age, or by acute oxygen deprivation. The events foreseen in these contradictory prophesies will obviously not all come to pass at the same time, if they come to pass at all. It seems desirable to make an attempt to sort fact from fiction and separate substantive knowledge from speculation.

Natural Climatic Fluctuations

In order to assess man's influence, we must first take a look at nature's processes.

The earth's atmosphere has been in a state of continuous slow evolution since the formation of the planet. Because of differences in the absorptive properties of different atmospheric constituents, the energy balance near the surface has been undergoing parallel evolution. Undoubtedly the greatest event in this evolution has been the emergence of substantial amounts of oxygen, photosynthetically produced by plants (1). The photochemical development of ozone in the upper atmosphere, where it forms an absorbing layer for the short-wave ultraviolet radiation and creates a warm stratum, is climatically also very important, especially for the forms of organic life now in existence.

* Originally published in *Science*, 1970, vol. 170, pp. 1265–8.

But for the heat balance of the earth, carbon dioxide (CO_2) and water vapor, with major absorption bands in the infrared, are essential constituents. They absorb a substantial amount of the dark radiation emitted by the earth's surface. The condensed or sublimated parts of the atmospheric water vapor also enter prominently into the energy balance. In the form of clouds they reflect incoming short-wave radiation from the sun, and hence play a major role in determining the planetary albedo. At night, clouds also intercept outgoing radiation and radiate it back to the earth's surface (2).

Over the past two decades Budyko (3) has gradually evolved models of the global climate, using an energy balance approach. These models incorporate, among other important factors, the incoming solar radiation, the albedo, and the outgoing radiation. Admittedly they neglect, as yet, nonlinear effects which might affect surface temperatures (4) but it seems unlikely that, over a substantial period, the nonlinear effects of the atmosphere–ocean system will change the basic results, though they may well introduce lags and superimpose rhythms. Budyko's calculations suggest that a 1.6 percent decrease in incoming radiation or a 5 or 10 percent increase in the albedo of the earth could bring about renewed major glaciation.

The theory that changes in the incoming radiation are a principal factor governing the terrestrial climate has found its major advocate in Milankovitch (5). He formulated a comprehensive mathematical model of the time variations of the earth's position in space with respect to the sun. This included the periodic fluctuations of the inclination of the earth's axis, its precession, and the eccentricity of its orbit. From these elements he calculated an insolation curve back into time and the corresponding surface temperature of the earth. He tried to correlate minima with the Pleistocene glaciations. These views have found considerable support in isotope investigations, especially of the $^{18}O/^{16}O$ ratio in marine shells (6) deposited during the Pleistocene. Lower ^{18}O amounts correspond to lower temperatures. Budyko and others (7) raise some doubts that

Milankovitch's theory can explain glaciations but admit that it explains some temperature fluctuations. For the last 1700 years there is also evidence that the ^{18}O content of Greenland glacier ice is inversely correlated to a solar activity index based on auroral frequencies (8). Again, low values of ^{18}O reflect the temperature at which the precipitation that formed the firn fell.

The fluctuations of externally received energy are influenced not only by the earth's position with respect to the sun but also by changes in energy emitted by the sun. Extraterrestrial solar radiation fluctuates with respect to spectral composition, but no major changes in total intensity have yet been measured outside the atmosphere. The occurrence of such fluctuations is indicated by a large number of statistical studies (9), but ironclad proof is still lacking. Such fluctuations are of either long or short duration. They have been tied to the solar activity cycle. Inasmuch as details are yet unknown, their effect on climate is at present one factor in the observed "noise" pattern.

In the specific context of this discussion, we are not concerned with the major terrestrial influences on climate, such as orogenesis, continental drift, and pole wanderings. But other, somewhat lesser, terrestrial influences are also powerful controllers of climate. They include volcanic eruptions that bring large quantities of dust and CO_2 into the air, and natural changes of albedo such as may be caused by changes in snow and ice cover, in cloudiness, or in vegetation cover (10). The fact that we have not yet succeeded in disentangling all the cause-and-effect relations of natural climatic changes considerably complicates the analysis of possible man-made changes.

The Climatic Seesaw

It was only a relatively short time ago that instrumental records of climate first became available. Although broad-scale assessments of climate can be made from natural sources, such as tree rings (11) or pollen associations, and, in historical times, from chronicles that list crop conditions or river freezes, this is tenuous evidence. But a

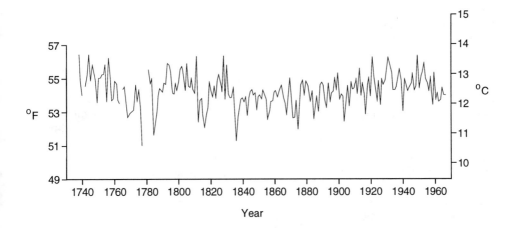

Figure 19.1 [*orig. figure 1*]   Annual temperatures for the eastern seaboard of the United States for the period 1738 to 1967 – a representative, reconstructed synthetic series centered on Philadelphia.

considerable number of instrumental observations of temperatures and precipitation are available for the period from the early 18th century to the present, at least for the Northern Hemisphere. These observations give a reasonably objective view of climatic fluctuations for the last two and a half centuries. This is, of course, the interval in which man and his activities have multi-

plied rapidly. These long climatic series are mostly from western Europe (*12*), but recently a series for the eastern seaboard of the United States has been reconstructed from all available data sources. In this series Philadelphia is used as an index location, since it is centrally located with respect to all the earlier available records (*13*). Figures 19.1 and 19.2 show the annual values for

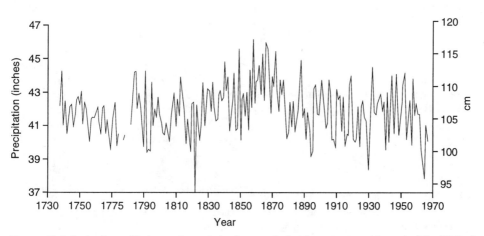

Figure 19.2 [*orig. figure 2*]   Annual precipitation totals for the eastern seaboard of the United States for the period 1738 to 1967 – a representative, reconstructed synthetic series centered on Philadelphia.

temperature and precipitation for a 230-year span; there are some minor gaps where the data were inadequate. These curves are characteristic of those for other regions, too. In particular they reflect the restlessness of the atmosphere. Many analysts have simply considered the variations to be quasi-random. Here I need only say that they do not reflect any pronounced one-sided trends. However, there are definite long or short intervals in which considerable one-sided departures from a mean are notable. On corresponding curves representing data for a larger area that encompasses most of the regions bordering the Atlantic, the major segments are those for the late 18th century, which was warm; the 19th century, which was cool; and the first half of the 20th century, in which there was a notable rising trend. This trend was followed by some cooling in the past two decades.

In the precipitation patterns, "noise" masks all trends, but we know that during a period in the middle of the last century there was considerably more precipitation than there is now. For shorter intervals, spells of drought alternate with high precipitation. Sometimes, for small areas, these can be quite spectacular. An example is the seasonal snowfall on Mount Washington, in New Hampshire; there the snowfall increased from an average of 4.5 meters in the winters of 1933–34 to 1949–50 to an annual average of 6 meters in the period 1951–2 to 1966–67 (14). Yet these values should not be taken as general climatic trends for the globe, or even for the hemisphere. Even if we take indices that integrate various climatic influences, we still cannot make categorical statements. Glacier conditions are typical in this group of indices. For example, the glaciers on the west coast of Greenland have been repeatedly surveyed since 1850. In consonance with temperature trends for lower latitudes, they showed their farthest advances in the 7th decade of the 19th century and have been retreating ever since (15). This pattern fits the temperature curves to the 1950 turning point, but, although glaciers in some regions of the world have been advancing since then, this is by no means true of all glaciers. The ques-

tion of whether these changes reflect (i) relatively short-term temperature fluctuations, or (ii) alterations in the alimenting precipitation, or (iii) a combination of these two factors remains unanswered.

Many of the shorter fluctuations are likely to be only an expression of atmospheric interaction with the oceans. Even if external or terrestrial impulses affect the energy budget and cause an initial change in atmospheric circulation, notable lag and feedback mechanisms involving the oceans produce pulsations which, in turn, affect the atmosphere (16). The oceans have a very large thermal inertia, and their horizontal motions and vertical exchanges are slow. Namias (17) has investigated many of the fluctuations of a few years' duration. He concluded, for example, that drought conditions on the eastern seaboard of the United States in the 1960's were directly affected by the prevailing wind system and by sea-surface temperatures in the vicinity but that the real dominant factor was a wind-system change in the North Pacific. Such teleconnections (relations among conditions in distant parts of the globe) complicate interpretations of local or even regional data tremendously. The worldwide effect of changes in the Pacific wind system is obvious from Namias's estimate that accelerations and decelerations cause large-scale breaks in the regime of sea-surface temperatures. These seem to occur in sequences of approximately 5 years and may cause temperature changes of 0.5°C over the whole North Pacific. Namias estimates that this can cause differences of 8×10^{18} grams in the annual amounts of water evaporated from the surface. The consequences for worldwide cloud and rain formation are evident. It is against this background that we have to weigh climatic changes allegedly wrought by man.

Carbon Dioxide

The fact that the atmospheric gases play an important role in the energy budget of the earth was recognized early. Fourier, and then Pouillet and Tyndall, first expressed the idea that these gases acted as a "greenhouse"

(18). After the spectrally selective absorption of gases was recognized, their role as climatic controls became a subject of wide debate. The capability of CO_2 to intercept long-wave radiation emitted by the earth was put forward as a convenient explanation for climatic changes. Arrhenius *(19)* made the first quantitative estimates of the magnitude of the effect, which he mainly attributed to fluctuating volcanic activity, although he also mentioned the burning of coal as a minor source of CO_2. The possibility that man-made CO_2 could be an important factor in the earth's heat balance was not seriously considered until Callendar *(20)*, in 1938, showed evidence of a gradual increase in CO_2 concentration in the earth's atmosphere. But it was Plass *(21)* who initiated the modern debate on the subject, based on his detailed study of the CO_2 absorption spectrum. The crucial question is, How much has CO_2 increased as a result of the burning of fossil fuels? It is quite difficult to ascertain even the mean amount of CO_2 in the surface layers of the atmosphere, especially near vegetation. There are large diurnal and annual variations. Various agriculturists have reported concentrations ranging from 210 to 500 parts per million. The daily amplitudes during the growing season are about 70 parts per million *(22)*. Nearly all early measurements were made in environments where such fluctuations took place. This, together with the lack of precision of the measurements, means that our baseline – atmospheric CO_2 concentrations prior to the spectacular rise in fossil fuel consumption of this century – is very shaky. Only since the International Geophysical Year have there been some regularly operating measuring points in polar regions and on high mountains and reliable data from the oceans which give some firm information on the actual increase *(23)*.

The best present estimate places the increase in atmospheric CO_2 since 1860 at 10 to 15 percent. This is hardly a spectacular change, but the rate of increase has been rising, and various bold extrapolations have been made into the 21st century. Much depends on the sinks for CO_2 which at present are not completely known. At present

concentrations, atmospheric O_2 and CO_2 stay in approximate equilibrium, through the photosynthetic process in plants. It is estimated that 150×10^9 tons of CO_2 per year are used in photosynthesis *(24)*. A corresponding amount is returned to the atmosphere by decay, unless the total volume of plant material increases *(25)*. This volume is one of the unknowns in the estimates of CO_2 balance. Perhaps satellite sensors can give some bulk information on that point in the future. The oceans are a major sink for CO_2. The equilibrium with the bicarbonates dissolved in seawater determines the amount of CO_2 in the atmosphere. In the exchange between atmosphere and ocean, the temperature of the surface water enters as a factor. More CO_2 is absorbed at lower surface-water temperatures than at higher temperatures. I have already pointed out the fact that surface-water temperatures fluctuate over long or short intervals; most of these ups and downs are governed by the wind conditions. The interchange of the cold deep water and the warm surface water through downward mixing and upwelling, in itself an exceedingly irregular process, controls, therefore, much of the CO_2 exchange *(26)*. Also, the recently suggested role of an enzyme in the ocean that facilitates absorption of CO_2 has yet to be explored. Hence it is quite difficult to make long-range estimates of how much atmospheric CO_2 will disappear in the oceanic sink. Most extrapolators assume essentially a constant rate of removal. Even the remaining question of how much the earth's temperature will change with a sharp increase in the CO_2 content of the atmosphere cannot be unambiguously answered. The answer depends on other variables, such as atmospheric humidity and cloudiness. But the calculations have been made on the basis of various assumptions. The model most widely used is that of Manabe and Wetherald *(27)*. They calculate, for example, that, with the present value for average cloudiness, an increase of atmospheric CO_2 from 300 to 600 parts per million would lead to an increase of 2°C in the mean temperature of the earth at the surface. At the same time the lower stratosphere would cool by 15°C. At the present rate of accumulation of CO_2 in the atmosphere, this doubling of the CO_2 would take about 400 years. The envisaged 2°C rise can hardly be called cataclysmic. There have been such worldwide changes within historical times. Any change attributable to the rise in CO_2 in the last century has certainly been submerged in the climatic "noise". Besides, our estimates of CO_2 production by natural causes, such as volcanic exhalations and organic decay, are very inaccurate; hence the ratio of these natural effects to anthropogenic effects remains to be established.

Dust

The influence on climate of suspended dust in the atmosphere was first recognized in relation to volcanic eruptions. Observations of solar radiation at the earth's surface following the spectacular eruption of Krakatoa in 1883 showed measurable attenuation. The particles stayed in the atmosphere for 5 years (28). There was also some suspicion that summers in the Northern Hemisphere were cooler after the eruption. The inadequacy and unevenness of the observations make this conclusion somewhat doubtful. The main exponent of the hypothesis that volcanic dust is a major controller of terrestrial climate was W. J. Humphreys (29). In recent years the injection into the atmosphere of a large amount of dust by an eruption of Mount Agung has renewed interest in the subject, not only because of the spectacular sunsets but also because there appears to have been a cooling trend since (30). The Mount Agung eruption was followed, in the 1960's, by at least three others from which volcanic constituents reached stratospheric levels: those of Mount Taal, in 1965; Mount Mayon, in 1968; and Fernandina, in 1968. Not only did small dust particles reach the stratosphere but it seems likely that gaseous constituents reaching these levels caused the formation of ammonium sulfate particles through chemical and photochemical reactions (31). The elimination of small particulates from the stratosphere is rela-

tively slow, and some backscattering of solar radiation is likely to occur.

As yet man cannot compete in dust production with the major volcanic eruptions, but he is making a good try. However, most of his solid products that get into the atmosphere stay near the ground, where they are fairly rapidly eliminated by fallout and washout. Yet there is some evidence that there has been some increase in the atmospheric content of particles less than 10^{-4} centimeter in diameter (32). The question is simply, What is the effect of the man-made aerosol? There is general agreement that it depletes the direct solar radiation and increases radiation from the sky. Measurements of the former clearly show a gradual increase in turbidity (33), and the same increase in turbidity has been documented by observations from the top of Mauna Loa, which is above the level of local contamination (34). From these observations the conclusion has been drawn that the attenuation of direct solar radiation is, in part at least, caused by backscattering of incoming solar radiation to space. This is equivalent to an increase in the earth's albedo and hence is being interpreted as a cause of heat loss and lowered temperatures (35). But things are never that categorical and simple in the atmosphere. The optical effects of an aerosol depend on its size distribution, its height in the atmosphere, and its absorptivity. These properties have been studied in detail by a number of authors (36). It is quite clear that most man-made particulates stay close to the ground. Temperature inversions attend to that. And there is no evidence that they penetrate the stratosphere in any large quantities, especially since the ban, by most of the nuclear powers, of nuclear testing in the atmosphere. The optical analyses show, first of all, that the backscatter of the particles is outweighed at least 9 to 1 by forward scattering. Besides, there is a notable absorption of radiation by the aerosol. This absorption applies not only to the incoming but also to the outgoing terrestrial radiation. The effectiveness of this interception depends greatly on the overlapping effect of the water vapor of the atmosphere. Yet the net effect of the man-made particulates seems to be that they lead to heating of the atmospheric layer in which they abound. This is usually the stratum hugging the ground. All evidence points to temperature rises in this layer, the opposite of the popular interpretations of the dust effect. The aerosol and its fallout have other, perhaps much more far-reaching, effects, which I discuss below. Suffice it to say, here, that man-made dust has not yet had an effect on global climate beyond the "noise" level. Its effect is puny as compared with that of volcanic eruptions, whose dust reaches the high stratosphere, where its optical effect, also, can be appreciable. No documented case has been made for the view that dust storms from deserts or blowing soil have had more than local or regional effects.

Dust that has settled may have a more important effect than dust in suspension. Dust fallen on snow and ice surfaces radically changes the albedo and can lead to melting (37). Davitaya (38) has shown that the glaciers of the high Caucasus have an increased dust content which parallels the development of industry in eastern Europe. Up until 1920 the dust content of the glacier was about 10 milligrams per liter. In the 1950's this content increased more than 20-fold, to 235 milligrams per liter. So long as the dust stays near the surface, it should have an appreciable effect on the heat balance of the glacier. There is fairly good evidence, based on tracers such as lead, that dusts from human activities have penetrated the polar regions. Conceivably they might change the albedo of the ice, cause melting, and thus pave the way for a rather radical climatic change – and for a notable rise in sea level. There has been some speculation along this line (39), but, while these dusts have affected microclimates, there is no evidence of their having had, so far, any measurable influence on the earth's climate. The possibility of deliberately causing changes in albedo by spreading dust on the arctic sea ice has figured prominently in discussions of artificial modification of climate. This seems technologically feasible (40). The consequences for the mosaic of climates in the lower latitudes have not yet been assessed. Present computer models of

world climate and the general circulation are far too crude to permit assessment in the detail necessary for ecological judgments.

All of the foregoing discussion applies to the large-scale problems of global climate. On that scale the natural influences definitely have the upper hand. Although monitoring and vigilance is indicated, the evidence for man's effects on global climate is flimsy at best. This does not apply to the local scale, as we shall presently see.

Extraurban Effects

For nearly two centuries it has been said that man has affected the rural climates simply by changing vast areas from forest to agricultural lands. In fact, Thomas Jefferson suggested repetitive climatic surveys to measure the effects of this change in land use in the virgin area of the United States (41). Geiger has succinctly stated that man is the greatest destroyer of natural microclimates (42). The changeover from forest to field locally changes the heat balance. This leads to greater temperature extremes at the soil surface and to altered heat flux into and out of the soil. Cultivation may even accentuate this. Perhaps most drastically changed is the low-level wind speed profile because of the radical alteration in aerodynamic roughness. This change leads to increased evaporation and, occasionally, to wind erosion. One might note here that man has reversed to some extent the detrimental climatic effects of deforestation in agricultural sectors, by planting hedges and shelter belts of trees. Special tactics have been developed to reduce evaporation, collect snow, and ameliorate temperature ranges by suitable arrangements of sheltering trees and shrubs (43).

The classical case of a local man-made climatic change is the conversion of a forest stand to pasture, followed by overgrazing and soil erosion, so that ultimately nothing will grow again. The extremes of temperature to which the exposed surface is subjected are very often detrimental to seedlings, so that they do not become established. Geiger pointed this out years ago. But not all grazing lands follow the cycle outlined above. Sometimes it is a change in the macroclimate that tilts the balance one way or another (44).

Since ancient times man has compensated for vagaries of the natural climates by means of various systems of irrigation. Irrigation not only offsets temporary deficiencies in rainfall but, again, affects the heat balance. It decreases the diurnal temperature ranges, raises relative humidities, and creates the so-called "oasis effect." Thornthwaite, only a decade and a half ago, categorically stated that man is incapable of deliberately causing any significant change in the climatic patterns of the earth. Changes in microclimate seemed to him so local and trivial that special instrumentation was needed to detect them. However, "Through changes in the water balance and sometimes inadvertently, he exercises his greatest influence on climate" (45).

What happens when vast areas come under irrigation? This has taken place over 62×10^3 square kilometers of Oklahoma, Kansas, Colorado, and Nebraska since the 1930's. Some meteorologists have maintained that about a 10 percent increase in rainfall occurs in the area during early summer, allegedly attributable to moisture reevaporated from the irrigated lands (46). Synoptic meteorologists have generally made a good case for the importation, through precipitation, of moisture from marine sources, especially the Gulf of Mexico. Yet ^3H determinations have shown, at least for the Mississippi valley area, that two-thirds of the precipitated water derives from locally evaporated surface waters. Anyone who has ever analyzed trends in rainfall records will be very cautious about accepting apparent changes as real until many decades have passed. For monthly rainfall totals, 40 to 50 years may be needed to establish trends because of the large natural variations (47).

This century has seen, also, the construction of very large reservoirs. Very soon after these fill they have measurable influences on the immediate shore vicinity. These are the typical lake effects. They include reduction in temperature extremes, an increase in humidity, and small-scale circulations of the land- and lake-breeze type, if the reservoir is large enough. Rarely do we have long records as a basis for comparing conditions before and after establishment of the reservoir. Recently, Zych and Dubaniewicz (48) published such a study for the 30-year-old reservoir of the Nysa Klodzka river in Poland, about 30 square kilometers in area. At the town of Otmuchow, about 1 kilometer below the newly created lake, a 50-year temperature normal was available (for the years 1881 to 1930). In the absence of a regional trend there has been an increase in the annual temperature of 0.7°C at the town near the reservoir. It is now warmer below the dam than above it, whereas, before, the higher stations were warmer because of the temperature inversions that used to form before the water surface exerted its moderating influence. It is estimated that precipitation has decreased, because of the stabilizing effect of the large body of cool water. Here, as elsewhere, the influence of a large reservoir does not extend more than 1 to 3 kilometers from the shore. Another form of deliberate man-controlled interference with microclimate, with potentially large local benefits, is suppression of evaporation by monomolecular films. Where wind speeds are low, this has been a highly effective technique for conserving water. The reduction of evaporation has led to higher water surface temperature, and this may be beneficial for some crops, such as rice (49).

The reduction of fog at airports by seeding of the water droplets also belongs in this category of man-controlled local changes. In the case of super-cooled droplets, injection of suitable freezing nuclei into the fog will cause freezing of some drops, which grow at the expense of the remaining droplets and fall out, thus gradually dissipating the fog. For warm fogs, substances promoting the growth or coalescence of droplets are used. In many cases dispersal of fog or an increase in visual range sufficient to permit flight operations can be achieved (50). Gratifying though this achievement is for air traffic, it barely qualifies as even a microclimatic change because of the small area and brief time scale involved. Similarly, the changes produced by artificial heating in orchards and vineyards to combat frosts hardly qualify as microclimatic changes.

Finally, a brief note on general weather modification is in order. Most of the past effort in this field has been devoted to attempts to augment rainfall and suppress hail. The results have been equivocal and variously appraised (51). The technique, in all cases, has been cloud seeding by various agents. This produces undoubted physical results in the cloud, but the procedures are too crude to permit prediction of the outcome. Thus, precipitation at the ground has been both increased and decreased (52). The most reliable results of attempts to induce rainfall have been achieved through seeding clouds forming in up-slope motions of winds across mountains and cap clouds (53). Elsewhere targeting of precipitation is difficult, and the effects of seeding downwind from the target area are not well known. No analysis has ever satisfactorily shown whether cloud seeding has actually caused a net increase in precipitation or only a redistribution. In any case, if persistently practiced, cloud seeding could bring about local climatic changes. But an ecological question arises: If we can do it, should we? This point remains controversial.

Attempts to suppress hail by means of cloud seeding are also still in their infancy. Here the seeding is supposed to achieve the production of many small ice particles in the cloud, to prevent any of them from growing to a size large enough to be damaging when they reach the ground. The seeding agent is introduced into the hail-producing zone of cumulonimbus – for example, by ground-fired projectiles. Some successes have been claimed, but much has yet to be learned before one would acclaim seeding as a dependable technology for eliminating this climatic hazard (54).

Hurricane modification has also been attempted. The objective is reduction of damage caused by wind and storm surges. Seeding of the outer-wall clouds around the eye of the storm is designed to accomplish this. The single controlled experiment that has been performed, albeit successfully in the predicted sense, provides too tenuous a basis for appraising the potential of this tech-nique (55). Here again we have to raise the warning flag because of the possibility of simultaneous change in the pattern of rainfall accompanying the storm. In many regions tropical storm rain is essential for water supply and agriculture. If storms are diverted or dissipated as a result of modification, the economic losses resulting from altered rainfall patterns may outweigh the advantages gained by wind reduction (56). As yet such climatic modifications are only glimpses on the horizon.

General Urban Effects

By far the most pronounced and locally far-reaching effects of man's activities on microclimate have been in cities. In fact, many of these effects might well be classified as mesoclimatic. Some of them were recognized during the last century in the incipient metropolitan areas. Currently the sharply accelerated trend toward urbanization has led to an accentuation of the effects. The problem first simply intrigued meteorologists, but in recent years some of its aspects have become alarming. Consequently the literature in this field has grown rapidly and includes several reviews summarizing the facts (57).

We are on the verge of having a satisfactory quantitative physical model of the effect of cities on the climate. It combines two major features introduced by the process of urbanization. They concern the heat and water balance and the turbulence conditions. To take changes in turbulence first, the major contributory change is an increase in surface roughness. This affects the wind field and, in particular, causes a major adjustment in the vertical wind profile so that wind speeds near the surface are reduced. The structural features of cities also increase the number of small-scale eddies and thus affect the turbulence spectrum.

The change in the heat balance is considerably more radical. Here, when we change a rural area to an urban one, we convert an essentially spongy surface of low heat conductivity into an impermeable layer with high capacity for absorbing and conducting heat. Also, the albedo is usually lowered.

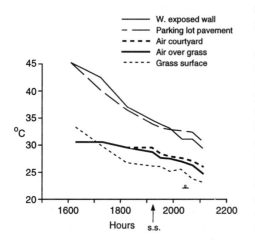

Figure 19.3 [*orig. figure 3*] A typical example of microclimatic heat island formation in incipient urbanization. The top two curves show radiative temperatures of wall and parking lot pavement on a clear summer evening (6 August 1968). The two middle curves show air temperatures (at elevation of 2 meters) in the paved courtyard and over an adjacent grass surface; from sunset (*s.s.*) onward, the courtyard is warmer than the air over the grass. The bottom (dashed) curve gives the radiative temperature of grass. The symbol at 2030 hours indicates the start of dew formation.

These radical changes in surface that accompany the change from rural to urban conditions lead to rapid runoff of precipitation and consequently to a reduction in local evaporation. This is, of course, equivalent to a heat gain – one which is amplified by radiative heat gain resulting from the lowering of the albedo. This heat is effectively stored in the stone, concrete, asphalt, and deeper compacted soil layers of the city. In vegetated rural areas usually more incoming radiation is reflected and less is stored than in the city. Therefore structural features alone favor a strongly positive heat balance for the city. To this, local heat production is added. The end result is what has been called the urban heat island, which leads to increased convection over the city and to a city-induced wind field that dominates when weather patterns favor weak general air flow.

Most of the features of the near-the-surface climatic conditions implied by this model have, over the years, been documented by comparisons of measurements made within the confines of cities and in their rural surroundings, mostly at airports. Such comparisons gave reasonably quantitative data on the urban effect, but some doubts remained. These stemmed from the fact that many cities were located in special topographic settings which favored the establishment of a city – such as a river valley, a natural harbor, or an orographic trough. They would by nature have a microclimate different from that of the surroundings. Similarly, airport sites were often chosen for microclimatic features favorable for aviation. Some of the uncertainties can be removed by observing atmospheric changes as a town grows. An experiment along this line was initiated 3 years ago in the new town of Columbia, Maryland. The results so far support earlier findings and have refined them (*58*).

Perhaps of most interest is the fact that a single block of buildings will start the process of heat island formation. This is demonstrated by air and infrared surface temperature measurements. An example is given in figure 19.3. The observations represented by the curves of figure 19.3 were made in a paved court enclosed by low-level structures which were surrounded by grass and vegetated surfaces. On clear, relatively calm evenings the heat island develops in the court, fed by heat stored in the daytime under the asphalted parking space of the court and the building walls. This slows down the radiative cooling process, relative to cooling from a grass surface, and keeps the air that is in contact with the surface warmer than that over the grass (*59*).

The heat island expands and intensifies as a city grows, and stronger and stronger winds are needed to overcome it (*60*). And although it is most pronounced on calm, clear nights, the effect is still evident in the long-term mean values. Figure 19.4 shows the isotherms in the Paris region, which is topographically relatively simple and

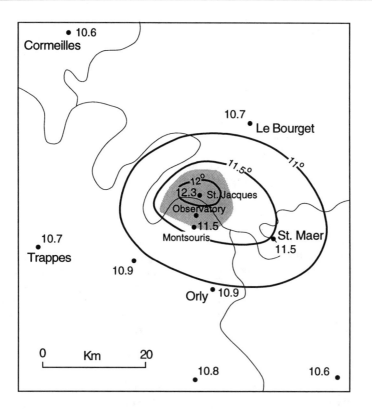

Figure 19.4 [*orig. figure 4*]   The urban heat island of Paris, shown by mean annual isotherms in degrees Celsius. The region is characterized by minimal orographic complexity. [After Dettwiller (*61*)].

without appreciable differences in elevation. A pronounced metropolitan heat island of about 1.6°C in the mean value can be seen. This is typical of major cities. In the early hours of calm, clear nights the city may be 6° to 8°C warmer than its surroundings. The Paris example is noteworthy because it has been demonstrated that the rise in temperature is not confined to the air but also affects the soil. It has been observed in a deep cave under the city, where temperatures have been measured for two centuries (*61*). Curiously enough, the cave temperature was once considered so invariant that the cave in question was proposed as one of the fixed points for thermometer scales. This artificially introduced trend in temperatures also plays havoc with the long-term temperature records from cities. They become suspect as guides for gaging the slow, natural climatic fluctuations.

Part of the rise in temperature must be attributed to heat rejection from human and animal metabolism, combustion processes, and air-conditioning units. Energy production of various types certainly accounts for a large part of it. In the urbanized areas the rejected energy has already become a measurable fraction of the energy received from the sun at the surface of the earth. Projection of this energy rejection into the next decades leads to values we should ponder. One estimate indicates that in the year 2000 the Boston-to-Washington megalopolis will have 56 million people living within an area of 30,000 square kilometers. The heat rejection will be about 65 calories per square centimeter per day. In winter this is about 50 percent, and in summer 15 percent, of the heat received by solar radiation on a horizontal surface (*62*). The eminent French geophysicist

J. Coulomb has discussed the implications of doubling the energy consumption in France every 10 years; this would lead to unbearable temperatures (63). It is one of a large number of reasons for achieving, as rapidly as possible, a steady state in population and in power needs.

An immediate consequence of the heat island of cities is increased convection over cities, especially in the daytime. That has been beautifully demonstrated by the lift given to constant-volume balloons launched across cities (64). The updraft leads, together with the large amount of water vapor released by combustion processes and steam power, to increased cloudiness over cities. It is also a potent factor in the increased rainfall reported from cities, discussed below in conjunction with air pollution problems. Even at night the heating from below will counteract the radiative cooling and produce a positive temperature lapse rate, while at the same time inversions form over the undisturbed countryside. This, together with the surface temperature gradient, creates a pressure field which will set a concentric country breeze into motion (65). A schematic circulation system of this type is shown in figure 19.5.

The rapid runoff of rainfall caused by the imperviousness of the surfaces of roads and roofs, as well as by the drainage system, is another major effect of cities. In minor rainfalls this has probably only the limited consequence of reducing the evaporation from the built-up area and thus eliminating much of the heat loss by the vaporization that is common in rural areas. But let there be a major rainstorm and the rapid runoff will immediately lead to a rapid rise of the draining streams and rivers. That can cause flooding and, with the unwise land use of flood plains in urban areas, lead to major damage. The flood height is linearly related to the amount of impervious area. For the 1- to 10-year recurrence intervals, flood heights will be increased by 75 percent for an area that has become 50 percent impervious, a value not at all uncommon in the usual urban setting. Observations in Hempstead, Long Island, have shown, for example, that, for a storm rainfall of 50 millimeters, direct runoff has increased from 3 millimeters in the interval from 1937 to 1943 to 7 millimeters in the interval from 1964 to 1966. This covers the time when the area changed from open fields to an urban community (66).

It is very difficult to document the

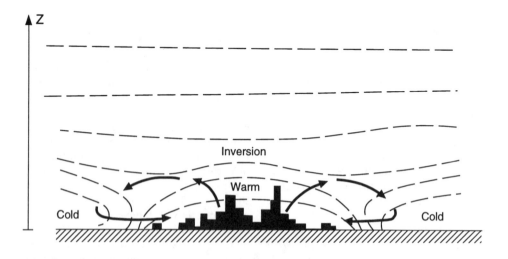

Figure 19.5 [*orig. figure 5*] Idealized scheme of nocturnal atmospheric circulation above a city in clear, calm weather. The diagram shows the urban heat island and the radiative ground inversions in the rural areas, a situation that causes a "country breeze" with an upper return current. (Dashed lines) Isotherms; (arrows) wind; Z, vertical coordinate.

decrease of wind speed over cities. Long records obtained with unchanged anemometer exposures at representative heights are scarce. Reasonable interpretations of available records suggest a decrease of about 25 percent from the rural equivalents. This is not unreasonable in the light of measurable increases in aerodynamic roughness. These are around 10 to 30 centimeters for meadows and cultivated fields and round 100 centimeters for woodland. There are several estimates for urban areas. I will give here a value calculated from the unique wind measurements on the Eiffel tower at a height of 316 meters, and from other wind records in the Paris region (67). These data yield values around 500 centimeters. They also suggest a decrease in wind at the top of the Eiffel tower from the interval 1890–1909 to the interval 1951–1960 of 0.4 meter per second, or 5 percent of the mean wind speed. In view of the height of this anemometer, this is quite a notable adjustment of the wind profile to the increase in terrain roughness.

Air Pollution Effects

Most spectacular among the effects of the city upon the atmospheric environment are those caused by air pollution. The catalogue of pollutants put into the air by man is long and has been commented upon in so many contexts that reference to the literature will have to suffice (68). Nor shall I dwell here on the special interactions of pollutants with the atmosphere in climatically and topographically specialized instances, such as the much investigated case of Los Angeles (69). I shall concentrate, instead, on the rather universal effects of pollutants on local climates.

Among these is the attenuation of solar radiation by suspended particulates. Although this affects the whole spectrum, it is most pronounced in the short wavelengths. The total direct radiation over most major cities is weakened by about 15 percent, sometimes more in winter and less in summer. The ultraviolet is reduced by 30 percent, on an average, and in winter often no radiation of wavelengths below 390

nanometers is received. The extinction takes place in a very shallow layer, as simultaneous measurements taken at the surface and from a tall steeple have shown (70).

Horizontally, the particulate haze interferes with visibility in cities. When shallow temperature inversions are present, the accumulation of aerosols can cause 80- or 90-percent reduction of the visual range as compared with the range for the general uncontaminated environment. The haze effect is accentuated by the formation of water droplets around hygroscopic nuclei, even below the saturation point. This is the more noteworthy because relative humidities near the surface are generally lower in cities than in the countryside. This is attributable partially to the higher temperatures and partially to the reduced evaporation. Nonetheless, fog occurs from two to five times as often in the city as in the surroundings. Fortunately, this seems to be a reversible process. Recent clean-up campaigns have shown that, through the use of smokeless fuels, considerable lessening of the concentration of particulates, and hence of fog and of the attenuation of light, can be achieved. In London, for example, with the change in heating practices, winter sunshine has increased by 70 percent in the last decade, and the winter visibilities have improved by a factor of 3 since the improvements were introduced (71).

I have alluded above to the increase in cloudiness over cities. It is likely that the enormous number of condensation nuclei produced by human activities in and around cities contributes to this phenomenon. Every set of measurements made has confirmed early assessments that these constituents are more numerous by one or two orders of magnitude in urbanized regions than in the country (72). Every domestic or industrial combustion process, principally motor vehicle exhaust, contributes to this part of the particulate. Independent evidence suggests that there is more rainfall over cities than over the surrounding countryside. But the evidence that pollutants are involved is tenuous. There is little doubt that the convection induced by the heat island can induce or intensify showers. This has been demonstrated for London, where apparently thundershowers yield 30 percent more rain than in the surrounding area (73). Orographic conditions would lead one to expect more showers in hilly terrain. This is not the case. Although this buoyancy effect is certainly at work, it does not stand alone: in some towns there are observations of precipitation increases from supercooled winter stratus clouds over urban areas. Some well-documented isolated cases of snow over highly industrialized towns suggest a cloud-seeding effect by some pollutants that may act as freezing nuclei (74). Also the rather startling variation of urban precipitation in accordance with the pattern of the human work week argues for at least a residual effect of nucleating agents produced in cities. The week is such an arbitrary subdivision of time that artificial forces must be at work. Observations over various intervals and in various regions indicate increased precipitation for the days from Monday through Friday as compared with values for Saturday and Sunday. These increases usually parallel the increase in industrialization, and, again, there is evidence for a more pronounced effect in the cool season (75).

Although most studies indicate that the increase in precipitation in urban areas is around 10 percent – that is, close to the limit of what could still be in the realm of sampling errors – some analyses have shown considerably larger increases in isolated cases. These instances have not yet been lifted out of the umbra of scientific controversy (76). But we should note here that some industrial activities and internal combustion engines produce nuclei that can have nucleating effects, at least on supercooled cloud particles. In the State of Washington in some regions that have become industrialized there is evidence of a 30-percent increase in precipitation in areas near the pulp mills over an interval of four decades (77). There are also incontrovertible observations of cloud banks forming for tens of kilometers in the plumes of power plants and industrial stacks. This is not necessarily associated with increased precipitation but raises the question of how far downwind

man's activities have caused atmospheric modifications.

In the absence of systematic three-dimensional observations, we have to rely on surface data. A recent study by Band (78) throws some light on the conditions. He found that, for a heat island 3°C warmer than its surroundings, a small but measurable temperature effect was still notable 3 kilometers to leeward of the town. Similarly, a substantial increase in the number of condensation nuclei was noted 3 kilometers downwind from a small town. In the case of a major traffic artery, an increased concentration of nuclei was measurable to 10 kilometers downwind. For a major city, radiation measurements have suggested that the smoke pall affects an area 50 times that of the built-up region. These values, which are probably conservative, definitely indicate that man's urbanized complexes are beginning to modify the mesoclimate.

As yet it is very difficult to demonstrate that any far-reaching climatic effects are the results of man's activities. If man-made effects on this scale already exist or are likely to exist in the future, they will probably be a result of the vast numbers of anthropogenic condensation and freezing nuclei. Among the latter are effective nucleating agents resulting from lead particles in automobile exhaust. These particles have become ubiquitous, and if they combine with iodine or bromine they are apt to act as freezing nuclei. Schaefer and others have pointed out that this could have effects on precipitation far downwind (79). These inadvertent results would lead either to local increases in precipitation or to a redistribution of natural precipitation patterns. They are, however, among the reversible man-made influences. As soon as lead is no longer used as a gasoline additive – which, hopefully, will be soon – the supply of these nucleating agents will stop and the influence, whatever its importance, should vanish promptly because of the relatively short lifetime of these nuclei.

Perhaps more serious, and much more difficult to combat, is the oversupply of condensation nuclei. Gunn and Phillips pointed out years ago that, if too many hygroscopic particles compete for the available moisture, cloud droplets will be small and the coalescence processes will become inhibited (*80*). This could lead to decreases in precipitation, a view that has recently been confirmed (*81*).

There remains one final area of concern: pollution caused by jet aircraft. These aircraft often leave persistent condensation trails. According to one school of thought, these artificial clouds might increase the earth's albedo and thus cause cooling. Although on satellite pictures one can occasionally see cloud tracks that might have originated from these vapor trails, they seem to be sufficiently confined, with respect to space and time, to constitute a very minute fraction of the earth's cloud cover. The other view of the effect of these vapor trails, which change into cirriform clouds, is that ice crystals falling from them may nucleate other cloud systems below them and cause precipitation. Any actual evidence of such events is lacking. And then we have the vivid speculations concerning weather modifications by the prospective supersonic transport planes. For some time military planes have operated at the altitudes projected for the supersonic transports. The ozone layer has not been destroyed, and no exceptional cloud formations have been reported. The water vapor added by any probable commercial fleet would be less than 10^{-9} of the atmospheric water vapor; thus, no direct influence on the earth's heat budget can be expected. At any rate, it seems that the sonic boom is a much more direct and immediate effect of the supersonic transport than any possible impact it may have on climate (*82*).

There is little need to comment on the multitude of schemes that have been proposed to "ameliorate" the earth's climate. Most of them are either technologically or economically unfeasible. All of them would have side effects that the originators did not consider. The new trend toward thinking in ecological terms would lead us to require that much more thoroughgoing analyses of the implications of these schemes be made than have been made so far before any steps are taken toward their implementation (*83*).

Summary

Natural climatic fluctuations, even those of recent years, cover a considerable range. They can be characterized as a "noise" spectrum which masks possible global effects of man-caused increases of atmospheric CO_2 and particulates. Local modifications, either deliberate or inadvertent, measurably affect the microclimate. Some artificial alterations of the microclimate are beneficial in agriculture. Among the unplanned effects, those produced by urbanization on local temperature and on wind field are quite pronounced. The influences on rainfall are still somewhat controversial, but effects may extend considerably beyond the confines of metropolitan areas. They are the result of water vapor released by human activity and of the influence of condensation and freezing nuclei produced in overabundance by motor vehicles and other combustion processes. Therefore it appears that on the local scale man-made influences on climate are substantial but that on the global scale natural forces still prevail. Obviously this should not lead to complacency. The potential for anthropogenic changes of climate on a larger and even a global scale is real. At this stage activation of an adequate worldwide monitoring system to permit early assessment of these changes is urgent. This statement applies particularly to the surveillance of atmospheric composition and radiation balance at sites remote from concentrations of population, which is now entirely inadequate. In my opinion, man-made aerosols, because of their optical properties and possible influences on cloud and precipitation processes, constitute a more acute problem than CO_2. Many of their effects are promptly reversible; hence, one should strive for elimination at the source. Over longer intervals, energy added to the atmosphere by heat rejection and CO_2 absorption remain matters of concern.

References and Notes

1   L. V. Berkner and L. S. Marshall, *Advan. Geophys.* 12, 309 (1967); S. I. Rasool, *Science* 157, 1466 (1967).
2   The climatic consequences of an original

single continent, continental drift, changing ocean size, and changing positions of the continents with respect to the poles are not discussed here.

3   M. I. Budyko, *Sov. Geogr.: Rev. Transl.* 10, 429 (1969); *J. Appl. Meteorol.* 9, 310 (1970). For a discussion and extension of Budyko's models, see W. D. Sellers, *ibid.* 8, 392 (1969); *ibid.* 9, 311 (1970).

4   For a recent review of the principal thoughts in this area, based primarily on work by C. E. P. Brooks (1951), W. D. Sellers (1965), and M. I. Budyko (1968), see H. L. Ferguson, *Atmosphere* 6, 133 (1968); *ibid.*, p. 145; *ibid.*, p. 151.

5   M. Milankovitch, "Canon of Insolation and the Ice-Age Problem," translation of *Kgl. Serbische Akad. Spec. Publ.* 132 (1941) by *Israel Program Sci. Transl.* (1969), *U.S. Dep. Comm. Clearing House Fed. Sci. Tech. Inform.*

6   C. Emiliani and J. Geiss, *Geol. Rundschau* 46, 576 (1957); C. Emiliani, *J. Geol.* 66, 264 (1958); *ibid.* 74, 109 (1966); *Science* 154, 851 (1966); W. S. Broecker, D. L. Thurber, J. Goddard, T. -L. Ku, R. K. Matthews, K. J. Mesolella, *ibid.* 159, 297 (1968).

7   M. I. Budyko, *Tellus* 21, 611 (1969); D. M. Shaw and W. L. Donn, *Science* 162, 1270 (1968).

8   J. R. Bray, *Science* 168, 571 (1970).

9   F. Baur, *Meteorol. Abhandl.* 50, No. 4 (1967).

10 For a recent review of the many factors causing climatic changes, see *Meteorol. Monogr.* 8, No. 30 (1968); for a divergent view on the problem, see L. R. Curry, *Ann. Ass. Amer. Geogr.* 52, 21 (1962); for factors involved in artificially induced changes, see H. Flohn, *Bonner Meteorol. Abhandl. No. 2* (1963).

11 H. C. Fritts, *Mon. Weather Rev.* 93, 421 (1965).

12 G. Manley, *Quart. J. Roy. Meteorol. Soc.* 79, 242 (1953); F. Baur, in Linke's *Meteorologisches Taschenbuch, Neue Ausgabe*, F. Baur, Ed. (Akademische Verlagsgesellschaft Geest und Portig, Leipzig, 1962), vol. 1, p. 710; Y. S. Rubinstein and L. G. Polozova, *Sovremennoe Izmenenie Klimata* (Gidrometeorolgicheskoe Izdatelstvo, Leningrad, 1966); H. H. Lamb, *The Changing Climate* (Methuen, London, 1966); H. von Rudloff, in *Europa seit dem Beginn der regelmässigen Instrumentenbeobachtungen (1670)* (Vieweg, Brunswick, 1967); H. E. Landsberg, *Weatherwise* 20, 52 (1967); M. Konček and K. Cehak, *Arch. Meteorol. Geophys. Bioklimatol. Ser. B Allg. Biol. Klimatol.* 16, 1 (1968); T. Anderson, "Swedish Temperature and Precipitation Records since the Middle of the 19th Century," *National Institute of Building Research, Stockholm, Document D4* (1970); for the Far East a particularly pertinent paper is H. Arakawa, *Arch. Meteorol. Geophys. Bioklimatol. Ser. B Allg. Biol. Klimatol.* 6, 152 (1964).

13 H. E. Landsberg, C. S. Yu, L. Huang, "Preliminary Reconstruction of a Long Time Series of Climatic Data for the Eastern United States," *Univ. Md. Inst. Fluid Dyn. Appl. Math. Tech. Note BN-571* (1968); for other assessments of climatic fluctuations in the United States, see also E. W. Wahl, *Mon. Weather Rev.* 96, 73 (1968); D. G. Baker, *Bull. Amer. Meteorol. Soc.* 41, 18 (1960).

14 C. W. Hurley, Jr., *Mt. Washington News Bull.* 10, No. 3, 13 (1969).

15 W. S. Carlson, *Science* 168, 396 (1970).

16 J. Bjerknes, *Advan. Geophys.* 10, 1 (1964); S. I. Rasool and J. S. Hogan, *Bull. Amer. Meteorol. Soc.* 50, 130 (1969); N. I. Yakovleva, *Izv. Acad. Sci. USSR, Atm. Ocean. Phys. Ser.* (American Geophysical Union translation) 5, 699 (1969).

17 J. Namias, in *Proc. Amer. Water Resources Conf. 4th* (1968), p. 852; *J. Geophys. Res.* 75, 565 (1970).

18 The term *greenhouse effect*, which has been commonly accepted for spectral absorption by atmospheric gases of long-wave radiation emitted by the earth, is actually a misnomer. Although the opaqueness of the glass in a greenhouse for long-wave radiation keeps part of the absorbed or generated heat inside, the seclusion of the interior space from advective and convective air flow is a very essential part of the functioning of a greenhouse. In the free atmosphere such flow is, of course, always present.

19 S. Arrhenius, *Worlds in the Making* (Harper, New York, 1908), pp. 51–54.

20 G. S. Callendar, *Quart. J. Roy. Meteorol. Soc.* 64, 223 (1938).

21 G. N. Plass, *Amer. J. Phys.* 24, 376 (1956).

22 W. Bischof and B. Bolin, *Tellus* 18, 155 (1966); K. W. Brown and N. J. Rosenberg, *Mon. Weather Rev.* 98, 75 (1970).

23 G. S. Callendar, *Tellus* 10, 253 (1958); B. Bolin and C. D. Keeling, *J. Geophys. Res.* 68, 3899 (1963); T. B. Harris, *Bull. Amer. Meteorol. Soc.* 51, 101 (1970); ESSA [*Environ. Sci. Serv. Admin.*] *Pam. ERLTM-APCL9* (series 33, 1970).

24 H. Lieth, *J. Geophys. Res.* 68, 3887 (1963).

25 E. K. Peterson, *Environ. Sci. Technol.* 3, 1162 (1969).

26 R. Revelle and H. E. Suess, *Tellus* 9, 18 (1957); H. E. Suess, *Science* 163, 1405 (1969); R. Berger and W. F. Libby, *ibid.* 164, 1395 (1969).

27 S. Manabe and R. T. Wetherald, *J. Atmos. Sci.* 24, 241 (1967).

28 G. J. Symons, Ed., *The Eruption of Krakatoa and Subsequent Phenomena* (Royal Society, London, 1888).

29 W. J. Humphreys, *Physics of the Air* (McGraw-Hill, New York, ed. 3, 1940), pp. 587–618.

30 R. A. Ebdon, *Weather* 22, 245 (1967); J. M. Mitchell, Jr. [personal communication and presentation in December 1969 at the Boston meeting of the AAAS] attributes about two-thirds of recent hemispheric cooling to volcanic eruptions.

31 A. B. Meinel and M. P. Meinel, *Science* 155, 189 (1967); F. E. Volz, *J. Geophys. Res.* 75, 1641 (1970).

32 In the 1930's I made a large number of counts of Aitken condensation nuclei [see H. Landsberg, *Mon. Weather Rev.* 62, 442 (1934); *Ergeb. Kosm. Phys.* 3, 155 (1938)]. These gave a background of ~100 to 200 nuclei per cubic centimeter. Measurements made in the last decade indicate an approximate doubling of this number [see C. E. Junge in *Atmosphärische Spurenstoffe und ihre Bedeutung für den Menschen* (1966 symposium, St. Moritz) (Birkhäuser, Basel, 1967)].

33 R. A. McCormick and J. H. Ludwig, *Science* 156, 1358 (1967).

34 J. T. Peterson and R. A. Bryson, *ibid.* 162, 120 (1968).

35 R. A. Bryson advocates this hypothesis. He states, in *Weatherwise* 21, 56 (1968): "All other factors being constant, an increase of atmospheric turbidity will make the earth cooler by scattering away more incoming sunlight. A decrease of dust should make it warmer." This remains a very simplified model, because "all other factors" never stay constant. See also W. M. Wendland and R. A. Bryson, *Biol. Conserv.* 2, 127 (1970). E. W. Barret in "Depletion of total short-wave irradiance at the ground by suspended particulates," a paper presented at the 1970 International Solar Energy Conference, Melbourne, Australia, calculates for various latitudes the depletion of radiation received at the ground because of dust. For geometrical reasons this is a more pronounced effect at higher than at lower latitudes. He therefore postulates that an order-of-magnitude increase in the amount of dust will redistribute the energy balance at the surface sufficiently to cause changes in the general circulation of the atmosphere.

36 W. T. Roach, *Quart. J. Roy. Meteorol. Soc.* 87, 346 (1961); K. Bullrich, *Advan. Geophys.* 10, 101 (1964); H. Quenzel, *Pure Appl. Geophys.* 71, 149 (1968); R. J. Charlson and M. J. Pilat, *J. Appl. Meteorol.* 8, 1001 (1969).

37 H. Landsberg, *Bull. Amer. Meteorol. Soc.* 21, 102 (1940); N. Georgievskii, *Sev. Morskoi Put. No. 13* (1939), p. 29; A. Titlianov, *Dokl. Vses (Ordena Lenina) Akad. Sel'skokhoz. Nauk Imeni V. I. Lenina* 6, No. 8, 8 (1941); A. I. Kolchin, *Les. Khoziaistvo* 3, 69 (1950); *Les i Step* 3, 77 (1951); G. A. Ausiuk, *Priroda (Moskva)* 43, No. 3, 82 (1954).

38 F. F. Davitaya, *Trans. Soviet Acad. Sci. Geogr. Ser. 1965 No. 2* (English translation) (1966), p. 3.

39 M. R. Block, *Paleogeogr. Paleoclimatol. Paleoecol.* 1, 127 (1965).

40 J. O. Fletcher, "The Polar Ocean and World Climate," *Rand. Corp.,* Santa Monica, Calif., *Publ. P-3801* (1968); "Managing Climatic Resources," *Rand. Corp.,* Santa Monica, Calif., *Publ. P-4000-1* (1969).

41 T. Jefferson, letter written from Monticello to his correspondent Dr. Lewis Beck of Albany, dated July 16, 1824.

42 R. Geiger, *Das Klima der bodennahen Luftschicht* (Vieweg, Brunswick, 1961), p. 503.

43 J. van Eimern, L. R. Razumova, G. W. Robinson, "Windbreaks and Shelterbelts," *World Meteorological Organ.,* Geneva, *Tech. Note No. 59* (1964); J. M. Caborn, *Shelterbelts and Windbreaks* (Faber and Faber, London, 1965).

44 I. A. Campbell, according to a news item in *Arid Land Research Newsletter No. 33* (1970), p. 10, studied the Shonto Plateau in northern Arizona, where he found that all gullies were stabilized, remaining just as they were 30 years ago. Yet there are now far more sheep in the area. He concluded that accelerated erosion there was caused by climatic variations and not by overgrazing.

45 C. W. Thornthwaite, in *Man's Role in Changing the Face of the Earth,* W. L. Thomas, Jr., Ed. (Univ. of Chicago Press, Chicago, 1956), p. 567.

46 L. A. Joos, "Recent rainfall patterns in the Great Plains," paper presented 21 October 1969 before the American Meteorological Society; F. Begemann and W. F. Libby,

Geochim. Cosmochim. Acta 12, 277 (1957).

47 In this context it is important to stress again the inadequacy of the ordinary rain gage as a sampling device. With about one gage per 75 square kilometers, we are actually sampling 5×10^{-10} of the area in question. But precipitation is usually unevenly distributed, especially when rain occurs in the form of showers. Then the sampling errors become very high. Even gages close to each other often show 10 percent differences in monthly totals. It takes, therefore, a long time to determine whether differences are significant or trends are real. This same caveat applies to analyses of rainmaking or to changes induced by effects of cities. This problem is often conveniently overlooked by statisticians unfamiliar with meteorological instruments and by enthusiasts with favorite hypotheses [see H. E. Landsberg, *Physical Climatology* (Gray, Dubois, Pa., ed. 2, 1966), p. 324; G. E. Stout, *Trans. Ill. Acad. Sci.* 53, 11 (1960)].

48 S. Zych and H. Dubaniewicz, *Zesz. Nauk. Univ. Lodz Riego Ser. II* 32, 3 (1969); S. Gregory and K. Smith, *Weather* 22, 497 (1967).

49 V. F. Pushkarev and G. P. Leochenko, *Sov. Hydrol. Select. Pap.* 3, 253 (1967); M. Gangopadhyaya and S. Venkataraman, *Agr. Meteorol.* 6, 339 (1969); R. Kapesser, R. Greif, I. Cornet, *Science* 166, 403 (1969).

50 W. B. Beckwith in "Human Dimensions of Weather Modification," *Univ. Chicago, Dep. Geogr. Res. Pap. No. 150* (1966), p. 195; B. A. Silverman, *Bull. Amer. Meteorol. Soc.* 51, 420 (1970).

51 "Weather and Climate Modification, Problems and Prospects," *Nat. Acad. Sci. Nat. Res. Counc. Publ. No. 1350* (1966); M. Neiburger, "Artificial Modification of Clouds and Precipitation," *World Meteorol. Organ., Geneva, Tech. Note No. 105* (1969); "Weather Modification, a Survey of the Present Status with Respect to Agriculture," *Res. Branch, Can. Dep. Agr., Ottawa, Publ.* (1970); M. Tribus, *Science* 168, 201 (1970).

52 L. Le Cam and J. Neyman, Eds., *Weather Modification Experiments* (Proceedings of the 5th Berkeley Symposium on Mathematical Statistics and Probability) (Univ. of California Press, Berkeley, 1967).

53 J. R. Stinson, in *Water Supplies for Arid Regions*, F. L. Gardner and L. E. Myers, Eds. (Univ. of Arizona Press, Tucson, 1967), p. 10; U.S. Department of the Interior, Office of Atmospheric Water Resources, Project

Skywater 1969 Annual Report, Denver (1970).

54 R. A. Schleusner, *J. Appl. Meteorol.* 7, 1004 (1968); "Metody vozdeistviia na gradovye protsessy," in *Vysokogornyi Geofiz. Trudy 11* (Gidrometeorologicheskoe Izdatelstvo, Leningrad, 1968).

55 R. C. Gentry, *Science* 168, 473 (1970).

56 G. W. Cry, "Effects of Tropical Cyclone Rainfall on the Distribution of Precipitation over the Eastern and Southern United States," ESSA [*Environ. Sci. Serv. Admin.*] *Prof. Pap. No. 1* (1967); A. L. Sugg, *J. Appl. Meteorol.* 7, 39 (1968).

57 H. E. Landsberg, in *Man's Role in Changing the Face of the Earth*, W. L. Thomas, Jr., Ed. (Univ. of Chicago Press, Chicago, 1956), p. 584; A. Kratzer, *Das Stadtklima*, vol. 90 of *Die Wissenshaft* (Vieweg, Brunswick, 1956); H. E. Landsberg, in "Air over Cities," *U.S. Pub. Health Serv. R. A. Taft Sanit. Eng. Center, Cincinnati, Tech. Rep. A 62-5* (1962); J. L. Peterson, "The Climate of Cities: A Survey of Recent Literature," *Nat. Air Pollut. Contr. Admin., Raleigh, N. C., Publ. No. AP-59* (1969).

58 P. M. Tag, in "Atmospheric Modification by Surface Influences," *Dep. Meteorol., Penn. State Univ., Rep. No. 15* (1969), pp. 1–71; M. A. Estoque, "A Numerical Model of the Atmospheric Boundary Layer," *Air Force Cambridge Res. Center, GRD Sci. Rep.* (1962); L. O. Myrup, *J. Appl. Meteorol.* 8, 908 (1969).

59 H. E. Landsberg, in "Urban Climates," *World Meteorol. Organ., Geneva, Tech. Note No. 108* (1970), p. 129.

60 T. R. Oke and F. G. Harnall, *ibid.*, p. 113.

61 J. Dettwiller, *J. Appl. Meteorol.* 9, 178 (1970).

62 R. T. Jaske, J. F. Fletcher, K. R. Wise, "A national estimate of public and industrial heat rejection requirements by decades through the year 2000 A.D.," paper presented before the American Institute of Chemical Engineers at its 67th National Meeting, Atlanta, (1970).

63 J. Coulomb, *News Report, Nat. Acad. Sci. Nat. Res. Counc.* 20, No. 3, 6 (1970).

64 W. A. Hass, W. H. Hoecker, D. H. Pack, J. K. Angell, *Quart. J. Roy. Meteorol. Soc.* 93, 483 (1967).

65 F. Pooler, *J. Appl. Meteorol.* 2, 446 (1963); R. E. Munn, in "Urban Climates," *World Meteorol. Organ., Geneva, Tech. Note No. 108* (1970), p. 15.

66 W. H. K. Espey, C. W. Morgan, F. D. Marsh, "Study of Some Effects of Urbanization on Storm Run-off from Small Watersheds,"

Texas Water Develop. Board Rep. No. 23 (1966); L. A. Martens, "Flood Inundation and Effects of Urbanization in Metropolitan Charlotte, North Carolina," *U.S. Geol. Surv. Water Supply Pap. 1591-C* (1968); G. E. Seaburn, "Effects of Urban Development on Direct Run-off to East Meadow Brook, Nassau County, Long Island, N. Y.," *U.S. Geol. Surv. Prof. Pap. 627-B* (1969).

67   J. Dettwiller, "Le vent au sommet de la tour Eiffel," *Monogr. Meteorol. Nat. No. 64* (1969).

68   See, for example, *Air Pollution*, A. C. Stern, Ed. (Academic Press, New York, ed. 2, 1968).

69   A. J. Hagen-Smit, C. E. Bradley, M. M. Fox, *Ind. Eng. Chem.* 45, 2086 (1953); J. K. Angell, D. H. Pack, G. C. Holzworth, C. R. Dickson, *J. Appl. Meteorol.* 5, 565 (1966); M. Neiburger, *Bull. Amer. Meteorol. Soc.*, 50, 957 (1969); in "Urban Climates," *World Meteorol. Organ., Geneva, Tech. Note No. 108* (1970), p. 248.

70   F. Lauscher and F. Steinhauser, *Sitzungsber. Wiener Akad. Wiss. Math. Naturw. Kl. Abt. 2a* 141, 15 (1932); *ibid.* 143, 175 (1934).

71   R. P. McNulty, *Atmos. Environ.* 2, 625 (1968); R. S. Charlson, *Environ. Sci. Technol.* 3, 913 (1969); R. O. McCaldin, L. W. Johnson, N. T. Stephens, *Science* 166, 381 (1969); C. G. Collier, *Weather* 25, 25 (1970); London Borough Association press release, quoted from UPI report of 14 Jan. 1970.

72   H. Landsberg, *Bull. Amer. Meteorol. Soc.* 18, 172 (1937).

73   B. W. Atkinson, "A Further Examination of the Urban Maximum of Thunder Rainfall in London, 1951–60," *Trans. Pap. Inst. Brit. Geogr. Publ. No. 48* (1969), p. 97.

74   J. von Kienle, *Meteorol. Rundschau* 5, 132 (1952); W. M. Culkowski, *Mon. Weather Rev.* 90, 194 (1962).

75   R. H. Frederick, *Bull. Amer. Meteorol. Soc.* 51, 100 (1970).

76   S. A. Changnon, in "Urban Climates," *World Meteorol. Organ., Geneva, Tech. Note 108* (1970), p. 325; B. G. Holzman and H. C. S. Thom, *Bull. Amer. Meteorol. Soc.* 51, 335 (1970); S. A. Changnon, *ibid.*, p. 337.

77   G. Langer, in *Proc. 1st Nat. Conf. Weather Modification*, Amer. Meteorol. Soc. (1968), p. 220; P. V. Hobbs and L. F. Radke, *J. Atmos. Sci.* 27, 81 (1970); *Bull. Amer. Meteorol. Soc.* 51, 101 (1970).

78   G. Band, "Der Einfluss der Siedlung auf das Freilandklima," *Mitt. Inst. Geophys. Meteorol. Univ. Köln* (1969), vol. 9.

79   V. J. Schaefer, *Science* 154, 1555 (1966); A. W. Hogan, *ibid.* 158, 800 (1967); V. J. Schaefer, *Bull. Amer. Meteorol. Soc.* 50, 199 (1969); State University of New York at Albany, *Atmospheric Sciences Research Center, Annual Report* 1969; J. P. Lodge, Jr., *Bull. Amer. Meteorol. Soc.* 50, 530 (1969); G. Langer, *ibid.* 51, 102 (1970).

80   R. Gunn and B. B. Phillips, *J. Meteorol.* 14, 272 (1957).

81   P. A. Allee, *Bull. Amer. Meteorol. Soc.* 51, 102 (1970).

82   G. N. Chatham, *Mt. Washington Observ. News Bull.* 11, No. 1, 18 (1970); P. M. Kuhn, *Bull. Amer. Meteorol. Soc.* 51, 101 (1970); F. F. Hall, Jr., *ibid.*, p. 101; V. D. Nuessle and R. W. Holcomb, *Science* 168, 1562 (1970).

83   P. Dansereau, *BioScience* 14, No. 7, 20 (1964); in *Future Environments of North America*, S. F. Darling and J. P. Milton, Eds. (Natural History Press, Garden City, N. Y., 1966), p. 425; R. Dubos, "A theology of the earth," lecture presented before the Smithsonian Institution, 1969; M. Bundy, "Managing knowledge to save the environment," address delivered 27 Jan. 1970 before the 11th Annual Meeting of the Advisory Panel to the House Committee on Science and Astronautics.

84   The work discussed here has been supported in part by NSF grants GA-1104 and GA-13353.

20

CLIMATE IMPACT OF INCREASING ATMOSPHERIC CARBON DIOXIDE*

J. Hansen, D. Johnson, A. Lacis, S. Lebedeff, P. Lee, D. Rind and G. Russell

Atmospheric CO_2 increased from 280 to 300 parts per million in 1880 to 335 to 340 ppm in 1980 (1, 2), mainly due to burning of fossil fuels. Deforestation and changes in biosphere growth may also have contributed, but their net effect is probably limited in magnitude (2, 3). The CO_2 abundance is expected to reach 600 ppm in the next century, even if growth of fossil fuel use is slow (4).

Carbon dioxide absorbs in the atmospheric "window" from 7 to 14 micrometers which transmits thermal radiation emitted by the earth's surface and lower atmosphere. Increased atmospheric CO_2 tends to close this window and cause outgoing radiation to emerge from higher, colder levels, thus warming the surface and lower atmosphere by the so-called greenhouse mechanism (5). The most sophisticated models suggest a mean warming of 2° to 3.5°C for doubling of the CO_2 concentration from 300 to 600 ppm (6–8).

The major difficulty in accepting this theory has been the absence of observed warming coincident with the historic CO_2 increase. In fact, the temperature in the Northern Hemisphere decreased by about 0.5°C between 1940 and 1970 (9), a time of rapid CO_2 buildup. In addition, recent claims that climate models overestimate the impact of radiative perturbations by an order of magnitude (10, 11) have raised the issue of whether the greenhouse effect is well understood.

We first describe the greenhouse mechanism and use a simple model to compare potential radiative perturbations of climate. We construct the trend of observed global temperature for the past century and compare this with global climate model computations, providing a check on the ability of the model to simulate known climate change. Finally, we compute the CO_2 warming expected in the coming century and discuss its potential implications.

* Originally published in *Science*, 1981, vol. 213, pp. 957–66.

Greenhouse Effect

The effective radiating temperature of the earth, T_e, is determined by the need for infrared emission from the planet to balance absorbed solar radiation:

$$\pi R^2 (1 - A)S_0 = 4\pi R^2 \sigma T_e \qquad (1)$$

or

$$T_e = [S_0(1 - A)/4\sigma]^{1/4} \qquad (2)$$

where R is the radius of the earth, A the albedo of the earth, S_0 the flux of solar radiation, and σ the Stefan-Boltzmann constant. For $A \sim 0.3$ and $S_0 = 1367$ watts per square meter, this yields $T_e \sim 255$ K.

The mean surface temperature is $T_s \sim$ 288K. The excess, $T_s - T_e$, is the greenhouse effect of gases and clouds, which cause the mean radiating level to be above the surface. An estimate of the greenhouse warming is

$$T_s \sim T_e + \Gamma H \qquad (3)$$

where H is the flux-weighted mean altitude of the emission to space and Γ is the mean temperature gradient (lapse rate) between the surface and H. The earth's troposphere is sufficiently opaque in the infrared that the purely radiative vertical temperature gradient is convectively unstable, giving rise to atmospheric motions that contribute to vertical transport of heat and result in $\Gamma \sim 5°$ to 6°C per kilometer. The mean lapse rate is less than the dry adiabatic value because of latent heat release by condensation as moist air rises and cools and because the atmospheric motions that transport heat vertically include large-scale atmospheric dynamics as well as local convection. The value of H is ~ 5 km at midlatitudes (where $\Gamma \sim 6.5°$C km^{-1}) and ~ 6 km in the global mean ($\Gamma \sim 5.5°$C km^{-1}).

The surface temperature resulting from the greenhouse effect is analogous to the depth of water in a leaky bucket with constant inflow rate. If the holes in the bucket are reduced slightly in size, the water depth and water pressure will increase until the flow rate out of the holes again equals the inflow rate. Analogously, if the atmospheric infrared opacity increases, the temperature of the surface and atmosphere will increase until the emission of radiation from the planet again equals the absorbed solar energy.

The greenhouse theory can be tested by examination of several planets, which provide an ensemble of experiments over a wide range of conditions. The atmospheric composition of Mars, Earth, and Venus lead to mean radiating levels of about 1, 6, and 70 km, and lapse rates of $\Gamma \sim 5°$, 5.5°, and 7°C km^{-1}, respectively. Observed surface temperatures of these planets confirm the existence and order of magnitude of the predicted greenhouse effect (Eq. 3). Data now being collected by spacecraft at Venus and Mars (12) will permit more precise analyses of radiative and dynamical mechanisms that affect greenhouse warming.

One-Dimensional Model

A one-dimensional radiative-convective (1-D RC) model (5, 13), which computes temperature as a function of altitude, can simulate planetary temperatures more realistically than the zero-dimensional model of Eq. 1. The sensitivity of surface temperature in 1-D RC models to changes in CO_2 is similar to the sensitivity of mean surface temperature in global three-dimensional models (6–8). This agreement does not validate the models; it only suggests that one-dimensional models can simulate the effect of certain basic mechanisms and feedbacks. But the agreement does permit useful studies of global mean temperature change with a simple one-dimensional model.

The 1-D RC model uses a time-marching procedure to compute the vertical temperature profile from the net radiative and convective energy fluxes:

$$T(h, t + \Delta t) =$$
$$T(h, t) + \frac{\Delta t}{c_p \rho}\left(\frac{dF_r}{dh} + \frac{dF_c}{dh}\right) \qquad (4)$$

where c_p is the heat capacity at constant pressure, ρ the density of air, h the altitude, and dF_r/dh and dF_c/dh the net radiative and convective flux divergences. To compute dF_r/dh the radiative transfer equation is integrated over all frequencies, using the temperature profile of the previous time

Table 20.1 [*orig. table 1*]   Equilibrium surface temperature increase due to doubled CO_2 (from 300 to 600 ppm) in 1-D RC models. Model 1 has no feedbacks affecting the atmosphere's radiative properties. Feedback factor f specifies the effect of each added process on model sensitivity to doubled CO_2; F is the equilibrium thermal flux into the ground if T_s is held fixed (infinite heat capacity) when CO_2 is doubled. Abbreviations: FRH, fixed relative humidity; FAH, fixed absolute humidity; 6.5LR, 6.5°C km^{-1} limiting lapse rate; MALR, moist adiabatic limiting lapse rate; FCA, fixed cloud altitude, FCT, fixed cloud temperature, SAF, snow/ice albedo feedback; and VAF, vegetation albedo feedback. Models 5 and 6 are based on f values from Wang and Stone (19) and Cess (20), respectively, and ΔT_s, of model 2.

Model	Description	ΔT_s (°C)	f	F (W m^{-2})
1	FAH, 6.5LR, FCA	1.22	1	4.0
2	FRH, 6.5LR, FCA	1.94	1.6	3.9
3	Same as 2, except MALR replaces 6.5LR	1.37	0.7	4.0
4	Same as 2, except FCT replaces FCA	2.78	1.4	3.9
5	Same as 2, except SAF included	2.5–2.8	1.3–1.4	
6	Same as 2, except VAF included	~3.5	~1.8	

step and an assumed atmospheric composition. The term dF_c/dh is the energy transport needed to prevent the temperature gradient from exceeding a preassigned limit, usually 6.5°C km^{-1}. This limit parameterizes effects of vertical mixing and large-scale dynamics.

The radiative calculations are made by a method that groups absorption coefficients by strength for efficiency (14). Pressure- and temperature-dependent absorption coefficients are from line-by-line calculations for H_2O, CO_2, O_3, N_2O, and CH_4 (15), including continuum H_2O absorption (16). Climatological cloud cover (50 percent) and aerosol properties (17) are used, with appropriate fractions of low (0.3), middle (0.1), and high (0.1) clouds. Wavelength dependences of cloud and aerosol properties are obtained from Mie scattering theory (14). Multiple scattering and overlap of gaseous absorption bands are included. Our computations include the weak CO_2 bands at 8 to 12 μm, but the strong 15-μm CO_2 band, which closes one side of the 7- to 20-μm H_2O window, causes ≥ 90 percent of the CO_2 warming.

Model Sensitivity

We examine the main processes known to influence climate model sensitivity by inserting them individually into the model, as summarized in table 20.1.

Model 1 has fixed absolute humidity, a fixed lapse rate of 6.5°C km^{-1} in the convective region, fixed cloud altitude, and no snow/ice albedo feedback or vegetation albedo feedback. The increase of equilibrium surface temperature for doubled atmospheric CO_2 is ΔT_s ~ 1.2°C. This case is of special interest because it is the purely radiative-convective result, with no feedback effects.

Model 2 has fixed relative humidity, but is otherwise the same as model 1. The resulting ΔT_s for doubled CO_2 is ~ 1.9°C. Thus the increasing water vapor with higher temperature provides a feedback factor of ~ 1.6. Fixed relative humidity is clearly more realistic than fixed absolute humidity, as indicated by physical arguments (13) and three-dimensional model results (7, 8). Therefore, we use fixed relative humidity in the succeeding experiments and compare models 3 to 6 with model 2.

Model 3 has a moist adiabatic lapse rate in the convective region rather than a fixed lapse rate. This causes the equilibrium surface temperature to be less sensitive to radiative perturbations, and ΔT_s ~ 1.4°C for doubled CO_2. The reason is that the lapse rate decreases as moisture is added to the air, reducing the temperature difference between the top of the convective region and the ground (ΓH in Eq. 3).

The general circulation of the earth's

atmosphere is driven by solar heating of the tropical ocean, and resulting evaporation and vertical transport of energy. The lapse rate is nearly moist adiabatic at low latitudes and should remain so after a climate perturbation. Thus use of a moist adiabatic lapse rate is appropriate for the tropics. But more stable lapse rates at high latitudes make the surface temperature much more sensitive to perturbations of surface heating (7, 8), and hence model 3 would underestimate the sensitivity there.

Model 4 has the clouds at fixed temperature levels, and thus they move to a higher altitude as the temperature increases (18). This yields $\Delta T_s \sim 2.8°C$ for doubled CO_2, compared to 1.9°C for fixed cloud altitude. The sensitivity increases because the outgoing thermal radiation from cloudy regions is defined by the fixed cloud temperature, requiring greater adjustment by the ground and lower atmosphere for outgoing radiation to balance absorbed solar radiation.

Study of Venus suggests that some clouds occur at a fixed temperature. The Venus cloud tops, which are the primary radiator to space, are at $H \sim 70$ km, where $T \sim T_e$. Analysis of the processes that determine the location of these clouds and the variety of clouds in the belts, zones, and polar regions on Jupiter should be informative. Available evidence suggests that the level of some terrestrial clouds depends on temperature while others occur at a fixed altitude. For example, tropical cirrus clouds moved to a higher altitude in the experiment of Hansen et al. (8) with doubled CO_2, but low clouds did not noticeably change altitude.

Models 5 and 6 illustrate snow/ice and vegetation albedo feedbacks (19, 20). Both feedbacks increase model sensitivity, since increased temperature decreases ground albedo and increases absorption of solar radiation.

Snow, sea ice, and land ice (ice sheets and glaciers) are all included in snow/ice albedo feedback. Snow and sea ice respond rapidly to temperature change, while continental ice sheets require thousands of years to respond. Thus a partial snow/ice albedo feedback is appropriate for time scales of 10 to 100 years. The vegetation albedo feedback was obtained by comparing today's global vegetation patterns with reconstruction of the Wisconsin ice age (20). Uncertainties in the reconstruction, the time scale of vegetation response, and man's potential impact on vegetation prevent reliable assessment of this feedback, but its estimated magnitude emphasizes the need to monitor global vegetation and surface albedo.

Model 4 has our estimate of appropriate model sensitivity. The fixed 6.5°C km⁻¹ lapse rate is a compromise between expected lower sensitivity at low latitudes and greater sensitivity at high latitudes. Both cloud temperature and snow/ice albedo feedback should be partly effective, so for simplicity one is included.

The sensitivity of the climate model we use is thus $\Delta T_s \sim 2.8°C$ for doubled CO_2, similar to the sensitivity of three-dimensional climate models (6–8). The estimated uncertainty is a factor of 2. This sensitivity (i) refers to perturbations about today's climate and (ii) does not include feedback mechanisms effective only on long time scales, such as changes of ice sheets or ocean chemistry.

Model Time Dependence

The time dependence of the earth's surface temperature depends on the heat capacity of the climate system. Heat capacity of land areas can be neglected, since ground is a good insulator. However, the upper 100 m of the ocean is rapidly mixed, so its heat capacity must be accounted for. The ocean beneath the mixed layer may also affect surface temperature, if the thermal response time of the mixed layer is comparable to the time for exchange of heat with deeper layers.

The great heat capacity of the ocean and ready exchange of continental and marine air imply that the global climate response to perturbations is determined by the response of the ocean areas. However, this response is affected by horizontal atmospheric heat fluxes from and to the continents. Ready exchange of energy between the ocean surface and atmosphere "fixes" the air temperature, and the ocean in effect removes from the atmosphere any net heat obtained from the continents. Thus the horizontal flux due to a climate perturbation's heating (or cooling) of the continents adds to the vertical heat flux into (or out of) the ocean surface. The net flux into the ocean surface is therefore larger than it would be for a 100 percent ocean-covered planet by the ratio of global area to ocean area, totaling ~ 5.7 W m^{-2} for doubled CO_2 rather than ~ 4 W m^{-2}. In a climate model that employs only a mixed-layer ocean, it is equivalent to use the flux ~ 4 W m^{-2} with the area-weighted mean land-ocean heat capacity.

The thermal response time of the ocean mixed layer would be ~ 3 years if it were not for feedback effects in the climate system. For example, assume that the solar flux absorbed by a planet changes suddenly from $F_0 \equiv \sigma T_0^4$ to $F_1 = F_0 + \Delta F \equiv \sigma T_1^4$, with $\Delta F \ll F_0$. The rate of change of heat in the climate system is

$$d(cT)/dt = \sigma T_1^4 - \sigma T^4 \qquad (5)$$

where c is heat capacity per unit area. Since $T_1 - T_0 \ll T_0$, the solution is

$$T - T_1 = (T_0 - T_1)e^{-t/t_{thr}} \qquad (6)$$

where

$$t_{thr} = c/4\sigma T_1^3 \qquad (7)$$

Thus the planet approaches a new equilibrium temperature exponentially with e-folding time t_{thr}. If the heat capacity is provided by 70 m of water (100 m for ocean areas) and the effective temperature is 255 K, t_{thr} is 2.8 years.

This estimate does not account for climate feedback effects, which can be analyzed with the 1-D RC model. Table 20.1 shows that the initial rate of heat storage in the ocean is independent of feedbacks. Thus the time needed to reach equilibrium for model 4 is larger by the factor $\sim 2.8°C/1.2°C$ than for model 1, which excludes feedbacks. The e-folding time for adjustment of mixed-layer temperature is therefore ~ 6 years for our best estimate of model sensitivity to doubled CO_2. This increase in thermal response time is readily understandable, because feedbacks come into play only gradually after some warming occurs.

It would take ~ 50 years to warm up the thermocline and mixed layer if they were rapidly mixed, or 250 years for the entire ocean. Turnover of the deep ocean, driven by formation of cold bottom water in the North Atlantic and Antarctic oceans with slow upwelling at low latitudes, is thought to require 500 to 1000 years (21), suggesting that the deep ocean does not greatly influence surface temperature sensitivity. However, there may be sufficient heat exchange between the mixed layer and thermocline to delay full impact of a climate perturbation by a few decades (6, 22, 23). The primary mechanism of exchange is nearly horizontal movement of water along surfaces of constant density (21).

Delay of CO_2 warming by the ocean can be illustrated with a "box diffusion" model (24), in which heat is stirred instantly through the mixed layer and diffused into the thermocline with diffusion coefficient k. Observed oceanic penetration by inert chemical tracers suggests that k is of order 1 square centimeter per second (2, 3, 24).

The warming calculated with the one-dimensional model for the CO_2 increase from 1880 to 1980 (25) is 0.5°C if ocean heat capacity is neglected (figure 20.1). The heat capacity of just the mixed layer reduces this to 0.4°C, a direct effect of the mixed layer's 6-year thermal response time. Diffusion into the thermocline further reduces the warming to 0.25°C for $k = 1$ cm^2 sec^{-1}, an indirect effect of the mixed layer's 6-year e-folding time, which permits substantial exchange with the thermocline.

The mixed-layer model and thermocline model bracket the likely CO_2 warming. The thermocline model is preferable for small climate perturbations that do not affect ocean mixing. However, one effect of warming the ocean surface will be increased vertical stability, which could reduce ocean warming and make the surface temperature response more like that of the mixed-layer case.

Lack of knowledge of ocean processes primarily introduces uncertainties about the time dependence of the global CO_2 warming. The full impact of the warming may be delayed several decades, but since

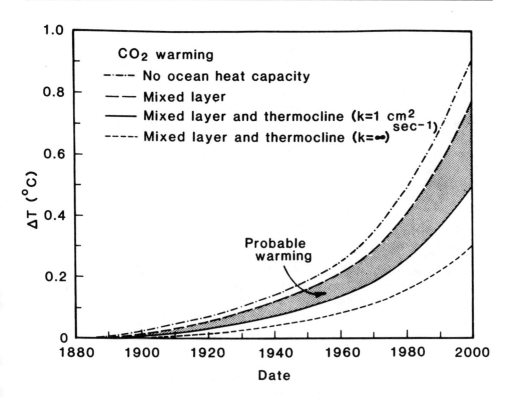

Figure 20.1 [*orig. figure 1*]   Dependence of CO_2 warming on ocean heat capacity. Heat is rapidly mixed in the upper 100 m of the ocean and diffused to 1000 m with diffusion coefficient k. The CO_2 abundance, from (*25*), is 293 ppm in 1880, 335 ppm in 1980, and 373 ppm in 2000. Climate model equilibrium sensitivity is 2.8°C for doubled CO_2.

man-made increases in atmospheric CO_2 are expected to persist for centuries (*1, 2, 6*), the warming will eventually occur.

Radiative Climate Perturbations

Identification of the CO_2 warming in observed climate depends on the magnitude of climate variability due to other factors. Most suspected causes of global climate change are radiative perturbations, which can be compared to identify those capable of counteracting or reinforcing the CO_2 warming.

A 1 percent increase of solar luminosity would warm the earth 1.6°C at equilibrium (figure 20.2) on the basis of model 4, which we employ for all radiative perturbations to provide a uniform comparison. Since the effect is linear for small changes of solar

luminosity, a change of 0.3 percent would modify the equilibrium global mean temperature by 0.5°C, which is as large as the equilibrium warming for the cumulative increase of atmospheric CO_2 from 1880 to 1980. Solar luminosity variations of a few tenths of 1 percent could not be reliably measured with the techniques available during the past century, and thus are a possible cause of part of the climate variability in that period.

Atmospheric aerosol effects depend on aerosol composition, size, altitude, and global distribution (*26*). Based on model calculations, stratospheric aerosols that persist for 1 to 3 years after large volcanic eruptions can cause substantial cooling of surface air (figure 20.2). The cooling depends on the assumption that the particles do not exceed a few tenths of a micrometer in size, so they do not cause greenhouse warming by blocking terrestrial radiation, but this condition is probably ensured by rapid gravitational settling of larger particles. Temporal variability of stratospheric aerosols due to volcanic eruptions appears to have been responsible for a large part of the observed climate change during the past century (27–30), as shown below.

The impact of tropospheric aerosols on climate is uncertain in sense and magnitude due to their range of composition, including absorbing material such as carbon and high-albedo material such as sulfuric acid, and their heterogeneous spatial distribution. Although man-made tropospheric aerosols are obvious near their source, aerosol opacity does not appear to have increased much in remote regions (31). Since the climate impact of anthropogenic aerosols is also reduced by the opposing effects of

absorbing and high-albedo materials, it is possible that they have not had a primary effect on global temperature. However, global monitoring of aerosol properties is needed for conclusive analysis.

Ground albedo alterations associated with changing patterns of vegetation coverage have been suggested as a cause of global climate variations on time scales of decades to centuries (32). A global surface albedo change of 0.015, equivalent to a change of 0.05 over land areas, would affect global temperature by 1.3°C. Since this is a 25 percent change in mean continental ground albedo, it seems unlikely that ground albedo variations have been the primary cause of recent global temperature trends. However, global monitoring of ground albedo is needed to permit definitive assessment of its role in climate variability.

High and low clouds have opposite

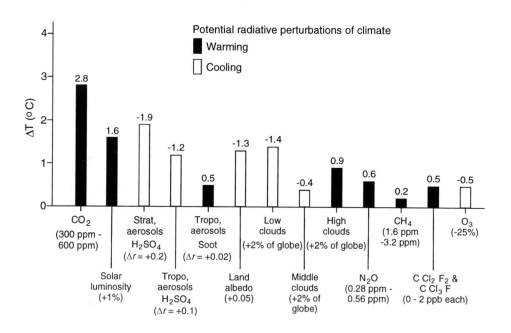

Figure 20.2 [orig. figure 2]   Surface temperature effect of various global radiative perturbations, based on the 1-D RC model 4 (table 20.1). Aerosols have the physical properties specified by (17). Dependence of ΔT on aerosol size, composition, altitude, and optical thickness is illustrated by (26). The $\Delta \tau$ for stratospheric aerosols is representative of a very large volcanic eruption.

effects on surface temperature (figure 20.2), high clouds having a greenhouse effect while low clouds cool the surface (14, 33). However, the nature and causes of variability of cloud cover, optical thickness, and altitude distribution are not well known, nor is it known how to model reliably cloud feedbacks that may occur in response to climate perturbations. Progress may be made after accurate cloud climatology is obtained from global observations, including seasonal and interannual cloud variations. In the meantime, some limits are implicitly placed on global cloud feedback by empirical tests of the climate system's sensitivity to radiative perturbations, as discussed below.

Trace gases that absorb in the infrared can warm the earth if their abundance increases (5, 34). The abundance of chlorofluorocarbons (Freons) increased from a negligible amount a few decades ago to 0.3 part per billion for CCl_2F_2 and 0.2 ppb for CCl_3F (35), with an equilibrium greenhouse warming of ~ 0.06°C. Recent measurement of a 0.2 percent per year increase of N_2O suggests a cumulative increase to date of 17 ppb (36), with an equilibrium warming of ~ 0.03°C. Tentative indications of a 2 percent per year increase in CH_4 imply an equilibrium warming < 0.1°C for the CH_4 increase to date (37). No major trend of O_3 abundance has been observed, although it has been argued that continued increase of Freons will reduce O_3 amounts (38). The net impact of measured trace gases has thus been an equilibrium warming of 0.1°C or slightly larger. This does not greatly alter analyses of temperature change over the past century, but trace gases will significantly enhance future greenhouse warming if recent growth rates are maintained.

We conclude that study of global climate change on time scales of decades and centuries must consider variability of stratospheric aerosols and solar luminosity, in addition to CO_2 and trace gases. Tropospheric aerosols and ground albedo are potentially significant, but require better observations. Cloud variability will continue to cause uncertainty until accurate monitoring of global cloud properties

provides a basis for realistic modeling of cloud feedback effects; however, global feedback is implicitly checked by comparison of climate model sensitivity to empirical climate variations, as done below.

Observed Temperature Trends

Data archives (39) contain surface air temperatures of several hundred stations for the last century. Problems in obtaining a global temperature history are due to the uneven station distribution (40), with the Southern Hemisphere and ocean areas poorly represented, and the smaller number of stations for earlier times.

We combined these temperature records with a method designed to extract mean temperature trends. The globe was divided by grids with a spacing not larger than the correlation distance for primary dynamical transports (41), but large enough that most boxes contained one or more stations. The results shown were obtained with 40 equal-area boxes in each hemisphere, but the conclusions are not sensitive to the exact spacing. Temperature trends for stations within a box were combined successively:

$$\frac{T_{1,n}(t)\,(n^*-1)\,T_{1,n}+T_n-\bar{T}_n+\bar{T}_{1,n}}{n^*} \qquad (8)$$

to obtain a single trend for each box, where the bar indicates a mean for the years in which there are records for both T_n and the cumulative $T_{1,n}$ and $n^*(t)$ is the number of stations in $T_{1,n}(t)$. Trends for boxes in a latitude zone were combined with each box weighted equally, and the global trend was obtained by area-weighting the trends for all latitude zones. A meaningful result begins in the 1880's, since thereafter continuous records exist for at least two widely separated longitudes in seven of the eight latitude zones (continuous Antarctic temperatures begin in the 1950's). Results are least reliable for 1880 to 1900; by 1900, continuous records exist for more than half of the 80 boxes.

The temperature trends in figure 20.3 are smoothed with a 5-year running mean to make the trends readily visible. Part of the

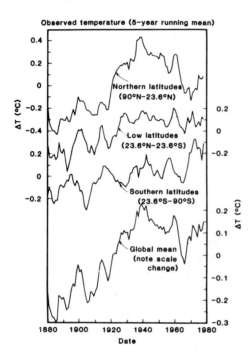

Observed temperature (5-year running mean)

Figure 20.3 [orig. figure 3] Observed surface air temperature trends for three latitude bands and the entire globe. Temperature scales for low latitudes and global mean are on the right.

noise in the unsmoothed data results from unpredictable weather fluctuations, which affect even 1-year means (42). None of our conclusions depends on the nature of the smoothing.

Northern latitudes warmed ~ 0.8°C between the 1880's and 1940, then cooled ~0.5°C between 1940 and 1970, in agreement with other analyses (9, 43). Low latitudes warmed ~ 0.3°C between 1880 and 1930, with little change thereafter. Southern latitudes warmed ~ 0.4°C in the past century; results agree with a prior analysis for the late 1950's to middle 1970's (44). The global mean temperature increased ~ 0.5°C between 1885 and 1940, with slight cooling thereafter.

A remarkable conclusion from figure 20.3 is that the global temperature is almost as high today as it was in 1940. The common misconception that the world is cooling is

based on Northern Hemisphere experience to 1970.

Another conclusion is that global surface air temperature rose ~ 0.4°C in the past century, roughly consistent with calculated CO_2 warming. The time history of the warming obviously does not follow the course of the CO_2 increase (figure 20.1), indicating that other factors must affect global mean temperature.

Model Verification

Natural radiative perturbations of the earth's climate, such as those due to aerosols produced by large volcanic eruptions, permit a valuable test of model sensitivity. Previous study of the best-documented large volcanic eruption, Mount Agung in 1963, showed that tropical tropospheric and stratospheric temperature changes computed with a one-dimensional climate model were of the same sign and order of magnitude as observed changes (45). It was assumed that horizontal heat exchange with higher latitudes was not altered by the radiative perturbation.

We reexamined the Mount Agung case for comparison with the present global temperature record, using our model with sensitivity ~ 2.8°C. The model, with a maximum global mean aerosol increase in the optical depth $\Delta\tau = 0.12$ (45), yields a maximum global cooling of 0.2°C when only the mixed-layer heat capacity is included and 0.1°C when heat exchange with the deeper ocean is included with $k = 1$ cm^2 sec^{-1}. Observations suggest a cooling of this magnitude with the expected time lag of 1 to 2 years. Noise or unexplained variability in the observations prevents more definitive conclusions, but similar cooling is indicated by statistical studies of temperature trends following other large volcanic eruptions (46).

A primary lesson from the Mount Agung test is the damping of temperature change by the mixed layer's heat capacity, without which the cooling would have exceeded 1.1°C (figure 20.2). The effect can be understood from the time constant of the perturbation and thermal response time of

a) Immediate response

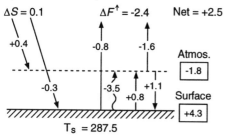

$\Delta S = 0.1$ $\Delta F^\uparrow = -2.4$ Net = +2.5

+0.4   -0.8   -1.6

Atmos.

-1.8

-0.3   -3.5   +1.1
              +0.8   Surface

+4.3

$T_S = 287.5$

b) A few months later

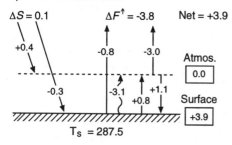

$\Delta S = 0.1$ $\Delta F^\uparrow = -3.8$ Net = +3.9

+0.4   -0.8   -3.0

Atmos.

0.0

-0.3   -3.1   +1.1
              +0.8   Surface

+3.9

$T_S = 287.5$

c) Many years later

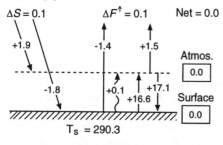

$\Delta S = 0.1$ $\Delta F^\uparrow = 0.1$ Net = 0.0

+1.9   -1.4   +1.5

Atmos.

0.0

-1.8   +0.1   +17.1
              +16.6   Surface

0.0

$T_S = 290.3$

Figure 20.4 [*orig. figure 4*]   Change of fluxes (watts per square meter) in the 1-D RC model when atmospheric CO_2 is doubled (from 300 to 600 ppm). Symbols: ΔS, change in solar radiation absorbed by the atmosphere and surface; $\Delta F^\uparrow$, change in outward thermal radiation at top of the atmosphere. The wavy line represents convective flux; other fluxes are radiative.

the mixed layer: $\Delta T \sim \{1 - \exp[(-1 \text{ year})/ (6 \text{ years})]\} \times 1.1°C \sim 0.17°C$, for the case $k = 0$. This large reduction of the climate response occurs for a perturbation that (unlike CO_2) is present for a time shorter than the thermal response time of the ocean surface.

Phenomena that alter the regional radiation balance provide another model test. Idso (*11*) found a consistent "empirical response function" for several such phenomena, which was 0.17°C per watt per square meter in midcontinent and was half as large on the coast. This response must depend on the rate of mixing of marine and continental air, since the phenomena occur on time scales less than the thermal relaxation time of the ocean surface. Thus, as one test of horizontal atmospheric transports, we read from our three-dimensional climate model (*8*) the quantities (solar insolation and temperature) that form Idso's empirical response function for seasonal change of insolation. Results ranged from 0.2°C W^{-1} m^2 in mid-continent, and about half that on the coast, to a value an order of magnitude smaller over the ocean, in agreement with the empirical response (*11*).

To relate these empirical tests to the CO_2 greenhouse effect, we illustrate the flux changes in the 1-D RC model when CO_2 is doubled. For simplicity we consider an instantaneous doubling of CO_2, and hence the time dependence of the response does not represent the transient response to a steady change in CO_2. The immediate response to the doubling includes (figure 20.4a): (i) reduced emission to space (−2.4 W m^{-2}), because added CO_2 absorption raises the mean altitude of emission to a higher, colder level; (ii) increased flux from atmosphere to ground (+ 1.1 W m^{-2}); and (iii) increased stratospheric cooling but decreased tropospheric cooling. The radiative warming of the troposphere decreases the "convective" flux (latent and sensible heat) from the ground by 3.5 W m^{-2} as a consequence of the requirement to conserve energy. There is a small increase in absorption of near-infrared radiation, the atmosphere gaining energy (+ 0.4 W m^{-2}) and the ground losing energy (− 0.3 W m^{-2}). The net effect is thus an energy gain for the planet (+ 2.5 W m^{-2}) with heating of the ground (+ 4.3 W m^{-2}) and cooling of the (upper) atmosphere (− 1.8 W m^{-2}). These flux changes are independent of feedbacks and are not sensitive to the critical lapse rate.

A few months after the CO_2 doubling

(figure 20.4b) the stratospheric temperature has cooled by ~ 5°C. Neither the ocean nor the troposphere, which is convectively coupled to the surface, have responded yet. The planet radiates 3.8 W m^{-2} less energy to space than in the comparison case with 300 ppm CO_2, because of the cooler stratosphere and greater altitude of emission from the troposphere. The energy gained by the earth at this time is being used to warm the ocean.

Years later (figure 20.4c) the surface temperature has increased 2.8°C. Almost half the increase (1.2°C) is the direct CO_2 greenhouse effect. The remainder is due to feedbacks, of which 1.0°C is the well-established H_2O greenhouse effect.

The greenhouse process represented in figure 20.4 is simply the "leaky bucket" phenomenon. The increased infrared opacity causes an immediate decrease of thermal radiation from the planet, thus forcing the temperature to rise until energy balance is restored. Temporal variations of the fluxes and temperatures are due to the response times of the atmosphere and surface.

Surface warming of ~3°C for doubled CO_2 is the status after energy balance has been restored. This contrasts with the Agung case and the cases considered by Idso (11), which are all nonequilibrium situations.

The test of the greenhouse theory provided by the extremes of equilibrium climates on the planets and short-term radiative perturbations is reassuring, but inadequate. A crucial intermediate test is climate change on time scales from a few years to a century.

Model versus Observations for the Past Century

Simulations of global temperature change should begin with the known forcings: variations of CO_2 and volcanic aerosols. Solar luminosity variations, which constitute another likely mechanism, are unknown, but there are hypotheses consistent with observational constraints that variations do not exceed a few tenths of 1 percent.

We developed an empirical equation that

fits the heat flux into the earth's surface calculated with the 1-D RC climate model (model 4):

$$F(t) = 0.018\Delta p/(1 + 0.0022\Delta p)^{0.6} - 17\Delta\tau - 1.5(\Delta\tau)^2 + 220\Delta S/S_0 - 1.5\Delta T + 0.033(\Delta T)^2 - 1.04 \times 10^{-4}\Delta p\Delta T + 0.29\Delta T\Delta\tau \qquad (9)$$

where $F(t)$ is in watts per square meter, p is the amount of CO_2 in parts per million above an "equilibrium" value (293 ppm), ΔS is the difference between solar luminosity and an equilibrium value S_0, $\Delta\tau$ is the optical depth of stratospheric aerosols above a background amount, and ΔT is the difference between current surface temperature and the equilibrium value for $\Delta p = \Delta S = \Delta\tau = 0$. Equation 9 fits the one-dimensional model results to better than 1 percent for $0 \le \Delta p \le 1200$ ppm, $0.98 \le \Delta S/S_0 \le 1.02$, and $\Delta\tau \le 0.5$. For the mixed-layer ocean model $T_s(t)$ follows from $dT_s/dt = F(t)/c_0$, where c_0 is the heat capacity of the ocean mixed layer per unit area. If the true mixed-layer depth is used to obtain c_0, $F(t)$ must be multiplied by $1/0.7$, the ratio of global area to ocean area. Diffusion of heat into the deeper ocean can then also be included by means of the diffusion equation with T_s as its upper boundary condition.

The CO_2 abundance increased from 293 ppm in 1880 to 335 ppm in 1980 (25), based on recent accurate observations, earlier less accurate observations, and carbon cycle modeling. The error for 1880 probably does not exceed 10 ppm (1, 2).

Volcanic aerosol radiative forcing can be obtained from Lamb's (27) dust veil index (DVI), which is based mainly on atmospheric transmission measurements after 1880. We convert DVI to optical depth by taking Mount Agung (DVI = 800) to have the maximum $\Delta\tau = 0.12$. The aerosol optical depth histories of Mitchell (47) and Pollack et al. (29), the latter based solely on transmission measurements, are similar to Lamb's. We use aerosol microphysical properties from (45). The error in volcanic aerosol radiative forcing probably does not exceed a factor of 2.

Solar variability is highly conjectural, so we first study CO_2 and volcanic aerosol

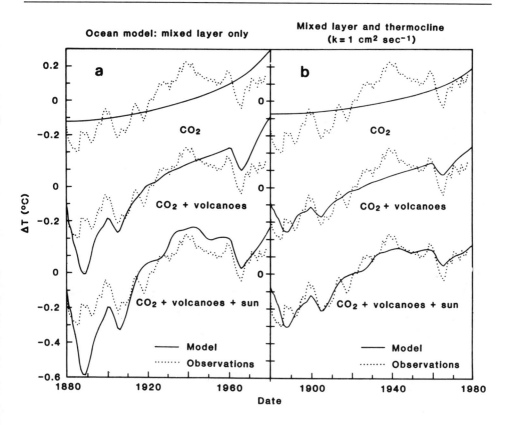

Figure 20.5 [*orig. figure 5*]   Global temperature trend obtained from climate model with sensitivity 2.8°C for doubled CO_2. The results in (a) are based on a 100-m mixed-layer ocean for heat capacity; those in (b) include diffusion of heat into the thermocline to 1000 m. The forcings by CO_2, volcanoes, and the sun are based on Broecker (*25*), Lamb (*27*), and Hoyt (*48*). Mean ΔT is zero for observations and model.

forcings and then add solar variations. We examine the hypothesis of Hoyt (*48*) that the ratio, r, of umbra to penumbra areas in sunspots is proportional to solar luminosity: $\Delta S/S_0 = f(r - r_0)$. Hoyt's rationale is that the penumbra, with a weaker magnetic field than the umbra, is destroyed more readily by an increase of convective flux from below. We take $f = 0.03$, which implies a peak-to-peak amplitude of ~0.4 percent for $\Delta S/S_0$ in the past century, or an amplitude of ~0.2 percent for the mean trend line. Taking S_0 as the mean for 1880 to 1976 yields $r_0 = 0.2$. The resulting $\Delta S/S_0$ has no observational corroboration, but serves as an example of solar variability of a plausible magnitude.

Radiative forcing by CO_2 plus volcanoes

and plus the sun both yield a temperature trend with a strong similarity to the observed trend of the past century (figure 20.5), which we quantify below. If only the heat capacity of the mixed layer is included, the amplitude of the computed temperature variations is larger than observed. However, mixing of heat into the deeper ocean with k = 1 cm^2 sec^{-1} brings both calculated trends into rough agreement with observations.

The main uncertainties in the climate model – that is, its "tuning knobs" – are (i) the equilibrium sensitivity and (ii) the rate of heat exchange with the ocean beneath the mixed layer. The general correlation of radiative forcings with global temperatures suggests that model uncertainties be constrained by requiring agreement with the observed temperature trend.

Therefore, we examined a range of model sensitivities, choosing a diffusion coefficient for each to minimize the residual variance between computed and observed temperature trends. Equilibrium sensitivities of 1.4°, 2.8°, and 5.6°C required $k = 0$, 1.2, and 2.2 cm^2 sec^{-1}, respectively. All models with sensitivities of 1.4° to 5.6°C provide a good fit to the observations. The smallest acceptable sensitivity is ~ 1.4°C, because it requires zero heat exchange with the deeper ocean. Sensitivities much higher than 5.6°C would require greater heat exchange with the deep ocean than is believed to be realistic (21, 22).

Radiative forcing by CO_2 plus volcanoes accounts for 75 percent of the variance in the 5-year smoothed global temperature, with correlation coefficient 0.9. The hypothesized solar luminosity variation (48) improves the fit, as a consequence of the luminosity peaking in the 1930's and declining into the 1970's, leaving a residual variance of only 10 percent. The improved fit provided

by Hoyt's solar variability represents a posteriori selection, since other hypothesized solar variations that we examined [for instance (49)] degrade the fit. This evidence is too weak to support any specific solar variability.

The general agreement between modeled and observed temperature trends strongly suggests that CO_2 and volcanic aerosols are responsible for much of the global temperature variation in the past century. Key consequences are: (i) empirical evidence that much of the global climate variability on time scales of decades to centuries is deterministic and (ii) improved confidence in the ability of models to predict future CO_2 climate effects.

Projections into the 21st Century

Prediction of the climate effect of CO_2 requires projections of the amount of atmospheric CO_2, which we specify by (i) the energy growth rate and (ii) the fossil fuel proportion of energy use. We neglect other possible variables, such as changes in the

Table 20.2 [orig. table 2]   Energy supplied and CO_2 released by fuels.

Fuel	Energy supplied in 1980*		CO_2 release per unit energy (oil = 1)	Airborne CO_2 added in 1980*		CO_2 added through 1980 (ppm)	Potential airborne CO_2 in virgin reservoirs† (ppm)
	$(10^{19}$ J)	(%)		(%)	(ppm)		
Oil	12	40	1	50	0.7	11	70
Coal	7	24	5/4	35	0.5	26	1000
Gas	5	16	3/4	15	0.2	5	50
Oil shale, tar sands, heavy oil	0	0	7/4	0	0	0	100
Nuclear, solar, wood, hydroelectric	6	20	0	0	0	0	0
Total	30	100		100	1.4	42	1220

* Based on late 1970's.   † Reservoir estimates assume that half the coal above 3000 feet can be recovered and that oil recovery rates will increase from 25 to 30 percent to 40 percent. Estimate for unconventional fossil fuels may be low if techniques are developed for economic extraction of "synthetic oil" from deposits that are deep or of marginal energy content. It is assumed that the airborne fraction of released CO_2 is fixed.

amount of biomass or the fraction of released CO₂ taken up by the ocean.

Energy growth has been 4 to 5 percent per year in the past century, but increasing costs will constrain future growth (1, 4). Thus we consider fast growth (~ 3 percent per year, specifically 4 percent per year in 1980 to 2020, 3 percent per year in 2020 to 2060, and 2 percent per year in 2060 to 2100), slow growth (half of fast growth), and no growth as representative energy growth rates.

Fossil fuel use will be limited by available resources (table 20.2). Full use of oil and gas will increase CO₂ abundance by < 50 percent of the preindustrial amount. Oil and gas depletion are near the 25 percent level, at which use of a resource normally begins to

be limited by supply and demand forces (4). But coal, only 2 to 3 percent depleted, will not be so constrained for several decades.

The key fuel choice is between coal and alternatives that do not increase atmospheric CO₂. We examine a synfuel option in which coal-derived synthetic fuels replace oil and gas as the latter are depleted, and a nuclear/renewable resources option in which the replacement fuels do not increase CO₂. We also examine a coal phaseout scenario: after a specific date coal and synfuel use are held constant for 20 years and then phased out linearly over 20 years.

Projected global warming for fast growth is 3° to 4.5°C at the end of the next century, depending on the proportion of depleted oil

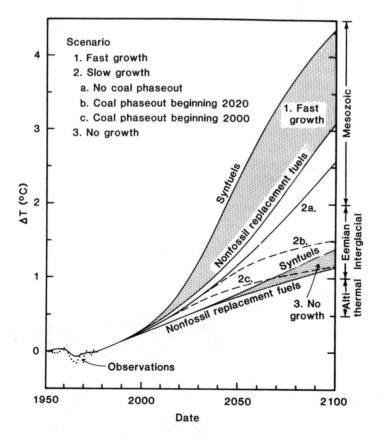

Figure 20.6 [*orig. figure 6*] Projections of global temperature. The diffusion coefficient beneath the ocean mixed layer is 1.2 cm² sec⁻¹, as required for best fit of the model and observations for the period 1880 to 1978. Estimated global mean warming in earlier warm periods is indicated on the right.

and gas replaced by synfuels (figure 20.6). Slow growth, with depleted oil and gas replaced equally by synfuels and nonfossil fuels, reduces the warming to ~ 2.5°C. The warming is only slightly more than 1°C for either (i) no energy growth, with depleted oil and gas replaced by nonfossil fuels, or (ii) slow energy growth, with coal and synfuels phased out beginning in 2000.

Other climate forcings may counteract or reinforce CO_2 warming. A decrease of solar luminosity from 1980 to 2100 by 0.6 percent per century, large compared to measured variations, would decrease the warming ~ 0.7°C. Thus CO_2 growth as large as in the slow-growth scenario would overwhelm the effect of likely solar variability. The same is true of other radiative perturbations; for instance, volcanic aerosols may slow the rise in temperature, but even an optical thickness of 0.1 maintained for 120 years would reduce the warming by less than 1.0°C.

When should the CO_2 warming rise out of the noise level of natural climate variability? An estimate can be obtained by comparing the predicted warming to the standard devi-

ation, σ, of the observed global temperature trend of the past century (50). The standard deviation, which increases from 0.1°C for 10-year intervals to 0.2°C for the full century, is the total variability of global temperature; it thus includes variations due to any known radiative forcing, other variations of the true global temperature due to unidentified causes, and noise due to imperfect measurement of the global temperature. Thus if T_0 is the current 5-year smoothed global temperature, the 5-year smoothed global temperature in 10 years should be in the range $T_0 \pm 0.1°C$ with probability ~ 70 percent, judging only from variability in the past century.

The predicted CO_2 warming rises out of the 1σ noise level in the 1980's and the 2σ level in the 1990's (figure 20.7). This is independent of the climate model's equilibrium sensitivity for the range of likely values, 1.4° to 5.6°. Furthermore, it does not depend on the scenario for atmospheric CO_2 growth, because the amounts of CO_2 do not differ substantially until after the year 2000. Volcanic eruptions of the size of Krakatoa or Agung may slow the warming, but barring

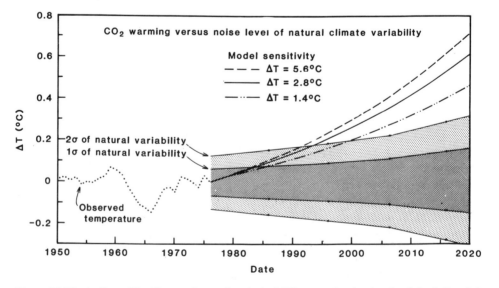

Figure 20.7 [*orig. figure 7*] Comparison of projected CO_2 warming to standard deviation (σ) of observed global temperature and to 2σ. The standard deviation was computed for the observed global temperatures in figure 20.3. Carbon dioxide change is from the slow-growth scenario. The effect of other trace gases is not included.

an unusual coincidence of eruptions, the delay will not exceed several years.

Nominal confidence in the CO_2 theory will reach ~ 85 percent when the temperature rises through the 1σ level and ~ 98 percent when it exceeds 2σ. However, a portion of σ may be accounted for in the future from accurate knowledge of some radiative forcings and more precise knowledge of global temperature. We conclude that CO_2 warming should rise above the noise level of natural climate variability in this century.

Potential Consequences of Global Warming

Practical implications of CO_2 warming can only be crudely estimated, based on climate models and study of past climate. Models do not yet accurately simulate many parts of the climate system, especially the ocean, clouds, polar sea ice, and ice sheets. Evidence from past climate is also limited, since the few recent warm periods were not as extreme as the warming projected to accompany full use of fossil fuels, and the climate forcings and rate of climate change may have been different. However, if checked against our understanding of the physical processes and used with caution, the models and data on past climate provide useful indications of possible future climate effects (51).

Paleoclimatic evidence suggests that surface warming at high latitudes will be two to five times the global mean warming (52–55). Climate models predict the larger sensitivity at high latitudes and trace it to snow/ice albedo feedback and greater atmospheric stability, which magnifies the warming of near-surface layers (6–8). Since these mechanisms will operate even with the expected rapidity of CO_2 warming, it can be anticipated that average high-latitude warming will be a few times greater than the global mean effect.

Climate models indicate that large regional climate variations will accompany global warming. Such shifting of climatic patterns has great practical significance, because the precipitation patterns determine the locations of deserts, fertile areas, and marginal lands. A major regional

change in the doubled CO_2 experiment with our three-dimensional model (6, 8) was the creation of hot, dry conditions in much of the western two-thirds of the United States and Canada and in large parts of central Asia. The hot, dry summer of 1980 may be typical of the United States in the next century if the model results are correct. However, the model shows that many other places, especially coastal areas, are wetter with doubled CO_2.

Reconstructions of regional climate patterns in the altithermal (53, 54) show some similarity to these model results. The United States was drier than today during that warm period, but most regions were wetter than at present. For example, the climate in much of North Africa and the Middle East was more favorable for agriculture 8000 to 4000 years ago, at the time civilization dawned in that region.

Beneficial effects of CO_2 warming will include increased length of the growing season. It is not obvious whether the world will be more or less able to feed its population. Major modifications of regional climate patterns will require efforts to readjust land use and crop characteristics and may cause large-scale human dislocations. Improved global climate models, reconstructions of past climate, and detailed analyses are needed before one can predict whether the net long-term impact will be beneficial or detrimental.

Melting of the world's ice sheets is another possible effect of CO_2 warming. If they melted entirely, sea level would rise ~ 70 m. However, their natural response time is thousands of years, and it is not certain whether CO_2 warming will cause the ice sheets to shrink or grow. For example, if the ocean warms but the air above the ice sheets remains below freezing, the effect could be increased snowfall, net ice sheet growth, and thus lowering of sea level.

Danger of rapid sea level rise is posed by the West Antarctic ice sheet, which, unlike the land-based Greenland and East Antarctic ice sheets, is grounded below sea level, making it vulnerable to rapid disintegration and melting in case of general warming (55). The summer temperature in its vicinity is about −5°C. If this temperature rises ~ 5°C, deglaciation could be rapid, requiring a century or less and causing a sea level rise of 5 to 6 m (55). If the West Antarctic ice sheet melts on such a time scale, it will temporarily overwhelm any sea level change due to growth or decay of land-based ice sheets. A sea level rise of 5 m would flood 25 percent of Louisiana and Florida, 10 percent of New Jersey, and many other lowlands throughout the world.

Climate models (7, 8) indicate that ~ 2°C global warming is needed to cause ~ 5°C warming at the West Antarctic ice sheet. A 2°C global warming is exceeded in the 21st century in all the CO_2 scenarios we considered, except no growth and coal phaseout.

Floating polar sea ice responds rapidly to climate change. The 5° to 10°C warming expected at high northern latitudes for doubled CO_2 should open the Northwest and Northeast passages along the borders of the American and Eurasian continents. Preliminary experiments with sea ice models (56) suggest that all the sea ice may melt in summer, but part of it would refreeze in winter. Even a partially ice-free Arctic will modify neighboring continental climates.

Discussion

The global warming projected for the next century is of almost unprecedented magnitude. On the basis of our model calculations, we estimate it to be ~ 2.5°C for a scenario with slow energy growth and a mixture of nonfossil and fossil fuels. This would exceed the temperature during the altithermal (6000 years ago) and the previous (Eemian) inter-glacial period 125,000 years ago (53) and would approach the warmth of the Mesozoic, the age of dinosaurs.

Many caveats must accompany the projected climate effects. First, the increase of atmospheric CO_2 depends on the assumed energy growth rate, the proportion of energy derived from fossil fuels, and the assumption that about 50 percent of anthropogenic CO_2 emissions will remain airborne. Second, the predicted global warming for a given CO_2 increase is based on rudimentary

abilities to model a complex climate system with many nonlinear processes. Tests of model sensitivity, ranging from the equilibrium climates on the planets to perturbations of the earth's climate, are encouraging, but more tests are needed. Third, only crude estimates exist for regional climate effects.

More observations and theoretical work are needed to permit firm identification of the CO_2 warming and reliable prediction of larger climate effects farther in the future. It is necessary to monitor primary global radiative forcings: solar luminosity, cloud properties, aerosol properties, ground albedo, and trace gases. Exciting capabilities are within reach. For example, the NASA Solar Maximum Mission is monitoring solar output with a relative accuracy of ~ 0.01 percent (57). Studies of certain components of the climate system are needed, especially heat storage and transport by the oceans and ice sheet dynamics. These studies will require global monitoring and local measurements of processes, guided by theoretical studies. Climate models must be developed to reliably simulate regional climate, including the transient response (58) to gradually increasing CO_2 amount.

Political and economic forces affecting energy use and fuel choice make it unlikely that the CO_2 issue will have a major impact on energy policies until convincing observations of the global warming are in hand. In light of historical evidence that it takes several decades to complete a major change in fuel use, this makes large climate change almost inevitable. However, the degree of warming will depend strongly on the energy growth rate and choice of fuels for the next century. Thus, CO_2 effects on climate may make full exploitation of coal resources undesirable. An appropriate strategy may be to encourage energy conservation and develop alternative energy sources, while using fossil fuels as necessary during the next few decades.

The climate change induced by anthropogenic release of CO_2 is likely to be the most fascinating global geophysical experiment that man will ever conduct. The scientific task is to help determine the nature of future climatic effects as early as possible. The required efforts in global observations and climate analysis are challenging, but the benefits from improved understanding of climate will surely warrant the work invested.

References and Notes

1   National Academy of Sciences, *Energy and Climate* (Washington D.C., 1977).
2   U. Siegenthaler and H. Oeschger, *Science* 199, 388 (1978).
3   W. S. Broecker, T. Takahashi, H. J. Simpson, T. H. Peng, *ibid.* 206, 409 (1979).
4   R. M. Rotty and G. Marland, *Oak Ridge Assoc. Univ. Rep. IEA-80-9(M)* (1980).
5   W. C. Wang, Y. L. Yung, A. A. Lacis, T. Mo, J. E. Hansen, *Science* 194, 685 (1976).
6   National Academy of Sciences, *Carbon Dioxide and Climate: A Scientific Assessment* (Washington, D.C., 1979). This report relies heavily on simulations made with two three-dimensional climate models (7, 8) that include realistic global geography, seasonal insolation variations, and a 70-m mixed-layer ocean with heat capacity but no horizontal transport of heat.
7   S. Manabe and R. J. Stouffer, *Nature (London)* 282, 491 (1979); *J. Geophys. Res.* 85, 5529 (1980).
8   J. Hansen, A. Lacis, D. Rind, G. Russell, P. Stone, in preparation. Results of an initial CO_2 experiment with this model are summarized in (6).
9   National Academy of Sciences, *Understanding Climate Change* (Washington D.C., 1975).
10 R. E. Newell and T. G. Dopplick, *J. Appl. Meteorol.* 18, 822 (1979).
11 S. B. Idso, *Science* 207, 1462 (1980); *ibid.* 210, 7 (1980).
12 *J. Geophys. Res.* 82 (No. 28) (1977); *ibid.* 85 (No. A13) (1980).
13 S. Manabe and R. T. Wetherald, *J. Atmos. Sci.* 24, 241 (1967).
14 A. Lacis, W. Wang, J. Hansen, *NASA Weather and Climate Science Review* (NASA Goddard Space Flight Center, Greenbelt, Md., 1979).
15 R. A. McClatchey et al., *U.S. Air Force Cambridge Res. Lab. Tech. Rep. TR-73-0096* (1973).
16 R. E. Roberts, J. E. A. Selby, L. M. Biberman, *Appl. Opt.* 15, 2085 (1976).
17 O. B. Toon and J. B. Pollack, *J. Appl. Meteorol.* 12, 225 (1976).
18 R. D. Cess, *J. Quant. Spectrosc. Radiat. Transfer* 14, 861 (1974).
19 W. C. Wang and P. H. Stone, *J. Atmos. Sci.* 37, 545 (1980).
20 R. D. Cess, *ibid.* 35, 1765 (1978).
21 G. Garrett, *Dyn. Atmos. Oceans* 3, 239 (1979); P. Müller, *ibid.*, p. 267.
22 S. L. Thompson and S. H. Schneider, *J. Geophys. Res.* 84, 2401 (1979).
23 M. I. Hoffert, A. J. Callegari, C. T. Hsieh, *ibid.* 85, 6667 (1980).
24 H. Oeschger, U. Siegenthaler, U. Schotterer, A. Gugelmann, *Tellus* 27, 168 (1975).
25 W. S. Broecker, *Science* 189, 460 (1975).
26 J. Hansen, A. Lacis, P. Lee, W. Wang, *Ann. N. Y. Acad. Sci.* 338, 575 (1980).
27 H. H. Lamb, *Philos. Trans. R. Soc. London Ser. A.* 255, 425 (1970).
28 S. H. Schneider and C. Mass, *Science* 190, 741 (1975).
29 J. B. Pollack, O. B. Toon, C. Sagan, A. Summers, B. Baldwin, W. Van Camp, *J. Geophys. Res.* 81, 1971 (1976).
30 A. Robock, *J. Atmos. Sci.* 35, 1111 (1978); *Science* 206, 1402 (1979).
31 W. Cobb, *J. Atmos. Sci.* 30, 101 (1973); R. Roosen, R. Angione, C. Klemcke, *Bull. Am. Meteorol. Soc.* 54, 307 (1979).
32 C. Sagan, O. B. Toon, J. B. Pollack, *Science* 206, 1363 (1979).
33 S. Manabe and R. F. Strickler, *J. Atmos. Sci.* 21, 361 (1964).
34 V. Ramanathan, *Science* 190, 50 (1975).
35 B. G. Mendonca, *Geophys. Monit. Clim. Change* 7 (1979).
36 R. Weiss, *J. Geophys. Res.*, in press.
37 R. A. Rasmussen and M. A. K. Khalil, *Atmos. Environ.* 15, 883 (1981).
38 F. S. Rowland and M. J. Molina, *Rev. Geophys. Space Phys.* 13, 1 (1975).
39 R. L. Jenne, *Data Sets for Meteorological Research* (NCAR-TN/IA-111, National Center for Atmospheric Research, Boulder, Colo., 1975); *Monthly Climate Data for the World* (National Oceanic and Atmospheric Administration, Asheville, N. C.).
40 T. P. Barnett, *Mon. Weather Rev.* 106, 1353 (1978).
41 S. K. Kao and J. F. Sagendorf, *Tellus* 22, 172 (1970).
42 R. A. Madden, *Mon. Weather Rev.* 105, 9 (1977).
43 W. A. R. Brinkman, *Quart. Res. (N. Y.)* 6, 335 (1976); I. I. Borzenkova, K. Ya. Vinnikov, L. P. Spirina, D. I. Stekhnovskiy, *Meteorol. Gidrol.* 7, 27 (1976); P. D. Jones and T. M. L. Wigley, *Clim. Monit.* 9, 43 (1980).

44 P. E. Damon and S. M. Kunen, *Science* 193, 447 (1976).

45 J. E. Hansen, W. C. Wang, A. A. Lacis, *ibid.* 199, 1065 (1978).

46 R. Oliver, *J. Appl. Meteorol.* 15, 933 (1976); C. Mass and S. Schneider, *J. Atmos. Sci.* 34, 1995 (1977).

47 J. M. Mitchell, in *Global Effects of Environmental Pollution*, S. F. Singer, Ed. (Reidel, Dordrecht, Netherlands, 1970), p. 139.

48 D. V. Hoyt, *Clim. Change* 2, 79 (1979); *Nature (London)* 282, 388 (1979).

49 K. Ya. Kondratyev and G. A. Nikolsky. *Q. J. R. Meteorol. Soc.* 96, 509 (1970).

50 R. A. Madden and V. Ramanathan [*Science* 209, 763 (1980)] make a similar comparison for the 60°N latitude belt. It is generally more difficult to extract the signal due to a global perturbation from a geographically limited area.

51 W. W. Kellogg and R. Schware, *Climate Change and Society* (Westview, Boulder, Colo., 1978).

52 CLIMAP Project Members, *Science* 191, 1131 (1976).

53 H. H. Lamb, *Climate: Present, Past and Future* (Methuen, London, 1977), vol. 2.

54 W. W. Kellogg in *Climate Change*, J. Gribbin, Ed. (Cambridge Univ. Press, Cambridge, 1977), p. 205; *Annu. Rev. Earth Planet. Sci.* 7, 63 (1979).

55 J. J. Mercer, *Nature (London)* 271, 321 (1978); T. Hughes, *Rev. Geophys. Space Phys.* 15, 1 (1977).

56 C. L. Parkinson and W. W. Kellogg, *Clim. Change* 2, 149 (1979).

57 R. C. Willson, S. Gulkis, M. Janssen, H. S. Hudson, G. A. Chapman, *Science* 211, 700 (1981).

58 S. H. Schneider and S. L. Thompson, *J. Geophys. Res.*, in press.

59 We thank J. Charney, R. Dickinson, W. Donn, D. Hoyt, H. Landsberg, M. McElroy, L. Ornstein, P. Stone, N. Untersteiner, and R. Weiss for helpful comments; I. Shifrin for several typings of the manuscript; and L. DelValle for drafting the figures.

21

ATMOSPHERIC METHANE: TRENDS OVER THE LAST 10,000 YEARS*

M. A. K. Khalil and R. A. Rasmussen

1 Introduction

Recent measurements have shown that the concentration of CH_4 is increasing in the earth's atmosphere (Rasmussen and Khalil, 1981a, b; Fraser et al., 1981; Blake et al., 1982; Khalil and Rasmussen, 1983). To a large extent this trend is probably caused by increasing agricultural and industrial activities linked to the growth of population. Rice paddy fields, cattle, production of oil and gas and urban areas are the main anthropogenic sources (Ehhalt and Schmidt, 1978; Khalil and Rasmussen, 1983). CH_4 is removed from the atmosphere principally by reacting with tropospheric OH radicals. There is some evidence that the concentrations of OH radicals have declined during the last century caused by increased levels of CO and CH_4 (Khalil and Rasmussen, 1985; Levine et al., 1985; Thomson and Cicerone, 1986). Decreased levels of OH may have contributed to the buildup of CH_4. Continued increases of CH_4 eventually may warm the earth by adding to the natural greenhouse effect causing global climatic changes (Wang et al., 1976; Lacis et al., 1981; Ramanathan et al., 1985).

As a result the past trends and the global cycle of CH_4 are of considerable interest (NRC, 1983, 1984).

The record of atmospheric CH_4 now extends back to more than 20,000 y. Systematic global measurements exist only for the last 10 y. Before then measurements had been taken sporadically by various scientists going back to the early 1960s when gas chromatographic techniques for measuring the concentrations of CH_4 came into common use. Most of the data, containing the record of atmospheric CH_4 for earlier times, are obtained by analyzing bubbles of ancient air preserved in polar ice. When all these data are put together a remarkably consistent and unified picture emerges that shows that the concentrations of CH_4 started increasing rapidly only about 100 or 200 y ago after being constant for perhaps 20,000 y or more. The rate of increase escalated, causing concentrations of CH_4 to double over the last century. From this record we estimate the trends of CH_4 over decades and longer times during the last several thousand years.

The plan of our paper is to first discuss the available data that span the last 10,000 y

* Originally published in *Atmospheric Environment*, 1987, vol. 21, pp. 2445–52.

(section 2). Section 3 contains the estimated trends based on the observed concentrations from ice cores and atmospheric measurements, and trends estimated using a global mass balance model. In the last section we discuss some of the implications of the findings and the remaining uncertainties.

2 A Review of the Methane Measurements and the Construction of the Composite Data Set

Measurements of CH_4 in samples of polar ice cores were first reported by Robbins et al. (1973). They found average concentrations of about 560 ppbv which were considerably lower than the 1400–1500 ppbv in the atmosphere during the early 1970s. However, Robbins et al. believed that the low concentrations of CH_4 may have been caused by the oxidation of CH_4 to CO in the bubbles by unknown processes and therefore did not represent the levels of CH_4 in the old atmosphere. Nearly 10 y later, based on a series of experiments, we provided evidence that the observations of Robbins et al. probably did reflect atmospheric levels several hundred years ago (Rasmussen et al., 1982). New experiments to extract CH_4 from polar ice cores showed that based on five samples, the concentrations of CH_4 were about 680 ppbv between 1000 and 3000 y ago (Khalil and Rasmussen, 1982). Soon afterwards Craig and Chou (1982) reported long term concentrations of CH_4 in polar ice cores representing air from as recent as the mid-19th century to more than 20,000 y before present. They concluded that methane increased for some 350 y before the present and more rapidly over the last several decades. These conclusions made it difficult to explain the increases of CH_4 by anthropogenic sources. However, Craig and Chou had obtained nine measurements in all and only five were from the last 350 y, leaving large gaps in the time series that affected their conclusions. These first three studies contained few data, but they established the feasibility of measuring CH_4 in polar ice cores, obtained the concentrations of CH_4 in the ancient atmosphere, and raised new questions about the causes of increasing CH_4.

The next two studies provided the most detailed information on the concentrations of CH_4 over the last 3000 y. Our study was designed to investigate the last 200–500 y in detail to determine when the increase of CH_4 began and whether there was evidence for natural changes in the concentration of CH_4 over the last 3000 y (Rasmussen and Khalil, 1984). Based on more than 80 analyses, we found that CH_4 concentrations in the ice cores did not change for thousands of years. Some 300 y ago CH_4 began to increase, as observed by Craig and Chou (1982); however the rate of increase was small and remained so until the last part of the 19th century, causing an insignificant net change. A rapid increase of CH_4 appeared to have started very recently, most likely over the last 100 y. From the ice available to us we were able to obtain only a few samples between 100 and 200 y old and none for more recent times. The study by Stauffer et al. (1985), consisting of about 30 measurements from a special core drilled in 1983–1984 at Siple Station in Antarctica, provided data during this period and closed the last remaining gap in the concentrations of CH_4 between the present and several hundred years in the past.

Most recently Pearman et al. (1986) have reported measurements of CH_4 from an Antarctic ice core. Their method of extracting air from the ice contaminated the CH_4 samples. However, they reported a few data after the contamination was eliminated. These data are included in our analysis.

The most recent concentrations of CH_4 in the ice cores are from around 1956 (Stauffer et al., 1985; Craig and Chou, 1982). The oldest direct measurements of CH_4 using gas chromatography are from 1962 (Rasmussen, unpublished data). Therefore, the ice core measurements and direct atmospheric measurements complement each other by completing the time series. Between 1962 and 1981 many sporadic atmospheric measurements have been reported. We have analysed these data and formed annual averages for use in this study (see also Khalil et al., 1987; Santanam, 1985). For the more

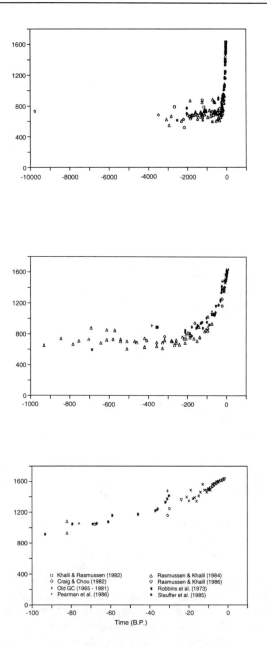

Figure 21.1 [*orig. figure 1*]   The atmospheric concentrations of CH_4 over the last 10,000 y (in ppbv). The concentrations before the 1960s are from analyses of ice cores and the more recent data are from direct atmospheric measurements.

Figure 21.2 [*orig. figure 2*]   The atmospheric concentrations of CH_4 over the last 1000 y (in ppbv). The figure shows the complementary relationship between different studies, the time when CH_4 concentrations began to increase rapidly, and the doubling of CH_4 over the last several hundred years.

Figure 21.3 [*orig. figure 3*]   Atmospheric CH_4 over the last century (in ppbv). The concentrations of CH_4 from polar ice cores and direct atmospheric measurements blend smoothly.

recent times between 1975 and 1985 we have adopted the concentrations reported by Rasmussen and Khalil (1986) based on systematic measurements taken at the South pole and the mid-northern latitudes. Characteristics of these data are similar to those obtained from the polar ice cores, thus making the composite time series more homogeneous.

The composite data base consists of all the data from these studies except the measurements from the Greenland ice sheet reported by Robbins et al. (1973). These data were excluded because of potentially large uncertainties in determining the ages of the air as discussed by Craig and Chou (1982). The different studies provide data at different times, fill critical gaps left by previous studies, and strengthen the data base during overlapping periods. Data during overlapping periods also show that there is general agreement on the absolute concentrations of CH_4 among the various investigators which justifies combining the data sets (see also Santanam, 1985).

The data, obtained from both poles, were corrected to represent the approximate global average concentration to make the ice core measurements comparable to each other and to the direct atmospheric measurements. Measurements from the South pole are increased by 5% and those from Greenland are decreased by 5%. Both direct atmospheric measurements and ice core measurements show that the concentrations from pole to pole vary by about 10% (Rasmussen and Khalil, 1984, 1986; Khalil and Rasmussen, 1983). For the direct atmospheric measurements reported by Rasmussen and Khalil (1986) an average of the concentrations in both hemispheres is taken to form a global average. The old GC data have also been adjusted to form global average concentrations.

The resulting data are shown in figure 21.1 containing the record of CH_4 over the last 10,000 y. The different symbols show how the various studies complement each other to produce a complete record of CH_4 from ancient times to the present. In figure 21.2 the concentrations of CH_4 are shown over the last 1000 y. This figure has several notable features. First, it shows more clearly the relationship of the data obtained in different studies. Second, it contains a detailed record of CH_4 concentrations over a long time with only short periods or gaps when data are not available and third, it shows that CH_4 began to increase only recently and its concentration has doubled compared to natural levels several hundred years ago. Figure 21.3 shows the concentrations over the last 100 y and demonstrates the smooth transition between the ice core and direct atmospheric measurements.

3 Trends of Methane

3.1 Estimates of trends

Trends of CH_4 over the last 7 y or so have been known from the systematic measurements reported by Khalil and Rasmussen (1982), Fraser et al. (1981) and Blake et al. (1982). Trends over longer time scales have not been determined except for those we reported earlier based on our own measurements of CH_4 in polar ice cores (Rasmussen and Khalil, 1984). The composite data base assembled here is much more detailed compared to the individual data sets and therefore it can be used to estimate trends more precisely and over shorter time spans than was possible before. We divided the measurements over 12 periods based partly on the amount of data available over each period. These periods are 1974–1984, 1962–1973, 1927–1956, 1900–1925, 1800–1900, 1700–1800, 1600–1700, 1300–1600, 800–1300, 300–800, 200 BC–300 AD and 1500 BC–200 BC. For certain years the data base contained more than one value. In such cases we averaged the reported concentrations to obtain a single representative value for each year when data existed. This method gives equal weight to each year for which concentrations have been reported. For the decade of 1974–1984 we have used only the data reported by Rasmussen and Khalil (1986). The average concentration for a given year is taken to be the average of the measured concentrations in January of that year and January of the next year.

Significant trends appear after 1700 AD

although the rates of increase are small, being around 1.5 ± 0.9 ppbv y^{-1} from 1700 to 1800 and 1.5 ± 1.2 ppbv y^{-1} from 1800 until 1900. The trends escalate to 2.3 ± 1.7 ppbv y^{-1} between 1900 and 1925, and reach 6.4 ± 3 ppbv y^{-1} between 1927 and 1956. Estimated trends between 1962 and 1973 are based on GC measurements by various groups. While the estimate appears accurate, it is highly uncertain. The rate of increase is 11 ± 11 ppbv y^{-1}. To further support the estimated rate of increase during this period we investigated the trend without using the sporadic GC data. We took 5-yearly averaged concentrations from analyses of ice cores that represent the most recent times between 1949 and 1956 (Craig and Chou, 1982; Stauffer et al., 1985; Pearman et al., 1986) and found the average concentration to be 1287 ± 45 ppbv. Next we took the 5 y of direct global measurements taken between 1979 and 1984; the average concentration was 1599 ± 19 ppbv (Rasmussen and Khalil, 1986). The difference is 312 ± 43 ppbv over about 30 y which amounts to 10.4 (± 1.6) ppbv y^{-1}. It is close to the estimated trend based on the GC data but the confidence limits of the total increase are considerably narrower. This calculation gives additional confidence in the trend between 1962 and 1974 that cannot be obtained from the sporadic GC measurements alone. Finally, in more recent times, we estimated the trend to be 17 ± 2.3 ppbv y^{-1} between 1974 and 1984. The rates of increase will be discussed further in section 3.2. The ± values are 90% confidence limits.

For earlier times, between 1700 AD and 1500 BC, there is no evidence for trends that are either statistically or environmentally significant. The rates of change, $dC/dt = R(T_1, T_2)$ for the periods between T_1 and T_2, are summarized as follows: $R(1600, 1700) = -2 \pm 2.3$ ppbv y^{-1}, $R(1300, 1600) = 0 \pm 0.5$ ppbv y^{-1}, $R(800–1300) = 0.1 \pm 0.2$ ppbv y^{-1}, $R(300, 800) = 0.2 \pm 0.2$ ppbv y^{-1}, $R(200 \text{ BC}, 300) = 0.2 \pm 0.3$ ppbv y^{-1}, $R(1500 \text{ BC}, 200 \text{ BC}) = 0 \pm 0.2$ ppbv y^{-1}.

From the measurements reported by Craig and Chou (1982), it appears that the concentrations of CH$_4$ may have remained constant for even longer times, perhaps as far back as 27,000 y BP; however, there are only two data during the period between 3500 y and 27,000 y ago.

3.2 Theoretical estimates based on population change

In an earlier paper we developed and discussed a theoretical mass balance model to explain our own measurements of CH$_4$ in polar ice cores (Khalil and Rasmussen, 1985). This model, without modification, can also explain both the concentrations and the trends in the composite data base developed in section 2.

We used the model to calculate the concentration of CH$_4$ as both the sources and sinks change over hundreds of years. The estimated present sources of CH$_4$ are divided into natural and anthropogenic components. The natural component is assumed to remain constant over the last 1000 y. The total annual anthropogenic emissions are scaled proportionately to the change of population between 1650 AD and the present. Before that time anthropogenic contributions are assumed to be small and stable. Since anthropogenic influences on the cycle of CH$_4$ are mainly from cattle, rice paddy fields, urban centers and the production of natural gas and oil, it is likely that they have increased with population. The population is estimated to have been about 0.5 billion people in 1650 AD, 1 billion in 1850, 2 billion in 1930 and more than 4 billion now (Ehrlich et al., 1977). Rapid increases of population began in the mid-17th century.

We also estimated that the concentrations of OH radicals have declined by 20% over the last century or so. The lifetime of CH$_4$ would therefore have changed from about 6 y to about 7.5 y. We estimated that approximately 20% of the annual removal of CH$_4$ from the atmosphere is by processes other than the reaction with OH radicals, including uptake by soils, other chemical reactions, and losses in the stratosphere (Khalil and Rasmussen, 1985). With these estimates of the sources and sinks the global mass balance model, $dC(t)/dt = S(t) - C(t)/\tau(t)$, was solved numerically. In this equation C represents the average concen-

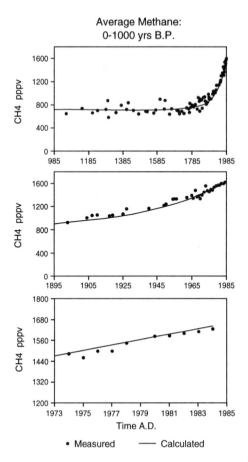

Figure 21.4 [orig. figure 4] Comparison of measured and calculated concentrations of CH$_4$ (in ppbv). Each point represents the average measured concentration for a year and is derived from the data shown in figure 21.1. The solid line is from a global mass balance calculation in which the sources affected by anthropogenic activities are increased in proportion to the increasing population and the concentrations of OH are reduced by 20% over the last century.

tration, S the emission rate and τ represents the atmospheric lifetime. The calculated concentrations of CH$_4$, discussed in our earlier paper (Khalil and Rasmussen, 1985), are shown in figure 21.4 over 1000 y, 100 y and 10 y scales. We derived the measured concentrations shown in figure 21.4 from those in figure 21.1 by forming a single

average concentration representative for each year when data exist, as described in section 2.

The agreement between the model and the measured concentrations of CH$_4$ can be described by %D(t) = 100% × [C (measured, t) – C (calculated, t)/C (measured, t) where %D(t) is the percent difference between measured and calculated concentrations at time t. Over the last 1000 y the average difference is 0.8% with a standard deviation of 8% which essentially reflects the variability of the measured concentrations and shows that the differences between the calculated and measured concentrations are insignificant. However, over shorter periods some systematic and statistically significant differences exist between observed and calculated concentrations but these are probably inconsequential. Over the last 10 y the % difference is –1% ± 0.5% or 16 ± 8 ppbv. The 1% difference between measured and calculated concentration is due partly to the effects of the last El-Nino which are not accounted for by the model calculations. During the El-Nino event of 1982–1983 concentrations of CH$_4$ fell below expected levels (Khalil and Rasmussen, 1986). It is also noteworthy that a difference of 1% constitutes very good agreement between measured and calculated values. Between the last 10 and 100 y the difference between observed and calculated concentrations is +2.4% ± 1.2% so that the observed concentrations are slightly higher than calculated using the mass balance model. At present there is no explanation for this small discrepancy.

Using the calculated concentrations we estimated the expected trends over the same periods for which we had evaluated the trends using observed concentrations. The results are summarized as follows: R(model, T$_1$, T$_2$) = the rate of change between times T$_1$ and T$_2$, calculated using the mass balance model. R(m, 1974, 1984) = 16 ppbv y^{-1}, R(m, 1962, 1973) = 13.7 ppbv y^{-1}, R(m, 1927, 1956) = 7 ppbv y^{-1}, R(m, 1904–1925) = 3.8 ppbv y^{-1}, R(m, 1800–1900) = 1.6 ppbv y^{-1}, R(m, 1700, 1800) = 0.4 ppbv y^{-1} and R(m, 1700, 1800) = 0.2 ppbv y^{-1}. In figure 21.5 we have summarized the estimated trends based on the

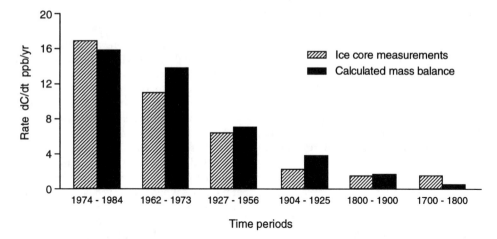

Figure 21.5 [*orig. figure 5*]   The trends of CH4 (ppbv y^{-1}). The rates of increase are calculated from the data and mass balance calculations shown in figure 21.2.

observed concentrations and the global mass balance model.

The model based on the population growth correctly predicts the concentrations over the last 10,000 y but more importantly, it correctly predicts the times when rapid increases of CH_4 are observed in the ice core data.

4   Discussion

We have compiled a history of the trends of CH_4 over the last 10,000 y based on composite data from several studies. The uncertainties we have reported represent the observed variability, however there are other uncertainties that cannot be quantified at present.

The age of the air in the ice cores is perhaps one of the most significant uncertainties. Although there are uncertainties in estimating the age of the ice, much larger uncertainties arise in relating the age of the air to the age of the surrounding ice. The uppermost layers of the ice sheets are firn and permeable to atmospheric gases. It takes several decades before the bubbles are formed in deeper ice and are sealed off from the influences of the atmosphere. This transition time may vary over the ice samples from the two poles and even over the ice sheet of the same pole. In some of the studies

this transition time is taken to be 90 y on average, but its variability or uncertainty is not known and it may be several decades (Craig and Chou, 1982; Rasmussen and Khalil, 1984). This uncertainty has its greatest effect on the most recent ice and the estimates of trends between 1900 and 1956 (see also, Schwander and Stauffer, 1984).

An ice core sample used for the measurement of CH_4 may contain air from several years, though we have associated a single year with each measurement. The core samples from deeper ice are likely to represent more years because annual layers are more compact and even ice at the same depth may contain bubbles with air from several different years. The number of years of CH_4 concentrations represented by each sample varies by an unknown amount.

In spite of these uncertainties the rates of increase and their temporal changes appear to coincide with the change and trends of population and may therefore be caused largely by industrial and agricultural activities associated with the production of food and energy for a growing population as discussed in section 3.2. The long period of constant concentrations places stringent constraints on the natural global emissions of CH_4 (Khalil and Rasmussen, 1985).

The composite data base strengthens the conclusion that the present high levels of

CH_4 are of anthropogenic origin. Figure 21.6 shows the gaps in time (in y) during which we have no data. It shows that over 300 y BP the gaps are smaller than 20 y and they become longer than 100 y only after about 1000 y BP. Data during the last 1000 y are so detailed that sizable natural fluctuations would not have gone undetected. The average concentration of CH_4 over the last century, obtained from the composite data, is about 1100 ppbv, which is more than any

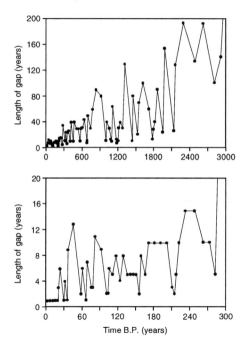

of the ice core measurements for air older than 100 y. Therefore, increases as large as those seen over the last century have not occurred for at least as long as 1000 y. For the times of 1000 y and earlier, whether CH_4 has undergone sizable natural fluctuations lasting decades or even centuries remains an open question. Figure 21.6 shows the periods that require further investigation.

All conclusions regarding the trends of CH_4, based on the ice core data, are predicted on the idea that the measured concentrations represent atmospheric levels at the times when the air is believed to have been sealed off in the bubbles. There is no proof of this assertion. However, the data, both collectively and from individual studies, show that CH_4 concentrations do not change in ice between recent times and more than 20,000 y ago. Even slow processes that remove CH_4 in the bubble can cause a significant decrease after hundreds or thousands of years. Therefore if CH_4 is removed in the bubbles or in the firn, the processes responsible must occur in the uppermost layers of the ice sheet.

The record of atmospheric CH_4 in polar ice cores shows the effects of anthropogenic activities on the cycle of CH_4. The relationship between population growth and the increase of CH_4 that we have shown in this study provides a basis for projecting the concentrations of CH_4 into the future when it may affect the global environment. We have reported the initial calculations of expected concentrations in the future (Khalil and Rasmussen, 1982, 1987).

Figure 21.6 [orig. figure 6] Gaps in the record of CH_4. The y-value is the number of years between the measurement at the x value and the measurement at the next later time. For example, in the lower panel, the graph shows that there is a measurement at 156 y BP and there is a gap of 8 y between it and the next measurement. The figure shows that there are detailed data over the last 1000 y. The length of the gaps in the data increases for older times, and there are only two measurements in ice older than about 3500 y.

Acknowledgments

We thank Martha Shearer and S. Sanatam for work on the published GC measurements between 1962 and 1981. This work was supported in part by grants from the National Science Foundation (No. ATM-8414020) and from the department of Energy (No. DE-FG06-85ER6031). Additional support was provided by the Biospherics Research Corporation and the Andarz Co.

References

Blake D. R., Meyer E. W., Tyler S. C., Makide Y., Montague D. C. and Rowland F. S. (1982) Global increase of atmospheric methane concentration between 1978 and 1980. *Geophys. Res. Lett.* 82, 477–480.

Craig H. and Chou C. C. (1982) Methane: record in polar ice cores. *Geophys. Res. Lett.* 9, 1221–1224.

Ehhalt D. H. and Schmidt U. (1978) Sources and sinks of atmospheric methane. *Pure appl. Geophys.* 116, 452–464.

Ehrlich P. J., Ehrlich A. H. and Holdren J. P. (1977) *Ecoscience: Population, Resources and Environment.* W. H. Freeman, San Francisco.

Fraser P. J., Khalil M. A. K., Rasmussen R. A. and Crawford A. J. (1981) Trends of atmospheric methane in the southern hemisphere. *Geophys. Res. Lett.* 8, 1063–1066.

Khalil M. A. K. and Rasmussen R. A. (1982) Secular trends of atmospheric methane. *Chemosphere* 11, 877–883.

Khalil M. A. K. and Rasmussen R. A. (1983) Sources, sinks and seasonal cycles of atmospheric methane. *J. geophys. Res.* 88, 5131–5144.

Khalil M. A. K. and Rasmussen R. A. (1985) Causes of increasing methane: depletion of hydroxyl radicals and the rise of emissions. *Atmospheric Environment* 19, 397–407.

Khalil M. A. K. and Rasmussen R. A. (1986) Interannual variability of atmospheric methane: possible effects of the El Nino-Southern oscillation. *Science* 232, 56–58.

Khalil M. A. K. and Rasmussen R. A. (1987) Trends of atmospheric methane: past, present and future. *Proceedings of Symposium on CO₂ and Other Greenhouse Gases*, Commission of European Communities, Brussels, Belgium, 3–5 Nov. 1986.

Khalil M. A. K., Rasmussen R. A., Shearer M. and Santanam S. (1987) An analysis of the GC atmospheric methane measurements: 1962–1981. In preparation.

Lacis A., Hanson J., Lee P., Mitchell T. and Lebedeff S. (1981) Greenhouse effect of trace gases: 1970–1980. *Geophys. Res. Lett.* 8, 1035–1038.

Levine J. S., Rinsland C. P. and Tennille G. M. (1985) The photochemistry of methane and carbon monoxide in the troposphere in 1950 and 1985. *Nature* 318, 254–257.

National Research Council (1983) *Changing Climate: Report of the Carbon Dioxide Assessment Committee.* National Academy Press, Washington D.C.

National Research Council (1984) *Global Tropospheric Chemistry, A Plan for Action.* National Academy Press, Washington D.C.

Pearman G. I., Etheridge D., de Silva F. and Fraser P. J. (1986) Evidence of changing concentrations of atmospheric CO_2, N_2O and CH_4 from air bubbles in Antarctic ice. *Nature* 320, 248–250.

Ramanathan V., Cicerone R. J., Singh H. B. and Kiehl J. T. (1985) Trace gas trends and their potential role in climate change. *J. geophys. Res.* 90, 5547–5566.

Rasmussen R. A. and Khalil M. A. K. (1981a) Atmospheric methane (CH_4): trends and seasonal cycles. *J. geophys. Res.* 86, 9826–9832.

Rasmussen R. A. and Khalil M. A. K. (1981b) Increase in the concentration of atmospheric methane. *Atmospheric Environment* 15, 883–886.

Rasmussen R. A. and Khalil M. A. K. (1984) Atmospheric methane in the recent and ancient atmospheres: concentrations, trends and interhemispheric gradient. *J. geophys. Res.* 89, 11599–11605.

Rasmussen R. A. and Khalil M. A. K. (1986) Atmospheric trace gases: trends and distributions over the last decade. *Science* 232, 1623–1624.

Rasmussen R. A., Khalil M. A. K. and Hoyt S. D. (1982) Methane and carbon monoxide in snow. *J. Air Pollut. Control Ass.* 32, 176–178.

Robbins R. C., Cavanagh L. A., Salas L. J. and Robinson E. (1973) Analysis of ancient atmospheres. *J. geophys. Res.* 78, 5341–5344.

Santanam S. (1985) A trend study of atmospheric methane 1965–1980 GC and polar ice core measurements. M.S. thesis, Oregon Graduate Center.

Schwander J. and Stauffer B. (1984) Age difference between polar ice and air trapped in its bubbles. *Nature* 311, 45–47.

Stauffer B., Fischer G., Neftel A. and Oeschger H. (1985) Increase of atmospheric methane recorded in Antarctic ice core. *Science* 229, 1386–1388.

Thompson A. M. and Cicerone R. J. (1986) Possible perturbations to CO, CH_4 and OH. *J. geophys. Res.* 91, 10853–10864.

Wang W. C., Yung Y. L., Lacis A. A., Mo T. and Hansen J. E. (1976) Greenhouse effect due to anthropogenic perturbations. *Science* 194, 685–690.

22

POSSIBLE CLIMATE CHANGE DUE TO SULPHUR DIOXIDE-DERIVED CLOUD CONDENSATION NUCLEI*

T. M. L. Wigley

It has been hypothesized that climate may be noticeably affected by changes in cloud condensation nuclei (CCN) concentrations, caused either by changes in the flux of dimethylsulphide (DMS) from the oceans[1,2] and/or by man-made increases in the flux of sulphur dioxide (SO_2) into the atmosphere.[3] When oxidized, the sulphur compounds produce non-sea-salt sulphate (n.s.s.-SO_4^{2-}) aerosols, which may act as CCNs. The CCN changes affect climate by altering the number density and size distribution of droplets in clouds, and hence their albedo. Here I am concerned primarily with the possible effects of SO_2. Because the increase in SO_2 emissions has been largely in the Northern Hemisphere, this raises the possibility of a cooling of the Northern Hemisphere relative to the Southern.[3] By comparing observed differences in hemispheric-mean temperatures with results from a simple climate model, one can place limits on the possible magnitude of any SO_2-derived forcing. The upper limit is sufficiently large that the effects of SO_2 may

have significantly offset the temperature changes that have resulted from the greenhouse effect.

Schwartz[3] has documented the likely increase in sub-micrometre-sized n.n.s.-SO_4^{2-} aerosol in the Northern Hemisphere, which he attributes largely to increased SO_2 fluxes. This is supported by Greenland ice-core analyses of sulphate,[4,5] which show a two- to threefold increase during the twentieth century, and is consistent with the estimated six- to sevenfold increase in global SO_2 flux this century,[6] more than 90% of which has been in the Northern Hemisphere. Furthermore, CCN concentrations over the north Atlantic are typically three or more times greater than those over southern oceans.[7] Cloud reflectivity differences have also been observed[8] in the vicinity of ship tracks, which could be partly a result of SO_2 effects.

Twomey et al.[7] and Charlson et al.[1] have estimated the effect of CCN changes on planetary albedo and radiation balance. For the cloud type most affected, stratiform

* Originally published in Nature, 1989, vol. 339, pp. 365–7.

water clouds, Charlson et al. (Fig. 1 and Table 1 of ref. 1) give results that may be expressed as

$$\Delta\alpha = -0.067 \, \Delta N/N \qquad (1)$$

where α is the TOA (top-of-atmosphere) albedo of the cloud averaged over the solar spectrum, and N is the cloud droplet number density (equation (1) assumes constant liquid-water content). A similar result has been given by Twomey et al. (ref. 7, Fig. 6b). (The example given in ref. 1 gives a proportionality constant of –0.053, but I have recalculated the constant from their Fig. 1. The uncertainty is about ±20%.) As $N \propto C^m$ where C is the CCN concentration and $m \approx 0.8$ (ref. 9), equation (1) becomes

$$\Delta\alpha = -0.053 \, \Delta C/C \qquad (2)$$

If the albedo change occurred in all suitable clouds over the Northern Hemisphere (such clouds cover roughly 45% of the global ocean area[1]), then the radiative balance for the hemisphere would be perturbed by an amount

$$\Delta Q_{NH} = -4.92 \, \Delta C/C \, (W \, m^{-2}) \qquad (3)$$

Unfortunately, the SO_2-induced change in CCN concentrations is unknown, and it is difficult even to guess at a reasonable value. Charlson et al.[1] assume $\Delta N/N = 0.3$. If increased SO_2 fluxes have caused a CCN change in the range $0.1 \leq \Delta C/C \leq 0.5$ since the late nineteenth century, then ΔQ_{NH} would lie in the range –0.5 to –2.5 W m^{-2} (or 1/0.6 times these values if averaged over only the Northern Hemisphere oceans). At the top end of this range, the perturbation is similar in magnitude to the forcing from changes in greenhouse-gas concentrations over the past few centuries (about 2.2 W m^{-2} at the top of the troposphere[10]).

Given the possibility of a significant SO_2-related forcing in the Northern Hemisphere, it is pertinent to ask the following questions: For a given forcing ΔQ_{NH}, what hemispheric temperature differential would be expected? What ΔQ_{NH} values are compatible with observed differences in changes in hemispheric-mean temperature? Could SO_2-related forcing have noticeably offset the greenhouse effect? These questions can

be answered by comparing modelled and observed differences in hemispheric-mean temperature. To do this, the climate model must include at least the damping effect of oceanic thermal inertia and must differentiate the two hemispheres. It is also important to use a reasonable parameterization of inter-hemispheric heat and mass exchanges, because although hemisphere-specific forcing will produce a response that is greater in the hemisphere concerned, the response will be far from negligible in the other hemisphere because of these exchanges.[11] In the model used here,[11] this is ensured by calibrating the land-sea and hemisphere-hemisphere exchange coefficients to give the correct seasonal cycles of temperature over the land and ocean areas of each hemisphere. The adequacy of the parameterization has been further tested by simulating volcanic eruptions in both hemispheres and comparing the results with hemispheric-mean temperature observations.[12]

The magnitude of the model's response to a given external-forcing change depends primarily on two factors, the climate sensitivity or gain[13] (determined here by the equilibrium global-mean warming produced by a doubling of the CO_2 concentration, ΔT_{2x}) and the rate of ocean mixing (controlled by the ocean's vertical diffusivity, κ). For any prescribed SO_2-derived forcing history, therefore, there will be a range of possible inter-hemispheric temperature differences and global-mean temperature changes because of uncertainties in ΔT_{2x} and κ.

For the forcing history, I have superimposed SO_2-derived forcing on the combined effect of changes in greenhouse-gas concentrations (CO_2, CH_4, N_2O and the CFCs) as given in ref. 10. Although other external forcings have been proposed,[14-17] greenhouse forcing is the only one that has a known history over the past century. The SO_2-derived forcing is here characterized by its 1985 value ($\Delta Q_{NH}(1985)$), for which I will consider a range of values, but the response may also depend on the history of ΔQ_{NH}. Although this is unknown, it must be related to the history of SO_2 flux changes, which show a relatively slow increase over 1850–1950 and a more rapid increase since then.[6] Another proxy for ΔQ_{NH} is the Greenland ice-core n.s.s.-SO_4^{2-} record,[4,5] which shows an increase over 1895–1915, little change to around 1940[4] or 1955,[5] an increase to 1970 and little change since then. Neither SO_2 flux nor n.s.s.-SO_4^{2-} will accurately portray changes in CCN, albedo or radiative forcing, because of possible saturation and nonlinear effects. I have investigated a number of different possible forcing histories, and all produce quantitatively similar results provided that they end at the same $\Delta Q_{NH}(1985)$ value. The particular history whose results are described below parallels the n.s.s.-SO_4^{2-} record in ref. 4 (their Fig. 2), beginning at zero in 1895, rising to a maximum in 1970 and then remaining constant. Although the relation of n.s.s.-SO_4^{2-} to CCNs has been used previously by Legrand et al.,[18] it can be considered only a rough approximation, not least because the relationship is known to be nonlinear.[19]

The key model results are the global-mean temperature change and the change in the temperature difference between the hemispheres. All model runs begin in equilibrium in 1765. The changes described below are all relative to values in 1900 in order to facilitate comparison with observations. The global-mean temperature change is

$$\delta \bar{T}(t) = \Delta \bar{T}(t) - \Delta \bar{T}(1900) \qquad (4)$$

and the change in the temperature difference between the hemispheres is

$$\delta T_{SN}(t) = (\Delta T_S(t) - \Delta T_N(t)) - (\Delta T_S(1900) - \Delta T_N(1900)) \qquad (5)$$

where Δ is the change since 1765, an overbar denotes a global average and subscripts N and S refer to the hemispheric averages. The SO_2-related forcing was applied only over the Northern Hemisphere ocean area, with magnitude scaled from ΔQ_{NH} by $1/0.6$. Results are shown in figure 22.1. Climate sensitivity is the main controlling model parameter, although this has little effect on $\delta T_{SN}(1985)$ for $\Delta T_{2x} \gtrsim 2$ °C. For the chosen forcing history, $\delta T_{SN}(1985)$ is virtually inde-

pendent of diffusivity. $\delta T_{SN}(1985)$ increases roughly linearly with increasing magnitude of $\Delta Q_{NH}(1985)$, and global-mean temperature change ($\delta \bar{T}(1985)$) decreases markedly as the magnitude of $\Delta Q_{NH}(1985)$ increases.

These results can be compared with observations, but there are uncertainties in such a comparison because of, on the observational side, changes in (or incomplete) data coverage, and because of, on statistical grounds, the existence of pronounced shorter-timescale (probably natural) fluctuations. Figure 22.2 shows δT_{SN} as a function of time. For the ocean areas, temperatures are based on observations of both sea surface temperatures (SSTs) and marine air temperatures (MATs) or night-time marine air temperatures (NMATs) from two different (but overlapping) data sources, the UK Meteorological Office marine data set[20] and the Comprehensive Ocean-Atmosphere Data Set (COADS).[21,22] For the land areas, data are from Jones et al.[23,24] Data from before about 1900 are less reliable than those after this date. The total trend over 1900–1985 is, with 95% confidence limits, 0.03 ± 0.13 °C (COADS MAT). 0.14 ± 0.14 °C (UKMO NMAT), 0.05 ± 0.12 °C (COADS SST) or 0.10 ± 0.12 °C (UKMO SST). The positive-trend values indicate cooling of the Northern Hemisphere with respect to the Southern. Autocorrelation has been accounted for in estimating the confidence limits. For comparison with the observations, figure 22.2 also shows a model result; specifically for $\Delta Q_{NH}(1985) = -1.0$ W m^{-2} with $\kappa = 1$ cm^2 s^{-1} and $\Delta T_{2x} = 3.0$ °C, which gives $\delta \bar{T}(1985) = 0.47$ °C and $\delta T_{SN}(1985) = 0.18$ °C (point C on figure 22.1).

The observations indicate a δT_{SN} trend over 1900–1985 of around 0.1 °C. This value is subject to considerable uncertainty, both statistical and observational, and could range anywhere between -0.1 °C and 0.3 °C. For any assumed overall trend, one can estimate the required value of $\Delta Q_{NS}(1985)$ using figure 22.1. This value depends on the assumed values of ΔT_{2x} and κ, and, provided no factors other than the greenhouse effect and SO_2 increases have contributed significantly to the century-timescale forcing, these quantities are

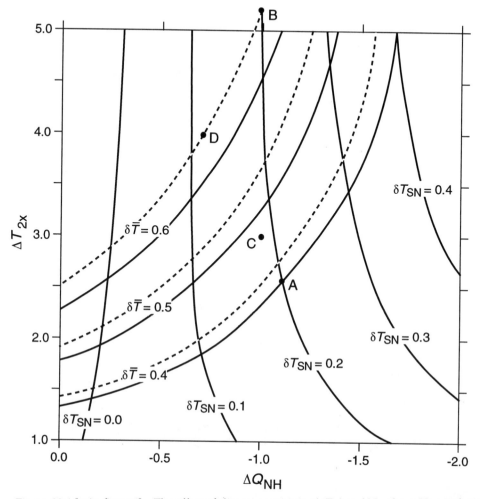

Figure 22.1 [*orig. figure 1*]   The effect of climate sensitivity (ΔT_{2x}) and Northern Hemisphere cloud-albedo-derived forcing ($\Delta Q_{NH}(1985)$) on global-mean temperature change between 1900 and 1985 ($\delta \bar{T}$; $\delta \bar{T}(1985)$ in text) and the relative warming of the hemispheres over this period (δT_{SN}; $\delta T_{SN}(1985)$ in text). Full lines show results for a diffusivity of 1.0 cm^2 s^{-1}, and dashed lines are for a diffusivity of 2.0 cm^2 s^{-1}. The effect of diffusivity on δT_{SN} is negligible. Points A–D are explained in text.

constrained, in turn, by the necessity for $\delta \bar{T}(1985)$ to be compatible with observations. From the late nineteenth century to 1985, global-mean temperature increased by about 0.5 °C, with an uncertainty of around ±0.1 °C (ref. 22).

An example showing the constraints on $Q_{NH}(1985)$ for given $\delta T_{SN}(1985)$ is given in figure 22.1. If $\delta T_{SN}(1985) = 0.2$ °C, $0.4 \leq$ $\delta \bar{T}(1985) \leq 0.6$ °C and $1 \leq \kappa \leq 2$ cm^2 s^{-1}, then $\Delta Q_{NH}(1985)$ must lie between the limits corresponding to points A and B in figure 22.1, specifically −1.0 to −1.1 W m^{-2}. Equation (1) implies that the required increase in average Northern Hemisphere marine CCN concentration is ~20%. For $\Delta Q_{NH}(1985) \sim -1$ W m^{-2}, the $\delta \bar{T}(1985)$ limits of 0.4 and 0.6 °C imply that ΔT_{2x} must be in the range

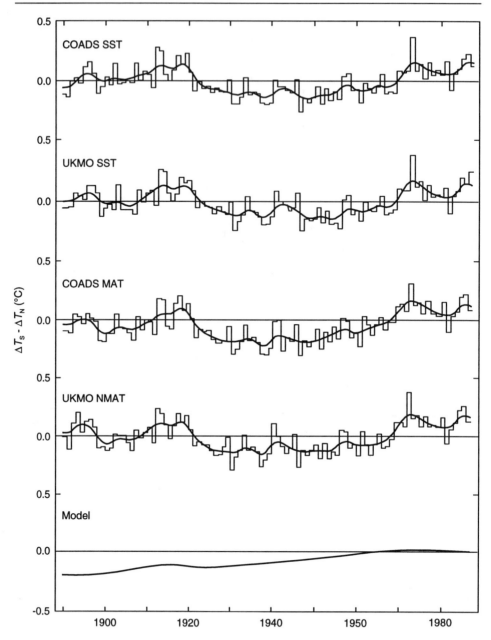

Figure 22.2 [*orig. figure 2*]   Differences between changes in Northern Hemisphere temperature (ΔT_N) and Southern Hemisphere temperature (ΔT_S). A positive value indicates that the Southern Hemisphere has warmed relative to the Northern Hemisphere. Marine data from two different sources are used, as described in the text. All values are relative to a 1950–1979 reference period. The lower panel shows modelled changes in δT_{SN}, with the 1985 value corresponding to point C in figure 22.1. (The zero point has been shifted to give zero mean over 1950–1979.) Abbreviations as in the text.

2.5–5.2 °C. For the estimated upper limit of $\delta T_{SN}(1985)$, 0.3 °C, the hemispheric-mean SO_2-derived forcing could be as high as –1.4 W m^{-2} (that is $\Delta C/C \approx 30\%$) and a ΔT_{2x} value close to 5 °C would be compatible with a global-mean warming of 0.5 °C.

The questions posed earlier can now be answered. Each 0.1 °C increase in the twentieth-century warming of the Southern Hemisphere relative to the Northern Hemisphere corresponds to a forcing of around –0.5 W m^{-2}, or a CCN increase of ~10%. Values of up to three times these amounts would still be compatible with observations. This suggests that SO_2-derived aerosols could have had a noticeable effect on global-scale temperatures. It should be noted, however, that there are alternative explanations for differing trends in hemispheric-mean temperatures (see refs 10 and 16, for example). Furthermore, even quite large CCN changes will fail to produce a statistically significant hemispheric temperature contrast. Given the noise level and the possibility of alternative explanations, detection of a SO_2-derived CCN effect would be exceedingly difficult.

The implications of even a modest value for the SO_2-derived forcing are of considerable importance. In the absence of such forcing (and assuming no long-term forcing other than that due to greenhouse-gas concentration changes), the implied ΔT_{2x} lies in the range 1.3–2.5 °C. These values are considerably below those given by most recent General Circulation Model (GCM) studies.[25] A value for $\Delta Q_{NH}(1985)$ of as little as –0.7 W m^{-2} (corresponding to $\Delta C/C \approx$ 14%) would admit a ΔT_{2x} value close to 4 °C, which is more in accord with these GCM results (point D on figure 22.1); but it is doubtful that the global-mean warming since 1900 is as high as the 0.6 °C required to give this ΔT_{2x} value. As the magnitude of the SO_2-derived forcing increases, this increasingly offsets the greenhouse-gas forcing and allows still larger values of ΔT_{2x}. Thus, the existence of SO_2-derived forcing may help to explain the apparent inconsistency between GCM results and observations.

An ironic twist arises if the SO_2-derived forcing is, after all, significant. The effects of SO_2 associated with acidic precipitation and urban pollution are clearly detrimental, and measures to reduce emissions are being implemented widely. However, if we were successful in halting or reversing the increase in SO_2 emissions, we could, as a by-product, accelerate the rate of greenhouse-gas-induced warming, so reducing one problem at the expense of increasing the rate of onset of another.

Acknowledgements
Supported by the US Department of Energy, Carbon Dioxide Research Division. Some of this work was carried out while the author was a visiting scientist at the National Center for Atmospheric Research (NCAR), Boulder, Colorado, USA. NCAR is sponsored by the NSF. Help from J. A. Coakley is gratefully acknowledged.

References
1   Charlson, R. J., Lovelock, J. E., Andreae, M. O. and Warren, S. G. Nature 326, 655–661 (1987).

2   Mészáros, E. Atmos. Envir. 22, 423–424 (1988).

3   Schwartz, S. E. Nature 336, 441–445 (1988).

4   Neftel, E., Beer, J., Oeschger, H., Zürcher, F. and Finkel, R. C. Nature 314, 611–613 (1985).

5   Mayewski, P. A. et al. Science 232, 975–977 (1986).

6   Möller, D. Atmos. Envir. 18, 19–27 (1984).

7   Twomey, S. A., Piepgrass, M. and Wolfe, T. L. Tellus 368, 356–366 (1984).

8   Coakley, J. A. Jr, Bernstein, R. L. and Durkee, P. A. Science 237, 1020–1022 (1987).

9   Pruppacher, H. R. and Klett, J. D. Microphysics of Clouds and Precipitation (Reidel, Dordrecht, 1978).

10 Wigley, T. M. L. Geophys. Res. Lett. 14, 1135–1138 (1987).

11 Wigley, T. M. L. and Raper, S. C. B. Nature 330, 127–131 (1987).

12 Sear, C. B., Kelly, P. M., Jones, P. D. and Goodess, C. M. Nature 330, 365–367 (1987).

13 Schlesinger, M. E. Clim. Dynam. 1, 35–51 (1986).

14 Hansen, J. et al. Science 213, 957–966 (1981).

15 Gilliland, R. L. Climatic Change 4, 111–131 (1982).

16 Gilliland, R. L. and Schneider, S. H. Nature 310, 38–41 (1984).

17 Vinnikov, K. Ya. and Groisman, P. Ya. Meteorol. i Gidrol. 1981 (11), 30–43 (1981).

18 Legrand, M. E., Delmas, R. J. and Charlson, R. J. *Nature* 334, 418–420 (1988).

19 Leaitch, W. R., Strapp, J. W., Isaac, G. A. and Hudson, J. G. *Tellus* 388, 328–344 (1986).

20 Folland, C. K., Parker, D. E. and Kates, F. E. *Nature* 310, 670–673 (1984).

21 Woodruff, S. D., Slutz, R. J., Jenne, R. L. and Steurer, P. M. *Bull. Am. met. Soc.* 68, 1239–1250 (1987).

22 Jones, P. D., Wigley, T. M. L. and Wright, P. B. *Nature* 322, 430–434 (1986).

23 Jones, P. D. et al. *J. Clim. appl. Met.* 25, 161–179 (1986).

24 Jones, P. D., Raper, S. C. B. and Wigley, T. M. L. *J. Clim. appl. Met.* 25, 1213–1230 (1986).

25 Schlesinger, M. E. and Mitchell, J. F. B. *Rev. Geophys.* 25, 760–798 (1987).

23

POSSIBLE CLIMATIC IMPACTS OF TROPICAL DEFORESTATION*

E. Salati and C. A. Nobre

1 Introduction

The conversion of tropical rainforests into a different vegetation, most notably pastures or annual crops, inevitably entails major changes in the ecosystem. There appears to be a consensus that this type of conversion changes the flora, the aquatic and land fauna, and the physico-chemical and biological characteristics of the soil and surface waters. There also appears to be agreement that qualitative and quantitative changes are caused in the biogeochemical cycles.

Yet, when analyzing the climatic impacts involved or associated with the conversion of tropical rainforests into pastures, the question becomes controversial. This is due to several factors, including in particular the difficulty of quantifying the components of the energy and water balances in the undisturbed and disturbed ecosystems, and the difficulty of developing climate models at the regional level that will permit reliable predictions based on the changes in land-use patterns to be introduced. Moreover, it has been generally accepted that the flora is a consequence of the climatic conditions in conjunction with the characteristics of soil and geomorphology. Therefore, it has been assumed that conversion of one type of

vegetation cover to another should not induce large-scale climatic changes. However, studies are producing evidence that for some ecosystems the present dynamic equilibrium of the atmosphere depends on the underlying vegetation and the present climate is the consequence of the interaction between the biosphere and the atmosphere (Salati, 1985; Shukla et al., 1990).

While the discussions on the subject proceed and the efforts to improve the measurements and develop more suitable models continue, the occupation by settlers, ranchers, and logging and mining operations press steadily deeper into the forest or into what remains thereof throughout the world (Setzer et al., 1988; Malingreau and Tucker, 1988; Repetto, 1988). Setzer et al. (1988) have shown that a sizeable area is burnt during the dry season in Amazonia every year ('queimadas') to give rise to settler's agricultural plots and grazing land in large cattle ranches. Their analysis was based on the satellite counting of satellite-detected fires. A large proportion of the area cleared was not primary forest but either secondary growth or crop residuals. A recent Landsat-based estimate of the total area of primary forest that has been cleared in Brazilian Amazonia up to 1989 (Fearnside

* Originally published in *Climatic Change*, 1991, vol. 9, pp. 177–96.

et al., 1990) puts this figure close to 400 000 km² (or about 10.5% of the area covered by forests), and the annual increase of deforested areas from 1988 to 1989 was about 27 000 km². This annual rate may indicate a decline in Brazilian Amazonian deforestation since previous estimates for the annual rate of deforestation were of the order of 35 000 km² or higher (Fearnside, 1989). At those deforestation rates most of the forests will be gone in less than 100 years.

The forest clearings are not contiguous but spread over large areas, mostly along the region's roads. Given the yet somewhat small percentage of total deforestation and the fact that it is scattered over a large area, one would not expect large changes in the basin-scale hydrological cycle to have been already detected. Indeed, a recent observational analysis of historical series of precipitation and streamflow in the Amazon basin (Rocha et al., 1989) has not shown any trend in the basin's hydrological cycle that could be attributed to changes in the vegetation cover.

2   Amazon Forest Micrometeorology

For subsequent comparisons, the micrometeorological data collected at the Ducke Forest Reserve, near Manaus, in Brazilian Amazonia, will be briefly reviewed. Research was carried out in a joint program by Brazilian and British researchers from 1983 through 1985. The data analysis has been published in Shuttleworth et al. (1984a, 1984b, 1985, and Shuttleworth, 1988) and summarized by Molion (1987).

The research area is covered by dense forest with trees measuring 35 m in average, with branches sometimes reaching 40 m. The micrometeorological observation tower is 45 m high. The solar energy reaching the ground was only 1.2% of that reaching the tree tops. The average albedo for this forest was 12%, varying with the solar zenith angle.

By measuring, during the dry season, vertical temperature and humidity profiles at the canopy level, the radiation budget above the canopy, and the fluxes of sensible and latent heating above the canopy, it was possible to conclude that for fine days 75% of net radiation goes into evaporating water and the remaining 25% is used to heat the air. For a daily average of 4.96 mm water equivalent of net radiation, 3.70 mm was used for evapotranspiration. Continuous measurements of throughfall, steamflow and rainfall at the top of the canopy by Lloyd and Marques (1988), during a period of more than 2 years, led to the estimate that interception loss (rainwater which is intercepted by leaves, stems, and trunks and directly reevaporated into the air) corresponds to about 10% of the measured rainfall at the top of the canopy. This estimate is somewhat smaller than previous estimates (Franken and Leopoldo, 1984) that interception loss accounts for up to 25% of rainfall. Shuttleworth (1988) points out that 'the spatial variability of precipitation throughfall beneath the canopy of Amazonian rainforest is very high and can result in large and systematic errors unless adequately sampled. This may well have contributed to the extreme variability of previous published results, and quite possibly towards an upward bias in reported interception ratio and canopy storage capacity.' Lloyd and Marques (1988) reported a value of 0.74 mm for the interception storage capacity for the experimental site forest at Ducke Forest Reserve.

Regarding the parameters connected with the wind structure in the surface layer, the following values were obtained: zero-plane displacement (d) was 25.3 ± 0.6 m, the roughness length (Z_o) was 5.0 ± 0.4 m, and the friction velocity (u^*) was 0.79 ± 0.13 m/s.

Shuttleworth (1988) developed an empirical model to estimate evaporation for the period September 1983 to September 1985. He used hourly-average meteorological measurements taken above the canopy and soil water tension averaged to a depth of 1 m. Based on this model he summarizes the 2 years and 1 month of micrometeorological data collection as follows: '... over the whole study period, approximately 10% of the rainfall was intercepted by the forest canopy, and this accounted for 20 to 25% of the evaporation. The remainder occurs as transpiration from the trees. Over the same period, about half the incoming precipitation is returned to the atmosphere as evaporation, a process which requires 90% of the energy input. These proportions exhibit some seasonal behaviour in response to the large seasonal variation in rainfall. The average evaporation over two years was within 5% of potential evaporation. Monthly average evaporation exceeds potential estimates by 10% during the wet months and fall below such estimates by at least this proportion in dry months.' Note that for fine days about 75% of the energy input is used for evapotranspiration; in contrast, when all days are taken into consideration, including wet days when evaporation of intercepted water is important, this amount increases to 90%.

For established pastures with grasses 0.6 m high the values for albedo range from 19% when the grass is green to 25% during the dry season. These values are about twice as large as typical albedoes for the rainforest. It means that the absorbed solar energy will likely be less over the areas converted to grass. In reality very little is known about the modifications of the surface microclimate for a grassy vegetation cover compared to the better studied microclimate of the rainforest surface. That has implications for the validation of GCM simulations of tropical deforestation (see section 5 for a review of recent simulations) since most parameters that describe the impoverished grassy vegetation of Amazonian pastures have yet to be measured. To address this gap in knowledge a new joint Brazilian-British micrometeorological experiment will be conducted for a 4-year period, starting in September 1990 in a cleared area in Amazonia, to study in detail the micrometeorological and soil changes associated with converting rainforest into pasture.

3 Water Balance in Small Watersheds

A number of studies have been made near Manaus, Amazon, to measure the various components of the water balance in a watershed. Continuous observations were made for more than two years and are summarized in Franken and Leopoldo (1984). The studies were conducted in two small watersheds.

a. *'Bacia Modelo' Watershed*. Located 80 km north of Manaus, this watershed comprises an area of 23.5 km^2 and is covered mainly by dense primary forest, with tree tops reaching 40 m in average, on a heavy yellow latosol. Precipitation, interception loss, and streamflow of *igarapés* (small forest streams) were measured from 2 February 1980 through 10 February 1981. Total precipitation during that period was 2089 mm; the interception loss was estimated at 534 mm, or 25.6% of precipitation, and the measured total runoff was 541 mm, or 25.9% of precipitation. From these measurements it was possible to estimate that the transpiration by plants was equal to 1014 mm. And the total evapotranspiration was 1548 mm, corresponding to 74.1% of precipitation.

b. *'Barro Branco' Watershed*. It is situated 26 km northeast of Manaus, at the Brazilian Institute for Amazonian Research's Ducke Forest Reserve. Ninety-five percent of the catchment area is covered by dense primary forest, 3% by managed forest, and 2% has been clear cut. The catchment area measures 1.3 km^2 and the soils are mainly yellow latosol. The water balance was measured during two different periods: from 29 September 1976 to 22 September 1977 and from July 1981 to 30 June 1982. The annual average values for the two periods studied shows that precipitation was 2293 mm, interception loss was 429 mm, or 18.7% of precipitation, runoff corresponds to 635 mm, or 27.7% and the transpiration was 1229 mm, or 53.6% of precipitation. Therefore, evapotranspiration was 1658 mm, corresponding to 72.3% of precipitation.

In the two areas studied, the measurements for the water balance components were similar and the evapotranspiration averaged 4.3 mm/day. Ribeiro et al. (1979), using monthly mean meteorological data for 1965–73, calculated potential evapotranspiration at 1536 mm/yr and 1075 mm/yr for real evapotranspiration. These results agree with those obtained by Franken and Leopoldo (1984). But, in view of the micrometeorological measurements reviewed in

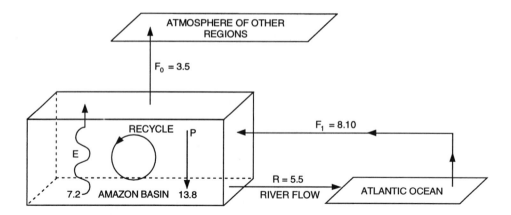

Figure 23.1 [*orig. figure 1*]   Schematic depiction of the water cycle in the Amazon Basin. E is evapotranspiration; P is precipitation; F_1 represents the amount of water vapor entering the region; F_o represents the amount of water vapor that leaves the region; and R is the flow of the Amazon River into the ocean. The fluxes are in units of 10^{12} m^3/yr.

the previous section, the measurements of interception loss in the two watersheds appear to be overestimated. That would explain in part why evapotranspiration in these calculations is about 20% higher than in the calculation of Shuttleworth (1988).

4   Water Balance of the Amazon Basin

The water balance of the Amazon basin is difficult to determine due to the lack of basic data systematically collected over time and space. However, using existing data through successive approximations, it was possible to quantify the fluxes involved.

Figure 23.1 is a schematic representation of the various fluxes involved in the water balance for the Amazon Basin. The fluxes were estimated according to the methods described below.

4.1   Aerological data

The daily radiosounding (upper air) data from the stations shown in figure 23.2 for the

years 1972–75 was analyzed. Calculations based on these data produced horizontal and vertical wind and humidity structures, precipitable water vapor, and the atmospheric horizontal moisture flux. For more details on these calculations see Salati et al. (1984a, 1984b); Marques et al. (1979a, 1979b, 1980a, 1980b); and Kagano (1979).

From these calculations it was possible to conclude that:

a. On the Atlantic coast and in the center of Amazonia winds are predominantly from the east at low levels. Figure 23.3 shows that most water vapor reaches the region coming from the Atlantic Ocean.

b. Water vapor fluxes decrease from east to west across the basin.

c. Precipitable water vapor in the region averages 35 mm,[1] with a seasonal variation of 10 mm. Therefore, the average water vapor stored in the atmosphere above the Amazon basin is of the order of 0.2×10^{12} t. The greenhouse absorption of outgoing longwave radiation by this significant mass of water vapor is largely what accounts for

[1] In the calculations of this average value of precipitable water vapor (PWV) data for Bogota and Lima were included. Because Bogota is on the Andes and Lima on an arid region, typical values of PWV for these two locations are substantially lower than a typical PWV for lowland Amazonia. Therefore the average PWV for the Amazon basin should be higher than 35 mm.

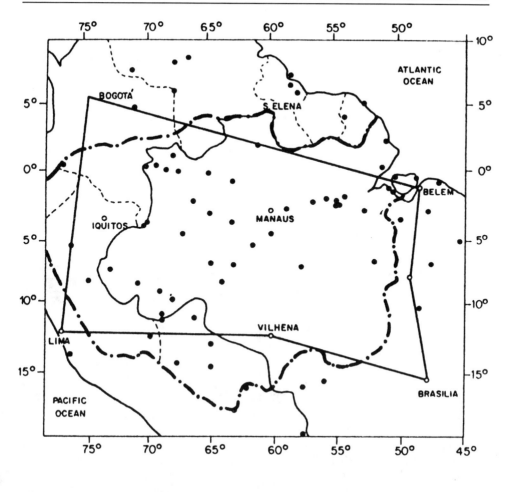

AMAZON RIVER BASIN LIMITS

Figure 23.2 [*orig. figure 2*]   Network of radiosounding stations used by Marques et al. (1980): Bogota, Belem, Carolina, Brasilia, Vilhena, Lima, Manaus.

the remarkable isothermal behavior observed in the region (low fluctuation of surface temperature between day and night).

d. Comparison of the seasonal cycle of the basinwide, vertically integrated atmospheric moisture divergence and the Amazon streamflow at Obidos (500 km from the mouth), show that the latter lags the former by approximately 3 months. Therefore, 3 months can be taken as a first-order estimate of the residence time for the water in the Amazon hydrological system (Marques et al., 1980a).

e. Recently, during the Global Atmospheric Experiment-Amazon Boundary Layer Experiment (ABLE-2B), conducted in Amazonia during April–May 1987, upper air data from 6 aerological stations (figure 23.4) were collected 4 times a day (00, 06, 12, and 18 UT) during a 1-month period (13 April–13 May 1987). Based on this data, water-balance estimates have been made for the area covered by the aerological stations (approximately 2.2 million km^2). Average precipitation in that area, collected in over 150 rain gauges and for that 1-month period, was 290 ± 20 mm; the calculated conver-

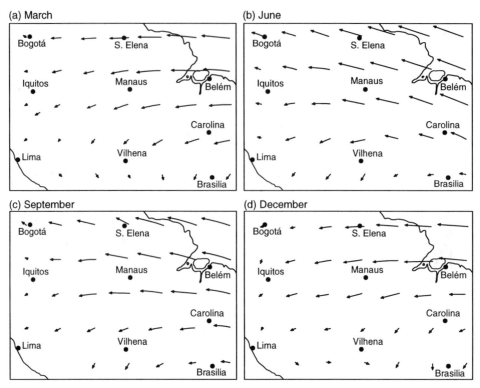

This length corresponds to a flux of 2000 g cm⁻¹s⁻¹

Figure 23.3 [*orig. figure 3*] Vector field values total moisture transport $\Omega = \Omega_\lambda + \Omega_\phi$. Mean monthly values for the 1972–1975 period obtained for the 5° lat. × 5° long. grid squares. Vector length of 1 cm corresponds to a flux of 2000 g cm⁻¹ s⁻¹.

gence of atmospheric water-vapor transport (total water-vapor transport entering the volume comprising the area encircled by the lines in figure 23.4 and the depth of the atmospheric column up to 20 km minus the total water-vapor transport exiting the volume) was 127 ± 12 mm. The variation of water vapor in that volume (storage term) during that 1-month period was –6 ± 1 mm, i.e., there was a slight drying from the beginning to the end of the period. The area and time averaged evapotranspiration can then be estimated as the residual term in the atmospheric water budget equation (storage term – convergence term = evapotranspiration – precipitation), resulting in 157 ± 32 mm of evapotranspiration, or about 54% of precipitation, which again shows the importance of water-vapor recycling in Amazonia.

That the rates of water-vapor recycling within the basin must be high also can be inferred indirectly by examining the values of precipitable water vapor (the numbers next to stations on the map of figure 23.4) in an east–west transect from Belem on the Atlantic coast westward to Tabatinga, which is about 2500 km inland. One can readily see that the values increase westward: 5.7 cm in Belem, 6.0 cm in Manaus, and 6.2 cm in Tabatinga. If oceanic water vapor were the main source of water vapor for precipitation, then one would expect precipitable water vapor to decrease as the air moves inland since rainfall would be extracting water vapor from the atmospheric column at a rate that would be higher than the rate of evapotranspiration into the column.

f. The estimated values of water-vapor

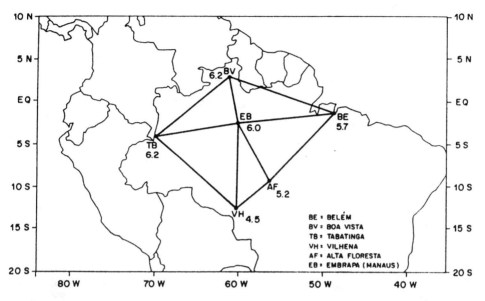

ABLE-2B- LARGE SCALE RAWINDSONDE NETWORK

Figure 23.4 [*orig. figure 4*]   Network of radiosounding stations used during the GTE-ABLE 2B Experiment (13 April–13 May 1987). The numbers next to the stations represent the mean precipitable water vapor content in cm for that 1-month period.

fluxes entering Amazonia on the Atlantic coast are not sufficient to explain the values of rainfall observed in the region, that is, the total influx of water vapor into the region is smaller than the total precipitation.

g. There is almost no data on the atmospheric exchange of water vapor between the Amazon basin and Orinoco basin to the north.

h. The inflow of oceanic water vapor from the Atlantic into the Amazon basin is assumed to be about 8–10 × 10^{12} t/y. This is the value with the highest uncertainty in the water balance of figure 23.1.

i. At the southern boundary of Amazonia the direction of water vapor fluxes is from north to south for almost the entire year. This shows that water vapor from Amazonia can influence water-vapor concentration in the atmosphere above the Brazilian Highlands.

4.2   Precipitation

In general precipitation is abundant along the Atlantic coast from the Guianas to about 3° S. In this coastal strip the total annual precipitation can be as high as 3500 mm. The main rainfall-producing mechanism is related to sea breeze-induced instability lines (bands of cumulus clouds aligned with the coastline). Often these lines propagate inland with phase speeds typically of the order of 10–15 m/s. On rare occasions they even reach the eastern foothills of the Andes. Their high frequency of occurrence throughout the year (over 250 days in average near the Equator) is the reason behind the 'traditional afternoon showers' at the Brazilian river port city of Belem on the mouth of the Amazon river. From the coast precipitation decreases inland to the east-southeast, reaching values as low as 1600 mm/yr around Santarem, about 700 km from the coast. Rainfall increases again westward of this relative minimum and a broad maximum (annual totals greater than 3000 mm) is found over western Amazonia in the Brazil–Colombia border area. A possible reason for this broad maximum is the proximity of this region to the Andes mountains to the west, which might provide the conditions for low-level convergence of the

Table 23.1 [*orig. table 1*]   Hydrological cycle of the Amazon Region; summary of the results obtained by different studies (adapted from Salati, 1987).

Research	Rainfall	Transpiration			Evapotranspiration			Runoff	
	mm	mm	%	mm/day	mm	%	mm/day	mm	%
Marques et al., 1980	2328[a]	–	–	–	1260 (r)	54.2	3.5	1068	45.8
	2328[b]	–	–	–	1000 (r)	43.0	2.7	1328	57.0
	2328[c]	–	–	–	1330 (p)	57.1	3.6	998	42.9
Villa Nova et al., 1976	2000[d]	–	–	–	1460 (p)	73.0	4.0	540	27.0
		–	–	–	1168 (r)	58.4	3.2	832	41.6
	2101[e]	–	–	–	1569 (p)	73.4	4.3	532	26.6
Molion, 1975	2379[f]	–	–	–	1146 (r)	48.2	3.1	1233	51.8
Ribeiro and Villa Nova, 1979	2478[g]	–	–	–	1536 (p)	62.0	4.2	942	38.0
		–	–	–	1508 (r)	60.8	4.1	970	39.2
IPEAN, 1978	2179[h]	–	–	–	1475 (r)	67.5	4.0	704	32.5
		–	–	–	1320 (r)	60.6	3.6	859	39.4
DMET, 1978	2207[i]	–	–	–	1452 (p)	65.8	4.0	755	34.2
		–	–	–	1306 (r)	59.2	3.6	901	40.8
Jordan and Heuveldop, 1981	3664[j]	1722	47.0	4.7	1905 (r)	52.0	5.2	1759	48.0
Leopoldo et al., 1981	2089[k]	1014	48.5	2.7	1542 (r)	74.1	4.1	541	25.9
Leopoldo et al, 1982	2075[l]	1287	62.0	3.5	1675 (r)	80.7	4.6	400	19.3
Shuttleworth, 1988	2636[m]	992	37.6	2.7	1320 (r)	50.0	3.6	–	–
ABLE-2B, 1987 (1 month)	290[n]	–	–	–	157 (r)	54.1	5.2	–	–

Observations: (r) – real evapotranspiration; (p) – potential evap.; [a] acrological method, applied for all Amazon basin, period 1972/1975; [b] idem, for the region between Belém and Manaus; [c] by Thornthwaite method, for the region between Belém and Manaus; [d] Penman method, mean for the period 1931/1960; [e] idem, for Manaus Region; [f] climatonomic method, for all Amazon Region, mean for the period 1931/1960; [g] water balance by Thornthwaite and Mather method for the Ducke Forest Reserve, mean for the period 1965/1973; [h] Thornthwaite method for all Amazon Region and estimated for a period over 10 years; [i] idem, for various periods; [j] water balance, with transpiration estimated by class A pan-evaporation for San Carlos Region; [k] 'Model Basin' water balance and [l] 'Barro-Branco' water balance (Ducke Forest Reserve); [m] Adaptation of Penman-Monteith for the period Sept. 1983–Sept. 1985; [n] aerological method applied to the Brazilian Amazon Basin during ABLE-2B, April 13–May 13, 1987.

airflow in that area. There is a secondary precipitation maximum in the southern portion of the basin (values as high as 2500 mm/yr). This region of maximum precipitation, which extends from southwestern Amazonia to the Atlantic, aligned on a NW–SE direction, is related to the interaction of frontal systems moving to tropical and subtropical latitudes with tropical convection over South America. This band of higher precipitation marks the northernmost position that frontal systems reach and where they show a tendency to become quasi-stationary.

The highest rainfall values in South America are found over localized areas on the eastern slopes of the Andes in Peru and on the western slopes in Colombia, where annual precipitation in excess of 5000 mm has been observed. These are caused fundamentally by mechanical uplifting of the prevailing low-level atmospheric flow by the Andean topographic barrier.

The mechanisms that explain the various precipitation maxima mentioned above are all apparently linked to either large-scale features (convergence provided by the Andes, frontal influences) or to other local and mesoscale forcings (topographic uplifting, diurnal land-sea temperature contrasts, etc). They do not appear to depend, to a first approximation, on the type of underlying vegetation. Yet, there is a wealth of observational evidence showing that evapotranspiration accounts for more than 50% of the precipitation (see table 23.1). This evidence suggests that the Amazonian forest is highly efficient in recycling water vapor back into the atmosphere. A different type of vegetation, such as grass, probably would not be as efficient in maintaining high evapotranspiration rates.

There is also a noticeable seasonal variation, with maximum rainfall in the Northern Hemisphere during July–August, and in the Southern Hemisphere during February–March (Salati and Marques, 1984). The average total annual precipitation for the Amazon basin has been estimated by various authors ranging from 2000 mm to 2400 mm. Recently Molion and Dalarosa (1990) have shown some evidence that these values might be underestimates because most rain gauges in Amazonia are located on the banks of large rivers and are possibly under the influence of a local diurnal river breeze circulation that would reduce precipitation over the river and its margins compared to forest areas distant from the river. With that in mind, a value of 2300 mm/yr was used for the water balance in figure 23.3, which corresponds to a flux of $13.8 \times 10^{12} \, m^3/yr$.

4.3   Evapotranspiration

Accurate measurements or estimates of real evapotranspiration for the various ecosystems present in Amazonia are not available presently. Evapotranspiration estimates can currently be obtained through the following methods: (a) aerological calculation for large areas; (b) water-balance calculations for small watersheds; (c) direct micrometeorological flux measurements; (d) isotopic dilution calculation; and (e) calculations based on the surface energy budget (Thornwaite, Penman, Monteith methods). Evapotranspiration estimates made by various authors for Amazonia are summarized in table 23.1.

For the Amazon basin as a whole, values ranging from 1146 mm to 1320 mm, corresponding to 48–54% of the precipitation, were obtained. In the water balance shown in figure 23.3, a value of 1200 mm/yr was used, corresponding to an annual input of 7.2×10^{12} t of water vapor from the surface into the atmosphere.

4.4   Recycling of water vapor

In view of the results found for water-vapor fluxes and precipitation, it is possible to conclude that water vapor has to recirculate in the region, i.e., total precipitation is larger than the total influx of water vapor into Amazonia. Supporting evidence also comes from the distributions of isotopes (0–18 and D) of rainwaters of different areas of Amazonia (Salati et al., 1979). Using current existing data it is estimated that 50–60% of rainfall originates from the recirculated water vapor through evapotranspiration.

5   Model Simulations of Amazonian Deforestation

Quantitatively estimating the effects that large changes in Amazonian ecosystems can have on the surface-energy and water budgets has been difficult because the equilibrium climate is determined by the momentum and energy exchanges at the Earth-atmosphere interface, interacting with complex dynamical processes in the atmosphere. Results of earlier model studies were generally inconclusive, and sometimes conflicting, about the regional (and global) climate changes following deforestation.

Table 23.2 [*orig. table 2*]   Summary of surface variables for control (C) and deforested (D) simulations averaged over 3 years for Amazonia (from: Lean and Warrilow, 1989).

Surface variable	C	D	
Evaporation (mm/d)	3.12	2.27	(−27.2%)
Precipitation (mm/d)	6.60	5.26	(−20.3%)
Soil moisture (cm)	16.13	6.66	(−58.7%)
Runoff (mm/d)	3.40	3.00	(−11.9%)
Net radiation (W/m^2)	147.3	126.0	(−14.4%)
Temperature (°C)	23.6	26.0	(2.4°C)
Sensible heat (W/m^2)	57.2	60.2	(+5.2%)
Bowen ratio	0.85	1.50	(+76.5%)

The models were of two types: either energy-box models (Lettau et al., 1979; Potter et al., 1975) or crude resolution GCMs (Henderson-Sellers and Gornitz, 1984). In general, the latter lacked both spatial resolution and an adequate treatment of the land-surface processes. For instance, their resolution was typically 10° long. × 5° lat., which would cause the whole of Amazonia to be represented by a few grid-points. Also their representation of evapotranspiration processes was based on simple parameterizations. In Henderson-Sellers and Gornitz (1984) runoff was proportional to soil moisture and to the precipitation. These parameterizations, including the 'bucket hydrology' parameterization, were inadequate to represent evapotranspiration processes over vegetated surfaces (Sellers et al., 1986) and make it difficult to represent the complex changes in soil hydrology following burning and land clearance. Henderson-Sellers (1987, her table 1, pp. 468–469) summarized the main results from these earlier model studies.

Realistic models of biosphere have only recently been developed that can be coupled with realistic models of the global atmosphere (Dickinson et al., 1986, Sellers et al., 1986). The pioneering work of assessing climate impacts of tropical deforestations using these novel coupled biosphere-atmosphere models was that of Dickinson and Henderson-Sellers (1988), hereafter referred to as DHS. In DHS the National Center for Atmospheric Research Community Climate Model (NCAR CCM), coupled to the Biosphere-Atmosphere-Transfer-Scheme

(BATS) of Dickinson et al. (1986), was used with a horizontal resolution of 7.5° long. × 4.5° lat. to study the effects of Amazonian deforestation. When the model's rainforests over Amazonia were replaced by degraded pasture, surface temperatures increased by 3–5 °C and evapotranspiration decreased over the region. The increase in surface temperature was attributed mostly to the decreased roughness length of the grass vegetation compared to that of forest and the reduction of evapotranspiration was mostly due to less absorbed solar radiation for grass given its higher albedo. Some difficulties were reported in the parameterization of incident solar radiation and of interception loss (Shuttleworth and Dickinson, 1989; Dickinson, 1989a, 1989b) that caused unrealistically high net radiation.

More recently two GCM simulations of tropical deforestation were conducted, one at the UK Meteorological Office (Lean and Warrilow, 1989, hereafter referred to as LW) and another at the Center for Ocean-Land-Atmosphere Interactions (COLA) (Shukla, Nobre, and Sellers, 1990, hereafter referred to as SNS). In LW the model's horizontal resolution was 3.75° long. × 2.5° lat. and all the model's vegetation north of 30° S in South America was replaced by grass. Although the total area in which the model's vegetation changed was almost twice that used in DHS and in SNS, their results were similar to those in DHS: surface temperature increased by 2.5 °C and evapotranspiration decreased for the pasture scenario compared to the forest one.

Additionally, it was found that simulated

Table 23.3 [*orig. table 3*]   Mean surface-energy budget for Amazonia. The data are 12-month mean (January–December) values. Values are in W/m^2, except for B and a which are non-dimensional, and T_s which is in °C. S is insolation; a is albedo; L_n is net upward longwave radiation; R_n is available radiative energy; E_t is transpiration plus soil evaporation; E_i is interception loss; E is evapotranspiration = $E_t + E_i$; H is sensible heating; G is ground heat flux; B is the Bowen ratio (H/E); and T_s is surface temperature (from: Shukla et al., 1990)

	S	$(1-a)S$	L_n	R_n	E_t	E_i	E	H	G	B	a	T_s
Control	233	204	−32	172	91	37	128	44	0	0.34	12.5	23.5
Deforestation	237	186	−40	146	64	26	90	56	0	0.62	21.6	26.0
Difference	+4	−18	−8	−26	−27	−11	−38	+12	0	−0.28	+9.1	+2.5

Table 23.4 [*orig. table 4*]   Mean water budget for Amazonia. The data are 12-month mean (January–December) values. Values E and P are in mm/yr; PW is in mm, P is total precipitation; E is evapotranspiration; and PW is precipitable water (from: Shukla et al., 1990)

	P	E	$(E-P)$	E/P	PW
Control	2464	1657	−807	0.67	37.7
Deforestation	1821	1161	−661	0.63	35.4
Difference	−642	−496	+146	−0.04	−2.3
Change (in percent)	−26.1	−30.0	+18.0	−5.9	−6.1

precipitation was reduced over Amazonia. As in DHS the increase on surface temperature was attributed to the decrease in roughness length. Table 23.2 (adapted from table 2 of LW) summarizes the main results of their study.

In SNS the COLA GCM, coupled to the Simple Biosphere Model (SiB) of Sellers et al. (1986), was used with a horizontal resolution of 2.8° long. × 1.8° lat. – i.e., the simulation with the highest horizontal resolution among the three studies – and the model's Amazonian tropical forests were replaced by degraded grass. The main results of SNS are summarized in tables 23.3 and 23.4 for the surface-energy and water balances in Amazonia, respectively (adapted from tables 1 and 2 of SNS), and described below.

Surface and soil temperatures were warmer by 1–3 °C in the deforested than in the control cases. The relative warming of the deforested land surface and the overlying air is consistent with the reduction in evapotranspiration and the lower surface

roughness length. The annual mean surface-energy budget (table 23.3) for Amazonia in the two simulations shows that absorbed solar radiation at the surface is reduced in the deforestation case (186 W/m^2) relative to the control case (204 W/m^2) because of the higher albedo (21.6%) for grassland compared to forest (12.5%). That plus the larger outgoing longwave radiation from the surface due to the higher surface temperature in the deforested case result in the amount of net radiative energy available at the surface for partition into latent and sensible heat flux being smaller in the deforested case (146 W/m^2) than in the control case (172 W/m^2). Also, as remarked in SNS, less precipitation is intercepted and reevaporated as the surface roughness and the canopy water-holding capacity of the pasture are relatively small. Furthermore, the transpiration rates are reduced due to the reduced soil moisture-holding capacity for the soils under pasture.

An interesting result was that the reduction in calculated annual precipitation (642 mm) was larger than the reduction in evapotranspiration (496 mm), as seen in table 23.4, which suggests that changes in the atmospheric circulation may act to reduce further the convergence of moisture flux in the region, a result that could not have been anticipated without the use of a dynamical model of the atmosphere, as noted in SNS. This, in turn, implies that runoff also decreased for the deforested case, a result also found in LW (table 23.2), since the decrease in precipitation was larger than the decrease in evapotranspiration.

Taken together the results of these three studies seem to suggest the existence of a significant sensitivity of the regional climate to the removal of the tropical forest. In general, the somewhat short period of integration in these studies precludes drawing conclusions on the significance of global climate changes or even climate changes in regions adjacent to Amazonia.

6   Discussion and Concluding Remarks

The conversion of tropical forested areas into pastures or other types of short vegetation will cause changes in the microclimate of the disturbed areas. If the size of the perturbed area is sufficiently large, even the regional climate may be altered. Depending on the scale of these alterations, they may cause climate changes at the global level and affect regions distant from the tropical forests.

6.1   Local changes in climate

Changes will occur in albedo and in energy and water balances. There will be a tendency toward less water infiltration and more runoff during rainy periods and less runoff during prolonged dry periods.

An important conclusion of the micrometeorological studies conducted at Ducke Reserve, near Manaus in central Amazonia (summarized in Shuttleworth, 1988) is that the annual flux of latent heat into the atmosphere is close to its potential value, that is, 20% smaller than the potential evapotranspiration during the dry season and about 10% above the evapotranspiration rate during the rainy season (implying a net transfer of sensible heating from the atmosphere into the canopy, i.e., negative Bowen ratio[2]).

Shuttleworth suggests that there might be a reduction of between 10 and 20% in the evapotranspiration for pastures as compared to the rainforest, mostly due to the higher albedo (thus, smaller available energy other things being equal) of grass compared to the albedo of tropical forests. That reduction, in turn, might cause rainfall to decrease by 10%, he suggested. Yet, this hypothetical scenario takes into account only changes in evapotranspiration due to changes in the available radiative energy. Important changes also would occur due to the decrease in surface roughness and at the soil level. Loss of top soil organic matter and soil fauna, compaction due to agricultural practices and overgrazing, and soil erosion may cause large changes in the physical and chemical characteristics of the predominantly clay soils of the Amazonian terra firme forest. Those changes likely would combine to reduce infiltration rates drastically, increase surface runoff during rainy periods, and decrease soil moisture in the shallower rooting zone of the grass vegetation primarily during the dry season. Decreased soil moisture availability also would contribute to reduce evapotranspiration.

Comparative measurements of the diurnal cycle of canopy and subsurface temperature at cleared and forested sites in Ibadan, Nigeria (Lawson et al., 1981), and in Surinam (Shulz, 1960) showed a large increase of soil (> 5 °C) and air (> 3 °C) temperatures for the cleared areas compared to the forested ones. Not being in the shade of a tall canopy, the diurnal fluctuation of

---

[2] It is not clear, though, how this condition could be maintained for a large area since a negative Bowen ratio would make the layer above the canopy more stable, therefore reducing the turbulent flux of latent heat from the canopy.

ground temperature and humidity deficit was much larger for the cleared sites in these two studies as well. Those changes in soil microclimate will have a profound effect on the biological, chemical, and physical processes in the top soil layer. Plants, animals, and microorganisms living in that layer will experience temperature, humidity deficit, and water stresses not present in the remarkably constant microclimate of the forest floor.

6.2 Regional climate changes

The summation of local climate change over a sufficiently large quasi-contiguous area (say larger than 1 million km^2) might change water-vapor transports and the water balance at a regional level with consequent changes in the energy balance. Climatic alterations and the scale at which they occur depend on the geographic location and its geomorphology. For instance, even small changes in the low-level wind regime on mountainous areas such as the Andean Cordillera can cause a large change in the temporal and geographical distribution of rainfall. It is not possible yet to predict accurately regional climate changes associated with the observed patterns of deforestation by means of climate model simulations. An important reason for such limitation is that when current climate models are integrated in a control mode, i.e., attempting to mimic the observed climate, they commonly fail to represent important aspects of the regional climate. One problem is, of course, resolution. It is expected that only when model resolution becomes of the order of 100 km (current climate model resolution is typically between 200 and 500 km) will the models probably capture the finer details of the regional climate. Yet, the results of recent climate model simulations of Amazonian deforestation, reviewed in the previous section, suggest the following changes at the regional level to be likely following extensive deforestation of tropical forests: increase in surface and soil temperature and in the diurnal fluctuation of temperature and specific humidity deficit, and a reduction of evapotranspiration and PBL moisture. In two of the three studies (LW

and SNS), yearly averaged precipitation and runoff decreased for Amazonia as a whole for the pasture vegetation compared to forest. The annual reduction in rainfall in these two simulations was larger than the corresponding reduction in evapotranspiration, thus explaining the reduction in runoff. It is likely, however, that runoff will increase following rainy periods, that is, runoff (and river streamflow) would be higher after deforestation during the rainy season and decrease during the dry season.

6.3 Global changes

Tropical forests contribute in many ways to maintain the present dynamical and chemical equilibrium of the atmosphere. Forests represent a carbon reservoir, both through their areal and root systems as well as through organic matter in the soil. Estimates indicate that tropical forests possess a reserve of carbon equivalent to 1.5–2.0 times the carbon store of CO_2 in the atmosphere. Therefore, conversion of forests into pastures will release CO_2 from the biosphere into the atmosphere, likely enhancing the greenhouse warming.

Forest burning associated with clearing processes for conversion into pastures also releases great quantities of particles and compound gases into the atmosphere. These particles cause changes in the atmosphere, especially in its chemical composition and energy balance.

To understand and predict any possible large-scale climate change due to tropical deforestation it is crucial to know to what extent the rainfall patterns will change when rainforests are converted into grasslands. It is well known that the tropical regions function as atmospheric heat sources through the release of latent heat of condensation in convective clouds. The heat so released drives large-scale tropical circulations (of the Hadley-Walker type) with ascending motion over the tropical regions, mostly over Amazonia, Tropical Africa, and the Indonesian-western Pacific region, and descending motion over the dry subtropics, primarily over the subtropical oceans. It is conceivable that a significant reduction in rainfall over Amazonia (say, greater than

20% reduction as the model simulations described in LW and SNS suggest) might have an effect in these tropical circulations. However, it is unclear what these changes would be and how they would manifest themselves in terms of climate changes in the Tropics, but away from the perturbed areas, and in the extra-tropics. Regarding the extra-tropics, it is interesting to note the suggestion by Paegle (1987) of a possible link between tropical convection and quasi-stationary features of the large-scale circulation over North America. He suggests that the westward shift of the subtropical jetstream from the east coast of North America in boreal winter to the west coast in spring and a concomitant westward shift of the North American long-wave trough may be linked to the seasonal northwestward migration of the area of rainfall maxima over Tropical South America from Central Amazonia in January–February to Central America in June–July.

Tropical forest areas also have a characteristic energy balance that contributes to the transport of energy as latent heat (water vapor) from the equatorial regions to those of greater latitude. This is particularly conspicuous in Central Brazil, southern Bolivia, Paraguay, and northern Argentina where, due to the generally southward low-level circulation, most of the water vapor present in those regions comes from Amazonia. Therefore, changes in atmospheric moisture in Amazonia due to deforestation might have an impact on the precipitation of the adjacent regions to the south.

So far we have focused our attention mostly on the Amazonian tropical forest. Can we say anything about climatic impacts arising from the removal of tropical forests in Equatorial Africa and Southeast Asia? It is likely that at the microclimate level the effects will be quite similar: higher near-ground temperatures and larger diurnal fluctuations of temperature and humidity deficit, increased runoff during rainy periods and decreased runoff during the dry season, decreased soil moisture, and, possibly, decreased evapotranspiration. The question whether there would be a significant change in precipitation is a complex one. For Southeast Asia large-scale changes in precipitation are less likely since the precipitation climate of that area of the western Pacific and Indian Oceans is controlled by large-scale features: on one hand, the precipitation distribution responds to the high sea surface temperatures (SST > 28 °C) that are conducive to large rates of evaporation besides a tendency for the low-level air to converge from areas of lower SST to areas with higher SST; these two factors enable cloud formation and high precipitation. On the other hand, land–sea heating contrast drives the monsoonal circulations of Southeast Asia. The monsoonal circulations account for the copious rainfall observed in that area.

In Africa there is, at least theoretically, the possibility that the removal of the tropical forest might influence the regional climate. A biophysical feedback mechanism as proposed by Charney et al. (1977) might cause an enhancement in aridification along the northern and southern boundaries of the forest. For reasons similar to the ones discussed in the earlier section, the changes in albedo, surface roughness, and soil moisture caused by replacement of forest by overgrazed pasture would result in decreased precipitation. That could, in turn, induce further clearings deeper into the forest. However, this question is not settled yet because interannual and longer-term rainfall variability in Tropical Africa is apparently also connected to planetary-scale phenomena, notably global SST distributions.

Finally, can we say anything on the ecological implications of the possibility of a future dryer and warmer climate in Amazonia following extensive deforestation? The decrease in precipitation suggested by the simulation studies for the deforested case is associated with a longer and more pronounced dry season. The authors in SNS remark that '[T]he lack of an extended dry season apparently sustains the current tropical forests, and therefore, a lengthening of the dry season could have serious ecological implications. Among other effects, the frequency and intensity of forest fires could increase significantly and the life-cycles of pollination vectors could be perturbed . . . Changes in the region's hydrological cycle and the disruption of complex plant-animal relations could be so profound that once the tropical forests were destroyed, they might not be able to re-establish themselves'. The authors then conclude that a 'complete and rapid destruction of the Amazon tropical rainforest could be irreversible'. Therefore, there might be a tendency of 'savannization' of Amazonia if one recalls that the savanna vegetation is naturally more adapted to withstand fire and a long (6 months or greater) dry season. Amazonia is surrounded to the south, east, and north by savanna-like vegetation. Any trend toward 'savannization' in Amazonia would likely be seen first in the transition forests straddled between the rainforest and the savanna because in those areas the dry season is usually longer than in the rainforest. This implies that any increase in the duration of the dry season in those regions might make it unsuitable for the reestablishment of the rainforest.

References

Brasil, Instituto de Pesquisas Espaciais (INPE): 1989, 'Avaliação da cobertura florestal na Amazonia Legal utilizando Sensoriamento Remoto Orbital', INPE, São José dos Campos, SP, Brazil.

Charney, J. G., Quick, W. J., Chow, S. H., and Kornfield, T. : 1977, 'A Comparative Study of the Effects of Albedo Change on Drought in Semi-Arid Regions', J. Atmos. Sci. 34, 1366–1388.

Departamento Nacional de Meteorologia (DMET): 1978, Balanço Hidrico de Brasil, DMET.

Dickinson, R. E. : 1989a, 'Modeling the Effects of Amazonian Deforestation on a Regional Surface Climate: A Review', Agric. and For. Meteorol. (in press).

Dickinson, R. E. : 1989b, 'Implications of Tropical Deforestation for Climate: A Comparison of Model and Observational Descriptions of Surface Energy and Hydrological Balance', Phil. Trans. R. Soc. Lond. B, (in press).

Dickinson, R. E., Henderson-Sellers, A., Kennedy, P. J., and Wilson, M. F. : 1986, Biosphere-Atmosphere Transfer Scheme (BATS) for the NCAR Community Climate Model, Tech. Note TN-275+STR, National Center for Atmospheric Research, Boulder, CO, U.S.A.

Dickinson, R. E. and Henderson-Sellers, A. : 1988, 'Modelling Tropical Deforestation: A Study of GCM Land-Surface Parameterizations', Q. J. R. Meteorol. Soc. 114, 439–462.

Fearnside, P. M. : 1987, 'Causes of Deforestation in the Brazilian Amazon', in R. E. Dickinson (ed.), The Geophysiology of Amazonia, Wiley, New York, pp. 37–61.

Fearnside, P. M., Tardin, A. T., and Meira Filho, L. G. : 1990, Deforestation Rate in Brazilian Amazônia, Instituto de Pesquisas Espaciais, S. José dos Campos, SP, Brazil, 8 pp.

Franken, W. and Leopoldo, P. R. : 1984, 'Hydrology of Catchment Area of Central Amazonian Forest Streams', in H. Sioli (ed.), The Amazon: Limnology and Landscape Ecology of a Mighty Tropical River and Its Basin, W. Junk, Dordrecht, The Netherlands.

Henderson-Sellers, A. : 1987, 'Effects of Change in Land Use on Climate in the Humid Tropics', in R. E. Dickinson (ed.), Geophysiology of Amazonia, John Wiley & Sons, New York, 526 pp.

Henderson-Sellers, A. and Gornitz, V. : 1984, 'Possible Climatic Impacts of Land Cover Transformations, with Particular Emphasis on Tropical Deforestation', Climatic Change 6, 231–258.

IPEAN: 1978, Instituto de Pesquisas Agropecuárias do Norte 1972, Zoneamento Agricola da Amazoñia, Bol. Téc., Belem, Brasil.

Jordan, C. F. and Heuveldop, J. : 1981, 'The Water Balance of an Amazonian Rain Forest', Acta Amazôn. 11, 87–92.

Kagano, M. T. : 1979, Um estudo climatológico e sinótico utilizando dados de radiossondagens, 1968–1976, de Manaus e Belém, INPE Report No. 1559–TDL/013, São José dos Campos, SP, Brazil.

Lawson, T. L., Lal, R., and Oduro-Afriyie, K. : 1981, 'Rainfall Redistribution and Microclimatic Changes over a Cleared Watershed', in R. Lal and E. W. Russel (eds.), Tropical Agriculture Hydrology, John Wiley & Sons, New York.

Lean, J. and Warrilow, D. A. : 1989, 'Climatic Impact of Amazon Deforestation', Nature 342, 311–413.

Leopoldo, P. R., Franken, W., Matsui, E., and Salati, E. : 1982a, 'Estimativa da evapotranspiração da floresta Amazônica de terra firme', Acta Amazôn. 12, 23–28.

Leopoldo, P. R., Matsui, E., Salati, E., Franken, W., and Ribeiro, M. N. G. : 1982b, 'Composição isotopica das precipitações e d'água do solo em floresta Amazônica do tipo terra firme na região de Manaus', *Acta Amazôn.* 12, 7–13.

Lettau, H., Lettau, K., and Molion, L. C. B. : 1979, 'Amazonia's Hydrologic Cycle and the Role of Atmospheric Recycling in Assessing Deforestation Effects', *Mon. Wea. Rev.* 107, 227–238.

Lloyd, C. R. and Marques, A. de O. : 1988, 'Spatial Variability of Throughfall and Steamflow Measurements in Amazonian Rain Forest', *Agric. and For. Meteorol.* 42, 63–73.

Malingreau, J. P. and Tucker, C. J. : 1988, 'Large-Scale Deforestation in the Southern Amazon Basin of Brazil', *Ambio* XVII(1), 49–55.

Marques, J., Santos, J. M., and Salati, E. : 1979a, 'O armazenaento atmosférico de vapor d'água sobre a região Amazônica', *Acta Amazôn.* 9, 715–721.

Marques, J., Santos, J. M., and Salati, E. : 1979b, 'O campo de fluxo de vapor d'água sobre a região Amazônica', *Acta Amazôn.* 9, 701–713.

Marques, J., Salati, E., and Santos, J. M. : 1980a, 'Cálculo da evapotranspiração real na bacia Amazônica através do método aerológico', *Acta Amazôn.* 10, 357–361.

Marques, J., Salati, E., and Santos, J. M. : 1980b, 'A divergência do campo do fluxo de vapor d'água e as chuvas na região Amazônica', *Acta Amazôn.* 10, 133–140.

Molion, L. C. B. : 1975, 'A Climatonomic Study of the Energy and Moisture Fluxes of the Amazonas Basin with Consideration of Deforestation Effects', Doctoral Thesis, University of Wisconsin.

Molion, L. C. B. : 1975, 'A Climatonomic Study of the Energy and Moisture Fluxes of the Amazonas Basin with Consideration of Deforestation Effects', Doctoral Thesis, University of Wisconsin.

Molion, L. C. B. : 1987, 'Micrometeorology of an Amazonia Rainforest', in R. E. Dickinson (ed.), *Geophysiology of Amazonia*, John Wiley & Sons, New York, 526 p.

Molion, L. C. B. and Dalarosa, R. : 1990, 'Pluviometria na Amazônia: São os dados confiáveis? *Climanálise* 5 (3), 38–42.

Potter, G. L., Ellsaesser, H. W., MacCracken, M. C., and Luther, F. M. : 1975, 'Possible Climatic Impact of Tropical Deforestation', *Nature* 258, 697–698.

Ribeiro, M. N. G. and Villa-Nova, N. A. : 1979, 'Estudos climáticos da Reserva Ducke, Manaus, Am.3. Evapotranspiração', *Acta Amazôn.* 9, 305–309.

Repetto, R. : 1988, *The Forest for Trees? Government Policies and the Misuse of Forest Resources*, World Resources Institute, Washington, D.C., U.S.A.

Rocha, H. R., Nobre, C. A., and Barros, M. C. : 1989, 'Variabilidade natural de longo prazo no ciclo hidrológico da Amazônia', *Climanálise* 4 (12), 36–43.

Salati, E. : 1985, 'The Climatology and Hydrology of Amazonia', in G. T. Prance (ed.), *Amazonia*, Pergamon, Oxford, England.

Salati, E. : 1987, 'The Forest and the Hydrological Cycle', in R. E. Dickinson (ed.), *Geophysiology of Amazonia*, John Wiley & Sons, New York, pp. 273–295.

Salati, E. and Marques, J. : 1984a, 'Climatology of the Amazon Region', in H. Sioli (ed.), *The Amazon: Limnology and Landscape Ecology of a Mighty Tropical River and Its Basin*, W. Junk, Dordrecht, The Netherlands.

Salati, E. and Vose, P. B. : 1984b, 'Amazon Basin: A System in Equilibrium', *Science* 225, 129–138.

Salati, E., Dall'Ollio, A., Matsui, E., and Gat, J. R. : 1979, 'Recycling of Water in the Amazon Basin: An Isotopic Study', *Water Resource Res.* 15, 1250–1258.

Sellers, P. J., Mintz, Y., Sud, Y. C., and Dalsher, A. : 1986, 'A Simple Biosphere Model (SiB) for Use within General Circulation Models', *J. Atmos. Sci.* 43, 505–531.

Setzer, A. W. and Pereira, M. C., Pereira, A. C., Junior, and Almeida, S. A. O. : 1988, *Progress Report on the IBDF-INPE 'SEQUE' Project, 1987*, TGech, Rep. No. INPE-45-34-RPE/565, Instituto de Pesquisas Espaciais, São José dos Campos, SP, Brazil.

Shukla, J., Nobre, C. A., and Sellers, P. : 1990, 'Amazonia Deforestation and Climate Change', *Science* 247, 1322–1325.

Shulz, J. P. : 1960, *Ecological Studies on Rain Forest in Northern Surinam*, North Holland, Amsterdam, 270 pp.

Shuttleworth, W. J. : 1988, 'Evaporation from Amazonian Rain Forest', *Proc. Roy. Soc. B.* 233, 321–346.

Shuttleworth, J. W. and Dickinson, R. E. : 1989, 'Comments on "Modelling Tropical Deforestation: A Study of GCM Land-Surface Parameterizations", by R. E. Dickinson and A. Henderson-Sellers', *Q. J. R. Meteorol. Soc.*, (in press).

Shuttleworth, W. J., Gash, J. H. C., Lloyd, C. R., Roberts, J., Marques Filho, A. O., Fish, G., Silva Filho, V. P., Ribeiro, M. N. G., Molion, L. C. B., Nobre, C. A., Sa, L. D. A., Cabral, O. M. R., Patel, S. R., and Moraes, J. C. : 1984a, 'Observations of Radiation Exchange above and below Amazon Forest', *Quart. J. Roy. Meteorol. Soc.* 110, 1163–1169.

Shuttleworth, W. J., Gash, J. H. C., Lloyd, C. R.,

Roberts, J., Marques Filho, A. O., Fish, G., Silva Filho, V. P., Ribeiro, M. N. G., Molion, L. C. B., Nobre, C. A., Sa, L. D. A., Cabral, O. M. R., Patel, S. R., and Moraes, J. C. : 1985, 'Daily Variations of Temperature and Humidity within and above Amazonian Forest', *Weather* 40, 102–108.

Villa Nova, N. A., Salati, E., and Matsui, E. : 1976, 'Estimativa da evapotranspiração na Bacia Amazônica, *Acta Amazôn.* 6, 215–228.

24

Large Losses of Total Ozone in Antarctica Reveal Seasonal ClOx/NOx Interactions[*]

J. C. Farman, B. G. Gardiner and J. D. Shanklin

Recent attempts[1,2] to consolidate assessments of the effect of human activities on stratospheric ozone (O_3) using one-dimensional models for 30° N have suggested that perturbations of total O_3 will remain small for at least the next decade. Results from such models are often accepted by default as global estimates.[3] The inadequacy of this approach is here made evident by observations that the spring values of total O_3 in Antarctica have now fallen considerably. The circulation in the lower stratosphere is apparently unchanged, and possible chemical causes must be considered. We suggest that the very low temperatures which prevail from midwinter until several weeks after the spring equinox make the Antarctic stratosphere uniquely sensitive to growth of inorganic chlorine, ClX, primarily by the effect of this growth on the NO_2/NO ratio. This, with the height distribution of UV irradiation peculiar to the polar stratosphere, could account for the O_3 losses observed.

Total O_3 has been measured at the British Antarctic Survey stations, Argentine Islands 65° S 64° W and Halley Bay 76° S 27° W, since 1957. Figure 24.1a shows data from Halley Bay. The mean and extreme daily values from October 1957 to March 1973 and the supporting calibrations have been discussed elsewhere.[4,5] The mean daily value for the four latest complete observing seasons (October 1980–March 1984) and the individual daily values for the current observing season are detailed in figure 24.1. The more recent data are provisional values. Very generous bounds for possible corrections would be ±30 matm cm. There was a changeover of spectrophotometers at the station in January 1982; the replacement instrument had been calibrated against the UK Meteorological Office standard in June 1981. Thus, two spectrophotometers have shown October values of total O_3 to be much lower than March values, a feature entirely lacking in the 1957–73 data set. To interpret

* Originally published in *Nature*, 1985, vol. 315, p. 207–10.

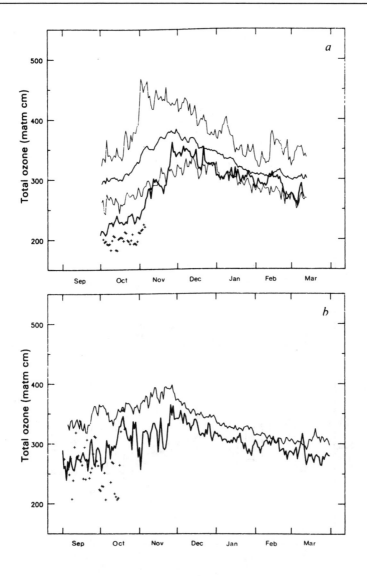

Figure 24.1 [orig. figure 1] Daily values of total O_3. a, Halley Bay: thin lines, mean and extreme values for 16 seasons, 1957–73; thick line, mean values for four seasons, 1980–84; +, values for October 1984. Observing season: 1 October to 13 March. b, Argentine Islands: as for Halley Bay, but extreme values for 1957–73 omitted. Observing season: 1 September to 31 March.

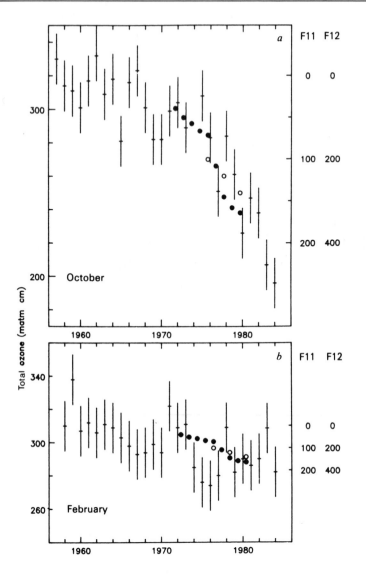

Figure 24.2 [*orig. figure 2*] Monthly means of total O_3 at Halley Bay, and Southern Hemisphere measurements of F-11 (O, p.p.t.v. (parts per thousand by volume) $CFCl_3O$ and F-12 (O, p.p.t.v. CF_2Cl_2). *a*, October, 1957–84. *b*, February, 1958–84. Note that F-11 and F-12 amounts increase down the figure.

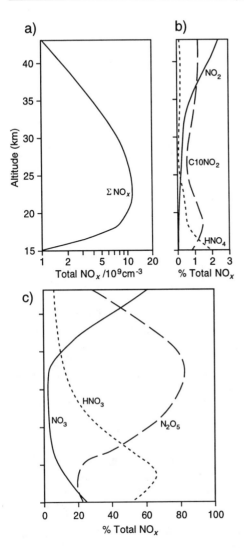

Figure 24.3 [*orig. figure* 3] NO_x during the polar night. *a*, Total NO_x cm^{-3}, from 15 to 43 km. *b*, NO_2, $ClONO_2$ and HNO_4 as percentages of total NO_x. *c*, NO_3, HNO_3 and N_2O_5 as percentages of total NO_x.

this difference as a seasonal instrumental effect would be inconsistent with the results of routine checks using standard lamps. Instrument temperatures (recorded for each observation) show that the March and October operating conditions were practically identical. Whatever the absolute error of the recent values may be, within the bounds quoted, the annual variation of total

O_3 at Halley Bay has undergone a dramatic change.

Figure 24.1b shows data from Argentine Islands in a similar form, except that for clarity the extreme values for 1957–73 have been omitted. The values for 1980 to the present are provisional, the extreme error bounds again being ±30 matm cm. The changes are similar to those seen at Halley Bay, but are much smaller in magnitude.

Upper-air temperatures and winds are available for these stations from 1956. There are no indications of recent departures from established mean values sufficient to attribute the changes in total O_3 to changes in the circulation. The present-day atmosphere differs most prominently from that of previous decades in the higher concentrations of halocarbons. Figure 24.2a shows the monthly mean total O_3 in October at Halley Bay, for 1957–84, and figure 24.2b that in February, 1958–84. Tropospheric concentrations of the halocarbons F-11 (CFCl$_3$) and F-12 (CF$_2$Cl$_2$) in the Southern Hemisphere[3] are also shown, plotted to give greatest emphasis to a possible relationship. Their growth, from which increase of stratospheric ClX is inferred, is not evidently dependent on season. The contrast between spring and autumn O_3 losses and the striking enhancement of spring loss at Halley Bay need to be explained. In Antarctica, the lower stratosphere is ~40 K colder in October than in February. The stratosphere over Halley Bay experiences a polar night and a polar day (many weeks of darkness, and of continuous photolysis, respectively); that over Argentine Islands does not. Figure 24.3 shows calculated amounts of NO_x in the polar night and the partitioning between the species.[6] Of these, only NO_3 and NO_2 are dissociated rapidly by visible light. The major reservoir, N_2O_5, which only absorbs strongly below 280 nm, should be relatively long-lived. Daytime levels of NO and NO_2 should be much less in early spring, following the polar night, than in autumn, following the polar day. Recent measurements[7] support these inferences. The effect of these seasonal variations on the strongly interdependent ClO$_x$ and NO_x cycles is examined below.

The O_3 loss rate resulting from NO_x and ClO_x may be written[8]

$$L=N+C=2\,k_2[O][NO_2]+2\,k_6[O][ClO] \quad (1)$$

L accounts for over 85% of O_3 destruction in the altitude range 20–40 km. At 40 km, N and C are roughly equal. Lower down, C decreases rapidly to 10% of L at 30 km, 3% at 20 km (refs 6, 8). Equation (1) is based on two steady-state approximations, (see table 24.1a for the reactions involved)

$$\psi = \frac{[NO_2]}{[NO]} \sim \frac{k_1[O_3]+k_4[ClO]}{k_2[O]+j_3} \quad (2)$$

and

Table 24.1 [*orig. table 1*]　Reaction list.

a　Governing ψ and χ (see text)

$NO + O_3 \rightarrow NO_2+O_2$	(1)
$NO_2+O \rightarrow NO+O_2$	(2)
$NO_2+h\nu \rightarrow NO+O$	(3)
$NO+ClO \rightarrow NO_2+Cl$	(4)
$Cl+O_3 \rightarrow ClO+O_2$	(5)
$ClO+O \rightarrow Cl+O_2$	(6)

b　Governing $[Cl+ClO]$

$HCl+OH \rightarrow Cl+H_2O$	(7)
$ClONO_2+h\nu \rightarrow ClO+NO_2$	(8)
$HOCl+h\nu \rightarrow Cl+OH$	(9)
$ClO+NO_2+M \rightarrow ClONO_2+M$	(10)
$ClO+HO_2 \rightarrow HOCl+O_2$	(11)
$Cl+CH_4 \rightarrow HCl+CH_3$	(12)
$Cl+HO_2 \rightarrow HCl+O_2$	(13)

c　$HCl+ClONO_2 \rightarrow Cl_2+HNO_3$ 　(14)

$$\chi = \frac{[Cl]}{[ClO]} \sim \frac{k_6[O]+k_4[NO]}{k_5[O_3]} \quad (3)$$

valid in daytime, with $[O]$ in steady state with $[O_3]$. Reaction (4) has a negative temperature coefficient, whereas reaction (1) has large positive activation energy,[9] with the result that ψ is strongly dependent on $[ClO]$ at low temperature, as shown in figure 24.4. $[ClO]$ is not simply proportional to total ClX, because $ClONO_2$ formation (reaction (10)) intervenes. Throughout the stratosphere, $\chi \ll 1$, so that $[ClO] \sim [Cl+ClO]$. From a steady-state analysis of the reactions given in table 24.1b,

$$[Cl+ClO] \sim \frac{k_7[HCl][OH]+j_8[ClONO_2]+j_9[HOCl]}{k_{10}[NO_2]+k_{11}[HO_2]+\chi(k_{12}[CH_4]+k_{13}[HO_2])} \quad (4)$$

Values of ψ, χ and $[Cl+ClO]$ obtained from equations (2), (3) and (4) are in good accord with full one-dimensional model results for late summer in Antarctica.[6] Neglecting seasonal effects other than those resulting from temperature and from variation of $[NO+NO_2]$, it is possible to solve simultaneously for $[NO_2]$ and $[ClO]$, and to derive L. Results are shown in table 24.2 as relaxation times,[8] $[O_3]/L$, for various conditions. The spring values (lines 2, 3 and 4) are highly dependent on ClX amount (compare columns *a* and *b*), the autumn values (line 1) much less so. At Argentine Islands, the sensitivity to ClX growth should resemble that seen in line 2, attributable solely to low temperature. Lines 3 and 4 show the enhanced sensitivity possible at stations

Table 24.2 [*orig. table 2*]　Relaxation times in days, $[O_3]/L$, for maximum chlorine levels 1.5 p.p.b.v. (*a*) and 2.7 p.p.b.v. (*b*) (1980).

Date	Altitude (km) Relative NO+NO₂	22 a	b	25.5 a	b	29 a	b	32.5 a	b	36 a	b	39.5 a	b	43 a	b
25 March	1	105	103	38	37	16	15	6.9	5.7	2.6	2.0	1.2	0.9	0.83	0.57
17 September	1	224	210	86	77	32	27	12.2	9.2	4.9	3.4	2.6	1.7	1.90	1.22
17 September	0.75	288	265	107	93	39	31	14.2	10.4	5.7	3.9	3.0	1.9	2.13	1.33
17 September	0.5	398	353	141	118	47	36	17.0	12.0	6.9	4.5	3.5	2.2	2.41	1.46

Noon values at 75.5° S, solar elevation 12.5°. Lower stratosphere at 230 K on 25 March; 190 K on 17 September. The altitudes shown apply to the summer temperature profile used in the model.[6]

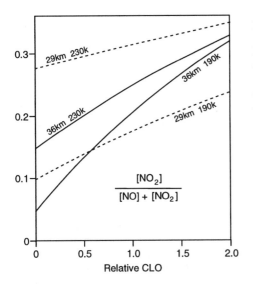

Figure 24.4 [*orig. figure 4*] [NO$_2$]/[NO+NO$_2$] has the status of an efficiency factor for O$_3$ destruction by the NO$_x$ cycle. In terms of the ratio ψ in the text, it is $\psi\,(1+\psi)$. The figure shows how this factor varies with [ClO] at 190 K and at 230 K, at altitudes of 29 and 36 km. Values of [O$_3$], [O], [ClO] and j_3 were taken from one-dimensional model results[6] (maximum chlorine 2.7 p.p.b.v.) for noon, 25 March at 75.5° S, solar elevation 12.5°. The abscissa is [ClO] relative to the model value. The rate-limiting reaction, (2) in table 24.1, for the NO$_x$ cycle has zero activation energy. Note how, nevertheless, O$_3$ was protected against destruction by NO$_x$ at low temperatures in a stratosphere with small amounts of ClX, but is losing this protection as ClX grows.

within the Antarctic Circle, such as Halley Bay, arising from slow release of [NO+NO$_2$] following the polar night. It remains to be shown how stable O$_3$ budgets were achieved with the relaxation times for the lower chlorine level (table 24.2*a*).

Much O$_3$ destruction is driven by visible light, but production requires radiation below 242 nm. On the dates shown (table 24.2), destruction persists for some 11 h, while, because of the long UV paths, production is weak (except around noon) at 29 km,

and is virtually absent below that altitude. Line 1 of table 24.2 then demands O$_3$ transport in autumn from the upper to the lower stratosphere, which is consistent with inferred thermally-driven lagrangian-mean circulations.[10] A mean vertical velocity of 45 m per day is in good accord with calculations of net diabatic cooling[11] and gives a realistic total O$_3$ decay rate in an otherwise conventional one-dimensional model.[6] The short relaxation times in the lower stratosphere in autumn are tolerable, with adequate transport compensating for lack of O$_3$ production.

In early spring, on the other hand, wave activity scarcely penetrates the cold dense core of the Antarctic polar vortex and with very low temperatures the net diabatic cooling is very weak.[11] Lagrangian transport in the vortex should then be almost negligible. (The virtual exclusion of Agung dust from the vortex supports this view.[5]) The final warming signals the end of this period of inactivity and is accompanied by large dynamically induced changes in O$_3$ distribution. However, before the warming, with low chlorine, total O$_3$ was in a state of near-neutral equilibrium, sustained primarily by the long relaxation times. With higher chlorine, relaxation times of the order seen in line 4, table 24.2, entail more rapid O$_3$ losses. With negligible production below 29 km and only weak transport, large total O$_3$ perturbation is possible. The extreme effects could be highly localized, restricted to the period with diurnal photolysis between polar night and the earlier of either the onset of polar day or the final spring warming. At the pole [NO+NO$_2$] rises continuously after the polar night, with the Sun. The final warming always begins over east Antarctica and spreads westwards across the pole. At Halley Bay the warming is typically some 14 days later than at the pole. Maximum O$_3$ depletion could be confined to the Atlantic half of the zone bordered roughly by latitudes 70 and 80° S.

Comparable effects should not be expected in the Northern Hemisphere, where the winter polar stratospheric vortex is less cold and less stable than its southern counterpart. The vortex is broken down, usually well before the end of the polar night, by major warmings. These are accompanied by large-scale subsidence and strong mixing, in the course of which peak O_3 values for the year are attained. Hence, sensitivity to ClX growth should be minimal if, as suggested above, this primarily results from O_3 destruction at low temperatures in regions where O_3 transport is weak.

We have shown how additional chlorine might enhance O_3 destruction in the cold spring Antarctic stratosphere. At this time of the year, the long slant paths for sunlight make reservoir species absorbing strongly only below 280 nm, such as N_2O_5, $ClONO_2$ and HO_2NO_2 relatively long-lived. The role of these reservoir species should be more readily demonstrated in Antarctica, particularly the way in which they hold the balance between the NO_x and ClO_x cycles. An intriguing feature could be the homogeneous reaction (table 24.1c) between HCl and $ClONO_2$. If this process has a rate constant as large as 10^{-16} cm^{3}s^{-1} (ref. 2) and a negligible temperature coefficient, the reaction would go almost to completion in the polar night, leaving inorganic chlorine partitioned between HCl and Cl_2, almost equally at 22km for example. Photolysis of Cl_2 at near-visible wavelengths would provide a rapid source of [Cl+ClO] at sunrise, not treated in equation (4). The polar-night boundary is, therefore, the natural testing ground for the theory of nonlinear response to chlorine.[1,2] It might be asked whether a nonlinear response is already evident (figure 24.2a). An intensive programme of trace-species measurements on the polar-night boundary could add greatly to our understanding of stratospheric chemistry, and thereby improve considerably the prediction of effects on the ozone layer of future halocarbon releases.

We thank B. A. Thrush and R. J. Murgatroyd for helpful suggestions.

References

1   Cicerone, R. J., Walters, S. and Liu, S. C. *J. geophys. Res.* 88, 3647–3661 (1983).
2   Prather, M. J., McElroy, M. B. and Wofsy, S. C. *Nature* 312, 227–231 (1984).
3   *The Stratosphere 1981. Theory and Measurements* (WMO Global Ozone Research and Monitoring Project Rep. 11 1981).
4   Farman, J. C., Hamilton, R. A. *Br. Antarct. Surv. Sci. Rep.* No. 90 (1975).
5   Farman, J. C. *Phil. Trans. R. Soc* B279, 261–271 (1977).
6   Farman, J. C., Murgatroyd, R. J., Silnickas, A. M. and Thrush, B. A. *Q. Jl. R. met. Soc.* (submitted).
7   McKenzie, R. L., Johnston, P. V. *Geophys. Res. Lett.* 11, 73–75 (1984).
8   Johnston, H. S., Podolske, J. *Rev. Geophys. Space Phys.* 16, 491–519 (1978).
9   *Chemical Kinetics and Photochemical Data for Use in Stratospheric Modelling, Evaluation No. 6* (JPL Publ. 83–62, 1983).
10 Dunkerton, T. *J. atmos. Sci.* 35, 2325–2333 (1978).
11 Dopplick, T. G. *J. atmos. Sci.* 29, 1278–1294 (1972).

25

CHLOROFLUOROCARBONS AND THE DEPLETION OF STRATOSPHERIC OZONE*

F. S. Rowland

Fourteen years have passed since Mario Molina and I first suggested, in an article in *Nature*, that chlorofluorocarbons are destroying the ozone layer in the earth's stratosphere. As the abstract of the article put it. "Chlorofluoromethanes are being added to the environment in steadily increasing amounts. These compounds are chemically inert and may remain in the atmosphere for 40–150 years, and concentrations can be expected to reach 10 to 30 times present levels. Photodissociation of the chlorofluoromethanes in the stratosphere produces significant amounts of chlorine atoms, and leads to the destruction of atmospheric ozone" (Molina and Rowland 1974; see also Rowland and Molina 1975a). According to this theory, now often described as the Rowland–Molina hypothesis, the release of atomic chlorine from the solar ultraviolet photolysis of molecules such as trichlorofluoromethane (CCl_3) and dichlorodifluoromethane (CCl_2F_2) triggers a catalytic chain reaction that causes significant depletion of the stratospheric ozone layer. The revelation in 1985 of a drastic reduction in ozone during the Antarctic springtime – the so-called Antarctic ozone

hole (figure 25.1) – was the most significant confirmation yet of the validity of the hypothesis. My research group has continued to explore atmospheric problems associated with chlorine chemistry, and in this article I will survey some of our findings during the past fourteen years.

The technological popularity of various chlorofluorocarbon molecules, known collectively as CFCs is based on their chemical inertness and volatility. Under trademarks such as Freon (Du Pont) and Genetron (Allied-Signal), they are used as aerosol propellants, refrigerants, cleaning solvents for electronic components, and foaming agents for plastics. Commercial users have long identified them through an arcane code in which the unit's digit is the number of fluorine atoms, the tens digit the number of hydrogen atoms plus one, and the hundreds digit the number of carbon atoms minus one. Thus, to take the four most commonly used, trichlorofluoromethane is CFC–11, dichlorodifluoromethane is CFC–12, trichlorotrifluoroethane (CCl_2FCClF_2) is CFC–113, and chlorodifluoromethane ($CHClF_2$) is CFC–22. Fourteen years ago,

* Originally published in *American Scientist*, 1989, vol. 77, pp. 36–45.

Figure 25.1 [*orig. figure 1*] The hypothesis that chlorofluorocarbons are destroying ozone in the earth's atmosphere, and in particular the layer in the upper stratosphere that protects the earth from dangerous solar ultraviolet radiation, has been confirmed by observations of a drastic decrease in ozone over Antarctica each spring during the late 1970s and 1980s (see figure 25.7). The so-called ozone hole appears as a large white area at the center of this picture taken by the total ozone mapping spectrometer aboard NASA's Nimbus 7 satellite on 5 October 1987, when the ozone loss reached nearly 60%.

chemists spoke simply of "fluorocarbons," but the problems with ozone depletion have caused them to distinguish chlorofluorocarbons, hydrofluorocarbons, and hydrochlorofluorocarbons.

The very lack of chemical reactivity which makes chlorofluorocarbon molecules commercially useful also allows them to survive unchanged in most applications and eventually emerge in gaseous form into the earth's atmosphere, where they persist for many decades. The major processes, or "sinks," by which most chemicals are removed from the atmosphere – photodissociation, rainout, and oxidation – are ineffective with the transparent, insoluble, nonreactive chlorofluorocarbons. The only important sink for trichlorofluoromethane and dichlorodifluoromethane is photodissociation in the mid-stratosphere by solar ultraviolet radiation with wavelengths shorter than 230 nm. These wavelengths, and indeed all wavelengths shorter than 293 nm, do not reach the lower atmosphere because they are intercepted and absorbed at higher altitudes, either by molecular oxygen, O_2, or by stratospheric ozone, O_3. The chlorofluorocarbons are subject to destruction only after they rise by random diffusion to altitudes of 25 to 40 km, above most of the molecular oxygen and ozone. The same stratospheric ozone which protects *Homo sapiens* and other living species from harmful ultraviolet radiation spreads its shield over the inanimate chlorofluorocarbons in the troposphere as well. Only in the harsh exposure of the mid-stratosphere do the chlorofluorocarbons forfeit this protection, absorb ultraviolet radiation, and release atomic chlorine.

The final escape route for these stratospheric chlorine atoms is a wandering descent back through the ozone layer towards ultimate deposition in rainfall as hydrochloric acid (HCl). Along the way, the chlorine oscillates chemically among two chain-reacting species, atomic chlorine and chlorine oxide (ClO):

$$Cl + O_3 \rightarrow ClO + O_2 \qquad (1)$$
$$ClO + O \rightarrow Cl + O_2 \qquad (2)$$

and several reservoir molecules, including hydrochloric acid, that temporarily suspend the reaction process (Stolarski and Cicerone 1974; Rowland and Molina 1975a). This so-called ClO chain is catalytic, because the chlorine reactant which initiates (1) is not permanently removed but reappears as a product from (2), ready to repeat the process again and again. The atmospheric odyssey to the final rainout requires a year or two, during which each chlorine atom destroys approximately 100,000 molecules of ozone. Ozone depletion by man attains significance on a global scale as the ultimate consequence of the yearly emission to the atmosphere of almost one million tons of chlorofluorocarbons, which translates into an ozone loss 10^5 times larger.

The attack on ozone can be understood in light of the mathematical truisms (odd + even = odd) and (odd + odd = even), which dominate two important classes of chemical processes in the stratosphere: reactions affecting "odd oxygen" – that is, pure oxygen with odd numbers of atoms, O_1 and O_3 – and the catalytic chain removal of ozone. The initial formation of ozone in the stratosphere results from solar ultraviolet bombardment of free molecular oxygen, which decomposes the oxygen into two separate atoms. The addition of each of these atoms to another diatomic oxygen molecule creates two ozone molecules. In contrast, the absorption of ultraviolet radiation by ozone with separation into atomic and molecular oxygen does not permanently remove ozone if the atomic oxygen immediately finds another diatomic molecule, recreating ozone. Instead, this two-step sequence converts solar energy into chemical bond energy and ultimately into heat, furnishing the source at 40 to 50 km for the temperature inversion layer that defines the stratosphere.

The depletion of stratospheric ozone involves processes which remove odd oxygen rather than just ozone. In odd-oxygen accounting, the interconversions of ozone and atomic oxygen represent zero change. The only direct method for transforming them into molecular oxygen – that is "even oxygen" – is the reaction of two odd species. While atomic oxygen and ozone do combine directly to produce molecular oxygen, most atmospheric removal of odd oxygen occurs through the catalytic chain reactions of (1) and (2) or through other analogous sequences.

The ClO_x chain is only one of several catalytic chain reactions driven by free radicals, chemical species with odd numbers of electrons, such as chlorine (17 electrons) and nitric oxide (15 electrons). One ozone molecule is removed in reaction (1), while the atomic oxygen which disappears in reaction (2) would otherwise have found a diatomic molecule of oxygen and reacted to create ozone; the net change in odd oxygen is − 1 for each step. Two other important stratospheric systems are the HO_x and NO_x chains, each of which can also accomplish in two steps the conversion of atomic oxygen and ozone into molecular oxygen without removing the catalyst. Removal of these odd electron species occurs only through chain termination, the combination of two radicals to form an even electron species – for example, chlorine oxide and nitrogen dioxide (NO_2) combining to form chlorine nitrate ($ClONO_2$). Such reservoir molecules can be broken into radicals again by ultraviolet radiation.

Detecting Chlorofluorocarbons

Calculations of stratospheric concentrations of trichlorofluoromethane and dichlorodifluoromethane furnished an important framework for many of our original conclusions. The simplest way to predict the vertical distribution of such trace constituents is a one-dimensional atmospheric model in which solar radiation of all wavelengths enters from the top and various molecules enter from the bottom through emission at the earth's surface. The rate of

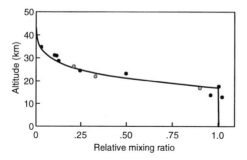

Figure 25.2 [*orig. figure* 2] That the chlorofluorocarbon trichlorofluoromethane (CCl_3F) reaches the stratosphere is shown in this graph. The curve represents the predictions of a one-dimensional atmospheric model used by the author and Mario Molina when they first suggested the hypothesis in 1974. Actual atmospheric samples collected by researchers from the National Center for Atmospheric Research and the National Oceanic and Atmospheric Administration are indicated respectively by dark gray and light gray dots. The data further demonstrate that the trichlorofluoromethane is photodissociated in the upper stratosphere.

upward mixing is adjusted to fit the known stratospheric profiles of other species such as methane (CH_4) and nitrous oxide (N_2O). Our 1974 predictions for the stratospheric concentrations of trichlorofluoromethane, based on laboratory observations of its properties, are shown in figure 25.2. (The mixing ratio – or mole fraction – is the fraction of all molecules present as each particular species, and is usually used in such comparisons instead of absolute concentrations in molecules/cm^3 because the total number of molecules decreases by a factor of 10 for about every 16-km increase in altitude.) Because no actual measurements of chlorine-containing molecules in the stratosphere had been reported when our first paper was submitted, the validity of our attempt to transfer laboratory observations was immediately called into question.

In 1975, samples collected in balloon-borne evacuated flasks demonstrated that trichlorofluoromethane reaches the stratosphere and that it photodissociates at an altitude of 20 to 35 km, as predicted in our calculations (Heidt et al. 1975; Schmeltekopf et al. 1975; see figure 25.2). These early experiments have been supplemented by abundant data for many chlorine-containing molecules at several latitudes (WMO-NASA 1986). A wide range of rates of decrease in mixing ratios with altitude has been predicted from the laboratory ultraviolet absorption cross sections and also confirmed in the stratosphere. One query still posed by laboratory-bound scientists is, How can molecules which are much heavier than air reach the stratosphere? (The molecular weight of trichlorofluoromethane is 137; that of air is 29.) Won't they just settle downward, as they do in the laboratory? The answer is that indoor experience with mass-dependent molecular diffusion is not applicable in the wind-driven atmosphere, in which mixing is dominated for altitudes to 100 km by the motions of very large air parcels. These moving masses of air carry along together gaseous molecules of all weights, allowing satisfactory calibration of the motions of trichlorofluoromethane with a molecule like methane, which is more than eight times lighter.

Further confirmation of stratospheric chlorine chemistry came through the positive identification of various other chemical states of chlorine, especially the carriers of the ClO_x chain reactions (NAS 1976a, 1976b). The presence of the chain-reacting chlorine oxide and the temporary reservoir hydrochloric acid soon confirmed the chemical reactivity of chlorine when released from the parent chlorofluorocarbons. For many scientists, the qualitative presence of chlorine oxide furnished definite proof of the ozone-depleting capabilities of chlorinated molecules in the upper stratosphere.

After stratospheric penetration and solar photodissociation of the chlorofluorocarbons had been confirmed, attention returned to the question of tropospheric sinks. Were our original evaluations of processes such as rainout, photodissociation, and oxidation correct in concluding that none would be important for the chlorofluorocarbons? After all, some methane reaches the upper stratosphere even though it has a strong tropospheric sink through reaction with hydroxyl (HO) radical. A number of specific possibilities for removal or destruction of trichlorofluoromethane and dichlorodifluoromethane were suggested, such as photolysis in the near-ultraviolet (293–400 nm) during absorption on grains of Sahara sand, degradation by reactions with atmospheric ions, and freeze-out on Antarctic snow. Subsequent investigations have demonstrated that none of these proposed sinks is important on a global scale (NAS 1976b, 1979a).

An estimate of the cumulative importance of all tropospheric sinks, including any not yet identified, can be derived from a comparison of the measured concentrations of trichlorofluoromethane and dichlorodifluoromethane in the atmosphere and the integrated amounts of each already released (NAS 1976b). The smaller the fraction surviving, the shorter the average atmospheric lifetime and the greater the probability of a significant tropospheric sink. Atmospheric lifetimes as short as 15 or 20 years were claimed on the basis of measurements near 50°N latitude,

converted into global estimates via the latitude concentration profile initially reported for trichlorofluoromethane, as shown in figure 25.3 (Jesson et al. 1977; Jesson 1982). However, this apparent distribution, with concentrations over England twice as large as those over South America, was quickly proved anomalous and presumably erroneous. Subsequent measurements by many research groups have indicated only about a 10 to 15% decrease in the concentration of trichlorofluoromethane from the Northern to Southern Hemispheres, as shown by the

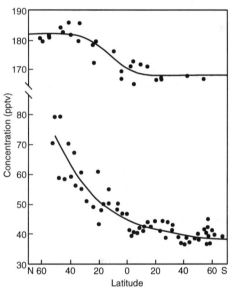

Figure 25.3 [orig. figure 3] Some scientists suggested that chlorofluorocarbons might be removed from the atmosphere by tropospheric sinks, processes such as rainout and oxidation. One report suggesting a relatively short lifetime for trichlorofluoromethane was based on calculations showing a steep decline in concentrations from the Northern to Southern Hemispheres (lower data points and curve). More recent measurements (upper data points and curve), however, have shown the calculations to be wrong, with a more even distribution of concentrations at a much higher level. The implication is a long lifetime for chlorofluorocarbons in the atmosphere.

1980 data in figure 25.3 (Rowland et al. 1985). The "missing" chlorofluorocarbons, whose presumed absence in the global integrations provided the basis for calculating short lifetimes had been there all along south of 30°N.

The steady accumulation of both trichlorofluoromethane and dichlorodifluoromethane has led to current estimates of atmospheric lifetimes of 75 years and 111 years respectively, consistent with removal by stratospheric ultraviolet radiation alone (Cunnold et al. 1986). Such lifetimes cannot be fixed with great precision because of possible errors involved in global assay of the atmospheric burden, in absolute calibration of instrument response, and in estimates of annual emission rates. The conclusion is clear, however, that the lifetimes approach a century or more, and lie toward the upper limits given in our 1974 paper. No rapid tropospheric removal processes exist for these molecules, or for other chlorofluorocarbons such as trichlorotrifluoroethane.

The total concentration of organochlorine in the atmosphere has grown substantially in the past two decades, as shown in figure 25.4 (Rowland 1988). Only methyl chloride, CH_3Cl, has an important natural source, and its current concentration of about 0.6 parts per billion by volume (ppbv) represents a reasonable upper limit to the total organochlorine in the atmosphere at the beginning of the twentieth century. The total concentration was no more than about 0.7 ppbv in 1950, increasing to about 1.0 ppbv in 1965, 1.7 ppbv in 1974, and about 3.5 ppbv in 1988 – an increase by at least a factor of five in the past 40 years, and a factor of two since we first warned of the threat to stratospheric ozone from chlorofluorocarbons. With continued emission at 1986 rates, the concentration will reach 5.0 ppbv by the end of the century.

We have noted from the beginning a sharp distinction in atmospheric behavior between fully halogenated chlorofluorocarbons – that is, chlorofluorocarbons that have no hydrogen – and their hydrogen-containing analogues. Tropospheric hydroxyl radicals can react with carbon-hydrogen bonds in methane and also with methylchloroform (CH_3CCl_3) and with chlorodifluoromethane ($CHClF_2$). Subsequent decomposition of the chlorinated fragments removes them from the atmosphere before they can reach the stratosphere. The threat to stratospheric ozone from such molecules is consequently greatly reduced. This distinction has made the hydrogenated chlorofluorocarbons an obvious source of substitute molecules for various technological uses (Rowland 1974; Rowland and Molina 1975a, 1975b).

Atmospheric lifetimes for the hydrogenated molecules can be estimated if

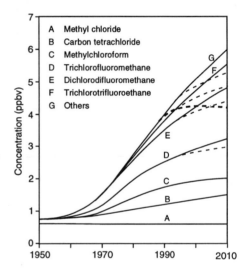

Figure 25.4 [*orig. figure* 4] The cumulative concentration of organochlorine in the atmosphere has grown substantially since the 1960s with the introduction of chlorofluorocarbons for a wide variety of commercial uses. The flat curve of the only organochlorine with an important natural source, methyl chloride (CH_3Cl), indicates the probable level of such substances in the atmosphere at the beginning of the century. Future levels include those projected from current levels of use (1), those that will result if the terms of the United Nations Environment Programme protocol of 1987 are met (2), and the cumulative level that would result from a complete phaseout over the next decade (3).

reaction with hydroxyl radicals is the primary sink and if a standard molecule with a known atmospheric lifetime is available. Such a standard requires both absolute assay in the atmosphere and known emission rates, and methylchloroform was early selected as the best available base for such calibrations (Rowland and Molina 1975b; Singh 1977). General agreement now exists that the atmospheric lifetime for methylchloroform is about six or seven years (Makide and Rowland 1981; Prinn et al. 1987). Most of the molecules now under discussion as technological and commercial substitutes for chlorofluorocarbons had already been identified a decade ago. One, chlorodifluoromethane, has been in widespread use for many years as the coolant in home air conditioners. Unfortunately, a six-year research hiatus after 1980 has left us with no really new substitute molecules available on the commercial market.

Greenhouse Gases

Solar radiation has its maximum emission in the visible wavelengths between 400 and 700 nm, consistent with the sun's surface temperature of about 5,600K. To balance the average 236 watts/m^2 it absorbs from the sun, the earth's atmosphere emits energy (in black body approximation) at wavelengths inversely proportional to the surface temperatures of the two bodies and therefore about 20 times longer – that is, in the infrared region at 8 to 14 μm. The quanta of energy in these wavelengths correspond to the energies of the internal vibrations of various polyatomic atmospheric molecules such as carbon dioxide, ozone, and water, and the wavelengths closely matching specific vibrational frequencies are strongly absorbed by the molecules. Such radiation is reemitted in all directions, greatly reducing emission to space. The existing atmospheric concentrations of these triatomic species are already sufficient to reduce the escape of infrared radiation by about 40% relative to an identical earth with no atmosphere. Earth has compensated by increasing its temperature from an estimated airless 254K to the present global average of 288K, and its total

infrared emission to 386 watts/m^2, thereby sending out 236 watts/m^2, most of it through the wavelengths not absorbed by these compounds. The reality of this "greenhouse effect" is quite evident from satellite readings of infrared deficits at those wavelengths corresponding to the vibrational frequencies of carbon dioxide, ozone, and water.

In current usage, however, "greenhouse effect" usually refers to an incremental increase in infrared absorption within the atmosphere because of increasing concentrations of various trace gases. Such atmospheric changes necessitate a further increase in the average temperature of the earth's surface in order that 236 watts/m^2 can still escape. Fifteen years ago, calculations of the greenhouse effect were directed solely toward carbon dioxide, the only trace gas then known to be increasing in concentration. Atmospheric concentrations of carbon dioxide have risen from about 315 parts per million by volume (ppmv) in 1957 to about 350 ppmv in 1988, largely through the burning of coal, oil, and natural gas, with a further contribution from tropical deforestation.

In 1975, Ramanathan pointed out that strong absorption features of molecules such as trichlorofluoromethane and dichlorodifluoromethane were located in optically transparent infrared regions of the atmosphere, and that incremental additions of chlorofluorocarbons are therefore far more effective than carbon dioxide, molecule for molecule, in retaining infrared radiation (Ramanathan 1975). The atmosphere is optically very thick toward the wavelengths of carbon dioxide, with the existing 350 ppmv already more than sufficient to absorb the radiation corresponding to its internal vibrational energies. Consequently, incremental increases in carbon dioxide add only a little per molecule to the total infrared absorption capability of the atmosphere. Of course, the 1,500 ppbv yearly increase in carbon dioxide is so much larger than increases of other trace gases that carbon dioxide still accounts for half or more of the incremental greenhouse increase. But yearly increases of 0.011 ppbv for trichlorofluoromethane and 0.018

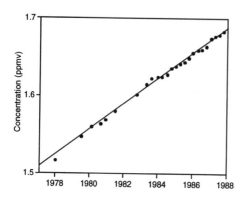

Figure 25.5 [*orig. figure 5*] Increases in both chlorofluorocarbons and methane contribute more per molecule to the greenhouse effect than do increases in carbon dioxide in the already carbon dioxide-rich atmosphere. Thus the monotonic increase in methane over the past ten years shown here has serious implications for global warming trends.

ppbv for dichlorodifluoromethane have substantial implications, because the chlorofluorocarbons are, per molecule, 10,000 times more efficient in absorbing infrared radiation than additional carbon dioxide in earth's existing carbon dioxide-rich atmosphere.

When we measured methane in the same tropospheric samples collected for measuring methylchloroform, we observed increases in seven different latitudes from both the Northern and Southern Hemispheres (Blake et al. 1982), in general confirmation of the increases reported from multiple daily measurements at Cape Meares, Oregon (Rasmussen and Khalil 1981). Our continued measurement of methane in air samples over the latitude range from Barrow, Alaska, to southern New Zealand and occasionally Antarctica has demonstrated a monotonic increase in the average global concentration of about 0.016 ppmv per year (Blake and Rowland 1988). Figure 25.5 shows this increase from 1.52 ppmv in early 1978 to 1.69 ppmv ten years later. Because the incremental greenhouse effect per molecule of methane is about 20 times that for carbon dioxide, the

contribution from methane is an important component of the total effect (Ramanathan et al. 1985).

The contributions of these and other trace gases such as nitrous oxide (N_2O) have been calculated to be approximately equivalent in importance to that of carbon dioxide – in effect, doubling the expected rate of increase for the average temperature of the earth's surface. Thus evaluation of the atmospheric consequences of continued emissions of chlorofluorocarbons must now include both ozone depletion *and* enhanced greenhouse warming.

Estimating Ozone Depletion

Our initial calculations of future chlorofluorocarbon concentrations were based on steady state conditions (yearly emissions = yearly losses), continuing one or more centuries into the future for atmospheric lifetimes of 40 to 150 years (Molina and Rowland 1974). We estimated the eventual losses in total ozone accompanying such increases in chlorofluorocarbon concentrations as a range between 7 and 13%. The 1975 report of the Federal Task Force on Inadvertent Modification of the Stratosphere estimated about 7% (IMOS 1975). The magnitudes of estimates have fluctuated over the succeeding years as atmospheric models have been adjusted to include more molecules, chemical reactions, and accurately measured rate constants and photoabsorption cross sections. For example, the semifinal draft of the first report by the National Academy of Sciences contained an estimate of 14% for steady state ozone depletion (Dotto and Schiff 1978), only to be delayed for six months for recalculation and reevaluation (to 7%) following our report of measurements of solar ultraviolet absorption cross sections for chlorine nitrate (Rowland et al. 1976).

There has been continuous agreement, however, that maximum ozone depletion from the ClO_x reactions (1) and (2) occurs in the upper stratosphere near 40 km (NAS 1976a, 1976b, 1979a, 1979b, 1982, 1984; WMO-NASA 1986). The ozone concentrations needed for (1) diminish rapidly above that altitude and the oxygen atoms required for (2) are quite scarce below it. The expected steady state depletions in the upper stratosphere have always been large, approaching 50%, being relatively unaffected by the various adjustments over the years in the input data for the atmospheric models.

The large changes in estimates of total ozone depletion have resulted chiefly from changing calculations of ozone concentrations in the lower stratosphere and troposphere. Here, polyatomic molecules shielded from shorter wavelength ultraviolet radiation by molecular oxygen and ozone in the stratosphere can survive for much longer periods, and the interactions among the HO_x, NO_x, and ClO_x chains are much more complex. An important effect of the ClO_x chain in the lower stratosphere is to tie up nitrogen dioxide as chlorine nitrate, preventing the removal of ozone by the NO_x analogues of (1) and (2) and partially offsetting the large ozone losses at higher altitudes.

By 1979, changes in model input data had increased the estimated steady state ozone depletion from 7% to 18% (NAS 1979a, 1979b). Then, a series of changes over the next several years acted to reduce the estimated ozone depletion (WMO-NASA 1986). Even the sign of the ozone variations in the lower stratosphere had altered, and the estimation of eventual total ozone loss decreased from 18% in 1979 to 5% in 1983 (NAS 1984).

During this period, too, the changing concentrations of organochlorine compounds were no longer treated in isolation in the atmospheric models. First, the effects from simultaneous increases in carbon dioxide – chiefly a reduction in upper-stratospheric temperatures, with accompanying shifts in the temperature-dependent rate constants for various chemical reactions – were introduced into the calculations. Then the increasing tropospheric concentrations of methane and nitrous oxide and increases in direct injection of nitrogen oxides from commercial subsonic aircraft were added. By the end of 1983, the Lawrence Livermore Laboratory one-dimensional atmospheric model was

providing estimates of 4.2% ozone depletion from increases in chlorofluorocarbon concentrations alone if all else was held constant, and even smaller eventual ozone losses for some of the optional scenarios (NAS 1984).

The succession of changes in the input data for the models between 1979 and 1983 not only sharply reduced ozone depletion estimates but also substantially lowered the calculated concentrations of hydroxyl radicals in the altitude range of 15 to 20 km, a region in which virtually no measurements were available for comparison or confirmation. The lifetime of hydrogen chloride, whose only reaction in the model was with hydroxyl radicals, became longer and longer, reaching several months in the computer output from the 1983 atmospheric models. The calculations typically did not include any "molecule–molecule" reactions but were restricted to about 160 different chemical and photochemical reactions, all of them involving solar photodissociation or chemical reactions of either atomic oxygen or free radicals. All were homogeneous gas-phase reactions involving only chemical species free to move as gases.

The absence of molecule–molecule reactions in the models was the result not of carelessness or conspiracy but rather of the general observation that molecule–molecule reactions are usually extremely slow in the gas phase, often not occurring at all in laboratory simulations. Consideration of possible atmospheric reactants with hydrogen chloride brought us back to the reservoir molecule chlorine nitrate:

$$HCl + ClONO_2 \rightarrow Cl_2 + HONO_2 \qquad (3)$$

By analogy, the hydrolysis of chlorine nitrate also seemed worth exploration:

$$H_2O + ClONO_2 \rightarrow HOCl + HONO_2 \qquad (4)$$

In 1984, Sato tried both of these experiments and found them to be exceptionally rapid. The infrared spectrum of gaseous chlorine nitrate disappeared within two seconds after the addition of hydrogen chloride and within minutes in separate experiments with water (Sato and Rowland 1984; Rowland et al. 1986).

A major unanswered question was whether the rapid reactions were homogeneous gas-phase reactions, or involved heterogeneous reactions, catalyzed by the surfaces of the reaction vessel, or were a combination of the two. None of the special inert surfaces normally used by chemists succeeded in slowing the reactions substantially. Subsequent experiments by others have demonstrated that the homogeneous gas-phase reactions of (3) and (4) are exceedingly slow and of no importance under atmospheric conditions (JPL 1987). The heterogeneous reactions have, however, proved to be crucial during the Antarctic springtime, as we will see.

Without waiting for the resolution of the homogeneous–heterogeneous dilemma, we asked Wuebbles and Connell to test the influence of the reactions on ozone depletion in the Livermore one-dimensional model. The results, as shown in figure 25.6, were startling: the addition of the reaction involving water at the rate constant indicated by Sato's experiments increased the estimated steady state ozone depletion almost six-fold, from 4.2% to 24%, while the addition of the hydrochloric acid reaction to the existing set of 160 reactions increased depletion to 31.7% (Wuebbles et al. 1984).

The problems inherent in modeling of heterogeneous reactions are intractable, especially when the surfaces involved are of uncertain composition and structure, as with the various kinds of stratospheric aerosols and clouds. Any consensus that heterogeneous reactions are significant in any part of the stratosphere almost automatically precludes modeling future ozone depletions, at least in that portion of the stratosphere. The recent discovery that such reactions contribute to extensive ozone depletion in the Antarctic polar vortex around the spring equinox casts severe doubt on any current predictions of ozone changes – the actual losses in the future are likely to be much more severe.

The Antarctic Ozone Hole

The ultimate proof of an ozone depletion theory is of course the observation of depletion. The first series of measurements of ozone began in the 1920s. It employed a

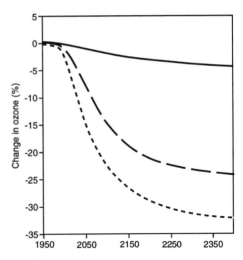

Figure 25.6 [*orig. figure 6*] Attempts to estimate the amount of ozone depletion resulting from the increase in chlorofluorocarbons in the atmosphere have yielded a wide range of values. In 1983 the Lawrence Livermore Laboratory's one-dimensional atmospheric model yielded a projection of a 4.2% decrease (1). Experiments conducted in 1984 with the same atmospheric model showed a 24% decrease in ozone with the addition of a reaction involving water and the reservoir molecule chlorine nitrate (2) and a 31.7% decrease from a reaction involving hydrogen chloride and chlorine nitrate (3).

ground-based ultraviolet spectrometer, now called a Dobson instrument after its first regular user. While several ground stations were operated in the 1930s, only the data set from Arosa, Switzerland, is continuous from then, with just five missing months since August 1931 (Dütsch 1984). A large number of new Dobson stations were established for the International Geophysical Year (1957–58), and the current ground-based network in the United States began operating in 1963.

Several Dobson stations were set up in Antarctica during the IGY, including one operated by the British Antarctic Survey at Halley Bay (76°S). The average October ozone measurements over Halley Bay drifted downward in the late 1970s and then plunged in the 1980s, as shown in figure 25.7 (Farman et al. 1985). These data were quickly supplemented by satellite data which showed the area of appreciable ozone loss – the so-called ozone hole (see figure 25.1) – extending over several million square miles, with deeper losses as the 1980s progressed (Stolarski et al. 1986).

A search for the possible causes of this springtime Antarctic ozone loss began, and the contending theories quickly sorted themselves into classes of chemical vs. dynamic and natural vs. anthropogenic. The questions raised have been decisively answered by scientific expeditions in 1986 and 1987 (NASA 1988), which demonstrated unequivocally that the ozone hole is the consequence of chlorine chemistry, abetted by the special conditions of Antarctic meteorology. The long polar night produces stratospheric temperatures as low as –90°C at 15 to 20 km altitude, allowing the formation of polar stratospheric ice clouds (McCormick et al. 1982); heterogeneous chemical reactions of the surfaces of these clouds convert hydrochloric acid and chlorine nitrate into molecular chlorine and hypochlorous acid (HOCl) by reactions (3) and (4) (Molina et al. 1987; Tolbert et al. 1987) and also sequester nitrogen oxides in the clouds as nitric acid (Crutzen and Arnold 1986; Toon et al. 1986). The first sunlight of approaching spring releases atomic chlorine, triggering ClO_x chain reactions (Solomon et al. 1986); with no nitrogen oxides to react with chlorine oxide, the chain reactions run essentially unhindered for five or six weeks into early spring, causing ozone depletions of 95% or more in the atmospheric layers at the altitudes of the stratospheric clouds and as much as 60% in total ozone over all altitudes.

The chlorine oxide dimer, ClOOCl, plays a critical role in the alternate ClO_x chain (Molina and Molina 1987). Two chlorine oxide radicals from (1) can react and then undergo photolysis to complete the sequence. This chain probably accounts for 75 to 85% of the rapid ozone depletion in the Antarctic vortex. Other chain routes of less importance involve hydrogen (Solomon et al. 1986) and bromine (McElroy et al. 1986).

Extremely high concentrations of chlorine oxide have been measured in the Antarctic polar vortex both from the ground and from the ER-2, a modified version of the U-2 spy plane, flying at 18 km altitude (Anderson et al. 1988). The striking correlation between high chlorine oxide and low ozone found in late September flights demonstrated their intimate relationship after a few weeks of spring sunlight. Concentrations of chlorine oxide as high as 1.3 ppbv far exceed the total

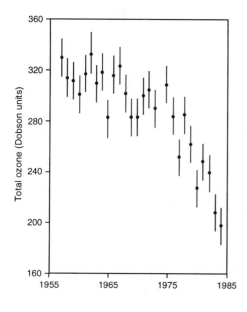

Figure 25.7 [*orig. figure 7*] The first published evidence of major ozone depletion came from measurements taken over a nearly 30-year period at the British Antarctic Survey base at Halley Bay. What appeared was a drastic decrease in ozone during the Antarctic spring commencing in the 1970s and intensifying in the 1980s (see figure 25.1). The data shown here – expressed in Dobson units, each of which is approximately equal to one part per billion by volume – are the monthly average of daily measurements in each October.

0.7 ppbv of chlorine available in the atmosphere as recently as 1965, demonstrating the critical role played by anthropogenic chlorine in the formation of the Antarctic ozone hole.

Global Trends

The search for global trends in ozone has been based largely on the accumulated data from the long-running Dobson stations, which show that major changes comparable to those found over Antarctica have not occurred elsewhere (NAS 1976b, 1979a, 1982, 1984; WMO-NASA 1986). Statistical evaluation through 1986 gave no indication of any trend in total global ozone significantly different from no change at all. The standard approach has been the so-called hockey-stick fit, in which an early period of no change is connected smoothly to a later period with a linear fit whose slope, plus or minus, is determined by least-squares analysis of the entire data set, after removal of the well-known seasonal cycle in concentrations.

These statistical trends have been expressed as best fits to a model embodying a linear yearly change in total ozone. The analyses have regularly included tests for contributions from other known geophysical cycles, such as the 11-year solar sunspot cycle and the 27-month quasi-biennial oscillation of wind directions in the equatorial stratosphere, as well as contributions from nuclear weapons tests in the atmosphere, which ended for the most part early in the 1960s. These statistical comparisons are basically tests of the predictions from one-dimensional models, which by their nature provide no information about possible latitudinal or seasonal variations in ozone losses. However, detailed analysis of the entire data set from Arosa for the period 1931 to 1986 (figure 25.8) demonstrated appreciable ozone loss in the winter and relatively little loss in the summer (Harris and Rowland 1986). This pattern was soon verified for several other Dobson stations, including those at Bismarck, North Dakota, and Caribou, Maine, and wintertime (December through March) losses have now

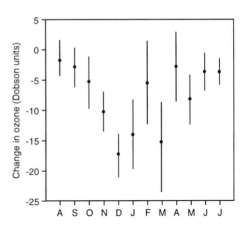

Figure 25.8 [orig. figure 8] Evidence of ozone depletion outside Antarctica has not been as easy to obtain. Data collected between 1931 and 1986 at the Dobson station at Arosa, Switzerland, indicate a less striking but nonetheless similar pattern of seasonal ozone loss. The monthly data points were generated from comparisons of average levels over the years 1931 to 1969 and 1970 to 1986.

been calculated at 18 of 19 stations north of 30°N latitude (Harris et al. 1988; NASA 1988).

These later evaluations allowed for the possibility of different trends in each month, in effect twelve monthly hockey sticks instead of the previous single yearly one. The evaluations were performed with data from each Dobson station and with data constructed from the averaged measurements in each of several latitude bands between 30°N and 64°N. Wintertime losses as high as 6.2% from 1970 to 1986 were calculated for the band from 53°N to 64°N, and 4.7% between 40°N and 52°N. Average annual losses ranged from 1.7% to 3% between 30°N and 64°N. The phenomenon of wintertime ozone loss in the northernmost quarter of the globe, already detectable by the end of 1980, was obscured in earlier analyses by the lumping together of data from all months. Moreover, the earlier analyses tended to weigh most heavily in the summer months, in which the standard deviation of the nautral ozone fluctuations was smallest.

The mechanisms of this ozone loss in the Northern Hemisphere are being tested with an experiment in Norway in January 1989. Although polar stratospheric clouds are scarcer in the Arctic because the polar vortex is not as constricted, reactions comparable to those over Antarctica seem quite likely, even if less persistent and not concentrated in an isolated air mass.

Events have been occurring on the political and regulatory front as well, perhaps in part in response to widespread public concern. The scientific discoveries of the mid-1970s led to a ban in the United States (announced in October 1976 and effective in 1978) on the use of chlorofluorocarbons as aerosol propellants and to similar restrictions in Canada and Scandinavia. The schedule of the proposed second phase of regulations in the United States covering all other uses of chlorofluorocarbons kept slipping in the late 1970s and was abandoned altogether in 1981. No additional regulations were seriously discussed until the Antarctic ozone hole brought a renewed sense of urgency. A United Nations Environment Programme convention to protect the ozone layer was signed in Vienna in 1985, and a protocol outlining proposed protective actions followed in Montreal in September 1987. The protocol specifies a 20% reduction from 1986 emissions of fully halogenated chlorofluorocarbons by 1994 and a further 30% reduction by 1999. The effect on the total amounts of organochlorine in the atmosphere is shown in figure 25.4. Even if the protocol were fully effective – and a large loophole exists for developing countries to go into production of chlorofluorocarbons – the reductions would have only a minor effect before the end of the century.

The unambiguous demonstration by the Antarctic expeditions that no hole would exist without chlorofluorocarbons and other anthropogenic organochlorine compounds has brought calls to strengthen the restrictive provisions of the protocol even before it has been ratified. The observation of ozone loss over the Northern Hemisphere has strengthened concern about global losses.

The pace of regulation depends greatly on the successful development of alternative substances, but the Environmental Protection Agency's call in September 1988 for a total phaseout of chlorofluorocarbons as aerosol propellants was accomplished swiftly in the United States and should go rapidly in Europe and Japan as well. An intensive international competition to create substitutes for the other technological uses is well under way.

In the meantime, the search is also intensifying for evidences of ozone depletion in the Arctic and in temperate latitudes, biological consequences of ozone loss, climatic alterations, and the onset of the greenhouse effect. Because the return of stratospheric chlorine to levels characteristic of the 1960s will require a few hundred years, both the chlorofluorocarbons and stratospheric ozone depletion will remain subjects for discussion, reminders of man's ability to alter nature on a global scale.

References

Anderson, I. C., W. H. Brune, M. I. Proffitt, W. Starr, and K. R. Chan. 1988. In situ observations of ClO in the Antarctic: Evidence for chlorine catalyzed destruction of ozone. NASA Polar Ozone Workshop, Aspen, CO.

Blake, D. R., and F. S. Rowland. 1988. Continuing worldwide increase in tropospheric methane, 1978 to 1987. *Science* 239:1129–31.

Blake, D. R., et al. 1982. Global increases in atmospheric methane concentrations between 1978–1980. *Geophys. Res. Lett.* 9:447–50.

Crutzen, P., and F. Arnold. 1986. Nitric acid cloud formation in the cold Antarctic stratosphere: A major cause for the springtime "ozone hole." *Nature* 324:651–55.

Cunnold, D. M. , et al. 1986. Atmospheric lifetime and annual release estimates for $CFCl_3$, and CF_2Cl_2 from 5 years of ALE data. *J. Geophys. Res.* 9:10797–817.

Dotto, L., and H. Schiff. 1978. *The Ozone War.* Doubleday.

Dütsch, H. U. 1984. An update of the Arosa ozone series to the present using a statistical instrument calibration. *Quart. J. Royal Meteorol. Soc.* 110:1079–96.

Farman, J. C., B. G. Gardiner, and J. D. Shanklin. 1985. Large losses of total ozone in Antarctica reveal seasonal ClO_x/NO_x interaction. *Nature* 315:207–10.

Harris, N., and F. S. Rowland. 1986. Trends in total ozone measurements at Arosa. Abstracts of fall meeting of the American Geophysical Union, San Francisco.

Harris, N., F. S. Rowland, R. Bojkov, and P. Bloomfield. 1988. Winter-time losses of ozone in high northern latitudes. NASA Polar Ozone Workshop, Aspen, CO.

Heidt, L. E., R. Lueb, W Pollock, and D. H. Ehhalt. 1975. Stratospheric profiles of CCl_3F and CCl_2F_2. *Geophys. Res. Lett.* 2:445–47.

IMOS (Federal Task Force on Inadvertent Modification of the Stratosphere). 1975. *Fluorocarbons and the Environment.* Council on Environmental Quality. Federal Council for Science and Technology.

Jesson, J. P. 1982. Halocarbons. In *Stratospheric Ozone and Man*, ed. F. A. Bower and R. B. Ward, vol. 2, pp. 29–63. CRC Press.

Jesson, J. P., P. Meakin, and L. C. Glasgow. 1977. The fluorocarbon-ozone theory. II. Tropospheric lifetimes – an estimate of the tropospheric lifetime of CCl_3F. *Atmos. Environ.* 11:499–508.

JPL. 1987. *Chemical Kinetics and Photochemical Data for Use in Stratospheric Modeling.* Evaluation no. 8. W. B. DeMore et al. Jet Propulsion Laboratory, Pasadena, JPL-87-41.

Makide, Y., and F. S. Rowland. 1981. Tropospheric concentrations of methylchloroform, CH_3CCl_3, in January 1978 and estimates of the atmospheric residence times for hydrohalocarbons, PNAS 78:5933–37.

McCormick, M. P., H. M. Steele, P. Hamill, W. P. Chu, and T. J. Swissler. 1982. Polar stratospheric cloud sightings by SAM II. *J. Atmos. Sci* 3:1387–97.

McElroy, M. B., R. J. Salawitch, S. C. Wofsy, and J. A. Logan. 1986. Reductions of Antarctic ozone due to synergistic interactions of chlorine and bromine. *Nature* 321:759–62.

Molina, L. T., and M. J. Molina. 1987. Production of Cl_2O_2 from the self-reaction of the ClO radical. *J. Phys. Chem.* 91:433–36.

Molina, M. J., and F. S. Rowland. 1974. Stratospheric sink for chlorofluoromethanes: Chlorine atom-catalysed destruction of ozone. *Nature* 249:810–12.

Molina, M. J., T-L. Tso, L. T. Molina, and F. C-Y. Wang. 1987. Antarctic stratospheric chemistry of chlorine nitrate, hydrogen chloride and ice: Release of active chlorine. *Science* 238:1253–57.

NAS (Committee on Impacts of Stratospheric Change). 1976a. *Halocarbons: Environmental Effects of Chlorofluoromethane Release.*

NAS (Panel on Atmospheric Chemistry). 1976b. *Halocarbons: Effects on Stratospheric Ozone.*

NAS (Panel on Stratospheric Chemistry and Transport). 1979a. *Stratospheric Ozone Depletion by Halocarbons: Chemistry and Transport.*

NAS (Committee on Impacts of Stratospheric Change). 1979b. *Protection against Depletion of Stratospheric Ozone by Chlorofluorocarbons.*

NAS (Committee on Chemistry and Physics of Ozone Depletion). 1982. *Causes and Effects of Stratospheric Ozone Reduction: An Update.*

NAS (Committee on Causes and Effects of Changes in Stratospheric Ozone). 1984. *Causes and Effects of Changes in Stratospheric Ozone: Update 1983.*

NASA. 1988. *Present State of Knowledge of the Upper Atmosphere 1988: An Assessment Report.* R. T. Watson and Ozone Trends Panel, M. J. Prather and Ad Hoc Theory Panel, and M. M. Kurylo and NASA Panel for Data Evaluation. NASA publ. 1208.

Prinn, R., et al. 1987. Atmospheric trends in methylchloroform and the global average for the hydroxyl radical. *Science* 238:945–50.

Ramanathan, V. 1975. Greenhouse effect due to chlorofluorocarbons: Climatic implications. *Science* 190:50–52.

Ramanathan, V., R. J. Cicerone, H. B. Singh, and J. T. Kiehl. 1985. Trace gas trends and their potential role in climate change. *J. Geophys. Res.* 90:5547–66.

Rasmussen, R. A., and M. A. K. Khalil. 1981. Atmospheric methane (CH_4); trends and seasonal cycles. *J. Geophys. Res.* 86:9826–32.

Rowland, F. S. 1974. Aerosol sprays and the ozone shield. *New Scientist* 64:717–20.

——. 1988. Testimony to Subcommittee on Hazardous Wastes and Toxic Substances. Senate Committee on Environment and Public Works, March 30.

Rowland, F. S., D. R. Blake, S. C. Tyler, and Y. Makide. 1985. Increasing concentrations of perhalocarbons, methylchloroform and methane in the atmosphere. In *Chemical Events in the Atmosphere and Their Impact on the Environment*, ed. G. B. Marini-Bettolo, pp. 305–34. Vatican: Pontifical Academy of Science.

Rowland, F. S., and M. J. Molina. 1975a. Chlorofluoromethanes in the environment. *Rev. Geophys. Space Phys.* 13:1–35.

——. 1975b. Stratospheric chemistry of chlorine compounds. Abstracts of 169th meeting of the American Chemical Society. Philadelphia. Phys. chem. abst. no. 70.

Rowland, F. S., H. Sato, H. Khwaja, and S. M. Elliott. 1986. The hydrolysis of chlorine nitrate and its possible atmospheric significance. *J. Phys. Chem.* 90:1985–88.

Rowland, F. S., J. E. Spencer, and M. J. Molina. 1976. Stratospheric formation and photolysis of chlorine nitrate. *J. Phys. Chem.* 80:2711–13.

Sato, H., and F. S. Rowland. 1984. Paper presented at the International Meeting on Current Issues in Our Understanding of the Stratosphere and the Future of the Ozone Layer, Feldafing, West Germany.

Schmeltekopf, A. L., et al. 1975. Measurements of Stratospheric CFCl₃, CF₂Cl₂ and N₂O. *Geophys. Res. Lett.* 2:393–96.

Singh, H. B. 1977. Atmospheric halocarbons. Evidence in favor of reduced average hydroxyl concentration in the troposphere. *Geophys. Res. Lett.* 4:101–04.

Solomon, S., R. R. Garcia, F. S. Rowland, and D. J. Wuebbles. 1986. On the depletion of Antarctic ozone. *Nature* 321:755–58.

Stolarski, R. S., and R. J. Cicerone. 1974. Stratospheric chlorine: A possible sink for ozone.

Can. J. Chem. 52:1610–15.

Stolarski, R. S., et al. 1986. Nimbus 7 satellite measurements of the springtime Antarctic ozone decrease. *Nature* 322:808–12.

Tolbert, M. A., M. J. Rossi, R. Malhotra, and D. M. Golden. 1987. Reaction of chlorine nitrate with hydrogen chloride and water at Antarctic stratospheric temperatures. *Science* 238:1258–60.

Toon, O. B., P. Hamill, R. P. Turco, and J. Pinto. 1986. Condensation of HNO₃ and HCl in the winter polar Stratospheres. *Geophys. Res. Lett.* 13:1284–87.

WMO-NASA. 1986. *Atmospheric Ozone 1985.* 3 vols. WMO Global Ozone Research and Monitoring Project. rep. no. 16.

Wuebbles, D. J., P. Connell, and F. S. Rowland. 1984. Paper presented at the International Meeting on Current Issues in Our Understanding of the Stratosphere and the Future of the Ozone Layer. Feldafing, West Germany.

26

Biomass Burning in the Tropics: Impacts on Atmospheric Chemistry and Biogeochemical Cycles*

P. J. Crutzen and M. O. Andreae

Biomass burning is widespread, especially in the tropics. It serves to clear land for shifting cultivation, to convert forests to agricultural and pastoral lands, and to remove dry vegetation in order to promote agricultural productivity and the growth of higher yield grasses. Furthermore, much agricultural waste and fuel wood is being combusted, particularly in developing countries. Biomass containing 2 to 5 petagrams of carbon is burned annually (1 petagram = 10^{15} grams), producing large amounts of trace gases and aerosol particles that play important roles in atmospheric chemistry and climate. Emissions of carbon monoxide and methane by biomass burning affect the oxidation efficiency of the atmosphere by reacting with hydroxyl radicals, and emissions of nitric oxide and hydrocarbons lead to high ozone concentrations in the tropics during the dry season. Large quantities of smoke particles are produced as well, and these can serve as cloud condensation nuclei. These particles may thus substantially influence cloud microphysical and optical properties, an effect that could

have repercussions for the radiation budget and the hydrological cycle in the tropics. Widespread burning may also disturb biogeochemical cycles, especially that of nitrogen. About 50 percent of the nitrogen in the biomass fuel can be released as molecular nitrogen. This pyrodenitrification process causes a sizable loss of fixed nitrogen in tropical ecosystems, in the range of 10 to 20 teragrams per year (1 teragram = 10^{12} grams).

The use of fire as a tool to manipulate the environment has been instrumental in the human conquest of Earth, the first evidence of the use of fires by early hominids dating back to 1 million to 1.5 million years ago (1). Even today, most human-ignited vegetation fires take place on the African continent, and its widespread, frequently burned savannas bear ample witness to this. Although natural fires can occur even in tropical forest regions (2, 3), the extent of fires has greatly expanded on all continents with the arrival of *Homo sapiens*. Measurements of charcoal in dated sediment cores have shown clear correlations between the rate of burning and

* Originally published in *Science*, 1990, vol. 250, pp. 1669–78.

human settlement (4). Pollen records show a shift with human settlement from pyrophobic vegetation to pyrotolerant and pyrophilic species, testimony to the large ecological impact of human-induced fires.

Natural fires have occurred since the evolution of land plants some 350 million to 400 million years ago and must have exerted ecological influences (5). In fact, high concentrations of black carbon in the Cretaceous-Tertiary boundary sediments suggest that the end of the age of the reptiles some 65 million years ago was associated with global fires that injected enormous quantities of soot particles into the atmosphere (6).

Today, the environmental impact of the burning of fossil fuels and biomass is felt throughout the world, and concerns about its consequences are prominent in the public's mind. Although the quantities of fossil fuels burned have been well documented, most biomass burning takes place in developing countries and is done by farmers, pioneer settlers, and housewives, for whom keeping records of amounts burned is not an issue. Biomass burning serves a variety of purposes, such as clearing of forest and brushland for agricultural use; control of pests, insects, and weeds; prevention of brush and litter accumulation to preserve pasturelands; nutrient mobilization; game hunting; production of charcoal for industrial use; energy production for cooking and heating; communication and transport; and various religious and aesthetic reasons. Studies on the environmental effects of biomass burning have been much neglected until rather recently but are now attracting increased attention (7). This urgent need has been recognized and will form an important element in the International Geosphere-Biosphere Programme (8).

In this article, we update quantitative estimates of the amounts of biomass burning that is taking place around the world and the resulting gaseous and particulate emissions and then discuss their atmospheric-chemical, climatic, and ecological consequences. Distinction should be made between net and prompt releases of CO_2. Net release occurs when land use changes take place by which the standing stock of biomass is reduced, for example, through deforestation. Biomass burning causes a prompt release of CO_2 but does not necessarily imply a net release of CO_2 to the atmosphere, as the C that is lost to the atmosphere may be returned by subsequent regrowth of vegetation. In either case, there is a net transfer of particulate matter and trace gases other than CO_2 from the biosphere to the atmosphere. Many of these emissions play a large role in atmospheric chemistry, climate, and terrestrial ecology.

Estimates of Worldwide Biomass Burning

In this section we derive some rough estimates of the quantities of biomass that are burned in the tropics through various activities, such as forest clearing for permanent use for agriculture and ranching, shifting cultivation, removal of dry savanna vegetation and firewood, and agricultural waste burning. In all cases, the available data are extremely scanty, allowing only a very uncertain quantitative assessment.

Clearing of forests for agricultural use. Two types of forest clearing are practiced in the tropics: shifting agriculture, where for a few years the land is used and then allowed to return to forest vegetation during a fallow period, and permanent conversion of forests to grazing or crop lands. In both cases, during the dry season, undergrowth is cut and trees are felled and left to dry for some time in order to obtain good burning efficiency. The material is then set on fire. The efficiency of the first burn is variable. Observations in forest clearings in Amazonia gave a burning efficiency of about 28% (9), similar to the value used by Seiler and Crutzen (10). This relatively low efficiency is due to the large fraction of the biomass that resides in tree trunks, only a small portion of which is consumed in the first burn. The remaining material may be left to rot or dry but is often collected and set on fire again. Adequate statistics are not available on how much of the original above-ground biomass is finally burned.

Taking reburn into account, we assume that in primary forests some 40% is combusted (9). For secondary forests, which have been affected by human activities and contain smaller sized material, we assume that 50% is burned.

According to Seiler and Crutzen (10), shifting agriculture (also called slash-and-burn agriculture, field-forest rotation, or bush-fallowing) was practiced by some 200 million people worldwide in the 1960s on some 300 million to 500 million ha, with an annual clearing of some 20 million to 60 million ha and a burning of 900 to 2500 Tg dm, that is, 400 to 1100 Tg of C; of this, 75% takes place in tropical secondary forest and the remainder in humid savannas (dm = dry matter; 1 g dm $\approx$ 0.45 g of C). Originally, shifting cultivators typically practiced crop and fallow periods of 2 to 3 and 10 to 50 years, respectively. Because of growing populations and lack of forest areas, fallow periods in many regions have shortened so much that the land cannot recover to the required productivity, which causes shifting agriculture to decline (11). On the other hand, in other regions it may still be expanding. According to Lanly (12), some 240 million ha were under traditional shifting agriculture by the end of the 1970s. On the basis of these statistics, Hao et al. (13) estimated that ~24 million ha are cleared annually for shifting cultivation in secondary forests. This clearing exposes ~2400 to 3000 Tg dm, that is, 1000 to 2400 Tg of C, to fire and thus leads to the release of 500 to 700 Tg of C. We combine the two ranges into an annual C release rate from shifting cultivation of between 500 and 1000 Tg. In traditional shifting agriculture, no net release of CO_2 to the atmosphere takes place because the forest is allowed to return to its original biomass density during the fallow period. The estimated rates, therefore, mainly represent prompt CO_2 release. However, because of overly frequent burning, the affected ecosystems often cannot recover to their original biomass, so that a net release of C to the atmosphere does result.

Permanent removal of tropical forests is currently progressing at a rapid rate. This process is driven by expanding human populations which require additional land, by large-scale resettlement programs, and by land speculation. The global rate of deforestation is subject to much uncertainty. The tropical forest survey of the Food and Agricultural Organization (FAO) of the United Nations for the latter part of the 1970s (12) has been the basis of several studies on net CO_2 release to the atmosphere. It now appears that the FAO statistics significantly underestimated deforestation rates, which, furthermore, may almost have doubled over the past decade (11, 14). As the earlier work on tropical deforestation was clearly based on questionable information, we feel that there is little point in reviewing it. Instead, we will estimate the consequences of deforestation activities for trace gas emissions, using the statistics assembled by Myers (11) and Houghton (14). This is the only available database that may be up to date and has also been adopted by the Intergovernmental Panel on Climate Change (15). Although newer data have now been assembled by the FAO, unfortunately they have not been released in time to be included in the present review. The net CO_2 release to the atmosphere due to deforestation from these sources still allows for the wide range of 1.1 to 3.6 Pg of C per year (14) [this range is given because of uncertainty regarding the areal extent of deforestation (11) and the original and successional biomass loadings (16)]. As about 60% of the total biomass is located below ground, including soil organic matter, this net release of CO_2 implies that 0.5 to 1.4 Pg of C per year of biomass are exposed to fire. As only 40 to 50% of the CO_2 release is through combustion (the rest is by microbial decomposition of organic matter), the resulting prompt release of CO_2 to the atmosphere would be in the range of 0.2 to 0.7 Pg of C per year.

Tropical savannas and brushland, typically consisting of a more or less continuous layer of grass interspersed with trees and shrubs, cover an area of about 1900 million ha (17). Savannas are burned every 1 to 4 years during the dry season with the highest frequency in the humid savannas (18). The extent of burning is increasing as a result of growing population pressures and more intensive use of rangeland. Although lightning may start some fires in savannas, most investigators are convinced that almost all are set by humans (4). Only dried grass, litter, weeds, and shrubs are burned; the larger trees of fire-resistant species suffer little damage.

Menaut (18) has estimated that in the West African savannas 45 to 240 Tg of C per year are burned. The total area of this savanna region is 227 million ha, including 53 million ha of Sahel semi-desert. No similarly detailed analysis on biomass burning has yet been attempted for other savanna regions. If we extrapolate to include all the savanna regions of the world (1900 million ha), we estimate that between 400 and 2400 Tg of C burn annually, and that most emissions are from the African continent. As, especially on the African continent, shifting cultivation also takes place in savanna regions, some double accounting could occur. The analysis by Seiler and Crutzen (10) indicates that a 30% correction may have to be applied to the above range, reducing it to 300 to 1600 Tg of burned C per year.

Fuel wood, charcoal, and agricultural waste. In the developing countries, fuel wood and agricultural waste are the dominant energy sources for cooking, domestic heating, and some industrial activities. It is difficult to estimate the amount of wood burned each year. The number given by FAO (19) for 1987, 1050 Tg dm, is certainly an underestimate because it includes only wood that is marketed. Scurlock and Hall (20) estimate that the annual per capita biofuel need (firewood, crop residues, dungcakes) is about 500 kg in urban and 1000 kg in rural regions, and that perhaps two-thirds of the rural energy use in China comes from agricultural waste. Altogether, they estimate that 14% of the global energy and 35% of the energy in developing countries is derived from biomass fuels, equivalent to 2700 Tg dm or 1200 Tg of C per year. Because of rapidly increasing populations in the developing world, this energy need is growing by several percent per year. An analysis of the situation in India (21), however, indicates a biofuel consumption of only 350 kg per capita per year in rural and 160 kg per capita per year in urban areas, adding up to a total consumption rate of 230 Tg per year for the Indian population of 760 million. About half of the biomass burned was firewood, the other half was mostly dung and crop residues.

It is clearly very difficult to extrapolate from this information. If the partitioning of biofuel between fuel wood and agricultural waste products derived for India were representative for the rest of the developing world, more than 1050 Tg dm of firewood and roughly an equal amount of agricultural waste products would be burned worldwide, that is, together at least about 950 Tg of C per year, 20% less than the 1200 Tg of C per year estimated by Scurlock and Hall (20). On the other hand, if the estimate of 230 Tg dm per year of biofuel burning for the Indian population is extrapolated to the total population in the developing world, then the amount is only about 600 Tg of C per year. Altogether we will assume a range of biofuel burning of 600 to 1200 Tg of C per year, with about equal contributions from firewood and agricultural waste products.

Burning of agricultural wastes in the fields, for example, sugar cane and rice straw, and stalks from grain crops, is another important type of biomass burning. The amount of residue produced equals about 1700 Tg dm per year in the developing world and a similar amount in the developed world (22). It is difficult to estimate what fraction of this waste is burned. Rice straw makes up 31% of the agricultural waste in the developing world, and, at least in Southeast Asia, burning of rice straw in the fields is the preferred method of waste disposal (23). Sugar cane residues account for about 11% of agricultural waste and are mostly disposed of by burning. We very tentatively guess that at least 25% of the agricultural waste, about 200 Tg of C per year, are burned in the fields. Summarizing from the uncertain information that is available, we estimate that yearly some 300 to 600 Tg of C of firewood and 500 to 800 Tg of C of agricultural wastes are burned in the developing world. In the industrial world

Table 26.1 [*orig. table 1*] Summary of the biomass exposed to fires, the total carbon released, the percentage of N to C in the fuel, and the total mass of N compounds released to the atmosphere by fires in the tropics.

Source or activity	Carbon exposed (Tg C/year)	Carbon released (Tg C/year)	N/C ratio (% by weight)	Nitrogen released (Tg N/year)
Shifting agriculture	1000–2000	500–1000	1	5–10
Permanent deforestation	500–1400	200–700	1	2–7
Savanna fires	400–2000	300–1600	0.6	2–10
Firewood	300–600	300–600	0.5	1.5–3
Agricultural wastes	500–800	500–800	1–2	5–16
Total	2700–6800	1800–4700		15–46

the corresponding figures are about one-tenth as large.

Prescribed burning and forest wildfires. It is interesting to compare the quantities of tropical biomass burned with those due to fires in temperate and boreal forests. Although individual wildfires may be large, because of fire-fighting efforts, the area burned per year is relatively small. Stocks (24) estimates that about 8 million ha of temperate and boreal forests are subject to wildfires each year.

Prescribed burning is commonly used for forest management. It serves mainly to reduce the accumulation of dry, combustible plant debris in order to prevent destructive wildfires. Because it is limited to North America and Australia and the area involved is only 2 million to 3 million ha per year (10), it has little impact on a global scale. Together some 150 to 300 Tg of C per year are burned by prescribed burning and wildfires, much less than through fires in the tropics.

Emissions to the Atmosphere

Table 26.1 summarizes the quantitative estimates of biomass burning in the tropics. We estimate that a total of 2700 to 6800 Tg of C are annually exposed to fires, of which 1800 to 4700 Tg of C are burned. The average chemical composition of dry plant biomass corresponds closely to the formula CH_2O. The nutrient element content varies with seasonal growth conditions; on a mass basis

it is relatively low: about 0.3 to 3.8% N, 0.1 to 0.9% S, 0.01 to 0.3% P, and 0.5 to 3.4% K (25). Consequently, although the emissions from biomass combustion are dominated by CO_2, many products of incomplete combustion that play important roles in atmospheric chemistry and climate are emitted as well, for example, CO, H_2, CH_4, other hydrocarbons, aldehydes, ketones, alcohols, and organic acids, and compounds containing the nutrient elements N and S, for example, NO, NH_3, HCN, and CH_3CN, SO_2, and COS. The smoke also contains particulate matter (aerosol) consisting of organic matter, black (soot) carbon, and inorganic materials, for example, K_2CO_3 and SiO_2. In table 26.2 we combine our estimates of global amounts of biomass burning with the emission ratios for various important trace species and derived global rates of pyrogenic emissions.

In spite of the large uncertainties, it is quite evident from table 26.2 that biomass burning results in globally important contributions to the atmospheric budget of several of the gases listed (26). Because much of the burning is concentrated in limited regions and occurs mainly during the dry season (July to September in the Southern Hemisphere and January to March in the Northern Hemisphere), it is not surprising that the emissions result in levels of atmospheric pollution that rival those in the industrialized regions of the developed nations. This comparison applies especially to a group of gases that are the main actors in atmospheric photochemistry: hydro-

Table 26.2 [*orig. table 2*] Estimates of emissions in teragrams of C, H₂, CH₃Cl, N, S, or aerosol mass per year (TPM, total particulate matter: POC, particulate organic carbon; EC, elemental carbon; K, potassium). Emission ratios for C and S compounds are in moles relative to CO_2: those for N compounds are expressed as the ratios of emission relative to the N content of the fuel; the emissions of TPM, POC, EC, and K are in grams per kilogram of fuel C. The emission ratios have been derived from information in (26, 29–31, 36 and 46). In calculating the ranges of total emissions, we used only half the ranges of total C emissions (2500 to 3900 Tg of C per year) and the emission ratios (for instance, 7.5 to 12.5% for CO). A similar procedure was followed for the N compounds.

Element or compound	Emission ratio	Emission from biomass burning	Total emissions from all sources
All C (from table 26.1)		1800–4700	
CO_2	≈90%	1600–4100	
CO	10 ± 5%	120–510	600–1300
CH_4	1 ± 0.6%	11–53	400–600
H_2	2.7 ± 0.8%	5–16	36
CH_3Cl	1.6 ± 1.5 x 10⁻⁴%	0.5–2	2
All N (from table 26.1)		15–46	
NO_x	12.1 ± 5.3%	2.1–5.5	25–60
RCN	3.4 ± 2.5%	0.5–1.7	>0.4
NH_3	3.8 ± 3.2%	0.5–2.0	20–60
N_2O	0.7 ± 0.3%	0.1–0.3	12–14
N_2	≤50%	≤11–19	100–170
SO_2	0.3 ± 0.15%	1.0–4.0	70–170
COS	0.01 ± 0.005%	0.04–0.20	0.6–1.5
TPM	30 ± 15 g/kg C	36–154	≈1500
POC	20 ± 10 g/kg C	24–102	≈180
EC	5.4 ± 2.7 g/kg C	6.4–28	20–30
K	0.4 ± 0.2 g/kg C	0.5–21	

carbons (for example CH_4), CO, and nitrogen oxides (NO_x). These gases have a strong influence on the chemistry of O_3, and OH, and thus on the oxidative state of the atmosphere. We will next discuss the most important emissions.

Carbon dioxide. Our estimates of the amount of biomass exposed to fire worldwide (2.7 to 6.8 Pg of C per year; table 26.1). and the resulting prompt CO_2 release to the atmosphere (1.8 to 4.7 Pg of C per year) are larger than earlier estimates (10, 13). They are 30 to 80% of the fossil fuel burning rate of 5.7 Pg of C per year (16).

We caution again that the prompt release of CO_2 to the atmosphere is not the same as the net CO_2 release from deforestation. The latter is estimated at 1.1 to 3.6 Pg of C per year (15). However, these figures need to be reduced somewhat, as a fraction of the burned biomass is converted into elemental C (charcoal), which is not subject to destruction by microbial activity (5, 10). There is hardly any information available on charcoal formation in fires. Fearnside and co-workers determined that in two forest clearings in Amazonia 3.6% of the biomass C exposed to the fires remained in the partially burned vegetation as elemental C (9). To this must be added the elemental C that is released in the smoke (27), so that the total elemental C yield may be about 4% of the C exposed to fire, or alternatively 14% of the C burned. From observation on a prescribed burn in a Florida pine forest (28), an elemental C yield of 5.4% (3.6 to 7.4%) of the C exposed or 9% (6 to 16%) of the C burned can be derived. From this limited data set we adopt charcoal yields of 5 and 10% of the C exposed or burned, respectively. When these elemental C yields are applied to the estimate of biomass burning given above, a range of elemental C production of 0.2 to 0.6 Pg of C per year can be deduced, which thus may reduce the range of net CO_2 emissions of 0.5 to 3.4 Pg of C per year. This correction is extremely tentative because of the paucity of measurements on elemental C production from forest fires and the total absence of data on yields from savanna fires or agricultural waste burning.

CO, CH_4 and other hydrocarbons, H_2, CH_3Cl. Figure 26.1 shows the sequence of the emission of CO_2 (maximum in the flaming stage), CO (maximum in the smoldering stage), and of various other gaseous products from experimental fires conducted in our laboratory. The fraction of CO emitted depends on the fire characteristics: hot flaming fires with good O_2 supply produce only a few percent, whereas smouldering fires may yield up to 20% CO (*29–31*). Therefore, CO may serve as a marker of the extent of smoldering combustion, so that emissions of gases from smoldering combustion can be better estimated on the basis of emission ratios relative to CO rather than relative to CO_2. Our estimates show very large emissions of CO, between 120 and 510 Tg of C per year. The estimated global source of CO is close to 1000 Tg of C per year (*32*); biomass burning is thus one of the main sources of atmospheric CO. Because about 70% of the OH

radicals in background air react with CO, biomass burning can substantially lower the oxidative efficiency of the atmosphere (which is mostly determined by the concentrations of OH), and thus can cause the concentrations of many trace gases to increase.

Methane contributes strongly to the atmospheric greenhouse effect, in this respect it has 20 to 30 times the efficiency per mole in the atmosphere of CO_2. It resides in

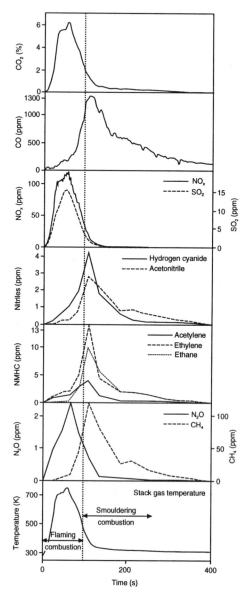

Figure 26.1 [*orig. figure 1*] Concentrations of CO_2, CO, and various other gases in the smoke from an experimental fire of Trachypogon grass from Venezuela as a function of time and the stack gas temperature. The dotted line separates the flaming phase from the smoldering phase. The flaming stage in this fire lasted for about 96 s. Concentrations are in percent by volume for Co_2, in volume mixing ratios (ppm) for the other species (1% = 10,000 ppm). Note that CO_2, NO_x, SO_2, and N_2O are mainly emitted in the flaming phase and the other gases in the smoldering phase: NMHC, nonmethane hydrocarbons.

the atmosphere long enough to enter the stratosphere. There, the oxidation of each molecule of CH_4 leads to the production of two molecules of H_2O; this process adds substantially to the stratospheric water vapor content. Because reaction with CH_4 also converts active Cl and ClO catalysts (which break down O_3) into inactive HCl, CH_4 plays a substantial role in stratosphere O_3 photochemistry. The pyrogenic emissions of CH_4, 11 to 53 Tg of C per year, may be about 10% of the global CH_4 source. On the basis of $^{13}CH_4$ isotope studies, the source of CH_4 from biomass burning was even estimated to be as large as 50 to 90 Tg per year (33), exceeding our estimated range of pyrogenic CH_4 emissions (table 26.2). Stevens et al. (34) indicate that the biomass burning source of CH_4 may have been increasing by 2.5 to 3 Tg per year during the past decade; this rate suggests that global biomass burning may have been increasing by as much as 5% per year, thus contributing strongly to the observed increases of CH_4 by almost 1% per year (32).

For H_2, biomass combustion may contribute 5 to 16 Tg per year. Its global sources and sinks have been estimated to be about 36 Tg per year, mostly due to fossil fuel burning (35). Biomass burning may thus make a significant contribution to the global source of H_2, which plays a role in stratospheric photochemistry.

Our estimated emission range of CH_3Cl, from 0.5 to 2 Tg of Cl per year by biomass burning, is large in comparison with its destruction rate of about 2 Tg per year by reaction with OH radicals (32). The photochemical breakdown of CH_3Cl is a significant source for active Cl in the stratosphere, so that it plays a role in O_3 depletion. CH_3Cl is often considered to be of natural origin, emanating from the ocean. This view needs to be reconsidered (26, 32).

Nitrogen gases. Emissions of NO from biomass burning may be in the range of 2 to 6 Tg of N per year, about 10 to 30% of the input from fossil fuel burning and comparable to the natural NO sources: lightning (2 to 10 Tg of N per year) and soil emissions (5 to 15 Tg of N per year) (32). Therefore, biomass burning contributes significantly to total NO emissions. For N_2O, on the other hand, we now estimate that pyrogenic emmissions are relatively small (0.1 to 0.3 Tg of N per year) (36), only a few percent of the global N_2O source of about 14 Tg of N per year (32). Earlier measurements of N_2O releases by biomass burning (29) have been biased by N_2O production in the collection devices.

The emissions of HCN and CH_3CN (together 0.5 to 1.7 Tg of N per year, and at a ratio of about 2:1) are significant, if not dominant, contributors to the atmospheric input rates of these compounds. The most important atmospheric sinks of HCN and CH_3CN, their reaction with OH, equals only 0.2 Tg of N per year for HCN and 0.02 to 0.2 Tg of N per year for CH_3CN (30). Consequently, other sinks must exist. Hamm and Warneck (37) proposed that these compounds may be taken up by the oceans. Another possibility is that they are consumed by vegetation, in which case they might serve as a minor source of fixed N. The atmospheric budget of NH_3 is not well known. Worldwide emissions are estimated to be in the range of 20 to 80 Tg of N per year (32, 38, 39), of which microbial release from animal excreta and soils make up the largest fraction. The pyrogenic source (0.5 to 2.0 Tg of N per year) is thus only a few percent of the global source.

An important outcome of the burning experiments at our laboratory (36) is that only about 25% of the plant N is emitted as NO, N_2O, NH_3, HCN, and CH_3CN. At most 20% of the N may be emitted as high molecular weight compounds, and about 10% of it is left in the ash. Recent measurements have shown that the remaining fraction, as much as 50% of fuel N, is emitted as N_2 (36). Thus biomass burning leads to pyrodentrification at a global rate of 10 to 20 Tg of N per year. This rate is 6 to 20% of the estimated terrestrial N fixation rate of 100 to 170 Tg of N per year (39) and therefore of potentially substantial significance. Most of the N loss occurs in the tropics, where it may lead to a substantial nutrient loss, especially from agricultural systems and savannas. Robertson and Rosswall (40) estimated that 8.3 Tg of N are emitted each year from West Africa into the atmosphere by burning, of which about 3.3 Tg could thus be N_2. This is almost three times their estimate of biological denitrification rate from the region.

Sulfur gases. In contrast to the N species, only relatively small amounts of SO_2 and aerosol sulfate are emitted. Biomass burning contributes only a few percent to the total atmospheric S budget, and only about 5% of the anthropogenic emissions. Still, because most of the natural emissions are from the oceans and most of the anthropogenic emissions are concentrated in the industrialized regions of the temperate latitudes, biomass burning could make a significant contribution to the S budget over remote continental regions, for example, the Amazon and Congo basins (41). Here, deposition may be enhanced five times because of tropical biomass burning.

Particles (smoke). Even though smoke is the most obvious sign of biomass burning, quantitative estimates on the amounts of particulate matter released are still highly uncertain. On the basis of an emission ratio of 30 g per kilogram of CO_2-C (27), we estimate that the emission of total particulate matter (TPM) is 36 to 154 Tg per year (table 26.2). This amount may appear to be minor compared to the total emission of particulate matter of the order of 1500 Tg per year. However, much of these emissions consists of large dust particles, which only briefly reside in the atmosphere. The smaller smoke particles are much more long-lived and more active in scattering solar radiation. The C content of smoke particles is about 66% (27), which is consistent with the notion that they consist mostly of partially oxygenated organic matter. This composition leads to an emission of about 30 to 100 Tg of particulate organic C, which would be about 15 to 50% of the organic C aerosol released globally (42). The content of black elemental C in smoke particles from biomass burning is highly variable. In smoldering fires it is as low as 4% (weight percent carbon in TPM), whereas in intensively flaming fires it can reach 40% (43). We use a value of 18%, based on our work in Amazonia (27). From this and the estimate for global TPM emissions of 36 to 154 Tg per year, we obtain a source estimated for black C aerosol of 6 to 30 Tg per year. This value already exceeds the earlier estimate of 3 to 22 Tg per year for the emission of black C from all sources (44).

Atmospheric Chemical Effects

Long-range transport of smoke plumes. The hot gases from fires rise in the atmosphere, entraining ambient air. Frequently, clouds form on the smoke plume and usually reevaporate without causing rain. When the plume loses buoyancy, it drifts horizontally with the prevailing winds, often in relatively thin layers, which can extend over a thousand kilometers or more. The height to which the smoke plumes can rise during the dry season is usually limited in the tropics by the trade wind inversion to about 3 km.

The further fate of the smoke-laden air masses depends on the large-scale circulation over the continent in which they originate. In tropical Africa, the plumes will usually travel in a westerly direction and toward the equator. As they approach the Intertropical Convergence Zone (ITCZ), vertical convection intensifies, destroys the layered structure, and causes the pyrogenic emissions to be distributed throughout the lower troposphere. Finally, in the ITCZ region, smoke and gases from biomass burning may reach the middle and upper troposphere, perhaps even the stratosphere. Air masses from the biomass burning regions in South America are usually moving toward the south and southeast, because of the effect of the Andes barrier on the large-scale circulation. Here again, they may become entrained in a convergence zone, the seasonal South Atlantic Convergence Zone (SACZ), which becomes established in austral spring, when biomass burning is abundant. Indeed, the data from the space-borne MAPS (Measurement of Air Pollution from Satellites) instrument typically show high concentrations of CO in the mid- and upper troposphere near the ITCZ and the SACZ (45).

Results of chemical measurements from satellites, space shuttle, aircraft, and research vessels indicate that pyrogenic emissions are transported around the globe. Soot C and other pyrogenic aerosol constituents have been measured during research cruises over the remote Atlantic and Pacific (46). High levels of O_3 and CO have also been observed from satellites over the tropical regions of Africa and South America, and large areas of the surrounding oceans (45, 47).

Photochemical smog chemistry. Biomass fires emit much the same gases as fossil fuel burning in industrial regions: CO, hydrocarbons, and NO_x, the starting ingredients for the formation of O_3 and photochemical smog. Once such a mixture is exposed to sunlight, hydrocarbons, including those naturally emitted by vegetation, are oxidized photochemically first to various peroxides, aldehydes, and so forth, then to CO. This CO is added to the amount directly emitted from the fires and is finally oxidized to CO_2 by reaction with OH. High concentrations of hydrocarbons and CO have been observed during the burning season in the tropics (27, 29, 48). In the presence of high levels of NO_x, as will be the case in the smoke plumes, the oxidation of CO and hydrocarbons is accompanied by the formation of O_3 (29, 48). The efficiency of O_3 formation, that is, the amount of O_3 formed per molecule of hydrocarbon oxidized, depends on the spread of the smoke plume and the chemical mix of hydrocarbons, NO_x, and O_3 present in the reaction mixture, and thus on the history of transport and mixing of the air mass (49). Increased concentrations of O_3 promote high concentrations of OH radicals and thus increase the overall photochemical activity of air masses affected by biomass burning. The effect may be enhanced further by the simultaneous emission of CH_2O [$\approx 2 \times 10^{-3}$ to 3×10^{-3} relative to CO_2 (31)], which is photolyzed in the tropical atmosphere within a few hours; this process leads in part to the production of HO_2 radicals via the formation of H and CHO. A similar effect may be caused by the photolysis of HONO, which may be emitted directly by the fires or formed by reactions on smoke particles (50).

High O_3 concentrations are produced in the plumes that extend over major parts of the tropical and subtropical continents during the dry season (27, 29, 48, 51). The highest concentrations, typically in the range from 50 to 100 ppb, are usually found in discrete layers at altitudes between 1 and 5 km, in accordance with the transport mechanisms of the burning plumes described above (figure 26.2). The concentrations at ground level are substantially lower and show a pronounced daily cycle with minima at night and maxima round midday. This cycle is controlled by the balance of O_3 sources and sinks: at night, O_3 consumption by deposition on the vegetation and reaction with hydrocarbons emitted by the vegetation and with NO emitted from soils reduce the concentration of O_3 near Earth's surface; during the day, these sinks are exceeded by photochemical O_3 formation and downward mixing of O_3-rich air. Also at ground level, O_3 volume mixing ratios in excess of 40 ppb are frequently measured during the dry season (48), similar to average values observed over the polluted industrialized regions of the eastern United States and Europe (52). Studies in temperate forest regions have linked such levels of O_3 pollution to damage to trees and vegetation, which has become widespread in Europe and North America (53). In view of the sharp increase of O_3 with altitude frequently observed in the tropics, the risk of vegetation damage by O_3 may be highest in mountainous regions, where O_3 concentrations above 70 ppb could be encountered. Ozone episodes with ground-level concentrations of 80 to 120 ppb must be expected to occur particularly during the dry season, when photochemically reactive air becomes trapped under the subsiding inversion layer (54). The regional ecological impact of high concentrations of phytotoxic O_3 on tropical vegetation and food production in the developing world is a matter of concern (55).

Perturbation of oxidant cycles in the troposphere. The global increase of tropospheric O_3, CO, and CH_4 concentrations, which is expected to continue in the future, is an indication of a fundamental change in the chemical behavior of the troposphere. Many gases, particularly hydrocarbons, are continuously emitted into the atmosphere from natural and anthropogenic sources. A buildup of these gases in the atmosphere is

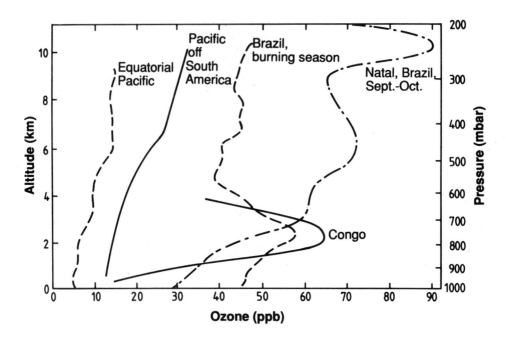

Figure 26.2 [*orig. figure 2*] Vertical profiles of O_3 in the tropical troposphere. The profile over the equatorial Pacific shows no influence from biomass burning, whereas the profile over the Pacific off South America suggests O_3 enhancement due to long-range transport from the tropical continents (*47*). The O_3 profiles over Brazil (*29*) and the Congo (*51*) show high O_3 concentrations at altitudes between 1 and 4 km due to photochemical production in biomass burning plumes. At higher altitudes, O_3 concentrations are also substantially enhanced, possibly also because of O_3 production by reactions in the effluents of biomass burning (*48*). [Adapted from (*47*) with permission of the author]

prevented by a self-cleaning mechanism, whereby these substances are slowly "combusted" photochemically to CO_2. The key molecule responsible for this oxidation process is OH. The reaction chains involved are such that OH is consumed when the concentration of NO_x is low. This is the normal condition of most of the unpolluted troposphere. On the basis of the observed increase of CO and CH_4, it has been suggested that global decreases in OH, the primary sink for CH_4 and CO, could lead through a feedback mechanism to a further increase in CO and CH_4, and that this situation could produce an unstable chemical condition (*56*). Injection of large amounts of NO_x from biomass burning and other anthropogenic activities may counteract this feedback, because hydrocarbon oxidation in the presence of elevated amounts of NO_x creates additional O_3 and OH. This counter-effect is, however, much more regionally limited because of the much shorter residence time of NO_x compared to that of CO and CH_4.

Model calculations (*57*) predict that a sixfold increase in regional OH concentrations in the boundary layer could occur as a result of deforestation and biomass burning in the tropics. There are two main reasons for this surprisingly large effect. One is that in the deforested regions NO_x is more easily ventilated to the atmosphere and a smaller portion reabsorbed in the less dense vegetation. The other is that removal of the trees eliminates the large emissions of isoprene (C_5H_8), which would normally react with and strongly deplete OH. We may, therefore, expect a strong enhancement of boundary layer O_3 and OH concentrations over tropical continental areas during the dry season, when vegetation is burned.

However, for the globe as a whole, it is likely that increasing CO and CH_4 emissions, with a large contribution from biomass burning, will lead to decreasing average OH concentrations and thereby to the buildup of the many gases that are removed from the atmosphere by reaction with OH. This change may be an extremely important development in global atmospheric chemistry.

Climatic and Ecological Effects

Climate change. With net global CO_2 emissions 1.1 to 3.6 Pg of C per year, the clearing of the tropical forests may be responsible for up to 20 to 60% of the greenhouse warming caused by the CO_2 emissions from fossil fuel burning. Biomass burning also releases another greenhouse gas, CH_4. In this case, biomass burning accounts for only about 10% of the global CH_4 sources, but probably for a greater fraction of the increase in global emissions (33). Estimates of the temporal trends of CH_4 source strengths from 1940 to 1980 (58) suggest that the pyrogenic contribution to the increase in CH_4 emissions over the time period is 10 to 40%.

The climatic effect of the smoke aerosols is beyond current understanding because of the complex nature of the interactions involved. Aerosols can influence climate directly by changing Earth's radiation balance. They reflect sunlight back into space. Smoke particles also contain black (elemental) C, which may strongly absorb sunlight and thus cause a heating of the atmosphere and less penetration of solar energy to Earth's surface. Such an effect has an influence on the heat balance of the lower troposphere; it results in less solar heating of the surface, warming of the atmosphere, and more stable meterological conditions. Robock (59) has shown that large daytime temperature drops can occur below smoke plumes from mid-latitude forest fires. Considering the great extent and expansion of tropical biomass burning, a widespread

effect of this kind may well have masked the expected greenhouse temperature rise during the dry season on the tropical continents.

Because the equatorial regions, particularly the Amazon Basin, the Congo Basin, and the area around Borneo, are extremely important in absorbing solar energy and in redistributing this heat through the atmosphere, any change affecting the operation of these "heat engines of the atmosphere" may be highly significant. A matter of considerable interest is the influence of submicrometer-sized pryogenic particles on the microphysical and optical properties of clouds and climate, an issue that has attracted considerable attention in connection with S emissions into the atmosphere (60, 61). Cloud droplets form on aerosol particles; these are called cloud condensation nuclei (CCN). The properties of the cloud depend on the number of available CCN: the more CCN, the more droplets that can form and the smaller the droplet size for a given amount of water. Clouds made up of smaller droplets reflect more sunlight back into space, and, because these clouds also are less likely to produce rain, cloud coverage also may increase. Because clouds are one of the most important controls on the heat balance of Earth, any large-scale modification of cloud properties is likely to have a strong impact on climate. Following proposals by Warner and Twomey (62) and by Radke et al (63), recent studies have shown that many of the submicrometer smoke particles produced by biomass fires can serve as CCN (64). Aged particles show enhanced CCN activity as their surfaces become coated with water-soluble materials, especially by uptake of HNO_3 and NH_3.

The pyrogenic production of smoke particles (40 to 150 Tg per year) is of the same magnitude as the input of sulfate particles from the anthropogenic emission of SO_2 from fossil fuel burning ($\approx$70 to 100 Tg of S per year worldwide) (65). On a molar basis and because of their larger surface to volume ratio, the emissions of pyrogenic particles may be even larger than those of sulfate aerosol. Consequently smoke particles, in addition to affecting the radiative properties of clouds and Earth's radiation balance, may also disturb the hydrological cycle in the tropics, with potential repercussions for regional and possibly global climate. Altogether, the climatic impact of biomass burning in the tropics may be impressively large. Recent general circulation model calculations by Penner et al. (66) indicate the possibility of a net change in Earth's radiation balance by -1.8 W/m^2, about equal, but opposite to the present greenhouse forcing.

The potential changes in precipitation efficiency add to the perturbation of the hydrological cycle in the tropics caused by deforestation and desertification. Tropical forests are extremely efficient in returning precipitation back to the atmosphere in the form of water vapor. There it can form clouds and rain again, and the cycle can repeat itself many times (67). A region such as the Amazon Basin can thus retain water (which ultimately comes from the ocean and will return there) for a long time and maintain a large standing stock of water. If the forest is replaced by grassland or, as is often the case, is converted into an essentially unvegetated surface by erosion and loss of topsoil, water runs off more quickly and returns through streams and rivers to the ocean, allowing less recycling. Beyond the unfavorable consequences that such large-scale changes in the availability of water will have on human activities, such a modification of the hydrological cycle may itself perturb tropical weather and maybe even climate (68). Furthermore, through the introduction of hotter and drier conditions, less evapotranspiration and precipitation, and a lengthening of dry season, there will be a much greater risk of and need for periodic fires (69). Together with changing biospheric emissions, the decrease in precipitation and cloudiness and changes in other meterological factors also have the potential to alter the chemistry of the tropical atmosphere in major ways.

Acid deposition. After acid rain had become a notorious environmental problem in Europe and North America, it came as a surprise to scientists to learn that it was also widespread in the tropics (table 26.3): acid rain has been reported from Venezuela (70), Brazil (71), Africa (72), and Australia (70, 73).

Table 26.3 [*orig. table 3*] Rainwater pH and acid deposition at some continental tropical sites and in the eastern United States.

Site	pH		Rainfall (cm)	Deposition (kg H+ ka/per year)	Reference
	Mean*	Range			
Venezuela					
San Eusebio	4.6	3.8–6.2	158	0.39	(70)
San Carlos	4.8	4.4–5.2			(70)
La Paragua	4.7	4.0–5.0			(70)
Brazil					
Manaus, dry season	4.6	3.8–5.0	240†	0.29	(71)
Manaus, wet season	5.2	4.3–6.1			
Australia					
Groote Eylandt	4.3				(73)
Katherine	4.8	4.2–5.4			(70)
Jabiru	4.3				(73)
Ivory Coast					
Ayame	4.6	4.0–6.5	179	0.41	(72)
Congo					
Boyele	4.4		185	0.74	(72)
Eastern United States	4.3	3.0–5.9	130	0.67	(76)

* Volume weighted. † Annual average.

In all instances, organic acids (especially formic and acetic acids) and nitric acid were shown to account for a large part of the acidity, in contrast to the situation in the industrialized temperate regions, where sulfuric acid and nitric acid predominate. It was originally thought that the organic acids were largely derived from natural, biogenic emissions, probably from plants (74). However, more recent evidence shows that acetic acid is produced directly by biomass burning and that both formic and acetic acid are chemically produced in the plumes (75). Nitric acid is formed photochemically from the NO_x emitted in the fires (57). Results from modeling the effects of biomass burning and a moderate amount of additional pollution, mostly connected with the activities related to logging and so forth, suggest that during the dry season pH values of ≈4.2 can be expected in the tropics as a consequence of the formation of nitric acid alone (57). For comparison, the mean pH in rain sampled throughout the eastern United States in 1980 was 4.3 (76) (table 26.3).

Acidic substances in the atmosphere can be deposited onto plants and soils either by rain and fog (wet deposition) or by the direct removal of aerosols and gases onto surfaces (dry deposition). In the humid tropics, wet deposition accounts for most of the deposition flux, whereas in the savanna regions, especially during the dry season, dry deposition dominates. Acid deposition has been linked to forest damage in Europe and the eastern United States (77). Acid deposition can act on an ecosystem through two major pathways: directly through the deposition of acidic aerosols and gases on leaves, or soil acidification. The danger of leaf injury is serious only at pH levels below 3.5, which is rarely encountered in the tropics (78), except perhaps in fog and dew. Nevertheless, the issue deserves some attention, as tropical forests may be inherently more sensitive to foliar damage than temperate forests because of the longer average leaf life of 1 to 2 years, which promotes cumulative damage.

Alterations of nutrient cycles and effects on soil degradation. Savanna and agricultural ecosystems are frequently deficient in N, P,

or S (79). When an area is burned, a substantial part of the N present in the ecosystem is volatilized. If this N were deposited again relatively nearby, this would cause no net gain or loss on a regional basis. If, however, as a result of fires, some 50% of the fuel N is emitted as N_2 (30, 36), a significant loss of nutrient N may result. In addition, long-range transfer of NO_x, NH_3, and nitriles to other ecosystems (savannas to tropical forests) depletes the fixed N reservoir of frequently burned ecosystems and thus provides a potential for long-term ecological effects. The budget of Robertson and Rosswall (40) for West African savannas implies that this loss of fixed N could deplete the fixed N load of these ecosystems in a few thousand years, a short time in comparison with the period during which humans have been present in these ecosystems. It may therefore be asked, to what extent enhanced N fixation can compensate for the loss of fixed N. This may indeed occur: laboratory research on tallgrass prairie soils has shown an enhancement of nonsymbiotic N_2 fixation after additions of available P in the ash from fires (80). Although the effects of biomass burning on the N cycle of the fire-affected ecosystems are most obvious, other nutrient elements, especially K, P, Mg, and S are also lost via smoke particles (81) in amounts that may have long-term, ecological consequences.

Regarding the C cycle, two issues appear to be of particular interest:

(1) The burial of pyrogenic charcoal residues that are not subject to microbial oxidation even over geological time scales, and that thus constitute a significant sink for atmospheric CO_2 and consequently a source for O_2 (10). Because the risk of fires increases with the growing atmospheric O_2 content (82), on geological time scales this may establish a positive feedback loop, which favors O_2 buildup in the atmosphere.

(2) The enhancement of biomass productivity of 30 to 60% or more after burning, despite the loss of nutrients, observed in some studies in humid savanna ecosystems (83). The results depend largely on burning practices, especially timing. Whether enhanced productivity may also increase the

pool of organic matter in the soil is unknown. Too little is yet known about the biogeochemical cycling of savanna ecosystems and changes thereof. It was discovered only recently that natural tropical grasslands may be much more productive than hitherto assumed, with productivity comparable to that of tropical forests (84). With a strong growth of the populations living in savanna regions, there will most likely be more frequent burning in these ecosystems. Significant effects on the global C cycle are possible, either through enhanced sequestering of C as charcoal and root-produced soil organic matter, if optimum burning practices are adopted (our speculation), or through loss of soil C by overly frequent fires and practices that lead to land degradation.

Ecosystems that are not burned, for example, remaining areas of intact rain forest, will receive an increased nutrient input. Studies of rainwater chemistry in the central Amazon Basin suggest that as much as 90% of the S and N deposited there is from external sources, and that long-range transport of emissions from biomass burning plays a major role (41). The long-term effects of such increasing inputs of nutrients to the rain forests, in combination with growing acid deposition and O_3 concentrations, are not known.

In addition to the immediate volatilization of N during the burns, enhanced microbial cycling of N in the soils occurs after fires. Emissions of NO and N_2O from soils at experimental sites in the temperate zone after burning were observed to be substantially higher than from soils at unburned sites (87). This effect persisted for at least 6 months after the fires. Following burning on a Venezuelan savanna site, enhancements in NO emissions by a factor of 10 were found for the 4 days during which the measurements were made (87). Other studies have also shown that the fluxes of NO_x from soils are enhanced after conversion from forests to grazing land (87), but in these studies the effect of burning was not isolated explicitly. However, in more extensive studies, Luizão et al. (88) did not observe enhanced N_2O fluxes on sites that were only burned and cleared but found a threefold enhanced

emission on 3 to 4-year-old pasture sites. According to these researchers, the enhanced emissions may be caused by increased input of oxidizable C from grass roots or rhizomes, or compaction of the soil surface by the cattle.

Although the above studies indicate that the emissions of trace gases increase after land disturbances, the total effect is complex and unclear. According to Robertson and Tiedje (89), denitrification (N_2 + N_2O production) is high in primary forests and at early successional sites but much lower at mid-successional sites. Studies by Sanhueza et al. (90) in a Venezuelan savanna and by Goreau and de Mello (91) on a cleared forest site during the dry season showed that forested areas may emit more N_2O than secondary grassland ecosystems derived by deforestation. Thus, although disturbed tropical forest ecosystems may initially emit more N_2O, this may only be temporary and in the long run less N_2O may be emitted. The issue is, therefore, unclear. Much long-term research is needed to elucidate the effects of biomass burning on nutrient cycling and especially on N volatilization in the tropics. This research is particularly important as there are indications that the main contributions to the total atmospheric N_2O source come from the tropics (92).

Conclusions

Our, still very uncertain, analysis of tropical biomass burning indicates emissions from about 2 to 5 Pg of C per year. In comparison, the present net release of CO_2 due to tropical land use change is estimated to range between 1.1 and 3.6 Pg of C per year (14). Significant amounts of C may be sequestered as charcoal, which may reduce the net release by 0.2 to 0.6 Pg of C per year. Because of the great importance of biomass burning and deforestation activities for climate, atmospheric chemistry, and ecology, it is clearly of the utmost importance to improve considerably our quantitative knowledge of these processes.

Biomass burning is a major source of many trace gases, especially the emissions of CO, CH_4 and other hydrocarbons, NO,

HCN, CH_3CN, and CH_3Cl are of the greatest importance. In the tropical regions during the dry season, these emissions lead to the regional production of O_3 and photochemical smog, as well as increased acid deposition with potential ecological consequences. On a global scale, however, the large and increasing emissions of CO and CH_4, the main species with which OH reacts in the background atmosphere, will probably lead to a decrease in the overall concentration of OH radicals and, therefore, to a decrease in the oxidation efficiency of the atmosphere. As the atmospheric lifetime of NO is only a few days, most of the atmosphere remains in an "NO-poor" state, where photochemical oxidation of CO and CH_4 leads to further consumption of OH. This in turn will enhance the atmospheric concentrations of CH_4 and CO, leading to a strong photochemical feedback.

Biomass burning is also an important source of smoke particles, a large amount (maybe all) of which act as CCN or can be converted to CCN by atmospheric deposition of hygroscopic substances. The amount of aerosols produced from biomass burning is comparable to that of anthropogenic sulfate aerosol. Through this process, the cloud microphysical and radiative processes in tropical rain and cloud systems can be affected with potential climatic and hydrological consequences.

An important recent finding is the substantial loss of fixed N that may be occurring because of biomass burning (pyrodenitrification). This loss appears to be of the greatest significance for savanna and agricultural ecosystems in the tropics and subtropics. The potential role of the savanna ecosystems in Earth's biogeochemical cycles deserves much more attention than it has been given so far. The savanna regions may play an important role in the global C cycle because of their large productivity, the potential interference of biomass burning with this productivity and the formation of long-lived elemental C. The geological importance of this C as a sink for atmospheric CO_2 (and source for O_2) should be explored.

References and Notes

1   W. Schüle, in *Fire in the Tropical Biota, Ecological Studies*, vol. 84 J. G. Goldammer Ed. (Springer-Verlag, Berlin, 1990), pp. 273–318; C. K. Brain and A. Sillen, *Nature* 336, 464 (1988).

2   J. G. Goldammer and B. Seibert, *Naturwissenschaften* 76, 51 (1989); E. F. Brünig, *Erdkunde* 23, 127 (1969); J. P. Malingreau, G. Stephens, L. Fellows, *Ambio* 14, 314 (1985).

3   R. L. Sandford, Jr., J. Saldarriaga, K. E. Clark, C. Uhl, R. Herrera, *Science* 227, 53 (1985).

4   R. Jones, *Annu. Rev. Anthropol.* 8, 445 (1979).

5   J. J. Griffin and E. D. Goldberg, *Science* 206, 563 (1979); *Environ. Sci. Technol.* 17, 244 (1983); J. R. Herring, thesis, University of California, San Diego (1977).

6   W. S. Wolbach, R. S. Lewis, E. Anders, *Science* 230, 167 (1985); W. S. Wolbach, I. Gilmour, E. Anders, C. J. Orth, R. R. Brooks, *Nature* 334, 665 (1988).

7   J. S. Levine, Ed., *Proceedings of the Chapman Conference on Global Biomass Burning* (MIT Press, Cambridge, MA, in press).

8   International Geosphere-Biosphere Programme, *A Study of Global Change of the International Council of Scientific Unions* (IGBP Secretariat, Royal Swedish Academy of Sciences, Stockholm, Sweden, 1990).

9   P. M. Fearnside, in (7).

10 W. Seiler and P. J. Crutzen, *Climatic Change* 2, 207 (1980).

11 N. Myers, *Deforestation Rates in Tropical Forests and Their Climatic Implications* (Friends of the Earth, London, 1989).

12 J. P. Lanly, *Tropical Forest Resources* (FAO, Forestry Pap. 30, Rome, 1982).

13 W. M. Hao, M. H. Liu, P. J. Crutzen, in *Fire in the Tropical Biota, Ecological Studies*, vol. 84, J. G. Goldammer, Ed. (Springer-Verlag, Berlin, 1990), pp. 440–462.

14 R. A. Houghton, *Climatic Change*, in press.

15 J. T. Houghton, G. J. Jenkins, J. J. Ephraums, Eds., *Climate Change, the Intergovernmental Panel on Climate Change Scientific Assessment* (Cambridge Univ. Press, New York, 1990).

16 S. Brown, A. J. R. Gillespie, A. Lugo, *For. Sci.* 35, 881 (1989).

17 B. Bolin, E. T. Degens, P. Duvigneaud, S. Kempe, in *The Global Carbon Cycle, SCOPE 13*, B. Bolin, E. T. Degens, S. Kempe, P. Ketner, Eds. (Wiley, Chichester, England, 1979), pp. 1–56.

18 J. C. Menaut, in (7).

19 *Yearbook of Forest Products 1987 (1976–1987)* (FAO, Rome, 1989).

20 J. M. O. Scurlock and D. O. Hall, *Biomass* 21, 75 (1990).

21 V. Joshi, in (7).

22 G. W. Barnard, in *Biomass and the Environment*, J. Pasztor and L. Kristoferson, Eds. (Westview, Boulder, CO, in press).

23 F. N. Ponnamperuma, *Organic Matter and Rice* (International Rice Research Institute, Los Banos, Philippines, 1984); A. Strehler and W. Stützle, in *Biomass*, D. O. Hall and R. P. Overend, Eds. (Wiley, Chichester, England, 1987), pp. 75–102.

24 B. J. Stocks, personal communication.

25 H. J. M. Bowen, *Environmental Chemistry of the Elements* (Academic Press, London, 1979).

26 P. J. Crutzen et al., *Nature* 282, 253 (1979).

27 M. O. Andreae et al., *J. Geophys. Res.* 93, 1509 (1988).

28 J. A. Comery, thesis, University of Washington (Department of Forestry), Seattle (1981).

29 P. J. Crutzen et al. *J. Atmos. Chem.* 2, 233 (1985).

30 J. M. Lobert, thesis, Johannes Gutenberg Universität, Mainz, Germany (1990); J. M. Lobert et al., in (7).

31 D. W. Griffith, W. G. Mankin, M. T. Coffey, D. E. Ward, A. Ribeau, in (7); W. R. Cofer III, J. S. Levine and E. L. Winstead, in (7); W. R. Cofer et al., *J. Geophys. Res.* 93, 1653 (1988).

32 World Meterological Organization, *Atmospheric Ozone* (Global Ozone Research and Monitoring Project Report 16, Geneva, Switzerland, 1985).

33 H. Craig, C. C. Chou, J. A. Welhan, C. M. Stevens, A. Engelkemeir, *Science* 242, 1535 (1988); P. D. Quay et al., *Global Biogeochem. Cycles*, in press.

34 C. M. Stevens, A. E. Engelkemeir, R. A. Rasmussen, in (7).

35 U. Schmidt, *Tellus* 26, 78, (1974).

36 J. M. Lobert, D. H. Scharffe, W. M. Hao, P. J. Crutzen, *Nature* 346, 552 (1990); Th. Kuhlbusch, thesis, Johannes Gutenberg Universität, Mainz, Germany (1990).

37 S. Hamm and P. Warneck, *J. Geophys. Res.*, in press.

38 I. E. Galbally, in *The Biogeochemical Cycling of Sulfur and Nitrogen in the Remote Atmosphere*, J. N. Galloway, R. J. Charlson, M. O. Andreae, H. Rodhe, Eds. (Reidel, Hingham, MA, 1985), pp. 27–53.

39 R. Söderlund and B. H. Svensson, in *Nitrogen, Phosphorus, and Sulphur*, B. H. Svensson and R. Söderlund, Eds. [*Ecol. Bull. Stockholm* 22, 23 (1976)].

40 G. P. Robertson and T. Rosswall, *Ecol. Monogr.* 56, 43 (1986).

41 M. O. Andreae and T. W. Andreae, *J. Geophys. Res.* 93, 1487 (1988); M. O. Andreae, R. W. Talbot, H. Berresheim, K. M. Beecher, *ibid* 95, 16987 (1990); M. O. Andreae et al., *ibid.*, p. 16813; M. O. Andreae et al., *ibid.*, in press.

42 R. A. Duce, *Pure Appl. Geophys.* 116, 244 (1978).

43 E. M. Patterson and C. K. McMahon, *Atmos. Environ.* 18, 2541 (1984).

44 R. P. Turco, O. B. Toon, R. C. Whitten, J. B. Pollack, P. Hamill, in *Precipitation Scavenging, Dry Deposition, and Resuspension*, H. R. Pruppacher, R. G. Semonin, W. G. N. Slinn, Eds. (Elsevier, Amsterdam, 1983), pp. 1337–1351.

45 H. G. Reichle, Jr., et al., *J. Geophys. Res.* 91, 10865 (1986); H. Reichle et al., *ibid.* 95, 9845 (1990).

46 M. O. Andreae, *Science* 220, 1148 (1983) — —, T. W. Andreae, R. J. Ferek, H. Raemdonck, *Sci. Total Environ.* 36, 73 (1984).

47 J. Fishman and J. C. Larsen, *J Geophys. Res.* 92, 6627 (1987).

48 V. W. J. H. Kirchhoff and E. V. A. Marinho, *Atmos. Environ.* 23, 461 (1989); V. W. J. H. Kirchhoff, E. V. Browell, G. L. Browell, *J. Geophys. Res.* 93, 15850 (1988); V. W. J. H. Kirchhoff, A. W. Setzer, M. C. Pereira, *Geophys. Res. Lett.* 16, 469 (1989); A. C. Delany, P. Haagensen, S. Walters, A. F. Wartburg, P. J. Crutzen, *J. Geophys. Res.* 90, 2425, (1985); V. W. J. H. Kirchhoff et al., *Geophys. Res. Lett.* 8, 1171 (1983); J. A. Logan and V. W. J. H. Kirchhoff, *J. Geophys. Res.* 91, 7875 (1986); V. W. J. H. Kirchhoff and R. A. Rasmussen, *ibid.* 95, 7521 (199).

49 R. B. Chatfield and A. C. Delany, *J. Geophys. Res.* 95, 18473 (1990).

50 A. Rondon and E. Sanhueza, *Tellus Ser. B* 41, 474 (1989).

51 V. W. J. H. Kirchhoff and C. A. Nobre, *Rev. Geofis.* 24, 95 (1986); B. Cros, R. Delmas, B. Clairac, J. Loemba-Ndembi, J. Fontan, *J. Geophys. Res.* 92, 9772 (1987); M. O. Andreae et al., *ibid.*, in press.

52 J. A. Logan, *J. Geophys. Res.* 90, 10463 (1985).

53 B. Prinz, in *Tropospheric Ozone*, I. S. A. Isaksen, Ed. (Reidel, Dordrecht, 1988), pp. 687–689.

54 B. Cros, R. Delmas, D. Nganga, B. Clairac, *J.*

Geophys. Res. 93, 8355 (1988).

55 P. M. Vitousek and P. A. Matson, *Biotropica*, in press.

56 P. J. Crutzen, in *The Geophysiology of Amazonia*, R. E. Dickinson, Ed. (Wiley, New York, 1987), pp. 107–130; N. D. Sze, *Science* 195, 673 (1977).

57 M. Keller, D. J. Jacob, S. C. Wofsy, R. C. Harriss, *Climatic Change*, in press.

58 H. J. Bolle, W. Seiler, B. Bolin, in *The Greenhouse Effect, Climatic Change, and Ecosystems, SCOPE 29*, B. Bolin, B. R. Döös, J. Jäger, R. A. Warrick, Eds. (Wiley, Chichester, England, 1986), pp. 157–203.

59 A. Robock, *Science* 242, 911 (1988).

60 R. C. Eagan, P. V. Hobbs, L. F. Radke, *J. Appl. Meterol.* 13, 553 (1974); S. Twomey, *J. Atmos. Sci.* 34, 1149 (1977); ——, M. Piepgrass, T. L. Wolfe, *Tellus Ser. B* 36, 356 (1984).

61 R. J. Charlson, J. E. Lovelock, M. O. Andreae, S. G. Warren, *Nature* 340, 437 (1989); T. M. L. Wigley, *ibid.*, 339, 365 (1989).

62 J. Warner and S. Twomey, *J. Atmos. Sci.* 24, 704 (1967).

63 L. F. Radke, J. L. Stith, D. A. Hegg, P. V. Hobbs, *J. Air Pollut. Control Assoc.* 28, 30 (1978).

64 F. Desalmand, R. Serpolay, J. Podzimek, *Atmos. Environ.* 19, 1535 (1985); F. Desalmand, J. Podzimek, R. Serpolay, *J. Aerosol Sci.* 16, 19 (1985); C. F. Rogers, B. Zielinska, R. Tanner, J. Hudson, J. Watson, in (7); L. F. Radke, D. A. Hegg, J. H. Lyons, P. V. Hobbs, R. E. Weiss, in (7).

65 C. F. Cullis and M. M. Hirschler, *Atmos. Environ.* 14, 1263 (1980); D. Möller, *ibid.*, 18, 19 (1984).

66 J. E. Penner, S. J. Ghan, J. J. Walton, in (7).

67 E. Salati and P. B. Vose, *Science* 225, 129 (1984).

68 J. Shukla, C. Nobre, P. Sellers, *ibid.* 247, 1322 (1990); P. Sellers, Y. Mintz, Y. C. Sud, A. Dalcher, *J. Atmos. Sci.* 43, 505 (1986); R. E. Dickinson and A. Henderson-Sellers, *Q. J. R. Meterol. Soc.* 114, 439 (1988); J. Lean and D. A. Warrilow, *Nature* 342, 411 (1989).

69 M. Fosberg, J. G. Goldammer, C. Price, D. Rind, in *Fire in the Tropical Biota, Ecological Studies*, vol. 84, J. G. Goldammer, Ed. (Springer-Verlag, Berlin, 1990), pp. 463–486.

70 U. Steinhardt and H. W. Fassbender, *Turrialba* 29, 175 (1979); J. N. Galloway, G. E. Likens, W. C. Keene, J. M. Miller, *J. Geophys. Res.* 87, 8771 (1982); E. Sanhueza, W. Elbert, A. Rondon, M. Corina Arias,

M. Hermoso, *Tellus Ser. B.* 41, 170 (1989).

71 M. O. Andreae, R. W. Talbot, T. W. Andreae, R. C. Harriss, *J. Geophys. Res.* 93, 1616 (1988).

72 J. P. Lacaux, J. Servant, J. G. R. Baudet, *Atmos. Environ.* 21, 2643 (1987); J. P. Lacaux et al., *Eos* 69, 1069 (1988).

73 P. J. Langkamp and M. J. Dalling, *Aust. J. Bot.* 31, 141 (1983); G. P. Ayers and R. W. Gillett, in *Acidification in Tropical Countries, SCOPE 36*, H. Rodhe and R. Herrera, Eds. (Wiley, Chichester, England, 1988), pp. 347–400.

74 R. W. Talbot, K. M. Beecher, R. C. Harris, W. R. Cofer, *J. Geophys. Res.* 93, 1638 (1988); W. C. Keene and J. N. Galloway, *ibid* 91, 14466 (1986).

75 G. Helas, H. Bingemer, M. O. Andreae, in preparation.

76 L. A. Barrie and J. M. Hales, *Tellus Ser. B* 36, 333 (1984).

77 F. H. Bormann, *BioScience* 35, 434 (1985).

78 W. H. McDowell, in *Acidification in Tropical Countries*, H. Rodhe and R. Herrera, Eds. (Wiley, Chichester, England 1988), pp. 117–139.

79 P. A. Sanchez, *Properties and Management of Soils in the Tropics* (Wiley, New York, 1976).

80 K. A. Eisele et al., *Oecologia* 79, 471 (1990).

81 E. Sanhueza and A. Rondon, *J. Atmos. Chem.* 7, 369 (1988).

82 A. J. Watson, thesis, Reading University, Reading, England (1978).

83 J. J. San Jose and E. Medina, in *Tropical Ecological Systems, Ecological Studies*, vol. 11, F. B. Golley and E. Medina, Eds. (Springer-Verlag, Berlin, 1975). pp. 251–264; D. Gillon, in *Tropical Savannas, Ecosystems of the World* 13, F. Bourliere, Ed. (Elsevier, Amsterdam, 1983), pp. 617–641.

84 D. O. Hall and J. M. O. Scurlock, *Ann. Bot. (London)*, in press.

85 I. C. Anderson, J. S. Levine, M. A. Poth, P. J. Riggan, *J. Geophys. Res.* 93 3893 (1988); I. C. Anderson and M. A. Poth, *Global Biogeochem. Cycles* 3, 121 (1989); J. S. Levine et al., *ibid.* 2, 445 (1988).

86 C. Johansson, H. Rodhe, E. Sanhueza, *J. Geophys. Res.* 93, 7180 (1988).

87 P. A. Matson, P. M. Vitousek, J. J. Ewel, M. J. Mazzarino, G. P. Robertson, *Ecology* 68, 491 (1987).

88 F. Luizão, P. Matson, G. Livingston, R. Luizão, P. Vitousek, *Global Biogeochem Cycles* 3, 281 (1989).

89 G. P. Robertson and J. M. Tiedje, *Nature* 336, 756 (1988).

90   E. Sanhueza,   W. M. Hao,   D. Scharffe,
      L. Donoso, P. J. Crutzen, *J. Geophys. Res.*, in
      press.
91   T. J. Goreau and W. Z. de Mello, *Ambio* 17,
      275 (1988).
92   R. G. Prinn et al., *J. Geophys. Res.*, in press.
93   We thank S. Brown, E. F. Bruenig, J. Clark,
      P. Fearnside, I. Y. Fung, D. W. Griffith,
      J. Goldammer, C. S. Hall, D. O. Hall, A. L.
      Hammond, W. M. Hao, R. Herrera, R. A.
      Houghton, M. Keller, V. W. H Kirchhoff,
      J. P. Lanly, J. Levine, J. Lobert, J. M. Logan,
      P. Matson, J. C. Menaut, N. Myers, Ph.
      Roberston,   C. F. Rogers,   B. J. Stocks,
      E. Sanhueza, and D. Schimel for comments.

27

ACID RAIN: A SERIOUS REGIONAL ENVIRONMENTAL PROBLEM*

G. E. Likens and F. H. Bormann

As part of the Hubbard Brook Ecosystem Study, we have monitored the chemistry of precipitation in north-central New Hampshire for about 11 years (1) and have found surprising acidity (2). Normally water in the atmosphere in equilibrium with prevailing CO_2 pressures will produce a pH of about 5.7 (3), but much stronger acids have recently been observed in rain and snow in the northeastern United States, with pH values as low as 2.1. The presence of these acids is presumably related to air pollution (2).

Current measurements and the few scattered observations during the past 11 years show that precipitation falling in northeastern United States is significantly more acidic than elsewhere in the United States (2, 4). For example, the annual mean pH, based upon samples collected weekly during 1970–71 and weighted proportionally to the amount of water and pH during each period of precipitation, was 4.03 at the Hubbard Brook Experimental Forest, New Hampshire; 3.98 at Ithaca, New York; 3.91 at Aurora, New York; and 4.02 at Geneva, New York. Measurements on individual rainstorms frequently showed values between pH 3 and 4 at all of these locations. Data from the National Center for Atmospheric Research included precipitation pH values as low as 2.1 in the northeastern United States during November 1964 (4). Summer rains are generally more acidic than winter precipitation (5).

The major cation is H^+, which accounts for 44 and 69 percent of the cations (milliequivalent basis) in the Ithaca and Hubbard Brook precipitation, respectively. Of the anions in Ithaca and Hubbard Brook precipitation, respectively, SO_4^{2-} represents 59 and 62 percent, NO_3^- contributes 21 and 23 percent, and Cl^- constitutes 20 and 14 percent.

Precipitation pH values from the northeastern United States are similar to those recorded over southern Sweden. More than 70 percent of the sulfur in the air in Sweden is thought to be anthropogenic, of which 77 percent has its source outside Sweden (6). The mean residence time for sulfur in the atmosphere is estimated to be 2 to 4 days (7). As a result, SO_2 may be transported more than 1000 km in this region before being deposited. Distant industrialized regions, such as England and the Ruhr Valley, are thought to be major contributors to

* Originally published in *Science*, 1974, vol. 184, pp. 1176–9.

Sweden's acid rain (8). Moreover, Odén's data show a striking increase in precipitation acidity in northeastern Europe over the past two decades, with the H⁺ concentration of rain in some parts of Scandinavia increasing more than 200-fold since 1956 (9).

We know of no long-term record of precipitation acidity for the United States. Recent data, since 1963, do not reveal any marked trends (figure 27.1), although extensive data on precipitation chemistry from Geneva, New York, provide indirect evidence of higher pH's in earlier years (10). Prior to 1940 pH was not measured, but

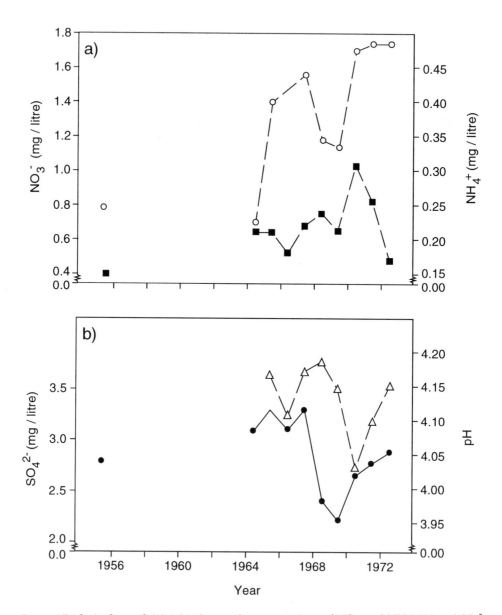

Figure 27.1 [*orig. figure 1*] Weighted annual concentrations of NO_3^- and NH_1^+ (A) and SO_1^{2-} and the pH (B) of precipitation at the Hubbard Brook Experimental Forest in New Hampshire. Values for 1955–1956 were estimated from Junge and Werby (35) and Junge (36).

methyl orange was used to indicate acidic or alkaline conditions in some of the samples collected in New York State. Since methyl orange changes color at pH 4.6, little exact information was obtained; but, significantly, no acidic reactions were recorded. In addition, Collison and Mensching (11) found relatively large amounts of HCO_3^- in precipitation at Geneva prior to 1930 (12). Because HCO_3^- does not coexist with stronger acids such as those found in today's rain, these data are consistent with the hypothesis that acid rain is a recent phenomenon in the northeastern United States.

Data on the sulfur content of rain and snow in New York State indicate that present-day precipitation contains about 70 percent less sulfur than that prior to 1950 (figure 27.2) (13). This drop in sulfur concentration at both Ithaca and Geneva is difficult

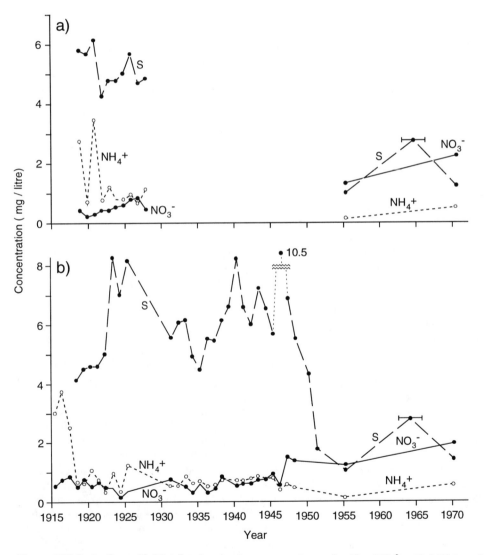

Figure 27.2 [orig. figure 2] Weighted annual concentrations of sulfur (SO_4^{2-} – S), NH_4^+, and NO_3^- in precipitation at Geneva, New York (A), and Ithaca, New York (B).

to reconcile with the proposed recent increase in acidity. High concentrations of sulfur in precipitation could reflect the presence of H_2SO_4. Since the pH of the precipitation apparently was not lowered prior to 1950, much of the sulfur must have existed as un-ionized particulate sulfate or as ionized forms neutralized by equivalent amounts of bases (14). Gorham (15) observed large amounts of calcium in smoke solids collected from chimneys in London and argued that such basic substances were more easily and quickly precipitated than acidic ones. His results and those of Overrein (16) indicated that various basic particles in smoke would neutralize acids and tend to fall out relatively near the pollution source. Overrein also found the lowest pH's some distance downwind from the combustion source, an indication that a surplus of acid-forming substances escaped neutralization and gave rise to acid rain (16).

It appears that much of the sulfur in early precipitation (before 1950) at Ithaca and Geneva resulted from the local combustion of coal (5). Prior to 1932, coal and wood were used almost exclusively as sources of fuel in Ithaca, although some manufactured coal gas was used. Natural gas was introduced into the Ithaca area late in 1932 but did not really replace coal as the prime fuel until 1950–1955 (17). The difference in the sulfur content between coal (high) and natural gas (low) is enormous, and thus this shift in use patterns is highly correlated with the decline in the sulfur content of precipitation in 1950–1955 (figure 27.2). The effect of coal burning on local precipitation chemistry also is indicated by a shift in seasonal characteristics. Before 1950 the wintertime concentrations of SO_4^{2-} in precipitation always exceeded the summer values (11, 12), whereas now maximum SO_4^{2-} concentrations occur during the summer (5).

Significantly, appreciable quantities of NO_x are normally produced in the combustion of natural gas. The relatively large increase in the NO_3^- content of precipitation since about 1945 (figures 27.1 and 27.2) may be correlated with the increased combustion of natural gas and motor fuels (5).

We suggest that the change in acidity occurred soon after 1950 because of a shift in the predominant form of sulfur in the atmosphere, and even though there was a decrease in the total sulfur content of the precipitation. When the major source of anthropogenic sulfur for the atmosphere was the combustion of coal, much of the sulfur was precipitated to the land near the combustion source in particulate form and as neutralized salts. Today, with the increasing combustion of fossil fuels, with mounting numbers of taller smokestacks fitted with precipitators to remove the larger particles (18), and with increasing combustion of fossil fuels other than coal, greatly increased quantities of SO_2 apparently are being introduced into the atmosphere, at least on a regional basis (19). Injected at heights of 60 to 360 m, the SO_2 may be dispersed over wide areas; and, in the absence of equivalent amounts of alkaline substances in the atmosphere, appreciable quantities of SO_2 are converted to acid. The consistently low pH of precipitation at rural sites in New England (2, 4), hundreds of kilometers from urban industrial centers, attests to the long-range dispersion of SO_2 and its secondary pollutant, H_2SO_4. It would appear, then, that these trends in fuel consumption, fuel preference, and pollution control technology (increasing the height of smokestacks and installing particle precipitators) have transformed local "soot problems" into a regional "acid rain problem."

The ecological effects of acid rain are as yet largely unknown, but potentially they are manifold and very complex. Effects may include changes in the leaching rates of nutrients from plant foliage, changes in the leaching rates of soil nutrients, effects on predator–prey relationships, acidification of lakes and rivers, effects on the metabolism of organisms, and the corrosion of structures.

It is believed that a reduction in forest growth in northern New England and in Scandinavia over the last two decades may be correlated with the concurrent acidification of precipitation (8, 20). Study of the direct effects of acid rain on plants is just beginning. Gordon (21) has observed that the emerging needles of several species of western pines, when inoculated with an atomized solution of H_2SO_4 at a pH <4.0, grew to only half the length of control needles. Wood and Bormann (22) noted spot necrosis and irregular development of leaf tissue on deciduous tree seedlings (yellow birch) 2 to 5 weeks old subjected to misting with aqueous H_2SO_4 at pH 3.0. The 2-week-old seedlings were more susceptible than the 5-week-old plants. In studies of tomatoes on the island of Hawaii, rain of low pH was found to decrease pollen germination and pollen tube growth and generally lower the quality and production of these plants (23). Toxic effects on plants may be associated with the increased proportions of HSO_3^- ions and undissociated H_2SO_3 at low pH (24).

Of equal concern are potentially numerous indirect effects (direct nutrient leaching or the erosion of the cuticle, providing ready access to pathogens, herbivores, or nutrient leaching) whereby the general vigor of plants may be reduced by acid precipitation and synergistically contribute with other impinging stresses to an adverse effect. In deciduous forests of the northeastern United States, the precipitation is generally most acidic during the growing season (25).

Increased input of acid to soil may lead to increased leaching of calcium and other nutrient elements [for example, see (8, 16)]. Although such losses would be small for some ecosystems and unlikely to result in any significant short-term damage to arable land, they do represent other added stresses to the ecosystem. During the growing season in forested areas, most of the acid rain impinges directly on the foliage where ion exchange occurs. Thus the rain that reaches the forest floor (soil) is far less acid than the incident precipitation (26).

The effect of acid precipitation on aquatic ecosystems may be large, particularly if the input ratio of direct precipitation to land drainage is high. A significant proportion of the decreasing trend in pH, observed during 1965–1970 in a large number of lakes and rivers in Scandinavia, has been attributed to acid rain (27). In Canada, numerous lakes west of the Sudbury smelters have increased in acidity by more than 100-fold

during the last decade (28); by 1970 some 33 of these lakes had pH values less than 4.5 and were termed "critically acid." Schofield (29) reports that the water chemistry of a relatively large clear-water, oligotrophic lake with a small drainage area in the Adirondack Mountains of New York has changed appreciably since 1938. In December 1938 the total alkalinity, expressed as $CaCO_3$, ranged from 12.5 to 20.0 mg/liter and the pH was 6.6 to 7.2, whereas during 1959–1960 the alkalinity ranged from 0 to 3.0 mg/liter and the pH was 3.9 to 5.8.

Serious fish mortality, particularly early age classes in salmonids, has been reported in Scandinavian rivers and lakes (8, 30) and Canadian lakes (28) and has been attributed directly to increased acidity from precipitation. Similar acidification of aquatic ecosystems must be occurring in other parts of the world, but it is unrecognized or is often confounded by other sources of pollution [for example, see (31)].

Damage to buildings, structures, and art forms by acid precipitation may be enormous (8, 32). The economic consequences of this damage are only now becoming apparent.

An estimate of total anthropogenic emissions for the United States indicates that this country is now annually injecting about 32 × 10^6 metric tons of SO_x into the atmosphere (33). Additional projections suggest that this amount will increase two- to fivefold by the year 2000 (33, 34). Development of nuclear-based power, of methods for extracting "clean" energy from coal and oil, and the passage of more stringent regulations governing SO_x emissions may slow this trend. In general, though, it seems safe to assume that, as long as energy demands mount and fossil fuels remain abundant and economically desirable, SO_x emissions will increase. Surely this will be true if air pollution standards are relaxed to meet energy needs during the "energy crisis." The extent to which the SO_2 concentration may ultimately rise is a matter for serious concern, for increased concentrations may promote still more serious and widespread acid rain with all that this outcome implies for the structure and function of both natural and man-manipulated ecosystems.

References and Notes

1 G. E. Likens, F. H. Bormann, R. S. Pierce, D. W. Fisher, in *Productivity of Forest Ecosystems*, P. Duvigneaud, Ed. (Proceedings of the Brussels Symposium, Unesco Publications, Paris, 1969): this apparently is one of the longest records of comprehensive precipitation chemistry for the United States.

2 G. E. Likens, F. H. Bormann, N. M. Johnson, *Environment* 14 (No. 2), 33 (1972).

3 E. Barrett and G. Brodin, *Tellus* 7, 251 (1955).

4 D. W. Fisher, A. W. Gambell, G. E. Likens, F. H. Bormann, *Water Resour. Res.* 4 (No. 5), 1115 (1968); F. J. Pearson, Jr., and D. W. Fisher, *U.S. Geol. Surv. Water Supply Pap. 1535P* (1971); A. L. Lazrus, B. W. Gandrud, J. P. Lodge, *Atmos. Environ.*, in press; A. L. Lazrus, personal communication.

5 G. E. Likens, *Cornell Univ. Water Resour. Mar. Sci. Cent. Tech. Rep. 50* (1972).

6 H. Rodhe, *Tellus* 24, 128 (1972).

7 E. Robinson and R. C. Robbins, final report for Stanford Research Institute Project PR-6755 to the American Petroleum Institute, New York, 1968 (mimeograph).

8 B. Bolin, Ed., "Report of the Swedish Preparatory Committee for the U. N. Conference on Human Environment" (Norstedt & Söner, Stockholm, 1971).

9 S. Odén, *Statens Naturvetensk. Forsk. Arsbok No. 1* (1968).

10 The first published data on the pH of precipitation in New York State that we know of is for 1963 from the Erie-Niagara Basin [R. J. Archer, A. M. LaSala, Jr., J. C. Kammerer, "Basin Planning Report EMB-4" (New York State Conservation Department, Water Resources Commission, Albany, 1968)]. The data in this report are the earliest data obtained in New York by the U.S. Geological Survey (R. J. Archer, personal communication). The pH values, ranging between pH 5.4 and 7.1 (median value, about 6.4), reported for rain at seven sites in 1963 are consistently higher than values published for New York State since. The pH of snow samples collected during 1964 and 1965 ranged between 3.0 and 6.8 (median value, 4.3).

11 R. C. Collison and J. E. Mensching, *N. Y. Exp. Stn. Geneva Tech. Bull. No. 193* (1932): "Bicarbonates have been found in considerable abundance in all samples of water collected at Geneva and vary from 12 to 92 pounds as HCO_3^- per acre annually" (p. 4); "at this station (Table 5) bicarbonates (HCO_3^- were present in all samples of rainwater analyzed both yearly and monthly . . . More bicarbonate was present during the summer months than in the winter months. As already noted all samples of rainwater collected contained this constituent" (p. 16). In the absence of data on pH, the large amounts of HCO_3^- in early rainwater samples at Geneva, New York (1919–1929) indicate much higher pH's than today. Indicator solutions (methyl orange and phenophthalein) were used to measure concentrations of HCO_3^- and CO_3^{2-} in the precipitation (p. 10), hence the above statements would clearly indicate that all samples would have had a pH>4.6. Only trace amounts of HCO_3^- are found in present-day precipitation samples at Geneva. The presence of HCO_3^-, therefore, would indicate that pH values in 1919–1929 were probably 5.7 or higher.

12 There is no record of pH measurements, HCO_3^- determinations, or methyl orange reactions for the precipitation samples collected at Ithaca, New York, during 1915–1952 [B. D. Wilson, *Soil Sci.* 11, 101 (1921); *J. Am. Soc. Agron.* 13, 226 (1921); *ibid.* 15, 453 (1923); *ibid.* 18, 1108 (1926); E. W. Leland, *Agron. J.* 44 (No. 4), 172 (1952)].

13 Data for Geneva and Ithaca are derived from various published sources; procedures for collection and analysis of samples have been evaluated in relation to observed trends in precipitation chemistry (5). Wilson (12) believed that, prior to 1923, the area adjacent to the precipitation collector at Ithaca was relatively free of smoke. However, in 1923 a large heating plant (Cornell University) was constructed within 1.6 km from the collection area and the sulfur content of precipitation (annual averages) was thereafter appreciably increased, at times more than doubled. At Geneva a branch railroad passed within 183 to 214 m of the precipitation collector and a coal-burning flour mill was also nearby (11). Various events, including the economic depression of the 1930's, the World War II industrial effort in the early 1940's, and the coal miners' strike in 1945, have undoubtedly affected the amounts of nitrogen and sulfur in precipitation.

14 An alkaline response to methyl orange indicator was observed in every precipitation sample collected in Tennessee during 1919–1921 in spite of high concentrations of sulfur in the samples [W. H. MacIntire and J. B. Young, *Soil Sci.* 15, 205 (1923)]. This was attributed to excess amounts of "lime" and other neutralizing bases in soot and fine dust found in precipitation samples. Moreover, the total amount of sulfur precipitated was about five times greater than the amount in water-soluble forms and was directly related to the amount of solid soot falling directly or with precipitation.

15 E. Gorham, *Geochim. Cosmochim. Acta* 7, 231 (1955).

16 L. N. Overrein, *Ambio* 1, 145 (1972).

17 By 1934 natural gas had completely replaced manufactured gas as a fuel source and was providing some 20 percent of the total fuel used by residential, commercial, and industrial users in the Ithaca area. By 1940, 30 percent of the fuel used in this area was natural gas; by 1950–1955 the proportion had increased to 70 percent and by 1965 to 85 percent (5).

18 J. T. Middleton, Ed., *Natl. Air Pollut. Control Adm. Publ. No. AP-50* (1969). A summary of studies showed that a major fraction (generally 80 percent or more) of urban atmospheric SO_4^{2-} was associated with particles smaller than $2\mu m$ in diameter.

19 In Greenland, obviously remote from major areas of fossil fuel combustion, the concentration of sulfur in glacial snows has increased since about 1960 [H. V. Weiss, M. Koide, E. D. Goldberg, *Science* 172, 261 (1971)]. We interpret this change as reflecting a general increase in the amount and distribution of SO_2 in the atmosphere over wide areas.

20 R. H. Whittaker, F. H. Bormann, G. E. Likens, T. G. Siccama, *Ecol. Monogr.*, in press.

21 C. C. Gordon (University of Montana, Missoula), Interim Report to the Environmental Protection Agency, 1972 (mimeograph).

22 T. Wood and F. H. Bormann, *Environ. Pollut.*, in press.

23 B. A. Kratky, personal communication.

24 K. Sundström and J. Hallgren, *Ambio* 2, 13 (1973).

25 The relation between SO_2, H_2SO_4, and relative humidity [the proportions shift toward the acid form as the relative humidity increases (18)], partially explains the increased acidity of precipitation during summer when the relative humidity is generally higher.

26 J. S. Eaton, G. E. Likens, F. H. Bormann, *J. Ecol.* 61 (No. 2), 495 (1973).

27 S. Odén and T. Ahl, *Sartryck Ur Ymer Årsbok* (1970), pp. 103–122.

28 R. J. Beamish and H. H. Harvey, *J. Fish. Res. Board Can.* 29 (No. 8), 1131 (1972); R. J. Beamish, *Water Res.* 8, 85 (1974).

29 C. L. Schofield, Jr., *Trans. Am. Fish. Soc.* 94 (No. 3), 227 (1965).

30 M. Grande and E. Snekvik, "Report of the Meeting of Representatives of the Norwegian Water Hydro-electricity Board" (Farmers' Association, Institute of Water Research, Fish and Wildlife Service, Oslo, 26 Feb. 1969); N. Johansson, E. Kihlstrom, A. Wahlberg, *Ambio* 2, 42 (1973); K. W. Jensen and E. Snekvik, *ibid.* 1, 223 (1972).

31 E. S. Selezneva, *Tellus* 24, 122 (1972).

32 B. Ross, *New York Times* (6 Feb. 1972), p. 1; see also the cover of the 31 August 1973 issue of *Science*.

33 J. H. Cavender, D. S. Kircher, A. J. Hoffman, *Nationwide Air Pollutant Emission Trends 1940–1970* (U.S. Environmental Protection Agency, Research Triangle Park, N. C., 1973).

34 A. J. Hoffman (personal communication) cautions that estimates in *Estimates of Possible Fivefold Increases in SO_x Emissions Based on 3-Year-Old Projection* (U.S. Environmental Protection Agency, Research Triangle Park, N. C., 1973) may be unrealistic.

35 C. E. Junge and R. T. Werby, *J. Meteorol.* 15, 417 (1958). *Air Chemistry and Radioactivity* (Academic Press, New York, 1963).

37 This is a contribution to the Hubbard Brook Ecosystem Study. Financial support was provided by the National Science Foundation and the U.S. Department of Interior, Office of Water Resources Research, through the Cornell University Water Resources and Marine Sciences Center. We thank T. Wood and C. Schofield for reading and commenting on the manuscript.

28

Decrease in Anthropogenic Lead, Cadmium and Zinc in Greenland Snows since the Late 1960s*

C. F. Boutron, U. Görlach, J.-P. Candelone, M. A. Bolshov and R. J. Delmas

More than twenty years ago, Patterson and co-workers[1] showed that evidence of lead concentrations in Greenland ice and snow had increased about 200-fold since ancient times. From their results, they concluded that more than 99% of this highly toxic metal in the global troposphere of the Northern Hemisphere originated from human activities in the mid 1960s – mainly from the use of alkyl-leaded petrol. At least in part because of this evidence, the United States and other countries limited the use of lead additives in petrol from about 1970. Here we report that, as a result of these policy initiatives, lead concentrations in Greenland snow have decreased by a factor of 7.5 over the past twenty years. We also show that over the same time period, cadmium and zinc concentrations have decreased by a factor of 2.5.

As many previous investigations of heavy metals in Greenland snows have been hampered by contamination problems (see recent reviews in refs 2–5), we first describe our sampling and analytical procedure in some detail. Samples were taken on 19–21 July 1989 in central Greenland, near Summit (72° 35'N, 37° 38' W, elevation 3,230 m, mean annual snow accumulation rate 21.5 g cm^{-2} yr^{-1}) as part of the european Eurocore program. The samples covered a continuous sequence of 22 years (1967–1989). They consisted of a snow core 10.5 cm in diameter and 10.7 m deep, which was drilled only for our heavy metal measurements, and of a few additional shallow samples. To minimize local contamination, the sampling site was selected several kilometers north of the Eurocore camp, in a clean area to which access was limited. The core was hand-drilled by operators wearing full clean-room clothing and shoulder-length polyethylene

* Originally published in *Nature*, 1991, vol. 353, pp. 153–6.

gloves, using a specially designed SIPRE-type all-plastic mechanical auger made of polycarbonate. The cutting blades, the extension rods and the T-shaped arm used to rotate the auger were also made of polycarbonate. Before being sent to Greenland, the auger, extension rods and all other items (also made of plastic) used during drilling operations had been extensively cleaned in our clean laboratory[6] by successive immersions in a series of 25% and 0.1% HNO_3 baths,[6] the last two washes being ultrapure 0.1% HNO_3 (ref. 7; from US National Institute of Science and Technology) diluted in ultrapure water.[6] The core sections (length ~50cm) were directly transferred by gravity into specially designed conventional polyethylene tube with threaded caps, which had also been extensively cleaned in the clean laboratory and were packed in multiple sealed acid-cleaned polyethylene bags. All samples were brought back frozen to France.

Despite the exceptional cleanliness of the field collection procedure, the possibility remained that significant, although probably very limited, heavy metal contamination might be present on the outside of the snow core. Each core section investigated was therefore subsampled through freshly prepared untouched surfaces, inside a laminar flow clean bench located inside a cold room, following a procedure similar to that described in ref. 8. From each section of initial length ~50cm, we could first obtain two ~20-cm-long core sections, with fresh untouched surfaces at both extremities, by breaking the original core section with a sharp-edged polyethylene splitting wedge (used only to initiate the crack) and a polyethylene hammer. The central part of each of these two sections is then extracted by hammering a home-made cylindrical polyethylene beaker (5 cm diameter, 20 cm long) into the centre of one of these untouched surfaces parallel to the axis of the core axis. Additional subsamples are also taken parallel to the axis at increasing distances from the centre of each section using narrower (1 cm diameter, 20 cm long) conventional polyethylene beakers. All items used during subsampling were extensively cleaned following the procedure described in detail in ref. 6.

Twenty-five ~20-cm-long sections were subsampled, representing about half of the total length of the core. Each subsample was analysed separately for Pb, Cd and Cu by graphite furnace atomic absorption spectrometry (GFAAS) after preconcentration by non-boiling evaporation[9] and for Zn, Al and Na by GFAAS without preconcentration. Some were also analysed for Cd by direct ultrasensitive laser-excited atomic fluorescence spectrometry (LEAFS),[10, 11] giving results which agree well with those from GFAAS (table 28.1). All results were carefully corrected for the blank contributions originating from the successive steps of the analytical procedure (these procedural blanks from the subsampling stage onwards

Table 28.1 [orig. table 1] Comparative determination of Cd in the central part of four sections of a 10.7–m snow core by graphite furnace atomic absorption spectrometry (after preconcentration by non-boiling evaporation), and by direct laser-excited atomic fluorescence spectrometry.

| | Measured Cd concentrations (pg per g) | |
| | GFAAS after | |
Depth (m)	preconcentration	LEAFS
1.90–2.03	1.3 ± 0.3	1.4 ± 0.8
3.32–3.47	2.0 ± 0.4	2.2 ± 1.2
6.40–6.52	3.3 ± 0.7	2.9 ± 1.6
8.77–8.87	3.9 ± 0.8	3.7 ± 2.1

always represent less than 10% of metals intrinsic to the samples). Precision of the data was estimated to be ~10–20%.

We investigated the variation of the measured concentrations from the outside to the centre of each core section in detail to determine whether the concentrations measured in the centre represented the original concentrations in the snow. In most cases, concentrations were fairly constant across the core, without any significant contamination on the outside. Significant, although rather limited contamination, especially of Cd and Cu was present in the outside layer of a few core sections, but the concentrations always reached a good plateau in the central parts of these core sections.

The measured variations in concentration of the heavy metals at Summit as a function of the age of the snow are shown in figure 28.1 for Pb (also shown are the Greenland Pb data for the period 5,500 BP to 1965, from refs 1, 12) and figure 28.2 for Cd, Zn and Cu. The dating is based on the seasonal variations of methanesulphonic acid (M. Legrand, personal communication) its precision is better than 0.5 yr. These profiles form comprehensive records of heavy metals in Greenland snows from the mid 1960s to present. Other published data on heavy metals for this period come from only a few samples covering very limited periods (a year at the most) collected at widely dispersed locations.[13–16]

The natural contribution to the heavy

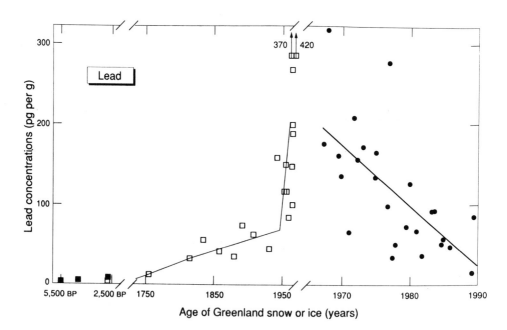

Figure 28.1 [*orig. figure 1*] Changes in Pb concentrations in Greenland ice and snow from 5,500 BP to present. Full dots (right) data from 1967 to 1989, from Summit, this work. Open squares (centre): 1753 to 1965, from north-west Greenland.[1] (the corresponding curve shown is the original one given in ref. 1). Full and open squares (left): 5,500 BP to 2,700 BP data from Camp Tuto and Camp Century respectively.[1,12] Although these data were obtained at widely dispersed locations it seems reasonable to put them together as the concentrations of major impurities such as Na, Al and sulphates are similar at these sites. After a great increase ~200-fold) from several thousand years ago to the mid 1960s, Pb concentrations have decreased rapidly (by a factor of ~7.5) during the past twenty years, mainly as a consequence of the fall in use of lead additives in gasolines.

metal from rock and soil dust in snow at Summit can be calculated from the Al concentrations (ranging from 1.4 to 17 ng per g) measured in our samples, using the mean metal/Al ratios in rock or soil. [17, 18] As illustrated in table 28.2 this contribution is very small (from ~2% down to ~0.2%) for Pb and Cd. For Zn, it is higher, but is always ≤20%. In contrast, the contribution for Cu is significant throughout the samples: for ~1/3 of the samples, it is close to the measured concentrations; for the others, it usually represents more than 50%, which indicates that Cu and Al values are significantly correlated. After correction for contribution from rock and soil dust, we used our Na concentration values (ranging from 1.7 to 19 ng per g) to estimate the natural heavy metal contribution from sea-salt spray from the mean metal/Na ratios in bulk sea water,[19] combined with the few published enrichment factors[20] relative to bulk sea water for these metals in sea-derived aerosols. This contribution is found to be negligible. The other possible natural contributions are from volcanoes, forest fires and continental or marine biogenic sources.[21] They cannot be evaluated from our data. But simple considerations using the best published estimates of the natural fluxes of metals from these sources to the global atmosphere[21] indicate that their contributions are probably negligible for Pb, Cd and Zn, but possibly significant for Cu. From this discussion, it appears that Pb, Cd and, to a lesser extent, Zn in snow from 1967–1989 at Summit are almost entirely derived from anthropogenic

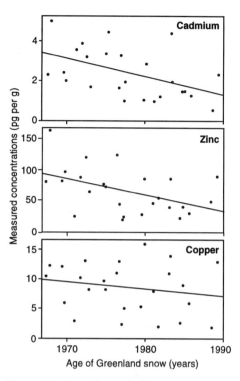

Figure 28.2 [*orig. figure 2*] Changes in Cd, Zn and Cu concentrations in snow at Summit, central Greenland, from 1967 to present. The least-square fits shown for each metal indicate that since 1967, Cd and Zn concentrations have decreased by a factor of ~2.5, probably as a consequence of the abatement policies for industrial emissions. Concentrations of Cu, which are mainly derived from natural sources, appear on the other hand to have remained fairly stable.

Table 28.2 [*orig. table 2*]   Crustal enrichment factors in snow cores from Summit, central Greenland.

Depth (m)	Year	*Crustal enrichment factor*[*]			
		Pb	Cd	Zn	Cu
0.05–0.25	1989	32	58	6.3	1.2
10.49–10.65	1967	230	186	19	3.0

* (Defined as metal/Al ratio in sample divided by metal/Al ratio in mean crustal material)[17] in snow from 0.05–0.25 m and 10.49–10.65 m (top and bottom sections of the snow core, respectively). Enrichment factors close to unity indicate that rock and soil dust make a large contribution to the measured metal concentrations. High enrichments indicate that rock and soil dust contribution is negligible.

sources, whereas a large fraction of Cu is natural. Similar conclusions have been reported previously[15] for these four metals in snow representing one year's accumulation (1983–84) from near Dye 3 in south Greenland.

As shown in figures 28.1 and 28.2, the concentrations of heavy metals at Summit from 1967 to 1989 vary widely. This is probably mainly because each data point represents a rather short and variable period (ranging from ~3 to 6 months) so that our data are certainly influenced by short-term changes such as seasonal effects.[1, 15] The least-square fits in figures 28.1 and 28.2, however, allow us to establish the general time trends for each metal. From 1967 to 1989, Pb concentrations are found to have strongly decreased (by a factor of ~7.5), while Cd and Zn have decreased less strongly (by a factor of ~2.5). The corresponding gradients (with confidence intervals at the 0.80 confidence level) are −7.95 ±1.86 pg per g per year for Pb, −0.090 ±0.035 pg per g per year for Cd and −2.58 ±1.08 pg per g per year for Zn. On the other hand, the very slight decrease observed for Cu (by a factor of ~1.3, with a gradient of −0.23 ± 0.16 pg per g per year) appears to be barely significant, which is consistent with the fact that Cu is probably mainly natural. This demonstrates that the concentrations in Cd and Zn in Greenland snow have decreased during the past two decades. For Pb, such a decrease was tentatively suggested by Wolff and Peel[15] to explain the low Pb concentrations measured near Dye 3 for snow laid down in 1983–84.

The observed decreases of Pb, Cd and Zn concentrations in central Greenland snows indicate that the pollution of the global troposphere of the Northern Hemisphere has significantly decreased during 1976–1989 for these three metals, and that this decrease was especially pronounced for Pb. The decrease for Pb follows the increase by a factor of ~200 from 5,500 BP to the mid 1960s (figure 28.1) which was reported by C. Patterson and co-workers.[1, 12] For Cd and Zn, the decrease probably also follows a previous increase, but this will need to be confirmed by obtaining reliable time series

for these two metals through the industrial period (reliable Greenland time series for heavy metals other than Pb are not available at present).

There has been a rapid fall in the use of Pb additives in gasolines during the past twenty years in the Northern Hemisphere, especially in North America and western Europe (refs 22–24; and J. O. Nriagu and J. M. Pacyna, personal communications). In the United States, for instance, the consumption of Pb additives fell by more than 90% between the late 1960s and present. Clearly, our data confirm that this fall has resulted in a very large decrease in the large-scale pollution of the troposphere of the Northern Hemisphere by this highly toxic metal.

For Cd and Zn, the situation is rather different. In contrast to Pb, for which only a small fraction of the anthropogenic emissions to the atmosphere originated from causes other than gasoline additives in the 1960s (ref. 25), anthropogenic Cd and Zn originated mainly from industrial processes such as the combustion of fossil fuels, nonferrous metal production, steel and iron manufacturing and incineration of refuse.[25] Our data indicate that the increasing efforts made in various Northern Hemisphere countries to reduce industrial emissions of Cd and Zn, and of other metals and pollutants (see, for example, refs 26, 27), have now resulted in a significant decrease in large-scale tropospheric pollution by these two metals.

References

1 Murozumi, M., Chow, T. J. and Patterson, C. C. *Geochim. cosmochim. Acta* 33, 1247–1294 (1969).

2 Peel, D. A. in *The Environmental Record in Glaciers and Ice Sheets* (eds Oeschger, H. and Langway, C. C. Jr) 207–223 (Wiley, Chichester, 1989).

3 Boutron, C. F. and Görlach, U. in *Metal Speciation in the Environment* (eds Broekaert, J. A. C., Gücer, S. and Adams, F.) 137–151 (Springer, Berlin, 1990).

4 Wolff, E. W. *Antarct. Sci.* 2, 189–205 (1990).

5 Boutron, C. F. in *National Perspectives on Global Metal Cycling* (eds Hutchinson, T. C., Gordon, C. A. and Meema, K.) 241–258 (Wiley Eastern, New Delhi, 1991).

6   Boutron, C. F. *Fresenius Z. analyt. Chem.* 337, 482–491 (1990).

7   Paulsen, P. J., Beary, E. S., Bushee, D. S. and Moody, J. R. *Analyt. Chem.* 60, 971–975 (1988).

8   Görlach, U. and Boutron, C. F. *J atmos. Chem.* (in the press).

9   Görlach, U. and Boutron, C. F. *Analyt. chim. Acta* 236, 391–398 (1990).

10   Apatin, V. M. et al. *Spectrochim, Acta* B44, 253–262 (1989).

11   Bolshov, M. A. et al. *Analyt. chim. Acta* (in the press).

12   Ng, A. and Patterson, C. C. *Geochim. cosmochim. Acta* 45, 2109–2121 (1981).

13   Herron, M. M., Langway, C. C., Weiss, H. V. and Cragin, J. H. *Geochim. cosmochim, Acta* 41, 915–920 (1977).

14   Boutron, C. F. *Geochim. cosmochim. Acta* 43, 1253–1258 (1979).

15   Wolff, E. W. and Peel, D. A. *Ann. Glaciol.* 10, 193–197 (1988).

16   Görlach, U. and Boutron, C. F. in *Heavy Metals in the Environment* (ed. Vernet, J. P.) 24–27 (Page, Norwich, 1989).

17   Taylor, S. R. *Geochim. cosmochim. Atca* 28, 1273–1285 (1964).

18   Bowen, H. J. M. *Environmental Chemistry of the Elements* (Academic, New York, 1979).

19   Duce, R. A., Arimoto, R., Ray, B. J., Unni, C. K. and Harder, P. J. *J. geophys. Res.* 88 5321–5342 (1983).

20   Weisel, C. P., Duce, R. A., Fasching, J. L. and Heaton, R. W. *J. geophys. Res.* 89, 11607–11618 (1984).

21   Nriagu, J. O. *Nature* 338, 47–49 (1989).

22   Trefry, J. H., Metz, S., Trocine, R. P. and Nelsen, T. A. *Science* 230, 439–441 (1985).

23   Boyle, E. A., Chapnick, S. D. and Shen, G. T. *J. geophys. Res.* 91, 8573–8593 (1986).

24   Dörr, H., Münnich, K. O., Mangini, A. and Schmitz, W. *Naturwissenschaften* 77, 428–430 (1990).

25   Nriagu, J. O. and Pacyna, J. M. *Nature* 333, 134–139 (1988).

26   Pacyna, J. M. in *Control and Fate of Atmospheric Trace Metals* (eds Pacyna, J. M. and Ottar, B.) 15–31 (Kluwer, Dordrecht, 1989).

27   Pacyna, J. M. *Emission Factors of Atmospheric Cd, Pb and Zn for Major Sources Categories in Europe in 1950 through 1985.* Rep. OR 30–91 (Norwegian Institute for Air Research, Lillestrøm, 1991).

Acknowledgements

We thank C. Rado for his participation in field sampling and S. Rudniev for assistance in LEAFS measurements. This work was supported by the Commission of European Communities and Switzerland within the framework of Cost 611 as part of the Eurocore program, the French Ministry of the Environment, the French–Soviet Commission and the University of Grenoble.

PART V

Biological Impacts

INTRODUCTION

The biosphere is the zone at the interface of the earth's crust, ocean and atmosphere where life is found. Part V looks at a selection of ways in which humans have transformed plant and animal life.

We begin by looking at the ways in which humans have modified habitats (the physical environment in which organisms live) and biomes (mixed communities of plants and animals occupying a major geographical area or at a continental scale). Chapter 29 looks in a rather general way at habitat, how it becomes fragmented, how this can lead to extinction and at what implications this has for the management of nature reserves. Chapters 30 and 31 concentrate on one particular biome – the great rainforests of the humid tropics. Chapter 30 considers some of the major environmental impacts while chapter 31 analyses the discrepancies that exist in published estimates of rates of rainforest loss.

Changes in the nature and extent of forests of the type mentioned in chapters 29–31 are some of the most pervasive ways in which organisms have been affected by human activities. However, there are other key habitats, which while perhaps of lesser extent, are nevertheless of very considerable ecological importance.

The first of these habitats is 'wetlands'. On a global basis the loss of wetland habitats (marshes, bogs, swamps, fens, mires, etc.) is a cause of considerable concern (Maltby, 1986; Williams, 1990). In all, wetlands cover about 6 per cent of the earth's surface (not far short of the total under tropical rain forest), so they are far from being trivial, even though they tend to occur in relatively small patches. More significant, however, is the fact they account for a disproportionately large amount – roughly one-quarter – of the earth's net primary productivity. They also have a very diverse range of organisms, and provide crucial wintering, breeding and refuge areas for wildlife. They are, however, under a range of threats. In chapter 32 Jesse Walker and colleagues try to assess the causes of wetland loss in Louisiana, USA. They look at natural and anthropogenic factors and point to the problems of establishing the causative ones.

The second of these habitat types is the coral reef. These are beautiful, diverse and productive ecosystems that have an importance that is out of all proportion to their areal extent. As Salvat (chapter 33) points out, they can be severely modified by natural events such as hurricanes and natural disease outbreaks. However, since the middle of the nineteenth century, the destruction and degradation of reefs has increased in frequency both spatially and temporally as a result of a multitude of human activities including runoff from land clearance, chemical pollution, mining, oil pollution, tourism and coastal construction projects. The causes of coral reef degradation, as with wetland loss, are complex, so that deciphering the impacts of natural and anthropogenic processes is far from easy. This is made evident by the discussion of two recent stress episodes suffered by reefs: predation by outbreaks of the Crown of Thorns Starfish (*Acanthaster Planci*) and degradation through so-called bleaching events.

The final habitat type to be included is

drylands and their margins. They cover around one-third of the earth's land surface and are coming under increasing development pressures. The degradation of such marginal areas is often termed desertification, and it may result from both natural climatic changes and from human activities. Since the beginning of the 1990s, however, the concept of desertification, while gaining public and political prominence, has become the subject of rather close and critical scrutiny, most notably by Thomas and Middleton (1994). In chapter 34, Tony Binns raises the question of whether the concept of desertification is a myth.

A second major facet of the human impact on the biosphere is the introduction of exotic species of fauna and flora and the breaking down of the old natural boundaries to the world's great faunal and floral realms. Introduced species can cause wholesale changes in the new environments to which they are transported, particularly if they spread explosively. Sometimes, as in the case of the introduction of Dutch Elm Disease to Britain or the migration of organisms from one ocean to another along ship canals – a process called Lessepsian Migration – the introduction is accidental. In the case of northern Australia (chapter 35) various large ungulates were introduced deliberately after colonization by European settlers in the nineteenth century. They have expanded their ranges and many of them have become feral. Their numbers are such that they are capable of achieving profound changes in the savanna woodlands, including devastating erosion.

Pollution is yet another major way in which human activities have modified the environment, and examples of both water (chapters 16, 17, 18), air (chapters 22, 26, 27, 28) and soil (chapters 7, 8) pollution have been given elsewhere in this volume.

Three more chapters are included in part V. Chapter 36 draws attention to how biocides can become magnified up the food chain so that egg-shells become thin and fragile. This in turn leads to egg breakage which means that successful breeding of hawks, falcons and the like becomes reduced. Populations then crash. Use of the many biocides implicated in egg-shell thinning is now subject to controls and the process of egg-shell thinning may have been reversed.

Another highly important pollution-related environmental change is 'Forest Decline', a process for which the German term *Waldsterben* is often used (chapter 37). This too is a highly complex issue in which a whole array of causative or triggering factors may be involved, some of which are natural (e.g. drought) and some of which are anthropogenic (e.g. acid deposition).

The arguments for and against each of the possible causative factors have been expertly reviewed by Innes (1987), who believes that in all probability most cases of forest decline are the result of the cumulative efforts of a number of stresses. He draws a distinction between predisposing, inciting and contributing stresses (p. 25):

Predisposing stresses are those that operate over long time scales, such as climatic change and changes in soil properties. They place the tree under permanent stress and may weaken its ability to resist other forms of stress. Inciting stresses are those such as drought, frost and short-term pollution episodes, that operate over short time scales. A fully healthy tree would probably have been able to cope with these, but the presence of predisposing stresses interferes with the tree's mechanisms of natural recovery. Contributing stresses appear in weakened plants and are frequently classed as secondary factors. They include attack by some insect pests and root fungi. It is probable that all three types of stress are involved in the decline of trees.

As with many environmental problems, interpretation of forest decline is bedevilled by a paucity of long-term data and detailed surveys. Given that forest condition oscillates from year to year in response to variability in climatic stress (e.g. drought, frost, wind throw) it is dangerous to infer long-term trends from short-term data (Innes and Boswell, 1990).

There may also be differences in causation in different areas. Thus while widespread forest death in eastern Europe may result from high concentrations of sulphur dioxide combined with extreme winter stress, this is a much less likely explanation in Britain, where sulphur dioxide concentrations have shown a marked decrease in recent years. Indeed, in Britain Innes and Boswell (1990, p. 46) suggest that the direct effects of gaseous pollutants appear to be very limited.

It is also important to recognize that some stresses may be particularly significant for a particular tree species. Thus, in 1987 a survey of ash trees in Great Britain showed extensive die-back over large areas of the country. Almost one-fifth of all ash trees sampled showed evidence of this phenomenon. Hull and Gibbs (1991) indicated that there was an association between die-back and the way the land is managed around the tree, with particularly high levels of damage being evident in trees adjacent to arable land. Uncontrolled stubble burning, the effects of drifting herbicides, and the consequences of excessive nitrate fertilizer applications to adjacent fields were seen as possible mechanisms. However, the prime cause of dieback was seen to be root disturbance and soil compaction by large agricultural machinery. Ash has shallow roots and if these are damaged repeatedly the tree's uptake of water and nutrients might be seriously reduced, while broken root surfaces would be prone to infection by pathogenic fungi.

Innes (1992, p. 51) also suggests that there has been some modification in views about the seriousness of the problem since the mid-1980s:

The extent and magnitude of the forest decline is much less than initially believed. The use of crown density as an index of tree health has resulted in very inflated figures of forest 'damage' which cannot now be justified. . . If early surveys are discounted on the basis of inconsistent methodology . . . then there is very little evidence for a large-scale decline of tree health in Europe.
. . . The term 'forest decline' is rather misleading in that there are relatively few

cases where entire forest ecosystems are declining. Forest ecosystems are dynamic and may change through natural processes.

He also suggests that the decline of certain species has been associated with climatic stress for as long as records have been maintained.

The ultimate consequences of habitat loss, habitat pollution and competition of resources, are extinction of species. It is possible that one great spasm of extinction, particularly perhaps of large animals and birds (mega-fauna), took place in Stone Age times. Alternatively it was caused by the severe climatic changes at the end of the Pleistocene. There are arguments in favour of both positions (see Goudie, 1993, pp. 125–31).

Another great spasm of species extinction and biodiversity loss may be occurring at the present time. Somebody who has repeatedly and cogently drawn attention to this issue is Norman Myers. According to Myers, writing in his influential *The Sinking Ark* (1979, p. 31) 'during the last quarter of this century we shall witness an extinction spasm accounting for 1 million species'. This is a considerable proportion of the estimated number of species living in the world today, for which Myers gives a figure of between 3 and 9 million. He has calculated that from AD 1600 to 1900 humans were accounting for the demise of one species every four years, that from 1900 onwards the rate increased to an average of around one per year, that at present the rate is about one per day, and that within a decade we could be losing one every hour. By the end of the century, our planet could lose anywhere between from 20 to 50 percent of its species. The need to maintain biodiversity has indeed become one of the crucial issues with which we must contend.

Because of the severity and urgency of the threat to biodiversity, it makes sense to identify those particularly significant areas, which Myers (chapter 38) terms 'hot-spots', where there are exceptional threats of destruction.

References

Hull, S. K., and Gibbs, J. N. 1991, Ash dieback: a survey of non-woodland trees. *Forestry Commission Bulletin*, 93: 32.

Innes, J. L. 1987, Air Pollution and Forestry. *Forestry Commission Bulletin*, 70.

Innes, J. L. 1992, Forest decline. *Progress in Physical Geography*, 16: 1–64.

Innes, J. L, and Boswell, R. C. 1990, Monitoring of forest condition in Great Britain 1989. *Forestry Commission Bulletin*, 94: 57.

Maltby, E. 1986, *Waterlogged Wealth. Why Waste the World's Wet Places?* London: Earthscan.

Myers, N. 1979, *The Sinking Ark: A New Look at the Problem of Disappearing Species.* Oxford: Pergamon Press.

Nunn, P. 1994. *Oceanic Islands.* Oxford: Blackwell Publishers.

Thomas, D. S. G., and Middleton, N. J. 1994, *The Myth of Desertification.* Chichester: Wiley.

Williams, M. 1990, *Wetlands: A Threatened Landscape.* Oxford: Basil Blackwell.

29

HABITAT FRAGMENTATION IN THE TEMPERATE ZONE*

D. S. Wilcove, C. H. McLellan and A. P. Dobson

In this chapter we examine three questions relating to habitat fragmentation in the temperate zone: (1) What is the effect of fragmentation on the species originally present in the intact habitat? (2) How does fragmentation lead to the loss of species? (3) For an already fragmented landscape, are there any guidelines for the selection and management of nature reserves? Here we shall set as our goal the long-term preservation of those species whose continued existence is jeopardized by habitat destruction. At the outset we note that this chapter is slanted towards vertebrate communities (especially birds) and forested habitats. Our bias reflects, in part, a bias in the existing literature. On the other hand, by virtue of their low population densities, birds and mammals are among the taxa most likely to disappear from isolated fragments (Wilcox, 1980).

Introduction

Fragmentation occurs when a large expanse of habitat is transformed into a number of smaller patches of smaller total area, isolated from each other by a matrix of habitats unlike the original. When the landscape surrounding the fragments is inhospitable to species of the original habitat, and when dispersal is low, remnant patches can be considered true "habitat islands," and local communities will be "isolates" (*sensu* Preston, 1962). If the matrix can support populations of many of the species from the original habitat, or if dispersal between patches is high, communities in the fragments will effectively be "samples" from the regional faunal or floral "universe" (Preston, 1962; Connor and McCoy, 1979). The process of isolate formation through fragmentation has been termed "insularization" by Wilcox (1980). The challenge to conservationists is to preserve as much of the species pool as possible within these fragments, in the face of continual habitat destruction.

Habitat fragmentation has two components, both of which cause extinctions: (1) reduction in total habitat area (which primarily affects population sizes and thus extinction rates); and (2) redistribution of the remaining area into disjunct fragments (which primarily affects dispersal and thus

* Originally published in M. E. Soulé (ed.), *Conservation Biology: The Science of Scarcity and Diversity*, (Sunderland, Mass.: Sinauer, 1986).

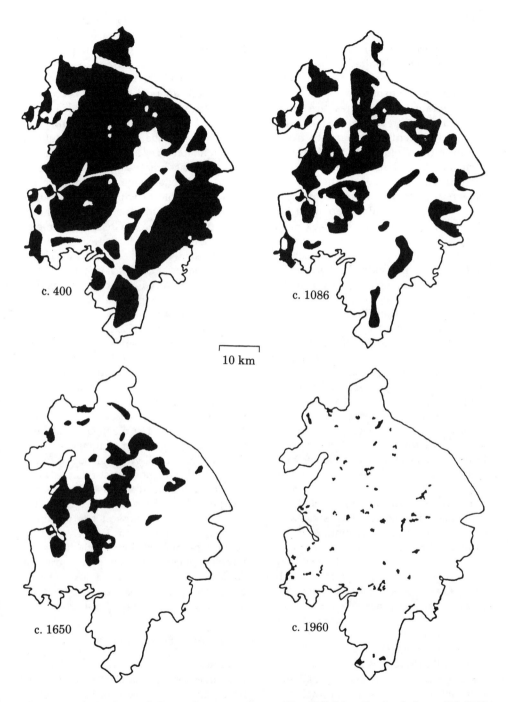

Figure 29.1 [*orig. figure 1*] Forest fragmentation in Warwickshire, England, from 400–1960 A.D. Forested areas are shown in black. (Redrawn from Thorpe 1978.)

immigration rates). Both Lovejoy et al. (1984) and Haila and Hanski (1984) stress the need to partition extinctions into those caused purely by habitat destruction and those in which insularization is an important additional component.

Temperate communities are widely believed to be more resistant to the effects of habitat fragmentation than are tropical communities. Temperate species tend to occur in higher densities, be more widely distributed, and have better dispersal powers than their tropical counterparts. These attributes should allow populations to persist in smaller patches of suitable habitat. Although local extinction rates may be high (due to high levels of population fluctuation and shorter individual lifespans; see Diamond, 1984a), high vagility can facilitate rapid recolonization from other fragments following extinction (Brown and Kodric-Brown, 1977).

On the other hand, one of the main reasons why habitat fragmentation seems less severe in the temperate zone is that most of the damage was done long before most people were aware of it. For example, in Great Britain reduction and fragmentation of the original forest cover began some 5000 years ago with permanent clearances by Neolithic farmers, and was well advanced by the time of the Norman Conquest in 1066 (figure 29.1). Species whose extinctions in Great Britain were certainly related to the destruction of the original forest (as well as other causes, especially hunting) include: brown bear (extinct by the time of the Norman Conquest), wild boar (18th century), wolf (18th century), goshawk (19th century; now reestablished in new conifer plantations), and capercaille (18th century; now reintroduced in new conifer plantations).

Much the same story can be told for the fauna of the deciduous forest of the eastern United States, where widespread forest destruction began with the arrival of European settlers (about 300 years ago) and reached a peak about the time of the Civil War. Here, too, a number of species vanished from the east as a result of habitat destruction combined with hunting. These include wolf (19th century), mountain lion (20th century, although a few persist in Florida), elk (19th century), passenger pigeon (20th century), and ivory-billed woodpecker (20th century). For both the British and American species, it is probably correct to say that even if they could be reintroduced to their former haunts, the outcome would be disappointing: suitable habitat no longer exists.

The species that survived this initial round of habitat fragmentation were the ones better able to withstand the human impact on the landscape, but by no means is the problem over. In Great Britain today, pressure on the land is so great that many of the "semi-natural" habitats which replaced the original forest are themselves severely reduced and fragmented. Examples include lowland heaths (Moore, 1962; Webb and Haskins, 1980), upland moors (Porchester, 1977; Parry et al., 1981) and calcareous grasslands (Blackwood and Tubbs, 1970; Jones, 1973). There is now a growing list of species characteristic of, or restricted to, these habitats whose decline or extinctions can be attributed at least in part to fragmentation (see Hawkesworth, 1974 for a general survey). In the United States, the continuing fragmentation of such habitats as old-growth Douglas fir forest in the Pacific northwest (Harris, 1984), deciduous forests in the east (Burgess and Sharpe, 1981; Whitcomb et al., 1981), and grasslands in the midwest has prompted concern for the continued survival of the species that inhabit them (including small whorled pogonia, greater prairie chicken, spotted owl, and Delmarva fox squirrel). Fragmentation remains the principal threat to most species in the temperate zone.

A Model of Fragmentation

Analyses of the effect of fragmentation (and guidelines for the design of nature reserves) have generally been based on the conceptual framework of island biogeography (Preston, 1962; MacArthur and Wilson, 1967; Soulé and Wilcox, 1980; Burgess and Sharpe, 1981). This theory suggests that the number of species on an oceanic island represents a

Table 29.1 [*orig. table 1*]   Scientific names of plants and animals mentioned in the chapter.

Plants	Dog's mercury: *Mercurialis perennis*
	Small whorled pogonia: *Isotria medeoloides*
Insects	Large blue butterfly: *Maculinea arion*
Amphibians	Red-spotted newt: *Notophthalmus viridescens*
Birds	American woodcock: *Scolopax minor*
	Blue-gray gnatcatcher: *Polioptila caerulea*
	Blue jay: *Cyanocitta cristata*
	Brown-headed cowbird: *Molothrus ater*
	Capercaille: *Tetrao urogallus*
	Common grackle: *Quiscalus quiscula*
	Goshawk: *Accipiter gentilis*
	Greater prairie chicken: *Tympanuchus cupido*
	Great spotted woodpecker: *Picoides major*
	Ivory-billed woodpecker: *Campephilus principalis*
	Kirtland's warbler: *Dendroica kirtlandii*
	Louisiana waterthrush: *Seiurus motacilla*
	Passenger pigeon: *Ectopistes migratorius*
	Spotted owl: *Strix occidentalis*
Mammals	Bobcat: *Lynx rufus*
	Brown bear: *Ursus arctos*
	Delmarva fox squirrel: *Sciurus niger cinereus*
	Eastern chipmunk: *Tamias striatus*
	Elk: *Cervus elaphus*
	Gray squirrel: *Sciurus carolinensis*
	Mountain lion: *Felis concolor*
	Opossum: *Didelphis virginiana*
	Raccoon: *Procyon lotor*
	Short-tailed weasel: *Mustela erminea*
	White-tailed deer: *Odocoileus virginianus*
	Wild boar: *Sus scrofa*
	Wolf: *Canis lupus*

balance, or dynamic equilibrium, between processes of immigration and extinction. The equilibrium number of species on an island depends upon the characteristics of the island – in particular, its size and isolation from potential sources of colonists – and the characteristics of the species themselves – in particular, their dispersal abilities and population densities.

We have recently (McLellan et al., 1986) developed a computer model that simulates the effects of habitat fragmentation on two pools of species with different minimum area requirements and dispersal abilities. This model has led to a number of insights regarding the extent of fragmentation that different species can tolerate.

The pattern of fragmentation of our hypothetical habitat is based largely on that of heathland in Dorset, England, as reported

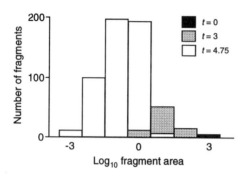

Figure 29.2 [*orig. figure* 2] The frequency distributions of fragment sizes at various stages of the fragmentation sequence. The top diagram corresponds to the original ($t = 0$), largely contiguous habitat. Fragmentation results in a larger number of smaller patches (lower diagrams) with a total area less than that of the original habitat. (From McLellan et al., 1986.)

by Moore (1962) and Webb and Haskins (1980). In our model, the original habitat is reduced from five extremely large tracts to an archipelago of over 450 fragments totalling 5 percent of the original area. For simplicity we have shown the total area of habitat decreasing linearly over time; in reality, the rate of destruction usually increases with time. The total number of fragments increases exponentially over time, reflecting a distribution increasingly skewed towards a large number of very small fragments (figure 29.2). At each stage of fragmentation, the remaining area of habitat is distributed among the growing number of fragments in a roughly lognormal fashion. There is no information on interfragment distances for the heathlands, so we have borrowed a pattern from another system, namely woodland in Cadiz township, Wisconsin. Sharpe et al. (1981) have shown that mean nearest-neighbor distances increased until about 10 percent of the original area was left, and then remained constant despite further losses.

Details of the Model

The two species pools were chosen to represent the extremes of susceptibility to fragmentation: one (resistant) pool consisted of species with good dispersal abilities and a low average proneness to local extinction. The other (susceptible) pool had species with poor dispersal abilities and a high average proneness to local extinction. The results therefore define qualitatively the range of patterns of species loss to be expected in habitats undergoing fragmentation. The fates of the P species in each pool were modeled using a species-by-species ("molecular"; see Gilpin and Diamond, 1981) formulation of the equilibrium model. The basic variable is the probability J_i that a given species i ($i = 1, 2 \ldots P$) occurs as a breeding population in a fragment. This probability, termed the "incidence" (Diamond, 1975a) of the species, increases with fragment area (A), due to increasing population size and thus decreasing chance of stochastic extinction, and decreases in distance (D) from a source of colonists due to decreasing frequency of immigration. Observed values of J_i plotted against area for a given distance, or vice versa, are called "incidence functions" (Diamond, 1975a; figure 29.3). Elaborations such as dependence of extinction rates distance (due to the "rescue effect;" see Brown and Kodric-Brown, 1977) and dependence of immigration rates on area (due, for example, to "passive sampling;" see Connor and McCoy, 1979) are not incorporated in this model.

This model has the general form

$$\frac{dJ_i}{dt} = I_i - E_i \qquad (1)$$

where I_i is the net rate at which unoccupied fragments are colonized by species i and E_i is the net rate at which the species becomes locally extinct. We chose to use a version of this general model first presented by Levins and Culver (1971):

$$\frac{dJ_i}{dt} = a_i J_i (1 - J_i) - b_i J_i \qquad (2)$$

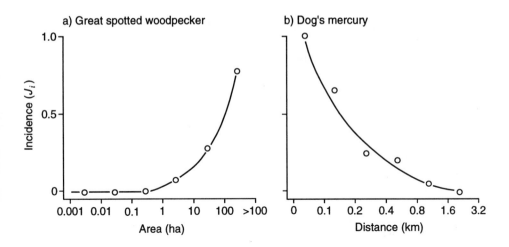

Figure 29.3 [*orig. figure 3*] Incidence functions for two species in forest fragments in Britain. (A) Incidence (J_i, the proportion of woods in each class occupied by a population) of the great spotted woodpecker plotted against woodland area. (From Moore and Hooper, 1975.) (B) Incidence of dog's mercury, a woodland herb, in secondary forest in Lincolnshire plotted against the distance to the nearest older forest containing the species. (From Peterken and Game, 1981.) Curves were drawn by eye.

where a_i is an instantaneous colonization rate per occupied fragment and e_i is an instantaneous local extinction rate. This model differs from the version described by Gilpin and Diamond (1981) in that the instantaneous rate of colonization of unoccupied fragments ($a_i J_i$) is a function of J_i rather than a constant. This is more appropriate for fragmented systems, where colonists must come from other (occupied) fragments rather than from some large, inviolate, "mainland" area. At equilibrium I_i and E_i are equal, and J_i is constant at the equilibrium incidence level J_i^*. For equation (2) this is

$$J_i^* = 1 - \frac{b_i}{a_i} \qquad (3)$$

Thus incidence is positive only when the colonization rate (a_i) exceeds the extinction rate (b_i).

Incidence functions were generated by specifying the dependence of a_i on distance and b_i on area. We used an exponential function for $a_i = f(D)$

$$a_i = c_i \cdot \exp \frac{-D}{D_i} \qquad (4)$$

where c_i is a colonization coefficient and $1/D_i$, the death rate per unit distance of migrants (Gilpin and Diamond, 1976). An inverse hyperbolic function was used for $b_i = f(A)$

$$b_i = e_i/A \qquad (5)$$

where e_i is an extinction coefficient (Gilpin and Diamond, 1976). Substituting equations (4) and (5) into equation (3) gives the expression for J_i^* as a function of area and distance

$$J_i^* = 1 - \left[\frac{e_i}{c_i} \cdot \exp \frac{D/D_i}{A} \right] \qquad (6)$$

When $D/D_i \simeq 0$ (i.e., in nonisolated fragments), equation 6 describes a hyperbolic function with increasing area, with the parameter e_i/c_i corresponding to the area at which incidence is zero. We distributed this "minimum area" lognormally with variance 1.0 in each of the species pools, with the susceptible pool having a mean value an order of magnitude greater than the resistant one. (For theoretical justification of the lognormal in such models see Gilpin and Armstrong, 1981, and for empirical evidence see Gilpin and Diamond, 1981.) For a given area, equation (6) describes an exponential function with the parameter D_i (the "mean dispersal distance;" see Gilpin and Diamond, 1976) corresponding to the distance required to reduce J_i to $1/e$ (36.8%) of its values at $D = 0$. We treated D_i unrealistically as constant among species in a given pool, assigning the susceptible pool a value half that given to the resistant pool.

Results of the model

The results of this exercise are illustrated in figure 29.4. Initially, when a large amount of habitat remains, mostly in large fragments, few or no species are lost from either pool. As fragmentation proceeds we eventually reach some critical level of reduction and fragmentation where species begin to die out. The susceptible pool loses species earlier and loses more species in total than does the resistant pool. When the resistant pool begins to lose species, it loses them very rapidly, because by this time the fragments are small and there is little habitat left.

Insularization causes extinctions over and above those expected through reduction in the total area of habitat. More species persist at equilibrium if the remaining habitat is concentrated into a single large patch rather than distributed over many small fragments (figure 29.4). We stress that the results in figure 29.4 are equilibrium patterns; depending on the relative time scales of habitat destruction and species' population dynamics, extinctions may either closely track the changing landscape patterns or lag behind them.

An important omission from the model is the explicit inclusion of population size as a variable (see Schoener, 1976; Williamson, 1981). Species' carrying capacities are assumed to be directly proportional to fragment size, and extinctions simply the result of demographic stochasticity. It seems more likely both that habitat heterogeneity means that the carrying capacity is not a simple function of area, and that factors like environmental stochasticity and population structure are important in determining extinction rates (see MacArthur and Wilson, 1967; Richter-Dyn and Goel, 1972; Leigh, 1975, 1981 for details; Diamond, 1984a; and

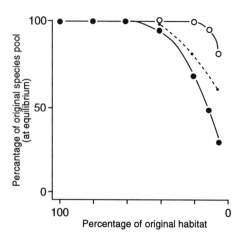

Figure 29.4 [*orig. figure 4*] The number of species remaining in each species pool as fragmentation proceeds. Closed circles show the pool of species with large area requirements and low vagility. Open circles show the species with less stringent area requirements. The small dots connected by the dashed line depict the proportion of the first pool that would be present when the habitat is minimally fragmented. (From McLellan et al., 1986.)

Gilpin and Soulé, Chapter 2 for a summary). Colonization will also be affected by habitat heterogeneity and population dynamics. If a fragment does not contain enough suitable habitat for a given species, establishment of a breeding population will not occur no matter how high the immigration rate. Similarly, abundant species will produce more colonists than scarce ones. Such factors will need to be included in more realistic (albeit more complex) models.

Despite these limitations, we believe the model provides a clear message: Even where most of the habitat has already been destroyed, subsequent fragmentation should be minimized, lest a rapid loss of species occur. Furthermore, insularization can cause extinctions independent of habitat reduction.

Mechanisms of Extinction

The above theoretical discussion has taken something of a black box approach to fragmentation. In this section, we shall focus on the proximate mechanisms of extinction. These include home range size, loss of habitat heterogeneity, effects of habitats surrounding fragments, edge effects, and secondary extinctions.

Home range

Some fragments will be smaller than the minimum home ranges or territories of certain species. This is often the case for large animals. For example, a single pair of ivory-billed woodpeckers may required 6.5–7.6 km^2 of undisturbed bottomland forest (Tanner, 1942). The European goshawk has a home range of appoximately 30–50 km^2 (Cramp and Simmons, 1979). Male mountain lions in the western United States may have home ranges in excess of 400 km^2 (Seidensticher et al., 1973). However, species often disappear from habitat fragments that far exceed their minimum home range sizes; mechanisms other than home range limitations must be operating in such cases.

Loss of habitat heterogeneity

One common consequence of fragmentation is a loss of habitat heterogeneity. Even a seemingly uniform expanse of habitat such as forest or grassland is, at some level of discrimination, really a mosaic of different habitats. Individual fragments may lack the full range of habitats found in the original block. Patchily distributed species or species that utilize a range of microhabitats are especially vulnerable to extirpation under these circumstances. An example of such a patchily distributed species is the Louisiana waterthrush of eastern North America. It nests and forages near open water, especially fast-moving streams (Chapman, 1907). Woodlots without open water do not provide suitable habitat for the waterthrush, and, not surprisingly, this bird is one of a number of forest songbirds that are rarely encountered in small woodlots (Robbins, 1980).

While the habitat requirements for most songbirds are far less obvious, they may nontheless play a major role in the response of these birds to fragmentation. Lynch and Whigham (1984) studied the bird communities and vegetation of 270 woodlots in Maryland. They discovered that structural or floristic characteristics of the vegetation strongly influenced the local abundance of each bird species. These vegetation characteristics, in turn, varied with the forest size. A qualitatively similar result was noted by Bond (1957) in forest fragments in southern Wisconsin. Although generalizations are risky, we might expect this problem of patchiness to be most acute for plants, and for insects that depend upon specific host plants.

When species require two or more habitat types, fragmentation may make it impossible for them to move between habitats. Karr (1982a) has attributed many of the extinctions of landbirds of Barro Colorado Island, Panama to just this mechanism. Within the temperate zone, this problem is likely to befall many kinds of organisms. The red-spotted newt is typical of a number of amphibians in having both a terrestrial and an aquatic stage. The terrestrial efts may remain ashore for up to three years, but eventually must return to the water to breed. Among birds, the blue-gray gnatcatcher in California moves from deciduous oak woodlands to chaparral and live oaks over the course of the breeding season (Root, 1967). Other temperate zone birds are also believed to make seasonal shifts in their ranging behavior (MacClintock et al., 1977). Unfortunately, too little is known about the behavior of most temperate zone animals to say with certainty what their habitat requirements are, or how these requirements may change seasonally. This lack of knowledge is always an obstacle to predicting the effects of fragmentation on individual species. Detailed information on habitat usage will be crucial to devising successful conservation programs for many species.

Effects of habitat between fragments

In the case of a true island, the ocean is an impassive barrier, and potential colonists will either traverse it successfully or perish in the attempt. In the case of a habitat fragment, the ocean has been replaced by a landscape of human dwellings or agricultural land. This landscape can also be a formidable barrier to colonists from the fragments. Unlike an ocean, however, a human-created landscape can contribute directly to the extinction of species within fragments. It does so by building up populations of animals that are harmful to species within the fragments. A good example of this problem comes from studies of forest-dwelling songbirds in forest fragments in the eastern United States.

Breeding populations of songbirds have been declining in small woodlots throughout the eastern United States since the late 1940s (Robbins, 1979; Whitcomb et al., 1981; Wilcove, 1985a). A number of factors have contributed to this decline, two of the most important being high rates of nest predation (Wilcove, 1985b) and brood parasitism by the brown-headed cowbird (Mayfield, 1977; Brittingham and Temple, 1983). In recent decades, the numbers of nest predators and cowbirds have increased greatly as a result of human-induced changes in the landscape.

Among the nest predators, blue jays, raccoons, and gray squirrels all occur in higher densities in suburban communities than in more natural habitats like forests (Flyger, 1970; Fretwell, 1972; Hoffman and Gottschang, 1977). Prior to the arrival of European settlers, the cowbird was largely confined to the grasslands of the midcontinent, where it followed the grazing mammals and ate the insects they stirred up. With the disruption of the eastern deciduous forest and the introduction of livestock, the cowbird spread throughout the eastern United States and Canada (Mayfield, 1977). More recently, the cowbird population in eastern North America has increased tremendously due to an increase in their winter food supply – waste grain in southern rice fields (Brittingham and Temple, 1983). The advent of mechanical harvesters has simultaneously increased the amount of land under rice cultivation and

the amount of waste grain. This range expansion and population increase has brought the cowbird in contact with populations of forest-dwelling songbirds, most of which lack behavioral defenses against cowbird parasitism (Rothstein, 1975; May and Robinson, 1985).

No habitat preserve is immune to the effects of human activity outside its borders, and wildlife managers must concern themselves with the ecological effects of land development outside the boundaries of protected areas. To quote Janzen (1983), "No park is an island."

Edge effects

Wildlife managers have long extolled the virtues of forest edge (see for example Dasmann, 1964, 1971; Yoakum and Dasmann, 1969; Burger, 1973), in a tradition dating back to the writings of Aldo Leopold (1933). Certainly a variety of game animals, including white-tailed deer and American woodcock, do well in edge habitats. But it is becoming increasingly clear that the forest edge has a strong negative impact on other members of the woodland flora and fauna (Hubbell and Foster, Chapter 10; Lovejoy et al., Chapter 12; Janzen, Chapter 13).

Ranney et al. (1981) believe that the seed rain into the cores of small woodlots is dominated by the seeds of the edge species. This may ultimately change the species composition of the woodlots, as the shade tolerant plants of the interior are replaced by shade intolerant forms from the edge. Such an effect would require the number of plants germinating to vary with the number of seeds set in the interior. Ranney et al. note that very small or irregularly shaped forest reserves may be unable to sustain populations of forest interior plants.

Field studies by Gates and Gysel (1978), Chasko and Gates (1982), and Brittingham and Temple (1983) have shown that the nesting success of songbirds is lower near the forest edges than in the interior (figure 29.5). This is because many nest predators (blue jay, American crow, common grackle, eastern chipmunk, short-tailed weasel, raccoon) and brood parasites (brown-

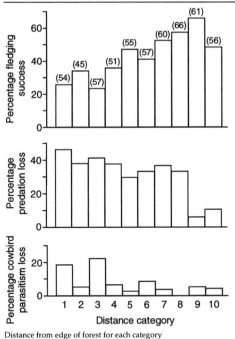

Distance from edge of forest for each category

(1) 0.0–0.82 m   (5) 6.87–10.06 m   (9) 46.25–65.59 m
(2) 0.83–2.19 m   (6) 10.07–14.18 m   (10) 65.59–123 m
(3) 2.20–4.34 m   (7) 14.19–26.74 m
(4) 4.35–6.86 m   (8) 26.75–46.24 m

Figure 29.5 [*orig. figure 5*] Nesting success of songbirds as a function of distance from the forest edge, based on a study in Michigan. Distance categories 1–10 are delimited at right, sample sizes are in parentheses. (From Gates and Gysel, 1978.)

the faunal effects of a forest edge would exceed the floral effects. Birds like crows, grackles, and cowbirds are not intolerant of the forest interior. Similarly, mammals like the raccoon, weasel, and chipmunk, while concentrating their activities near the forest edge, will also frequent the forest interior (Whitaker, 1980).

One consequence of these observations deserves special emphasis: if 600 m is taken as a liberal estimate of the faunal edge effect, then circular reserves smaller than 100 ha will contain no true forest interior. In the case of forest songbirds, this finding suggests that reserves should contain *at least* several hundred ha of uninterrupted forest. In fact, far larger areas may be needed to ensure the long term survival of these birds (Whitcomb et al., 1976).

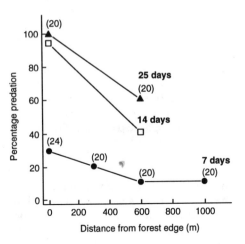

Figure 29.6 [*orig. figure 6*] Percentage of experimental nests preyed upon as a function of distance from the forest edge. Nests are wicker, open-cup baskets containing fresh quail eggs. Solid circles are the predation rates after 7 days, open squares are after 14 days, and closed triangles are after 25 days. The numbers in parentheses are the numbers of experimental nests. The data suggest that the edge-related increase in predation may extend from 300–600 m inside the forest. (From Wilcove, 1985a.)

headed cowbird) occur in higher densities around forest edges (Bider, 1968; Robbins, 1980; Whitcomb et al., 1981; Brittingham and Temple, 1983).

For management purposes, it is important to know how far into the forest the influence of the edge is felt. Studies by Ranney (1977) and Wales (1972) show that the major vegetational changes caused by the edge extend only 10–30 m inside the forest, depending on whether the edge has a northerly or a southerly exposure. However, by placing artificial nests at varying distances from the edge, Wilcove (1985a) has shown that the edge-related increase in predation may extend from 300–600 m inside the forest (figure 29.6). It should not be surprising that

Secondary extinctions

Fragmentation often disrupts many of the important ecological interactions of a community, including predator–prey, parasite–host, and plant–pollinator relations, and mutualisms (Gilbert, 1980; Terborgh and Winter, 1980). The disruption of these interactions may lead to additional extinctions, sometimes referred to as "secondary extinctions." Typically, these secondary extinctions are associated with the decay of complex tropical communities, but they are certainly not unknown in the temperate zone. For example, small woodlots in the eastern United States support few, if any, large predators like mountain lions, bobcats, and large hawks or owls that may regulate populations of smaller, omnivorous species like raccoons, opossums, squirrels, and blue jays (Matthiae and Stearns, 1981; Whitcomb et al., 1981). These omnivores, in turn, prey upon the eggs and nestlings of the forest songbirds. As noted earlier, the rate of nest predation in small woodlots is very high, and this may be one reason why songbird populations have declined. A similar explanation has been invoked to explain some of the avian extinctions on Barro Colorado Island, Panama (Terborgh, 1974).

A more complicated example involves the extinction of the large blue butterfly in Britain (Thomas, 1976; Ratcliffe, 1979). This butterfly has a remarkable life history in that the larvae must develop within the nests of the red ant *Myrmica sabuleti*. The large blue was brought to the brink of extinction when land development and reduced grazing by livestock eliminated the open areas it required. The remaining populations vanished through a complex chain of events. An epidemic of myxomatosis in the mid-1950s depressed rabbit populations, and as a result many of the sites became overgrown with scrub. The *Myrmica sabuleti* ants were unable to survive in the overgrown areas, and their decline meant the end of the large blue.

We suspect that as more data on fragmentation are gathered, secondary extinctions will prove a common occurrence in temperate communities. The prevention of these extinctions will require synecological studies involving threatened species, coupled with active management of preserves.

Guidelines for Temperate Zone Reserves

Although blanket prescriptions for the design of nature reserves (Wilson and Willis, 1975; Diamond, 1975b) have come under criticism in recent years (examples include Simberloff and Abele, 1976; Abele and Connor 1979; Higgs and Usher, 1980, Game, 1980; Margules et al., 1982; Boeklen and Gotelli 1984), we believe that the theory of island biogeography provides a useful framework within which more detailed studies of particular cases can be planned. In this final section, we focus on three questions:

1 How much of the available habitat must be set aside as reserves, and in what distribution of sizes?
2 Should reserves be clustered together in close proximity to each other, or spread out over a broad area?
3 What is the optimum shape for reserves?

It is important to realize that the "correct" answers to these questions may depend very much on the scale of the conservation effort, because local, regional, and national conservation operations usually operate under very different budget constraints and spatial scales. Once again, we note that our perspective on these questions is rather ornithocentric.

How much and how large?

A characteristic pattern in habitats undergoing fragmentation is an increasingly skewed distribution of fragment sizes as the total area of habitat declines. In general, a large proportion of the remaining area of highly fragmented habitats should be targeted for protection in order to avert (or at least minimize) the biotic collapse which models suggest can occur in such systems. All other things being equal, priority should go to the largest remaining fragments, for several reasons. First, as emphasized in our model, different species have different area requirements, and the large fragments will often be the only refuge for species which exist at low densities (such as top predators and large herbivores) or who are habitat specialists whose requirements are only satisfied in large areas. Second, the large fragments may well serve as sources of immigrants for marginal populations in neighboring small fragments. If many species are maintained in these small fragments by the "rescue effect" (Brown and Kodric-Brown, 1977), then the small fragments do not represent a viable reserve strategy on their own (although they may be useful in an integrated regional strategy; see below). Third, the trend will always be for large fragments to be eroded unless protected. Because of the cost involved, the responsibility for acquiring and managing these large reserves must rest primarily with national conservation organizations.

The foregoing discussion is not meant to denigrate the value of small reserves. Indeed, their selection emerges as a logical strategy when one considers the different levels of organization and scale at which conservation policy is determined (McLellan et al., 1986). We have argued above that the largest fragments of threatened habitats should generally be obtained as reserves by *national* conservation organizations. However, in a heterogeneous environment these reserves may not encompass all of the habitat variation (and thus all of the characteristic biota) present in the ecosystems concerned. Thus, we suggest the primary task for conservation organizations operating on a *regional* scale should be to distribute their funds for land acquisition among a series of medium-sized reserves designed to capture this variation. The optimal trade-off between capturing more habitat heterogeneity (by purchasing several smaller reserves) and maintaining viable populations of area-sensitive species (by purchasing fewer larger reserves) will have to be determined by detailed studies of each particular system (Simberloff and Abele, 1982). Conservation on a *local* scale, as in a township, operates under the tightest

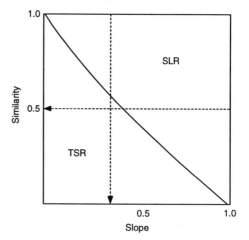

Figure 29.7 [*orig. figure 7*] This diagram shows how the similarity in species composition (Jaccard's coefficient) between two equal-sized small reserves and the slope of the (log-log transformed) species–area relationship determine whether or not more species will be preserved in one large area or two smaller areas of the same total area. Above the solid line, a single large reserve (SLR) holds more species; below it two small reserves (TSR) hold more species. The dotted lines and shaded area define the expected parameter space when the SLR is small in absolute terms, and all of the reserves are geographically close (i.e. when conservation is operating on a local scale). (After Higgs and Usher, 1980.)

budgetary constraints and thus the most restricted size range for possible reserves (all of which will be small in absolute terms). However, it is possible to state with some confidence that the best strategy at this scale is to go for single "large" reserves rather than several (very) small ones. There are two main reasons: (1) The slope (z) of the species–area relationship is normally steep (>0.35) at small areas where fragments contain a small proportion (<0.25) of the species pool (Martin, 1981); (2) The similarity in species composition among the small local reserves will usually be very high (>>0.5) because of their physical proximity and likely similarity in habitat, and because the effective species pool which can colonize them may be considerably less than the regional pool due to minimum area effects. Higgs and Usher (1980) have shown that under these circumstances more species will be contained in single large reserves (figure 29.7).

For each major habitat type within a given region, the result of applying these hierarchical strategies would be a small number of large national reserves, a network of medium-sized regional reserves, and a large number of small local reserves. With sufficient integration between organizational levels, primarily in regard to reserve placement (see below), a composite strategy such as this might be adequate to ensure the long-term persistence of those target species not already extirpated by fragmentation.

How close?

It will often be impractical to speak of clustering large national reserves such as national parks and forests, since they can be in widely separated regions of the country. On a local level, there may be great benefit to placing reserves close to each other. The large national preserves can serve as sources of colonists for the smaller local preserves, which themselves may serve reciprocally as stepping stones. These benefits will accrue only to the more vagile organisms such as birds (Lynch and Whigham, 1984), bats (Wilcox, 1984), and those species able to pass through the variety of habitats in the surrounding landscape (many temperate zone mammals). In terms of linking reserves, the value of corridors *per se* is debatable. They are unlikely to reduce the isolation of two distant reserves, and dispersal might occur anyhow if the reserves are close (Frankel and Soulé, 1981). More useful are land use practices which allow populations of many target species to exist at least marginally in the surrounding habitat. These populations can then diffuse into the reserves.

Reserve shape

Diamond (1975b) and Wilson and Willis (1975) have recommended that reserves be as nearly circular in shape as possible. The stated reason is to minimize dispersal distances within a reserve (but see Game, 1980). In the case of temperate zone forest reserves, we may add a second reason – to minimize the proportion of forest edge to forest interior. (By similar reasoning, clearings should not be allowed within the forest. If clearings must be created, they should be placed as close to the edge as possible and clustered together.)

Blouin and Connor (1985) have produced a detailed statistical analysis of data for oceanic islands which suggests that island shape is unimportant in determining the species composition of the islands studied. However, the analysis misses the point about application of the theory of island biogeography to the management of species in nature reserves. Essentially, circularity in the shape of forest fragments may be advocated purely to diminish the impact of edge effects. This is unlikely to be important in oceanic island habitats, where interactions between species across the border of the island are very infrequent. Indeed, at the risk of overgeneralizing, we suggest that the optimal shape for *any* habitat reserve is circular, so as to minimize contact between the protected interior and the surrounding habitat.

Management

Finally, we believe that over the long run virtually all temperate zone reserves will require active management to prevent or overcome the ecological imbalance created by fragmentation or human activity. Good reserve design will lessen but rarely eliminate the need for management (Gilbert, 1980). Such management may take several forms, including controlled treatment of the vegetation to preserve particular successional stages (open country for the large blue butterfly); the elimination of foreign species (wild boars in the Great Smoky Mountains National Park); or the culling of populations of "nuisance" animals (cowbirds in the breeding grounds of the Kirtland's warbler). Conservationists must realize that the battle is not over once the land has been saved. Indeed, it has just begun.

Suggested Readings

Burgess, R. L. and D. M. Sharpe (eds.). 1981. *Forest Island Dynamics in Man-Dominated Landscapes*. Springer Verlag, New York.

Harris, L. D. 1984. *The Fragmented Forest*. Univ. of Chicago Press, Chicago.

Verner, J., M. Morrison and C. J. Ralph (eds.). 1986. *Modeling Habitat Relationships of Terrestrial Vertebrates*. Univ. of Wisconsin Press, Madison.

Wilcox, B. A. 1980. Insular ecology and conservation. In *Conservation Biology: An Evolutionary-Ecological Perspective*, M. E. Soulé and B. A Wilcox (eds.). Sinauer Associates, Sunderland, MA.

30

THE EFFECTS OF DEFORESTATION IN AMAZONIA*

H. Sioli

An exposé on the effects of deforestation in Amazonia has to start with an introduction into the most important peculiarities or principles of the ecology of the region. I shall restrict myself to the high-forest-covered *terra firme* of the Amazonian lowlands.

1 Ecological Peculiarities

The first scientist to recognize and outline the most striking basic ecological characteristics of that area was the Swiss Hans Bluntschli who stated that '. . . wind and plain, forest and water act intrinsically together. We perceive that everything in Amazonia must come under their influence, from the smallest living being to the activity and behaviour of mankind . . .' (Bluntschli, 1921: 51, translated by H. S.). As a comparative anatomist, Bluntschli observed with eyes schooled in structural interrelations that the forest and water 'act intrinsically together', depend on each other and are mutual expressions of one another.

Chemical poverty including lack of nutrients

Bluntschli, however, did not make any chemical analysis of water and evidently did not know that Katzer had already discovered the surprising chemical pureness and weakness of most Amazonian waters (Katzer, 1903). Otherwise, that keen observer would have discovered deeper and more revealing interrelationships. Systematic research on Amazonian water chemistry started only after 1945 and still continues. It confirmed Katzer's early discovery and extended it to by far the largest areas of the whole forest-covered Amazonian region: the 'hylaea' of Alexander von Humboldt (Sioli, 1950, 1954, 1955; Schmidt, 1972; Furch, 1976; Junk and Furch, 1980). The only exceptions to that general rule are a few small local zones with peculiar geology and lithology (Sioli, 1963). It is especially true of the strip of land of the Andean foothills with its projections along the whitewater rivers and the alluvial lands of the *várzeas*. Here we find chemically richer waters, signs of richer, more fertile soils, and conditions completely different from those

* Originally published in *Geographical Journal*, 1985, vol. 151, pp. 197–203.

of the *terra firme*. Most Amazonian waters, however, prove to be so pure and poor in electrolytic content that they are best compared to distilled water of low quality and almost equal to rainwater.

In a climate as humid as that of the Amazonian hylaea, and under a mature forest cover (i.e., with a biomass constant over a long period of time) all soluble substances are liberated from the soil by weather processes and are being washed out. Since they are nowhere retained or accumulated, as would happen with an increasing mass of vegetation, they must appear in the ground, spring and creek waters. But since these waters are as chemically pure and poor as I have described, we must conclude that the soils are correspondingly poor and do not contain appreciable quantities of substances, including nutrients for plant growth, that could be mobilized by weathering. By implication this means that those soils are extremely infertile.

Chemical analyses of soils (Camargo, 1948; Sombroek, 1966; Falesi, 1972) confirmed the conclusion drawn from the water chemistry.

These results enable us to understand one fact or principle which is basic for the forest ecosystem of by far the greatest part of the Amazonian hylaea: the poorness in nutrients of its soils, or in other words their infertility.

Recycling of nutrients

Despite their infertility, an exuberant forest stands on the Amazonian soils. At first glance this seems a paradox and must be clarified. The answer is that the living part of that forest ecosystem, the forest vegetation itself, responds to that challenge by a strictly closed circulation of nutrients within the living biomass (Stark, 1969; Klinge and Fittkau, 1972). This constant recycling of the same nutrient molecules through generations of forest organisms must be understood as one of the basic principles in the functioning of the Amazonian forest system.

This principle alone does not, however, explain how the forest manages to realize and maintain its closed nutrient cycle.

Studying it in detail, we find that nature has applied all possible means to achieve that aim: to reduce to a minimum the loss of nutrients by leaks from the cycle. There is no appreciable humus layer on top of the Amazonian forest soil. Instead, the root system of the trees is shallow, usually restricted to the uppermost 20–30 cm, and on average three times as dense as that of a temperate forest (Klinge, 1973a). This indicates that the nutrients are never temporarily retained or stored in dead matter on top of this soil (from which they might easily be leached by the high rainfall) but are rapidly reabsorbed from the forest litter – and from the droppings and remains of animals – which rapidly decomposes in the hot humid climatic conditions. Stark (1969) was the first researcher to discover that the decomposition of dead forest litter, and hence the recycling of nutrients within the forest system, is performed by the fungi. Herrera et al. (1978) were even able to show fungi with one end of their hyphae inside a decomposing litter leaf and the other end inside a living root. And Singer (1984) recently found specific details and conditions of the mycorrhiza.

The intensity of nutrient recycling in the Amazon forest has also been shown by W. Franken (verbal communication) by comparing the chemical composition of rainwater, of water falling from the canopies, of stem runoff, and of groundwater in a forest near Manaus. While the rainwater was chemically as poor as would be expected in an area far from industrial civilization, the canopy water and the stem runoff were very rich in dissolved matter including nutrients, whereas the groundwater was practically as poor as the rainwater. This means that the dense mat of roots of the forest trees acts as a highly effective filter. It retains all dissolved substances from the canopy water and the stem runoff, and immediately recycles them back into the living trees without losing any into the groundwater, creeks and rivers which would finally carry them into the ocean to be lost forever from the forest ecosystem (see also Klinge and Fittkau, 1972).

The chemical richness of the subcanopy

water and the stem runoff is explained, not only by the leaching effect of rainwater on the leaves and epiphytic and epiphyllic plants of the forest trees, but far more importantly by the fact that probably the greatest part of Amazon forest fauna lives in or on the canopies of the trees: insects, birds and up to the coatis and monkeys. They not only feed but also defecate there and their 'manure' is washed down by the rains. The fauna is thus perfectly integrated into the general nutrient recycling process within the life of the forest as a whole: it simply adds a lateral extension to it.

Tightly closed and rapid recycling is all the more essential for the forest's existence, since *terra firme* soils consist of up to 80 per cent or even more of fine quartz, the rest being kaolinite as the only clay mineral present (Irion, 1976, 1978). Because of the low absorption capacity of kaolinite, the soil would never retain nutrients, even those liberated by decomposition of inorganic and organic matter. The first rains would extract such nutrients from the soil and carry them beyond the forest ecosystem into the groundwater, creeks and finally into the sea.

Protection against soil erosion

Good percolation capacity of the soil under high forest cover reduces surface runoff to practically zero, and the dense canopies of high forest inhibit surface erosion. The dense surface mat of tree roots not only recovers nutrients for the living vegetation: it also immediately soaks up a high percentage of fallen rainwater to be evapotranspired again by the evergreen latifolious trees and quickly returns it to the atmosphere, where it condenses and falls once again as new rain. Salati et al. (1978, 1979) found that about 50 per cent of rains in Amazonia consist of water recycled within the region. The pluvial climate of Amazonia favours uninterrupted growth of vegetation through its annual average rainfall of some 2500 mm and its absence of arid season (Koppen types Af and Am). But it depends on that portion of the rainwater which is recycled thanks to the existence of latifolious forest with its high evapotranspiration capacity. That is the final implication of Bluntschli's early observation.

We thus see why recycling processes are essential for the existence and survival of the forest ecosystem in Amazonia.

Diversity of the forest ecosystem

Once these recycling processes have overcome the basic difficulty of perpetuating the Amazon forest ecosystem – that is the lack of a steady supply of fresh nutrients – nature has developed here the most diverse ecosystem we know on earth (see Fittkau, 1973; Klinge, 1973b). The greatest number of plant and animal species we are aware of (estimated at between 1.5 and 2 million species) divides the big, general nutrient cycle into an immense number of subcycles. Indigenous man in his native culture is perfectly adapted to and integrated in that system. The diversity of species may be taken as yet another 'principle' of the Amazon forest ecosystem.

The enormous number of plant and animal species means not only that there is an equally vast number of ecological niches occupied and utilized by those species or nutrient subcycles intercalated into the major closed cycle. It also means that these plant and animal species, together with the abiotic environmental conditions, by their mutual interaction and feedback form a single complex that includes an even greater number of homeostatic cycles, from very tiny to large ones, all of which interact and influence one another. At the same time, they all depend on the fact that the greatest homeostatic cycle of that system – that of the tightly closed cycle which results from the cooperation of the smaller ones – functions with a minimum of loss. The whole system may be loosely compared to the circulation of an organism in which there are blood vessels of the most diverse sizes, from capillaries to the main arteries.

This extremely complex, diverse and rich Amazon biota has developed over millions of years, uninterrupted by glacial periods or general aridification. It represents the greatest genetic pool and reserve we know on earth, and is at the same time the basis of and the starting material for future biotic evolution.

2 Effects of Deforestation

With these basic ecological peculiarities or principles of Amazon *terra firme* and its ecosystem in mind we may attempt a survey of some consequences to be predicted when current 'development' projects are undertaken. All of them interfere with the Amazon forest and involve large-scale deforestation, almost all for export purposes.

Circulation and reserves of nutrients

Deforestation interrupts the tightly closed recycling of nutrients within the ecosystem. The inorganic components of the biomass, particularly the nutrients, are liberated by burning the felled forest and are contained in the ashes – since they are not fixed in the soils, as already shown. Most are then washed out by the first rains and are removed from the environment by surface runoff from the denuded areas or through the groundwater. The rest are absorbed by crops or through grasses by cattle, and are exported with the harvest or the beef.

The word 'colonialism' is used nowadays as a synonym for the exploitation or pauperization of one region for the benefit or enrichment of another. Deforestation and export of the resulting products can only be termed neo or endocolonialism. I must also mention that cattle ranches, which have already destroyed tens of thousands of km² of forest and transformed them into artificial steppe, need relatively few personnel and are thus of no advantage to the local population.

The result of such activity is impoverished soils, deprived of the stock of nutrients contained in the former living biomass. Carrying capacity of artificial pastures in a former hylaean section of the Belém–Brasilia highways decreased from 0.9–1 head of cattle on young pastures to only 0.3 head after some six years (H. O'R. Sternberg, verbal communication). Nutrients are one basis of constant biological productivity. Their irreplaceable loss reduces and limits not only harvests but also biological production.

Surface erosion and compaction of the soil

The soils, sediments and thick weathering layers of the Amazon *terra firme* are easily eroded. With deforestation, these soils and layers are deprived of the protection of the dense forest canopy and exposed to the direct impact of heavy rains and tropical thunderstorms. The substitution of forest by short-cycle crops or by planted grasses removes the former protection. In the case of planted pasture, we must also consider that there is no tropical grass perfectly resistant to trampling by cattle. Besides, during the dry seasons, pasture grasses dry up, the sod is exposed still more, and the bare soil is strongly heated, destroying its microflora. There is a danger that the region may develop into a new 'dust bowl' – dust, formerly unknown in Amazonia, is already common on the new roads during the dry season. Also, formerly porous forest soil is compacted by deforestation (Schubart, 1977). Through general and especially slope clearing, deforestation increases surface runoff. With the onset of the first heavy rains of the next rainy season, the unprotected dust-dry soil will be eroded with unprecedented violence and washed away into the creeks and rivers.

Sandification

Sandification on the bare surface layer of the soil is another effect. The direct impact of heavy raindrops causes what can be called 'selective erosion', washing away the finer clay particles while the coarser and heavier sand is left behind. This effect can also be observed on small areas left without protective vegetation for a longer time. The most striking example of that type of selective erosion is the famous 'sandy campos' behind Santarém which the famous British naturalist Henry Walter Bates (1892) described in the last century. With time, 'sandification' proceeds deeper into the soil, thereby reducing its water-retaining capacity. This new edaphic condition is especially hostile to young seedlings of forest trees and finally prevents the regrowth of a forest.

Climatic change

Climatic change will be one of the most serious effects of large-scale deforestation in Amazonia. Small and temporary clearings (*roças* of the native people, Indians as well as *caboclos*) are mere pinpricks in a coherent forest, which soon heal and cicatrize after abandonment and do not noticeably affect the general recycling of nutrients nor reduce the generally high evapotranspiration of the whole forest. But the effect on it of great clearings will be worse, the more extensive the deforestation. With the reduction of rainwater re-evaporating from the original forest, the rate of its recycling will also diminish. Evapotranspiration of steppe is estimated to be only a third that of latifolious evergreen forest, and the availability of water for evapotranspiring steppe plants is reduced in the same proportion. The consequence will be that total annual rainfall will decrease considerably when a certain percentage of Amazon forest has been destroyed, and the seasonality of rainfall will become more pronounced. This will probably have a disastrous effect on the survival of spared forest areas which are intended as 'nature reserves' or the like.

With longer and more severe dry seasons, the supply of rainwater to the generally surface root systems of the forest trees will be interrupted for longer periods. Since the groundwater level of Amazon *terra firme* is usually very deep and is not reached by the roots of most trees, these depend on having only short periods without rain, just the relatively dry seasons called 'summer' in Amazonia. Through climatic change, it is inevitable that the planned limited forest reserves will automatically also die.

Rivers

More intense surface runoff and soil erosion, and seasonal changes in the pluvial climate on the regime, will have obvious effects on the sediment load and sedimentation processes of the rivers. We must certainly expect more sudden and higher floods, lower water levels during the dry season, greater turbidity and increased bottom freight, and partial silting of the river beds in unpredictable places.

Global consequences

Large-scale deforestation of Amazonia must also be taken into consideration. Determination of the biomass of the Amazon high forest (undertaken in a subregion near Manaus that is poor in nutrients and with a noticeably weakly developed forest) has revealed that the dry vegetable matter amounts to about 500 tons per hectare, which corresponds to about 250 metric tons of carbon fixed in a steady-state equilibrium (Klinge, 1976). Extrapolated for the whole Amazonian hylaea of nearly 5×10^6 km^2, this means that about 115×10^9 tons of carbon are retained in the forest matter. That is of the order of nearly 20 per cent of the carbon of the entire atmosphere's CO_2.

Removal of that forest biomass and replacement with artificial steppe (pasture) or short-cycle crops means that at most 20 per cent of the C-content of the former forest biomass will be fixed in the new vegetation. The rest will be oxidized, by burning or by rotting etc., and will enter the atmosphere as carbon dioxide. Experience during recent decades indicates that about half that additional input of CO_2 into the atmosphere will disappear into the sink of the oceans. There would be thus a net increase of about 8 per cent in the CO_2 content of the global atmosphere if the entire Amazon forest were replaced by a much less voluminous, manmade vegetation. Together with the increase of about 16 per cent of CO_2 in the atmosphere already observed since the last century, that new addition would contribute to the greenhouse effect of the CO_2 in the atmosphere. (For more details see Sternberg, in press.)

Trying to relate the existence or disappearance of the Amazon forest to the O_2 content of the atmosphere, or to hold that forest responsible for up to 50 per cent (!) of the oxygen production on earth, is naturally nonsense and invented only for propaganda reasons.

The annihilation of the enormous number and diversity of species of organisms which make up the unique Amazonian forest ecosystem will probably have a most incisive effect on all life on earth and its future. Not only will its present diversity, richness and beauty be impoverished, but a very high percentage of the global genetic stock and reserves will be irretrievably exterminated by large-scale deforestation and with it the basis for their future evolutionary potential will be cut off.

'Life' on earth does not consist only of the metabolic processes occurring in the organisms or in the flux rates of matter and energy through the ecosystems. Perhaps the most striking essentials of life are its diversity and polychromy. That peculiarity ranges from the number of possible combinations of amino acids in protein molecules (compared to which the mass of the whole universe expressed in the lightest hydrogen atoms is infinitely small) (Eigen, 1979) to the insight that 'the purpose of the world is not the best possible general welfare but richness in destinies' (Kleinschmidt, 1930, translated by H. S.).

With this statement we leave the realm of pure classic natural sciences and enter another spiritual dimension of such qualities as beauty and, more essentially, ethics. In this context let me quote the American anthropologist R. A. Rappaport who said 'knowledge will never replace respect in the attitude of men vis-à-vis ecological systems, since the latter are so complex that sufficient understanding of their content and structure may never be at hand to permit predictions about results of their actions' (Rappaport, 1969, cit. from Sternberg, 1980).

These words are clear cut, but they are not observed by modern 'development' schemes. These envisage and take for granted only hoped-for quantitative production, mostly for export goods. They fail to consider the losses which will occur when the Amazon forest ecosystem, evolved over millions of years and maintained by its native peoples in their adapted cultures, is destroyed within a few decades.

The question thus arises: what is behind the risky projects whose proponents show no 'respect' for that ecosystem, but aim only to exploit Amazonia?

Life, in an ecological sense, can be defined as the mutual interaction between an organism and its environment, from which a new function unit results (see Sioli, 1973). 'Development' is the interaction by the organism, man, on the environment, the Amazon forest ecosystem. Since we have hitherto considered only the response of the environment, we must now look at man who performs that interaction in order to understand the whole process in truly ecological terms. Only then can we comprehend what is happening in that part of the tropics.

It is only too obvious that the 'development' ideas come from non-Amazonian people, be they in Europe, North America, Japan, or in the modern centres of southern Brazil. They come from a mentality evolved in temperate climates and under different geographical, historical and social conditions, culminating in the present 'highly developed' industrial–commercial civilization (see Sioli, 1972). The plans are conceived and elaborated solely to serve that civilization all over the world. They are not for the local population of Amazonia (whether Indians, caboclos or other Amazonized inhabitants), and are not intended to safeguard the basis of these peoples' future. This is nothing but a new wave of conquests now sweeping over the last vast tract of land spared by the original and subsequent European conquests.

What is needed is to care about the local population, which is also multiplying, and about its needs and its future. Current projects of giant enterprises for export, and of mega-colonization by foreign settlers, are obnoxious to the people of Amazonia and to future generations.

Instead, a really rational and lasting utilization of Amazonia must benefit her true inhabitants and must be based on strict observance of the well-known ecological peculiarities of that country. That means the reverse of the present ideas. The diverse Amazonian forest must be maintained as a continuum, in space and time. Inside it, population centres of limited size and number should be built as 'islands' of human civilization separated from each other by the forest. As much as possible should be recycled, with just enough produced for export to enable the people to purchase undeniably necessary goods. As a basic concept for such true development, in a humane sense for a growing number of people, without destroying the extremely fragile landscape, its diversity and sustainable bio-productivity, the ideas of Schumacher's *Small is Beautiful* (1973) may be taken as a guideline. Some groping experiments in that direction have already started. May they be successful and triumph over the destructive objectives of instant profits for foreign regions and peoples!

References

Bates, H. W. 1982 *The Naturalist on the River Amazon*. London: John Murray.

Bluntschli, H. 1921 Die Amazonasniederung als harmonischer Organismus. *Geogr. Zeitung* 27: 49–67.

Camargo, F. C. de 1948 Terra e colonização no antigo e novo Quaternário na Zona da Estrada de Ferro de Bragança, Estado do Pará, Brasil. *Bol. Mus. Paraense E. Goeldi. Belém* 10: 123–47.

Eigen, M. 1979 Zeugen der Genesis – Versuch einer Rekonstruktion der Urformen des Lebens aus ihren in den Biomolekülen hinterlassenen Spuren. In: *Jahrbuch der Max-Planck Gesellschaft*. Göttingen: Vandenhoeck and Ruprecht.

Falesi, I. C. 1972 *Solos da Rodovia Transamazonica*. Instituto de Pesquisas Agropecuárias do Norte (IPEAN). No. 55 Boletim Técnico, Belém.

Fittkau, E.-J. 1973 Artenmannigfaltigkeit amazonischer Lebensräume aus ökologischer Sicht. *Amazoniana* 4: 321–40.

Furch, K. 1976 Haupt- und Spurenelementgehalte Zentralamazonischer Gewässertypen (Erste Ergebnisse). *Biogeographica* 7: 27–43.

Herrera, R. et al. 1978 Amazon ecosystems: their structure and functioning with particular emphasis on nutrients. *Intersciencia* 3(4): 223–32.

Irion, G. 1976 Mineralogisch-geochemische Untersuchungen an der pelitischen Fraktion amazonischer Oberböden u. Sedimente. *Biogeographica* 7: 7–25.

—— 1978 Soil infertility in the Amazonian Rain Forest. *Naturwissenschaften* 65: 515–19.

Junk, W. F., and Furch, K. 1980 Water chemistry and macrophytes of creeks and rivers in southern Amazonia and the central Brazilian Shield. *Proc. Vth Int. Symp. Tropical Ecology, Kuala Lumpur*: 771–96.

Katzer, Fr. 1903 *Grundzüge der Geologie des unteren Amazonasgebietes (des Staates Pará in Brasilien)*. Leipzig; Max Weg.

Kleinschmidt, O. 1930 *Naturwissenschaft u. Glaubenserkenntnis*. Berlin: Martin Warneck.

Klinge, H. 1973a Root mass estimation in lowland tropical rainforests of Central Amazonia. Brazil: I. Fine root masses of a pale yellow latosol and a giant humus podsol. *Trop. Ecol.* 14: 29–38.

—— 1973b Struktur u. Artenreichtum des Zentralamazonischen Regenwaldes. *Amazoniana* 4: 283–92.

—— 1976 Bilanzierung von Hauptnährstoffen im Ökosystem tropischen Regenwald (Manaus): Vorläufige Daten. *Biogeographica* 7: 59–77.

Klinge, H., and Fittkau, E. J. 1972 Filterfunktionen im Ökosystem des zentralamazonischen Regenwaldes. *Mitt. Deutsch Bodenkundl. Ges.* 16: 130–35.

Salati, E. et al. 1978 Origem e distribuição das chuvas na Amazônia. *Intersciencia* 3(4): 200–06.

—— 1979 Recycling of water in the Amazon basin: an isotopic study. *Water Resources Research* 15 (5): 1250–58.

Schmidt, G. 1972 Chemical properties of some waters in the tropical rainforest region of Central Amazonia. *Amazoniana* 3: 199–207.

Schubart, H. O. R. 1977 Critérios ecológicos para o desenvolvimento agricola das terra-firmes da Amazônia. *Acta Amazonica* 7: 559–67.

Singer, R. 1984 The role of fungi in Amazonian forests and in reforestation. In: *The Amazon, Limnology and landscape ecology of a mighty tropical river and its basin*. Dordrecht: Dr. W. Publishers.

Sioli, H. 1950 Das Wasser im Amazonasgebiet. *Forsch. u. Fortschr.* 26: 274–80.

—— 1954 Beiträge zur regionalen Limnologie des Amazonasgebietes: II. Der Rio Arapiuns. *Arch. f. Hydrobiol.* 49: 448–518.

—— 1955 Beiträge zur regionalen Limnologie des Amazonasgebietes: III. Über einige Gewässer des oberen Rio Negro Gebietes. *Arch. f. Hydrobiol.* 50: 1–32.

—— 1963 Beiträge zur regionalen Limnologie des Amazonasgebietes: V. Die Gewässer der Karbonstreifen Unteramazoniens (sowie einige Angaben über Gewässer der anschliessenden Devonstreifen) *Arch. f. Hydrobiol.* 59: 311–50.

—— 1972 Ökologische Aspekte der technisch-kommerziellen Zivilisation u. ihrer Lebensform. *Biogeographica* 1: 1–13.

—— 1973 Introduction into the Problem: The Situation of Modern Civilisation in the Light of the Ecological Aspect of Life. In: *Ökologie u.*

Lebenschutz in Internationaler Sicht: Ecology and Bioprotection. International conclusions. Freiburg: Verlag Rombach.

Sombroek, W. G. 1966 *Amazon Soils*. Wageningen: Centre f. Agricult. Public. and Documentation.

Stark, N. M. 1969 Direct nutrient cycling in the Amazon Basin. *11° Simpósio y Foro de Biologia*

Tropical Amazonica: 172–77.

Sternberg, H. O'R. 1980 Amazonien: Integration u. Integrität. In: *Integration in Lateinamerika*: 293–322. Munich: Wilhelm Fink Verlag.

—— in press O. *Pulmão Verde*. Regional Latin-American Conference of the IGU, Rio de Janeiro, 1982.

31

RATES OF DEFORESTATION IN THE HUMID TROPICS: ESTIMATES AND MEASUREMENTS*

A. Grainger

There has been serious international concern for at least 20 years about the rapid rate at which tropical forests, particularly those in the humid tropics, are being cleared or otherwise modified. This concern has intensified recently owing to the contribution which carbon dioxide emitted from biomass burning after deforestation is making to global climate change caused by the greenhouse effect. Despite this, current estimates of tropical forest areas and deforestation rates are still very inaccurate, for we have continued to rely on subjective estimates and been too slow to use the full potential of the remote-sensing technology at our disposal to monitor this major component of global environmental change. This paper stresses the need to base estimates on measurements, reviews estimates of forest areas and deforestation rates in the humid tropics made since 1970, proposes a set of criteria to evaluate them, and suggests how the present situation could be improved by better monitoring.

Types of Tropical Forests

The Tropics is a large region situated between the Tropics of Cancer and Capricorn containing a diversity of climates and vegetation types. It is important when making or comparing estimates of tropical forest areas and deforestation rates to know which types of forests are included. The all-embracing term 'tropical forest' is divided most conveniently for statistical purposes into closed forest and open forest. Tropical closed forest has a closed canopy and includes forests in both the humid tropics (tropical moist forest) and dry tropics. Tropical open forest includes a diversity of open (savanna) woodlands, mainly in drier tropical areas, where trees are widely scattered and there is no continuous closed canopy, but it is difficult to establish a strict dividing line between closed and open forests in terms of canopy coverage (Lanly, 1982). Not all forest in dry tropical areas is open: about one-tenth of all tropical closed forest is in dry areas, so this term should not

* Originally published in *Geographical Journal*, 1993, vol. 159, pp. 33–44.

be used interchangeably with tropical moist forest.

Tropical moist forest is generally used to refer to all closed forest in the humid tropics (Sommer, 1976) and includes two main forest types, tropical rain forest and tropical moist deciduous forest. Tropical rain forest is found in permanently humid areas, i.e. those with at most only a limited seasonality in rainfall distribution, and tropical moist deciduous forest – also called monsoon forest – in areas with a distinct dry season. Each of these two main forest types can be divided into numerous subtypes (Whitmore, 1990), but until 1992 available data did not allow us even to estimate the separate areas of tropical rain forest and tropical moist deciduous forest, so the estimates reviewed here will refer to all tropical moist forest.

Deforestation

There is often confusion about what constitutes 'deforestation'. Usually this is only a matter of terminology, but it can lead to problems when comparing different estimates. Deforestation has been defined by this author as 'the temporary or permanent clearance of forest for agriculture or other purposes' (Grainger, 1984) and this corresponds to the definition used by UN agencies (Lanly, 1982). The key word is 'clearance': if forest is not cleared then deforestation does not take place, according to this definition. Deforestation results when forest is replaced by another land use, including different types of shifting and permanent agriculture as well as non-agricultural uses such as mining.

However, there are other human impacts which do not cause deforestation. The second main impact on tropical moist forest, for example, is selective logging, but as usually only two to 10 good-sized trees are removed from a hectare of forest, which may contain between 400 and 700 trees greater than 10 centimetres in diameter, this does not lead to forest clearance and so does not constitute deforestation as defined here. Clearfelling would cause deforestation but is rarely practised in the humid tropics,

unlike the situation in temperate forests. Selective logging should therefore be treated as a separate phenomenon. To encompass it and other lesser impacts requires a wider term, e.g. 'degradation', which could be defined as 'the temporary or permanent deterioration in the density or structure of vegetation cover or its species composition' (Grainger, 1993). Deforestation represents one extreme of degradation and is easy to monitor by satellite sensors in comparison with selective logging.

Measurements and Estimates

Forests cover huge areas in the Tropics, so it is easy to appreciate the difficulties involved in assessing the current extent of tropical moist forest and how rapidly this is changing. Many estimates are still based on the subjective judgement of experts supposedly knowledgeable about the area concerned, rather than on remotely-sensed measurements. When evaluating estimates the distinction between those based on measurements and those relying on a high degree of expert judgement is therefore a crucial one.

In previous centuries and the early part of this century, tropical forest areas had to be estimated by a combination of ground inspection, maps and subjective judgement. The situation changed when aerial photography became established as a standard forest inventory tool, but photographing all the forests in a country and then interpreting the many photographs collected is very time-consuming. In the humid tropics bad weather and heavy cloud cover are additional constraints on national aerial photographic forest surveys, which have therefore tended to be fairly infrequent, commonly at intervals of not less than every ten years and often longer. This is also a major handicap when trying to estimate historical deforestation rates.

However, in 1972, the first earth resources remote-sensing satellite, later called Landsat 1, was launched by the US National Aeronautics and Space Administration (NASA). For the first time this gave the potential for continuous monitoring of forests and other earth resources. Satellite remote sensing in the visible and infra-red parts of the spectrum could not avoid problems due to high cloud cover, and still required ground checking to provide valid results, but it was no longer necessary to mount a large number of flights each time a national forest survey was needed, as the satellites collected data continuously. Image interpretation was also speeded up: larger areas were included on a single satellite image and as data were stored in digital form interpretation could be assisted by computers. Since 1972, a growing number of satellites have been launched and sensor resolution has improved. There are also a number of weather satellites operated by the US National Oceanographic and Atmospheric Administration (NOAA) which carry lower resolution (AVHRR) sensors but have the advantages of an even larger area covered in a single image and daily overflights of each area, compared with a return time of more than two weeks for a Landsat satellite. Cloud problems can be overcome with radar sensors, carried either on aircraft (side-looking airborne radar was used to survey forests in Brazilian Amazonia in the early 1970s) or on the new generation of radar satellites, which include the European Space Agency's ERS–1, soon to be followed by ERS–2 and a Japanese satellite, JRS–1 (Lillesand and Kiefer, 1987).

The technology therefore exists to make frequent large-area forest measurements and monitor the state of the world's forests on a continuous basis (Grainger, 1984). Global assessments of tropical forest areas, since 1980, have indeed made increasing use of satellite data for some countries, but not for all, and the number of estimates of deforestation rates made in this way is even smaller. People assume that because the technology is available all estimates make use of remote-sensing measurements, but this is not the case. For even though satellites collect data continuously this does not mean that the data are interpreted continuously, or even periodically. So it is very important when evaluating estimates to find if they were based on measurements or simply rely on expert assessment.

Criteria for Evaluating Estimates

Lack of measurements can lead to major differences between alternative estimates of deforestation rates. People wanting to know which is the most reliable estimate are therefore placed in a quandary. To help overcome these problems this section presents a list of nine criteria, summarized in table 31.1, to use when comparing the relative merits of different estimates of tropical forest areas and deforestation rates. Generally speaking, the most recent estimate is not always the best, and like should be compared with like. Variations occur even if measurements are

Table 31.1 [*orig. table I*] A check list for assessing the reliability of estimates of deforestation rates.

1	Who made the estimate: is it a primary or secondary estimate?
2	What type of change in forest cover is estimated? How is deforestation defined?
3	To which type(s) of tropical forest does the estimate refer?
4	Is the estimate based on measurements or subjective judgement?
5	If the former, what form of remote sensing was used?
6	What sensor was used and what was its resolution?
7	What were the dates when the measurements were made?
8	Was the whole country/region surveyed?
9	If not, was a statistical sampling methodology used?
10	Was a remotely-sensed survey compared with a map?
11	If so, is it known upon what measurements the map was based?
12	Is the figure an estimate of an historical change or a projection of a future change?

made and the criteria will help in those circumstances too.

1   Who made the estimate? Estimates of national deforestation rates may be derived from government statistics, or from studies by remote-sensing scientists either in the country itself or overseas. At the global level there are, so far, only two sources of primary estimates, i.e. those made by a detailed examination of the full range of available national data, including specially commissioned surveys as necessary. These are the UN Food and Agriculture Organization (FAO) – often in partnership with the UN Environment Programme (UNEP) – and an independent specialist, Norman Myers. All other global estimates, such as those made by the World Resources Institute (1990), have until now been derived from these estimates and are referred to here as secondary estimates. The reliability of secondary estimates is much less and they should be treated to greater scrutiny.

A common error on the part of those making secondary estimates has been to use estimates of national areas of 'forests and woodlands' published annually in the *FAO Production Yearbook* series to estimate trends in deforestation rates. These highly aggregated figures normally merely report statistics released by the governments concerned, without detailed checking and clarification by forest inventory experts to ensure homogeneity associated with estimates made by the FAO Forestry Department. The latter previously undertook regular world forest inventory studies beginning in the early 1950s but suspended its work in the middle of preparing the 1973 inventory because of dissatisfaction at the quality of the data submitted to it by governments: the low frequency of national forest inventories has already been referred to (Persson, 1974). The 1980 FAO/UNEP Tropical Forest Resource Assessment (Lanly, 1981) was actually the first comprehensive survey of tropical forest areas made by the FAO Forestry Department since 1965. The implications for the reliability of forest data issued by other departments of FAO are obvious.

2   To what type of change in forest cover does the estimate refer? If the focus is on 'deforestation', how is the term defined by the author of the estimate? Definitions are very important. Deforestation, as defined above, is a clear physical change in forest cover and one that is easily measurable by remote-sensing techniques. Lesser impacts, such as the degradation caused by selective logging, are more diffuse and harder to measure. If the estimated rate of 'deforestation' includes these lesser impacts it will be much greater than an estimate of the rate of clearance alone.

Norman Myers used the term 'conversion' in his report for the US National Research Council to refer not only to deforestation but also selective logging and the clearance of forest fallow vegetation on shifting cultivation plots (the latter had been previously cleared of forest before cultivation) (Myers, 1980). Myers clearly stated what his much larger figure referred to, but many people subsequently saw it as representing an alternative estimate of the rate of deforestation and the dispute took some years to be resolved.

3   What type of tropical forest is referred to in the estimate? The diversity of tropical forest types was referred to above, and it is understandable that the deforestation rate for tropical forest (which includes open forests) is greater than that for tropical closed forest, which in turn exceeds that for tropical moist forest. It is surprising how often these three deforestation rates are confused. For example, the deforestation rate for tropical closed forest is frequently quoted as though it referred to tropical moist forest. No separate estimates of tropical rain forest area or deforestation rate was made until 1992.

4   Was the estimate based on specified measurements or on the subjective judgement of an expert? All estimates must come from somewhere: if they are based on measurements and the measuring technique is quoted – increasingly the case with recent estimates – their reliability is easier to assess. It should not be assumed that because an

estimate is published by a large international organization it must be accurate.

5 If measurements were made, what remote-sensing technique was employed – aerial photography, airborne radar, visible/infra-red satellite or radar satellite? What was the resolution of the sensor? How were the remote sensing data interpreted? Measurements made by the medium- to high-resolution sensors on the Landsat and SPOT satellites are more reliable than those made by low-resolution AVHRR sensors on NOAA weather satellites. This particularly applies to deforestation rates, because coarse resolution sensors cannot detect the many small forest clearances of shifting cultivators. These can also be missed when estimates rely on visual interpretation of photographic versions of satellite images from medium- to high-resolution sensors, rather than computer-assisted analysis of the original digital data.

6 If measurements were made, was the whole forest area of a country surveyed or just a part of it? In the case of a partial coverage, how large was this and was a statistically-based sampling methodology used? If not, then extrapolation to the whole country could be questionable.

7 What were the dates of the measurements? Deforestation involves a reduction in forest area over a given period, so two sets of measurements of forest areas are required at the initial and final years of that period. Ideally the dates of measurements should match in those years. Sometimes one or both sets of measurements are spread over a number of years, perhaps because heavy cloud cover meant that good quality images of all areas could not be collected in the same year. This reduces reliability, as found with a Brazilian estimate referred to below, and needs to be taken into account.

8 Were the two forest area estimates, on which the deforestation rate was based, both derived from remote-sensing measurements? It is still common for one estimate of forest area based on a recent remote-sensing survey to be compared with another estimated from an earlier forest map, perhaps dating back to the 1960s or 1970s. Such maps are assumed to be reliable and objective and the measurements on which the map was based are not quoted. Although usually derived from aerial photographic surveys this might not have been the case, and varying degrees of subjectivity would have been involved on the part of those who drew the maps. The date of the original survey should be ascertained, as a map supposedly referring to 1965, for example, might have simply been redrawn from a 1955 map (based on the last full aerial photographic survey) using subjective judgement to make the necessary adjustments.

9 Does the estimate refer to a change that has already happened or is it a projection of a change that might occur in the future? This may sound a rather trivial question but FAO often publishes both estimates and projections, the latter based not on measurements but on the use of mathematical models or subjective forecasts. The two types of figures must not be confused. Unfortunately a number of secondary estimates have used FAO projections as though they referred to actual historical changes. It was common in the late 1980s, for example, for projections made by FAO/UNEP in 1981 (Lanly, 1981) concerning possible deforestation rates in 1981–85 to be regarded as the 'most recent' estimates of deforestation rates, even though they were published in the same report as a comprehensive set of estimates of actual deforestation rates for 1976–80.

Many of these criteria could be used to evaluate forest area estimates as well. It is not suggested that each global estimate, whether of forest areas or deforestation rates, will contain sufficient information to allow all the questions to be answered. But the amount of supplementary information supplied continues to improve and even if only some of the questions can be answered this should be sufficient to make a good assessment of reliability.

Table 31.2 [*orig. table II*] Estimates of tropical moist forest area (million ha).

Source	Date	Area
Persson*	1974	979
Sommer	1976	935
Myers*	1980	972
Grainger	1983	1081
Myers	1989	800
FAO	1992	1282

* Derived from source data by Grainger (1984).

Estimates of Tropical Forest Areas

Sommer (1976) The first person to specifically estimate the area of tropical moist forest was Adrian Sommer in 1976. In a report for FAO he listed 65 countries as being in the humid tropics and estimated their total forest area as 935 million hectares in the 1970s (Sommer, 1976). National forest areas given in two other surveys, by Reidar Persson (1974) and Norman Myers (1980), were used by this author to produce similar secondary estimates of 979 million hectares and 972 million hectares respectively (table 31.2) (Grainger, 1984).

FAO/UNEP (1981) The 1980 FAO/UNEP Tropical Forest Resource Assessment, published in 1981, was the first of a new generation of surveys of this kind. It contained a much more disaggregated set of data on different types and conditions of forests, made greater use of remote-sensing surveys, applied a uniform forest classification to all countries, corrected forest area

estimates, to the same year (1980), and listed its data sources (including types and dates of measurements where known). It gave the total area of tropical closed forest, in 1980, as 1201 million hectares and that of open forest as 734 million hectares (table 31.3) (Lanly, 1981).

Grainger (1983) The present author produced a secondary estimate of 1081 million hectares for tropical moist forest alone, based on these FAO/UNEP data for 55 of the countries listed by Sommer (1976). Best available estimates were used from other sources for another six countries, together with Puerto Rico and Vanuatu, none of which were included in Lanly (1981) (Grainger, 1983). (To enable a comprehensive estimate to be made, best available data from countries not listed by FAO/UNEP were used by the author for Australia, Fiji, New Caledonia, Solomon Islands and Puerto Rico [Myers, 1980a]; Reunion and Vanuatu [Persson, 1974]; and Hawaii [Nelson and Wheeler, 1963]). Over half of all tropical moist forest was shown to be in Latin America, a quarter in Asia-Pacific and the rest in Africa (table 31.4).

Comparing the 1980 estimate with those made in the 1970s suggests that tropical moist forest area actually increased over that decade, which is impossible. The problem results from the relative inaccuracy of the earlier estimates. Because the 1980 figure relied on remote-sensing surveys made since 1970 for just under three-fifths of the total tropical moist forest area we must assume it was the most reliable figure then available. But it was not without its faults,

Table 31.3 [*orig. table III*] Areas of tropical closed forests, tropical moist closed forests and tropical dry open woodlands in 1980 (million ha).

	Closed Forest All[a]	Open Moist[b]	Woodland[a]
Africa	217	205	486
Asia-Pacific	306	264	31
Latin America	679	613	217
Total	*1201	*1081	734

NB * Totals are not the sum of regional figures due to rounding.
Sources: [a]Lanly (1981); [b]Grainger (1983), based on Lanly (1981).

Table 31.4 [*orig. table IV*] Estimates of regional tropical moist forest area (million ha).

	Persson	*Sommer*	*Myers*	*Grainger*	*Myers*	*FAO*
	*1974**	*1976*	*1980**	*1983*	*1989*	*1992*
Africa	183	175	156	205	152	405
Asia-Pacific	249	254	290	264	211	180****
Latin America	547	506	527	613	416	687
Total***	979	935	972	1081	800**	1282

*	Derived from source data by Grainger (1984).
**	Regional figures are totals of individual countries listed by Myers. Totals was adjusted by him for the whole biome-type.
***	Total may not always equal sum of regional figures due to rounding.
****	Asia only.

and the lack of cited measurements for the remaining two-fifths of all tropical moist forest means that it cannot be considered a reliable baseline for comparison with later estimates (Grainger, 1984), which was the intention at the start of the project (FAO, 1975).

Myers (1989) In 1989 Norman Myers estimated the total area of tropical moist forest as 800 million hectares (table 31.4), extrapolating from an aggregate of estimates for 34 countries accounting for 97 per cent of all tropical moist forest (Myers, 1989).

FAO (1990, 1991) On the other hand, in the following year FAO released an interim figure as part of its forthcoming 1990 Forest Resource Assessment, putting the area of 'tropical moist forest' at 1282 million hectares. As this was even larger than the 1980 figure for all tropical closed forest it must be assumed that it either referred to that or to the sum of closed forest and forest fallow in the humid tropics (FAO, 1990). A second interim estimate published the year afterwards put the total area of tropical forest (i.e. closed and open, humid and dry) in 87 countries at 1714 million hectares. This was rather more comparable with the corresponding figure of 1935 million hectares in the 1980 assessment, which covered 76 countries (FAO, 1991). (For the purpose of comparison FAO [1991] released a revised estimate of 1882 million hectares for the total area of tropical forest in the same 76 coun-

tries in 1980.) The 1991 estimate was disaggregated by region but not by forest type. However, in the following year FAO released interim figures of 1282 million hectares for the area of tropical moist forest and 656 million hectares for the area of tropical rain forest in 1990 (FAO, 1992). These suggest that Myers understated the actual area in 1989.

We might question the reliability of both recent reports: Myers indicated a larger amount of forest clearance during the 1980s than would be expected from earlier estimates of deforestation rates: a loss of 200–300 million hectares from the Grainger (1983) figure compared with up to 100–150 million hectares even if deforestation rates were much higher in the 1980s (see below). Meanwhile, FAO's interim estimates all imply yet another 'increase' in tropical moist forest area. It seems best to reserve judgement until FAO's final report is published. Whatever the correct figure for 1990 may be, clearly much inaccuracy remains so it would be unwise to use trends in such global area estimates to estimate mean annual deforestation rates over the last 20 years.

Estimates of Deforestation Rates

Sommer (1976) A number of primary estimates have been made of deforestation rates in the humid tropics (table 31.5). The earliest of these, 11–15 million hectares per annum, was published by Adrian Sommer in his report for FAO mentioned above. During

Table 31.5 [*orig. table V*] Estimates of rates of deforestation in the humid tropics (million ha per annum).

Source	Date	Period	Total	Notes
Sommer	1976	1970s	11–15	15 commonly quoted
Myers	1980	1970s	7.5–20	7.5 a later revision
Grainger	1983	1976–80	6.1	cf 7.3 for all tropics
Myers	1989	late 1980s	14.2	–
FAO	1992	1980s	12.2	–

the second half of the 1970s it formed the basis for the commonly accepted rate of 15 million hectares per annum, though the figure that Sommer gave was actually 'at least 11 million hectares per annum'. This was derived by calculating the average percentage deforestation rate for just 13 countries (Bangladesh, Colombia, Costa Rica, Ghana, Ivory Coast, Laos, Madagascar, Malaysia, Papua New Guinea, the Philippines, Thailand, Venezuela and North Vietnam) for which estimates of deforestation rates were available and then extrapolating the percentage to all tropical moist forest. These countries accounted for 16 per cent of all tropical moist forest. The sources of these estimates were not given (Sommer, 1976).

Myers (1980) The next primary estimate emerged from a report made by Norman Myers for the US National Research Council and published in 1980. The report focused on examining, in a mainly qualitative way, the scale and causes of all impacts on forests in the leading countries in the humid tropics,

and Myers used the term 'conversion' to refer to all impacts 'from marginal modification to fundamental transformation'. He only gave estimates of deforestation rates for 13 countries (Brazil, Burma, Colombia, Indonesia, Laos, Liberia, Madagascar, Nicaragua, Papua New Guinea, Peru, the Philippines, Thailand and Zaire), but the sum of these rates (7.8 million hectares per annum) can be extrapolated to give a rate for all tropical moist forest of 11 million hectares per annum, similar to Sommer's estimate (Grainger, 1984). Myers did not list these individual rates in a table or use them to make a rigorous estimate of the total rate of deforestation or conversion. He merely stated (on the penultimate page of the report) that it was possible that Sommer's estimate was significantly low, and that a better estimate of the rate at which forests were disappearing could be over 21 million hectares per annum (Myers, 1980). While this figure seemed to refer only to deforestation, it was later revised by Myers and colleagues to give an annual loss of 7.5 million hectares of closed forest and 14.5

Table 31.6 [*orig. table VI*] Annual rates of deforestation for tropical closed forests and tropical moist closed forests (1976–80) and tropical dry open woodlands (projections 1981–85) (million ha per annum).

	Closed Forest	Open	
	All[a]	*Moist[b]*	*Woodland[c]*
Africa	1.3	1.2	2.3
Asia-Pacific	1.8	1.6	0.2
Latin America	4.1	3.3	1.3
Total	7.3*	6.1	3.8

NB * Totals are not the sum of regional figures due to rounding.

Sources: [a]Lanly (1981); [b]Grainger (1983), based on Lanly (1981); [c]Lanly (1982).

Table 31.7 [*orig. table VII*] Three estimates of regional rates of deforestation in the humid tropics for the periods stated (million ha per annum).

	Grainger (1983) 1976–80	Myers (1989) late 1980s	FAO (1992) 1980s
Africa	1.2	1.6	3.0
Asia-Pacific	1.6	4.6	2.6
Latin America	3.3	7.7	6.6
Total	6.1	14.2*	12.2

*Regional figures are total of individual countries listed by Myers. Total was adjusted by him for the whole biome-type.

million hectares of forest fallow (Melillo et al., 1985).

FAO/UNEP (1981) In the following year an estimate of 7.3 million hectares per annum for the deforestation rate for all tropical closed forest in 1976–80 was published in the 1500-page report of the FAO/UNEP Tropical Forest Resource Assessment Project (Lanly, 1981). Here deforestation referred strictly to forest clearance. Usually only the summary of this work is cited (Lanly, 1982), but the full report is very impressive and, as mentioned above, represented a wholly new direction in this field. For the first time, mean national deforestation rates were listed for 76 tropical countries for the same period (1976–80). Also included was a projection of 7.5 million hectares per annum for the deforestation rate for all tropical closed forest in 1981–85. Lanly (1982) gave projections for 1981–85 of 3.8 million hectares per annum for the deforestation rate of tropical open forest (for which no 1976–80 figure was given) (table 31.6) and 4.4 million hectares per annum for the area of tropical closed forest logged.

Grainger (1983) By selecting the FAO/UNEP rates for 55 of the 65 countries listed by Sommer (1976) as constituting the humid tropics, the present author produced a secondary estimate of 6.1 million hectares per annum for the deforestation rate for tropical moist forest alone during 1976–80 (tables 31.7 and 31.8). This represented an annual decline of 0.56 per cent in total forest area. Just over half of all deforestation took place in Latin America and a quarter in Asia-

Pacific. Rates for the remaining countries were not listed by FAO/UNEP and reliable alternative estimates were not available. But since the 55 countries contained 99.5 per cent of all tropical moist forest the total was used unadjusted (Grainger, 1983, 1984).

The accuracy of this figure was probably quite low, because only for six countries had deforestation rates been estimated wholly or partly on the basis of remote-sensing surveys. In another seven countries local estimates of unverifiable accuracy had been used. The remaining rates had been estimated by FAO staff using best available information.

The FAO/UNEP report was controversial at the time for suggesting that deforestation rates had halved compared with Sommer's previous estimate, but FAO's international standing ensured that its 1976–80 estimates eventually became widely accepted. Nevertheless, a polarization developed between it and Myers' global estimate. Some scientists assumed that the actual rate lay between 7.3 and 20 million hectares per annum until Myers finally resolved the matter and concluded that the part of his global figure which actually referred to deforestation was just 7.5 million hectares per annum (Melillo et al., 1985).

Myers (1989) Two other primary estimates have appeared since then. Norman Myers' second report, in 1989, claimed that the deforestation rate for tropical moist forest had doubled to 14.2 million hectares per annum. He gave national deforestation rates for 24 individual countries and for a further ten countries in two regional groups, Central

Table 31.8 [*orig. table VIII*] Two estimates of national rates of deforestation of tropical moist forest (thousand ha per annum).

	Grainger 1976–80	Myers late 1980s
Africa		
Cameroon	80	200
Central African Republic	5	na
Congo	na	70
Equatorial Guinea	15	na
Gabon	27	60
Ghana	15	na
Guinea	na	na
Guinea-Bissau	15	na
Ivory Coast	310	250
Liberia	41	na
Madagascar	165	200
Nigeria	285	400
Sierra Leone	6	na
Uganda	10	na
Zaire	165	400
Asia-Pacific		
Brunei	7	na
Burma	89	800
India	na	400
Indonesia	550	1200
Kampuchea	15	50
Laos	120	100
Malaysia	240	480
Papua New Guinea	21	350
Philippines	100	270
Thailand	325	600
Vietnam	60	350
Latin America		
Belize	9	b
Bolivia	65	150
Brazil	1360	5000
Colombia	800	650
Costa Rica	60	b
Cuba	2	na
Dominican Republic	2	na
Ecuador	300	300
El Salvador	4	b
French Guiana	64	a
Guatemala	1	b
Guyana	3	a
Honduras	53	b
Nicaragua	97	b
Panama	31	b
Peru	160	350
Surinam	3	a
Venezuela	125	150

[a] Total deforestation rate for French Guiana, Guyana and Surinam given as 50 000 ha per annum (cf. 70 000 ha per annum in 1976–80).

[b] Total deforestation rate for Belize, El Salvador, Honduras, Nicaragua, Costa Rica and Panama given as 300 000 ha per annum (cf. 255 000 ha per annum in 1980).

na indicates data not available.

America and the Guyanas (Guyana, French Guiana and Surinam). The overall deforestation rate for these countries (which contained 97 per cent of all tropical moist forest) in the 1980s was 13.8 million hectares per annum, and Myers extrapolated it to give a rate of 14.2 million hectares per annum for all tropical moist forest (table 31.7). Just over half of all deforestation took place in Latin America as before, but Asia-Pacific's share had now increased to about one-third (Myers, 1989).

How reliable was this estimate? Did the deforestation rate really double? (Myers had clearly accepted the lower figure as correct for the late 1970s.) It is difficult to make a definitive evaluation because while Myers gave various references to his data sources he did not specify in detail which if any measurement techniques had been used and to what extent national estimates depended on the expert judgement of local experts or of Myers himself.

The criteria listed earlier in the paper cannot therefore be applied to the whole report, but it is possible to evaluate the deforestation rates given for Brazil and three Asian countries. The rate for Brazilian Amazonia was estimated as 5 million hectares per annum, which happened to be a rough mean of two different estimates: 1.7 million hectares per annum for 1978–88 made in 1989 by one group at the Brazilian National Space Agency (INPE) (quoted in da Cunha, 1989), and 8.1 million hectares per annum for 1987 made by another INPE group (Setzer et al., 1988).

The higher figure was actually a serious overestimate, relying on an image collected by the low resolution AVHRR sensor on a NOAA satellite that was used in the study to measure not changes in forest area but the number of fires in Amazonia, and it was later disowned by INPE (M. Barboso, pers. comm., 1990). The lower figure, obtained by comparing imagery from the high resolution Thematic Mapper (TM) sensor on a Landsat satellite with an earlier complete survey in 1978 based on imagery from Landsat's medium resolution multispectral scanner (MSS), was, on the other hand, an underestimate. The 1988 survey covered only half of

Brazilian Amazonia and just three-quarters of the images were from 1988, with the remainder from previous years. Visual, rather than computer-assisted image interpretation was used, missing many small clearances. But that the lower estimate was closer to the true figure was shown the following year, again by an INPE group in conjunction with an independent expert, Philip Fearnside of the National Institute for Research in Amazonia (INPA). They gave an estimate of 2.1 million hectares per annum for the average rate between 1978 and 1989 by comparing the 1978 MSS survey with a fully-comprehensive survey of Amazonia in 1989 using TM imagery. They also showed that only eight per cent of all forests in the region had been cleared by 1989 (Fearnside et al., 1990).

Myers' estimates for Indonesia, Malaysia and the Philippines were all at least twice the 1976–80 rates. One reason for this was that he assumed that selective logging was so intensive there, that it led to deforestation. However, this is open to question, for as Sayer and Whitmore (1991) have commented: 'No one who has either personally examined heavily logged rain forests or studied the research data on their recovery . . . is likely to agree with Myers that they are destroyed by timber extraction alone'. If Myers' assumption does not hold and Brazil's deforestation rate is also cut to 2.1 million hectares per annum the total rate falls to just over ten million hectares per annum. A detailed examination of estimates for other countries could reduce it even more.

FAO (1990, 1991) FAO expects to publish a new comprehensive assessment of tropical forest areas and deforestation rates in early 1993. In 1990 it released an interim estimate of 16.8 million hectares per annum for the deforestation rate for tropical moist forest in the 1980s (FAO, 1990). As this was greater than the Myers' estimate, doubts about the latter placed a question mark over this one too. The doubts were accentuated for there was known to be heavy reliance on extrapolations and modelling rather than measurements alone.

In 1991 a second interim estimate of 16.9 million hectares per annum was released, which was clearly stated to refer to all tropical forest, both closed and open (FAO, 1991) and so was equivalent in scope to the aggregate rate of 11.1 million hectares per annum in the 1980 assessment (Lanly, 1981), rather than 7.3 million hectares per annum for all tropical closed forest. In the following year FAO disaggregated this figure to give a deforestation rate of 12.2 million hectares per annum for tropical moist forest, of which 4.9 million hectares per annum was in tropical rain forest (table 31.7) (FAO, 1992). However, given the interim nature of this estimate it is probably best to wait until FAO produces its final estimate of deforestation rates for the 1980s before making a final judgement.

The Need for Better Global Monitoring

Given the immense importance of tropical deforestation, it is most unsatisfactory that estimates of deforestation rates are so inaccurate and that such a low priority has been given to monitoring deforestation using the full potential of remote-sensing technology at our disposal. The result of such uncertainty is that policy-makers are often faced with widely varying alternative estimates and no basis to choose between them, which can easily lead them to be very cautious about taking action to control deforestation. Similarly, scientists studying the greenhouse effect and other aspects of contemporary global environmental change have to use estimates of deforestation rates made without the rigorous measurement and error estimation normally expected of scientific data. Some studies of the amount of carbon dioxide emissions from tropical deforestation (Houghton et al., 1985) have tried to cope with the large disparity between different estimates by employing an alternative scenario approach, but this is far from ideal. The situation is even more worrying because the data requirements of scientists, and the policy-makers whom they advise, continue to grow. So the need now is not just for national deforestation rates but for them to be disaggregated spatially by forest type and biomass density. There is no guarantee that such data will be available even by the year 2000.

It is surely time for the world to undertake methodical measurement of deforestation rates by continuous monitoring, something which scientists would regard as fundamental, as opposed to the making of intermittent, subjective estimates, which until recently has been regarded as largely acceptable by the international forestry and environmental community. Yet we seem to be curiously reluctant to intensively monitor terrestrial global environmental change. Since the human race is now capable of changing global climate, in addition to other lesser environmental depredations, proper monitoring would seem to be a duty, not an optional extra, and an inescapable part of rational global environmental management. As long ago as 1972, the UN Conference on the Human Environment, held in Stockholm, recommended the establishment of a continuous global monitoring system for all the world's forests, but 20 years later we are still waiting for an effective system for just the tropical moist forests.

For the most part the obstacles to setting up such a system are institutional rather than technical (Grainger, 1984). The first concerns the nature of the role of UN agencies. For example, it has taken considerable time for the FAO, which is responsible for forestry data within the UN system, to make the transition from merely acting as a repository for national statistics to being an active monitor of world forest resources in its own right. Its 1990 Tropical Forest Resource Assessment marks another important step forward from the 1980 assessment, which only undertook special remote-sensing surveys of forest areas for which appropriate data could not be obtained from other sources. Yet even the 1990 assessment will, we understand, rely heavily on the use of low-resolution satellite imagery and modelling, and there is no assurance that the FAO will maintain an active monitoring role in future.

There is also reticence about adopting

continuous monitoring. Given the rapid rate of change in tropical forest cover, the speedy regrowth of forest after temporary clearance, and the limitations placed on satellite monitoring by heavy cloud cover, a continuous monitoring programme is a vital operational requirement, and ideally it would be good to monitor deforestation rates on an annual basis. There are certainly no data limitations in this regard, for satellites are collecting images on a daily basis, although coping with the large number of images involved would require efficient management. Unfortunately, continuous monitoring seems rather foreign to UN practice, and instead we have to make do with projects that only report once every ten years. The team which undertook the 1980 Tropical Forest Resource Assessment was disbanded after its report was published and a new team had to be assembled for the 1990 assessment. There is no guarantee that it will stay together afterwards either.

Even if the concept of continuous monitoring were accepted there are still differences of opinion as to how it should be conducted. One view, which certainly seems to be the dominant one at the FAO, is that in the long term monitoring should be primarily the responsibility of national governments, with UN agencies only collating the results. This decentralized approach has the virtues of being sensitive to justifiable concerns that global monitoring impinges on national sovereignty, and of recognizing the importance of good ground data collection to assist in remote-sensing image interpretation. But so far only a few tropical countries, such as Brazil, the Philippines and Thailand, have the technical and institutional capability to monitor their forests on a frequent basis, and while national remote-sensing capabilities are improving all the time it is likely that for the next ten years at least these will be focused on providing data to meet national development priorities rather than those needed for global monitoring. A decentralized approach is certainly feasible in principle, given a common monitoring framework were agreed, but in practice variation in the kind of data collected and in the periods

monitored would make it difficult to ensure homogeneous monitoring output.

In a centralized programme, on the other hand, a single organization, possibly with the support of regional centres, would be responsible for collecting and interpreting satellite imagery and other remote sensing data, supplemented by national ground data collection and cooperative interpretation by local scientists. This would not lose the advantage of local involvement and data collection, and indeed many tropical countries would benefit greatly from the data collected locally and globally.

A centralized approach could monitor more frequently, and produce more homogeneous global data than a decentralized approach. It could also begin operations within the space of a few years, whereas with a decentralized approach we might have to wait a decade or more even for reliable data on forest areas and deforestation rates.

A centralized monitoring programme appears the best pragmatic choice but there are four other reasons why it is attractive. First, scientists need data on where deforestation is happening so this may be linked with the type of forest being cleared, its biomass density and its biodiversity characteristics. So it is important for a monitoring programme not only to produce frequent output in the form of simple deforestation rates but also global maps of deforestation events in digital form, suitable for use in geographic information system (GIS) modelling by associated programmes monitoring changes in, for example, terrestrial biomass carbon and biodiversity. A centralized monitoring system would be much better able to do this.

Second, such data will be of more than scientific benefit. They will also be needed by the new global environmental institutions established to oversee implementation of the Climate Change Convention and Biodiversity Convention agreed at the UN Conference on Environment and Development (UNCED) in 1992, and their successors. If tropical countries are to be paid large sums of money in 'carbon credits' for protecting tropical rain forest – either to reduce carbon emissions or allow continuing carbon uptake in regenerating forest – or planting more forest to increase carbon storage, then an independent and comprehensive monitoring programme will be needed to ensure value for money. Other international financial transfers may be anticipated for conserving tropical rain forest biodiversity and to support sustainable development in tropical countries. In each of these cases too, the institution responsible for financial control will demand effective global monitoring.

Third, monitoring tropical deforestation on a global scale is not just about comparing satellite images of forests taken in one year with those a few years later. A wide range of new techniques and methodologies will have to be developed to handle the large number of images involved so that monitoring can be undertaken as efficiently and cheaply as possible, e.g. by combining images from a range of sensors of different resolutions. A centralized monitoring programme would have the organizational weight to establish a substantial applied research programme for this purpose. It would also have the flexibility needed to respond quickly to policy-makers' needs for new kinds of data.

Fourth, deforestation of tropical moist forest is only one of the major components of global environmental change in need of frequent monitoring. Desertification is also happening on a huge scale and we know even less about this than tropical deforestation (Grainger, 1990a). A global environmental monitoring system is needed to monitor this and other phenomena in an integrated way so the interdependences between them can be identified and the complex links with climate change, in particular, can be better understood. A centralized continuous monitoring programme for the tropical moist forests could provide the foundation for such a comprehensive monitoring system.

Who will undertake such monitoring? UN agencies are ideally suited to global tasks like this. If the FAO does not wish to continue with an active global monitoring role then this author has suggested a number of other options. First, UNEP could take over responsibility for monitoring through its Global Environment Monitoring System (GEMS), which could also feed data directly into its Global Resource Inventory Database (GRID). Second, the UN could decide to set up a new organization entirely devoted to global monitoring. But if the UN does not act fast, then, since satellite imagery is freely available to anyone who can afford to buy it, another international organization, such as a non-governmental environmental organization or a scientific research institute, could well step in and take the initiative the UN has forfeited (Grainger, 1990b). IUCN, for example, has had an impressive mapping programme for the tropical rain forests under way for several years (Collins, 1990) and a number of research groups in the USA, in particular, are already interpreting satellite imagery of large areas of the tropics as part of studies related to global climate change.

Conclusions

This paper has reviewed the main estimates of forest areas and deforestation rates for tropical moist forest made since 1970 and provided a set of criteria to evaluate their relative reliability. Although the reliability of estimates of global forest area has increased during this period, as a result of greater use of satellite imagery, there is still considerable uncertainty, as a result of which it is not feasible to use trends in global forest area as a basis for determining the overall deforestation rate. In addition, we still lack estimates of the area of tropical rain forest as such. The actual magnitude of deforestation rates for tropical moist forest are even more uncertain, as there is still an overwhelming reliance on subjective expert assessments rather than remote-sensing measurements. The final results of the latest FAO study of tropical forest areas and deforestation rates are awaited with interest.

It has been argued that the only sure way to provide reliable data on the status and trends in tropical forest resources is to establish a continuous monitoring system which places the emphasis on actual measurements, making use of the full range of remote-sensing techniques at our disposal and local data collection. Such a system was called for by the UN Conference on the Human Environment as far back as 1972. The UN Conference on Environment and Development, held in Rio last year, should have reinforced this recommendation as a matter of urgency but did not. Given the very limited achievements of that conference it is important that the UN keeps global monitoring at the top of its action agenda, undertaking internal consultations on the division of future responsibilities for global monitoring, and initiating a widespread consultation process to resolve other key issues, such as the role of national sovereignty and the best way to link global monitoring with new international environmental institutions and scientific research groups. Given the rate at which deforestation is taking place in tropical moist forest there is no time to waste. Establishing a continuous global monitoring system for the tropical forests, and later for other major components of global environmental change, is not only a scientific necessity. It is a moral imperative too.

References

Collins, N. M. (ed.) 1990 *The last rain forests*. A Mitchell Beazley World Conservation Atlas. Mitchell Beazley, in association with IUCN, London.

Da Cunha, R. P. 1989. Deforestation estimates through remote sensing: the state of the art in the Legal Amazonia. In *Proceedings of a conference on Amazonia: facts, problems and solutions*, 31 July–2 August 1989, University of São Paulo, Brazil: 205–38.

FAO, 1975 Formulation of a tropical forest cover monitoring project. UN Food and Agriculture Organization/UN Environment Programme, Rome.

——, 1990. Interim report on Forest Resources Assessment 1990 Project. Item 7 of the Provisional Agenda, Committee on Forestry, Tenth Session, 24–28 September 1990. UN Food and Agriculture Organization, Rome.

——, 1991 Second interim report on the state of tropical forests. Presented to the Tenth World Forestry Congress, Paris, September, 1991. Forest Resources Assessment 1990 Project, FAO, Rome.

——, 1992 The forest resources of the tropical zone by main ecological regions. Presented to the UN Conference on Environment and Development, Rio de Janeiro, June 1992. Forest Resources Assessment 1990 Project, FAO, Rome.

Fearnside, P. M., Tardin, A. T. and Filho, L. G. M. 1990 *Deforestation rate in Brazilian Amazonia*. Brasilia: National Secretariat of Science and Technology.

Grainger, A. 1983 Improving the monitoring of deforestation in the humid tropics. In Sutton, S. L., Whitmore, T. C. and Chadwick, A. C. (eds) *Tropical rain forest-ecology and management*: 387–95. Oxford: Blackwell Scientific Publications.

——, 1984 Quantifying changes in forest cover in the humid tropics: overcoming current limitations. *J. Wld For. Resource Mgmt* 1: 3–63.

——, 1990a *The threatening desert: controlling desertification*. London: Earthscan Publications.

——, 1990b Overcoming institutional constraints on the global monitoring of natural resources. In *Proc. International conference on global natural resource monitoring and assessments: preparing for the 21st century*, Venice, 24–30 September 1989: 1408–15. American Society for Photogrammetry and Remote Sensing, Bethesda, Maryland.

——, 1993 *Controlling tropical deforestation*. London: Earthscan Publications.

Houghton, R. A., Boone, R. D., Melillo, J. M., Palm, C. A., Woodwell, G. M., Myers, N., Moore, B. and Skole, D. L. 1985 Net flux of carbon dioxide from tropical forests in 1980. *Nature* 316: 617–20.

Lanly, J. P. (ed.) 1981 *Tropical Forest Resources Assessment Project (GEMS): tropical Africa, tropical Asia, tropical America*, 4 Vols. Rome: FAO/UNEP.

——, 1982 *Tropical forest resources*. FAO Forestry Paper No. 30. Rome: FAO.

Lillesand, T. K. and Kiefer, R. W. 1987 *Remote sensing and image interpretation*. Chichester: John Wiley.

Melillo, J. M., Palm, C. A., Houghton, R. A., Woodwell G. M. and Myers, N. 1985 Comparison of two recent estimates of disturbance in tropical forests. *Environ. Conserv.* 12: 37–40.

Myers, N. 1980 *Conversion of tropical moist forests*. Washington DC: US National Research Council.

—, 1989 *Deforestation rates in tropical forests and their climatic implications*. London: Friends of the Earth (UK).

Pearson, R. 1974 *World forest resources*. Research

Notes No. 17. Stockholm: Department of Forest Survey, Royal College of Forestry.

Sayer, J. A. and Whitmore, T. C. 1991 Tropical moist forests: destruction and species extinction. *Biol. Conserv.* 55: 199–213.

Setzer, A. W., Pereira, M. C., Pereira, A. C. Jr and Almeida, S. A. O. 1988 Relatorio de Atividades do Projeto IBDF–INPE 'SEQUE'– ano 1987. INPE-4534-RPE/565. Instituto de Pesquisas Espaciais, São Paulo.

Sommer, A. 1976 Attempt at an assessment of the world's tropical forests. *Unasylva* 28: 5–25.

Whitmore, T. C. 1990 *An introduction to tropical rain forests*. Oxford: Clarendon Press.

World Resources Institute, 1990, *World Resources 1990–91*. New York: Basic Books

32

WETLAND LOSS IN LOUISIANA*

H. J. Walker, J. M. Coleman, H. H. Roberts and R. S. Tye

Introduction

In the third and fourth decades of this century, R. J. Russell (1936) and H. N. Fisk (1944) reported that Louisiana was losing its wetlands and that the state's coastal marshes were rapidly changing composition. However, little attention was paid to such statements until 1970 when research by Gagliano and van Beek again focused on the problem. Subsequently, a wetland loss rate of 130 km²/year for the State as a whole has been estimated by Gagliano (1981). In many sections of coastal Louisiana, especially the modern Mississippi River delta, rates of land loss are exceptionally high (figure 32.1) and in some areas rates are accelerating (figure 32.2).

Wetland erosion results from the interaction of the many physical, chemical, and biological processes that operate in the natural environment and, in more recent times, the processes induced by the human utilization of this and adjacent environments. All of these processes operate at different scales, in both time and space; some can be manipulated by man, others are not amenable to his control.

The Setting

The Louisiana wetlands, with an area of 15.8×10^3 km² (Ringold and Clark, 1980, p. 88) are primarily a product of the Mississippi River which presently drains an area of 3.34×10^6 km². They are only the latest in a series of wetlands that have more or less continuously occupied the prograding coastal zone since the Cretaceous in this part of North America. The large volumes of sediment transported by the Mississippi River and its ancestral systems have created the Gulf Coast Geosyncline. This geosyncline, one of the world's largest, is composed of a number of subsurface sedimentary sequences that were laid down in localized depocenters. Each shift in depocenter location brought about changes in the shoreline and in the size of associated wetlands. Overall, however, there has been a net gain in sediment and the long-term pattern has been shoreline progradation. Nonetheless, the patterns of coastal advance and retreat have been complex; they varied with changes in sediment input, climate, sea level, and the location of the depocenters.

The most recent fluctuations in wetland position and size are those that occurred with the numerous changes in sea level that

* Originally published in *Geografiska Annaler*, 1987, vol. 69A, pp. 189–200.

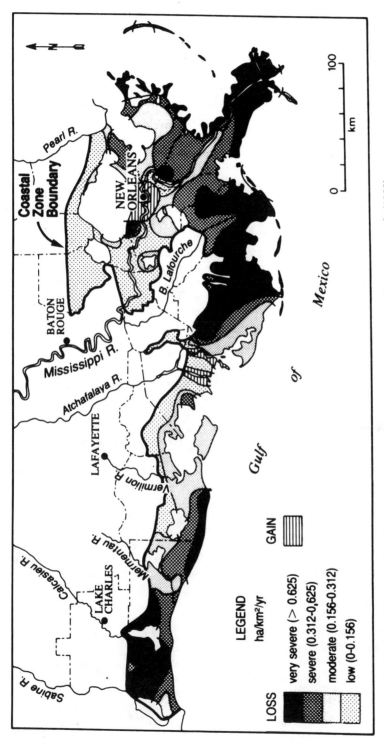

Figure 32.1 [*orig. figure 1*] Rates of land loss in coastal Louisiana. After van Beek and Meyer-Arendt (1982).

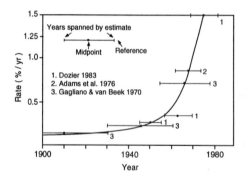

Figure 32.2 [*orig. figure* 2] The acceleration of the rate of wetland loss in the Mississippi River delta. After Dozier (1983).

accompanied the waxing and waning of continental glaciers during the Pleistocene. Marsh deposits representative of low-stand wetlands have been recovered by coring bottom sediments in water-depths of as much as 200 m off the Louisiana and Texas shoreline (Coleman, et al. 1986.)

Causes of Wetland Loss

The major processes contributing to the erosion of the present-day wetlands are 1. geologic, 2. catastrophic, 3. biologic, and 4. human. Each of these categories contains a number of second- and third-order processes some of which are especially important to Louisiana's wetland loss.

Geologic Factors

The three most important geologic factors are those related to 1. universal and localized sea-level change, 2. subsidence and compaction, and 3. changes in the location of the depositional centers within the deltaic environment.

1. *Sea-Level Change*: The total volume of water in the lithosphere–atmosphere–hydrosphere system is believed to have remained fairly constant throughout the last billion years. However, the proportions of water resident in each of these spheres varies with time. Further, interplay between these spheres (e.g., plate tectonics and climate) is often reflected in an erratic sea-level record (Vail et al. 1977). Sea-level

variation may occur on a worldwide (eustatic) scale or it may be of local significance only. For example, localized downwarping occurs in Louisiana because of sediment and water loading on coastal sediments. Occasionally, local processes override the global trend resulting in regional high and low levels that may, at times, be out of phase with eustatic high and low levels.

Although the response of the oceans to climatic change is the most important factor influencing short-term sea-level variation, coastal Louisiana is highly vulnerable to those shorter-term changes in water level caused by hurricanes, cold-front passages, and river flooding. Increases in sea level caused by these phenomena may range from a few cm to several m; durations may last from a few hours to several weeks.

Late Pleistocene and Holocene sea levels. During the last glacial advance (Late Wisconsin; ca 30,000 years BP) sea levels are estimated to have been from 100 to 140 m lower than at present (Fisk and McFarlan 1955; Milliman and Emergy 1968; Dillon and Oldale 1978). Subsequent to the onset of sea-level rise about 17,000 years ago, the locus of river deposition progressively shifted landward. Fisk and McFarlan (1955) describe four subaqueous terraces that mark the positions of deltaic deposition. The terraces at −183, −122, and −61, and −30 m are evidence that transgression occurred sporadically. Eustatic sea-level rise was at times balanced by sediment input which resulted in stabilizing the position of the shoreline. The rapid rate of sea-level rise that occurred during the Late Pleistocene and Early Holocene had decreased drastically by 5000 years BP. About the same time the present Chenier Plain and Deltaic Plain began to form.

Recent and projected sea level. Numerous attempts have been made to quantify the present rate of sea-level rise, but, owing to highly variable regional controls and an inability to produce a reliable and representative data base, current estimates of the eustatic sea-level rise range from 1.2 to 3.0 mm/year (Kraft 1971; Nummedal 1983).

Concern over present rates of rise in sea

level have prompted the publication of numerous projections of future rises. Most of these projections are tied to estimates of the effect increased quantities of carbon dioxide and other gases in the atmosphere will have on air temperature, and in turn on glacial melting and oceanic water expansion. A study by the United States Environmental Protection Agency states that a global rise in sea level of between 1.44 and 2.17 m by 2100 is likely but that a rise as low as 0.56 m or as high as 3.45 m cannot be ruled out. Further, it reports that "along most of the Atlantic and Gulf Coasts of the United States, the rise will be 18 to 24 cm more than the global average" (Hoffman et al. 1983, p. vi).

2. *Subsidence and Compaction:* Subsidence occurs naturally in Louisiana on both regional and local scales as a result of processes ranging from the down-warping of the earth's crust in response to thermal cooling and sediment loading to the rapid compaction of unconsolidated sediments. During much of the Quaternary the major depocenter of the Gulf Coast Geosyncline was offshore of southwest Louisiana. However, in the Holocene this center shifted eastward causing an increase in the sediment input to the area off south and southeast Louisiana (Suter 1986, p. 27). This shift was accompanied by an increase in the rates of subsidence in the area (Trahan 1982).

Geosynclinal subsidence. The depocenters, where sediments from the ancestral Mississippi River accumulated to great thicknesses, are genetically linked to "down-to-the-Gulf" fault systems that roughly parallel the present northern Gulf shoreline. Many of the faults are active today adding to the regional subsidence in Louisiana. In addition, the lateral and vertical flowage of thick salt beds (Worzel and Burk 1979) that underlie deposits of the ancestral Mississippi River, as well as the modern delta, also affect regional subsidence.

Sediment loading and compaction. The dominantly fine grained and highly organic sediments of Louisiana's coastal plain are subjected to three processes immediately after deposition. All three increase the rate of subsidence and include:

1   Primary consolidation – a reduction in the volume of the soil mass owing to dewatering under a sustained load. The load is transferred from the interstitial water to the soil particles.

2   Secondary compression – a decrease in soil volume associated with the rearrangement of constituent particles, and

3   Oxidation of organic matter – reduction of soil volume as bio-chemical reactions occur that cause organic matter to decompose into its mineral constituents (Terzaghi 1943).

These processes of sediment compaction associated with loss of interstitial water, particle rearrangement, and oxidation are fundamental properties of all sediment deposition. However, in areas where sedimentation rates are high, where sediments contain large amounts of water, and where the organic content is high, these processes are extremely active and contribute significantly to land loss. Although depositional events associated with a delta lobe progradation often only take about 1000 to 1500 years to complete (Kolb and van Lopik 1958; Frazier 1967), sediment thickness may exceed 100 m (figure 32.3). Such short depositional histories for these thick deposits suggest that normal processes of compaction will not have time to fully consolidate the sediments.

On a regional scale, this point of view can be supported by comparing long-term water level records. In dynamic areas of sedimentation such as in the Mississippi River delta, compaction and subsidence causes the mean water level (as recorded by a gage) to increase relative to the rise in sea level caused by eustatic processes. In such areas, water-level rise over even short periods of time is significantly higher than in areas having less subsidence and compaction (figure 32.4). The rate of water-level rise at Eugene Island, Louisiana, of 1.61 cm/year, includes eustatic sea-level rise. If a eustatic sea-level rise of 0.12 cm is subtracted, the rate of compaction and subsidence at the Louisiana site is 1.49 cm/year. This amount becomes especially significant when

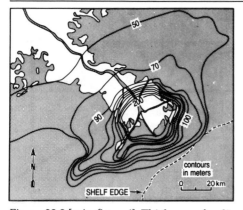

Figure 32.3 [*orig. figure 4*] Thickness of sediments in the Balize delta lobe. After Coleman (1976).

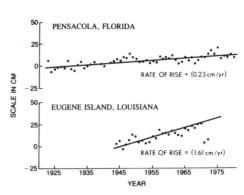

Figure 32.4 [*orig. figure 5*] Water-level gage records from Florida and Louisiana. After Coleman et al. (1986).

compared to the vertical accretion rate of marsh deposits (inland marsh accretion is 0.6–0.75 cm/year as determined by CS 137 dating, (DeLaune et al. 1978, p. 532).

On a local scale, the thickness of recently deposited sediments over a more consoli-

dated surface can affect subsidence and compaction rates. Three cores taken across an old Pleistocene alluvial valley wall show that the subsidence in the area of thick Recent sediment fill is about four times as

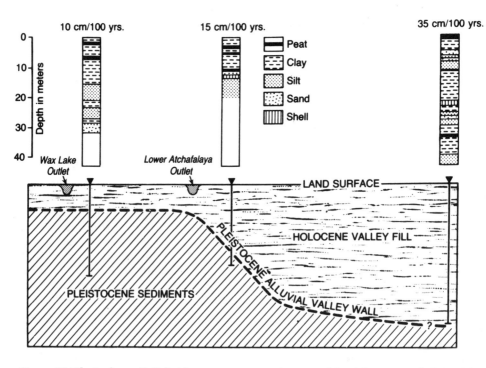

Figure 32.5 [*orig. figure 6*] Subsidence rates across the central Louisiana coastal plain. After Roberts (1985).

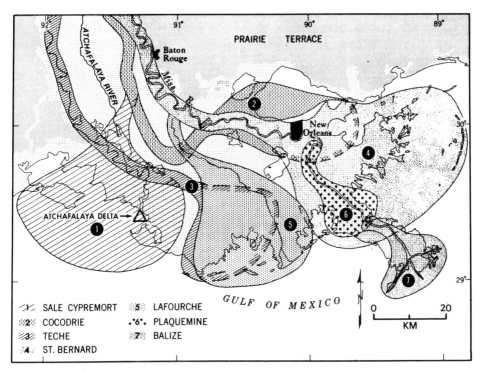

Figure 32.6 [*orig. figure 7*] Major delta lobes that have constructed the Holocene Mississippi River deltaic plain. Note the location of the most recent lobe in the Mississippi River delta complex, the Atchafalaya delta (A). After Kolb and van Lopik (1958).

great as in the area of thin Recent sediment cover (figure 32.5). Although such data are not available from many areas in the Louisiana coastal marshes, it is certain that the thickness of the Holocene varies considerably and that subsidence and compaction must vary accordingly.

3. *Changes in Deltaic Sites of Deposition*: Sea-level changes during the Pleistocene caused alternating entrenchment and infilling of coastal river systems, especially the Mississippi River system. When sea level was low, rivers entrenched; when high, entrenched valleys were infilled. Valley infilling deprived low sea-level coasts of valuable sediment that results in the formation of coastal marshes. In addition to river entrenchment and infilling, changes in sea level also affect the sites of deltaic deposition that form broad coastal marshes. During falling sea level, sites of deposition shift

seaward. During rising sea-level, these sites shift landward and coastal marshes suffer inundation.

At different times in the past the area of coastal wetlands was governed by the locus of deposition and position of major delta lobes. The presence of numerous delta lobes now buried beneath continental shelf deposits points out the role that sea level and subsidence play in controlling the total area of coastal marshes. From the last low sea-level stand to the present, it is estimated that Louisiana's coastal marshes have decreased in area by approximately 40 to 50% (Coleman et al 1986). If submergence of the coast had not occurred along the Louisiana shoreline, many of these older deltaic lobes would still be present and the wetlands would be much more extensive.

The latest phase of the Quaternary cycle, characterized by relative stability of climates and relatively small changes in sea level, began approximately 5000 to 6000 years ago. It was during this period of time that 7 major deltaic lobes formed (figure 32.6). The result of the building and subsequent abandonment of these lobes was the construction of a deltaic coastal plain with an area of 28,568 km², of which only 23,900 km² is now exposed above the sea surface (Coleman 1977).

The most recent of the delta lobes is the Balize or bird-foot delta (figure 32.3 and 32.6), which has required approximately 800 years to form. This lobe has nearly completed its depositional cycle. In the recent past, a new distributary, the Atchafalaya River (figure 32.6), began to capture part of the Mississippi's water and sediment discharge and to form a new delta (Van Heerden and Roberts 1980; Wells et al. 1982). During the present period of delta switching, large volumes of sediment are being delivered to and trapped in the alluvial valley of the Atchafalaya River. In consequence, some of the sediment that normally would nourish the coastal marshes

is no longer reaching the coast, a condition that severely impacts the rate of coastal marsh loss.

During delta progradation, broad coastal marshes are formed. However, once a river begins to abandon its major site of deposition, the unconsolidated mass of sediment continues to be subjected to those processes associated with marine reworking and subsidence (figure 32.7). These processes combine to cause inundation of the marshes and lead eventually to complete transgression by the sea (Scruton 1960). Thus, over a relatively short period of geologic time, land gain and loss are to a large extent a function of the delta cycle stage. The initial phase of delta progradation is characterized by rapid formation of coastal marshes, while deterioration of coastal marshes is associated with the abandonment phase of a delta lobe (figure 32.8).

Similar cycles, although on smaller areal and at shorter temporal scales, occur during delta growth in response to crevassing along the delta's distributaries. Crevassing permits sediment to flow into the bay that

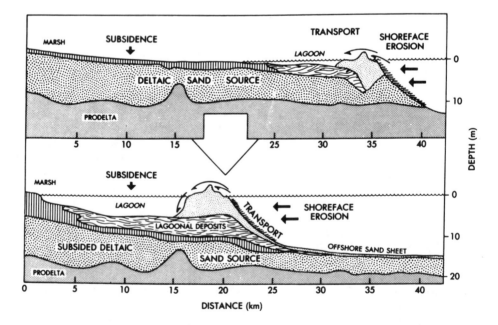

Figure 32.7 [orig. figure 8] Subsidence and coastal barrier erosion after delta abandonment. After Penland and Boyd (1982).

TRANSGRESSIVE MISSISSIPPI DELTA BARRIER MODEL

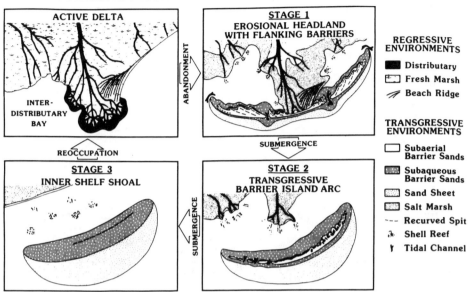

Figure 32.8 [*orig. figure 9*] Model for the evolution of an abandoned Mississippi River delta. After Penland and Boyd (1981).

borders the levee. The bay is eventually filled and the original levee break sealed. Subsequent subsidence of the bay fill leads to inundation by marine water and eventual reversion to a bay environment. The length of time for the completion of one bay-fill cycle varies with a number of factors but generally takes about 100 to 150 years (figure 32.9).

Catastrophic Factors

Among the many catastrophic events (tsunami, landslides, floods . . .) that impact shorelines, by far the most important in Louisiana is the hurricane. Hurricanes most likely to strike Louisiana form between 5 degrees and 15 degrees N Latitude in the Atlantic Ocean during summer and fall. The degree of impact is controlled largely by hurricane size, speed, and path as well as by slope of the continental shelf and orientation of the shoreline. Wind is especially important in that it helps determine the intensity of the waves and, along with low atmospheric pressure, the height of superelevated

water levels or storm surge. The maximum winds and therefore highest surge levels and most intense waves occur to the right of the hurricane's center. The dominant impact on coastal morphology is through erosion on the shoreline, marshes, and coastal dunes. As the elevation of water level rises, increased areas of coast are exposed to erosion and overtopping of barriers is facilitated.

Hurricane Frederick (13 September 1979) provides an example of positioning and storm surge heights. Dauphine Island, Alabama, located to the right of Hurricane Frederick's landfall, was overtopped by a 3.6 m storm tide. Maximum shoreline retreat along the oceanward part of the barrier was 40 m; average retreat was 15 m (Nummedal et al. 1980; Boyd and Penland 1981; Kahn and Roberts 1982). In addition, erosion of as much as 25 m occurred along the lagoonal shoreline. This extensive erosion has been attributed to the increased scour caused by water flowing across the barrier into the Mississippi Sound (Nummedal et al. 1980).

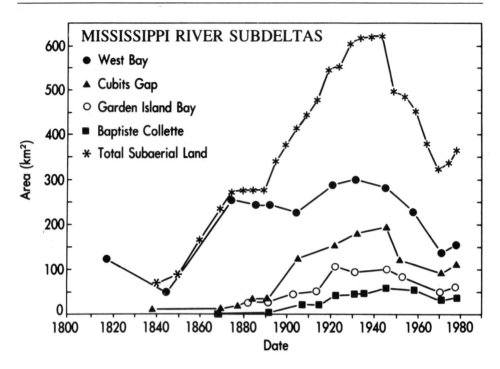

Figure 32.9 [*orig. figure 10*] Growth and deterioration of four bay fills within the Mississippi River delta. Growth rates range from 0.8 to 2.7 km²/year; deterioration rates from 1.0 to 4.1 km²/year. After Wells et al. (1982).

To the left of Frederick's path are located the Chandeleur Islands where the storm surge was only 1.3 m. Nonetheless, these barriers, which are a relatively sediment-starved chain of islands, were severely eroded (30-m retreat) and flattened. Frederick reopened numerous channels that had filled since first formed or reformed by Hurricane Camille in 1969 (Kahn 1980; Nummedal et al. 1980; Kahn and Roberts 1982). Much of the sand from the subaerial part of the Chandeleur's beaches was removed and transported into the Sound where it was deposited as washover lobes.

In addition to direct erosion, storm surges introduce large volumes of saltwater into the wetland system causing major change in vegetative growth. The longterm impact of saltwater inundation is not well understood but it is believed that it initiates changes that may increase wetland erosion rates (Gosselink 1984).

Biologic Factors

The Louisiana coastal zone is composed primarily of four main wetland types; saline marsh, brackish marsh, fresh marsh, and swamp forest. Their distribution, to a large extent, is dependent on the salinity tolerance of numerous plant species. Plants that tolerate salty Gulf waters form a narrow band along the coast. Inland of this salt marsh are the brackish water species which grade inland into totally freshwater species (Chabreck 1982). All of these wetland types are undergoing loss although at different rates. Presently, the brackish marsh is eroding more rapidly than the others (Craig et al. 1979).

The major condition for marsh maintenance is that marsh growth equals or exceeds decomposition including peat formation and relative sea-level rise caused by eustacy, subsidence, and compaction (figure 32.10).

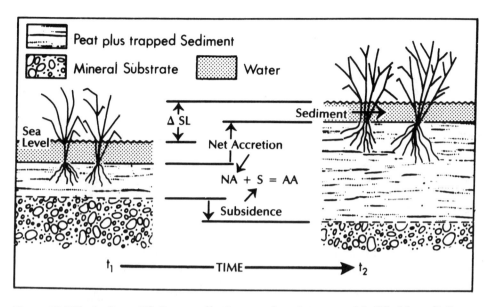

Figure 32.10 [*orig. figure 11*] Factors affecting marsh maintenance. Modified from DeLaune and Smith (1984).

Rates of vertical accretion are known from Caesium 137 profiles and from marker horizons that have been buried since formation (Salinas et al. 1986). Growth rates vary with plant type and position in relation to streams and the shoreline. The rate of accretion is about 1.4cm/year for stream-side settings and about 0.60–0.75 cm/year for inland marshes (table 32.1). When these rates are compared with present-day water-level changes, it is apparent that Louisiana's marshes are in a critical state: For example, the rate of rise being recorded by the water level gage at Eugene Island, Louisiana (figure 32.5) is 1.61 cm/year, a value that is exceeded only by the marsh accretion rate of

1.69 cm/year recorded at one brackish marsh location (table 32.1). If, as is being predicted by numerous investigators (Hoffman et al. 1983), the eustatic rate of sea-level rise continues to increase, marshland deterioration will accelerate.

Degradation due to biotic organisms also contributes to wetland loss. Invertebrates, which are exceptionally common, and large consumers such as snow geese, muskrats, and nutria, all cause marsh destruction. However, data on the extent to which these organisms contribute to marsh loss are lacking (Gosselink 1984; O'Neil 1949).

Table 32.1 [*orig. table 1*] Accretion rates for selected marshes (cm/yr). After Hatton 1981.

| Marsh Type | Location | | | |
| | Streamside | | Inland | |
	mean	range	mean	range
Fresh	1.06		0.65	0.31–0.69
Intermediate	1.35	1.30–1.40	0.64	0.38–1.06
Brackish	1.40	1.06–1.69	0.59	0.38–0.81
Salt	1.35		0.75	

Human Factors

Although the wetland habitat has been used by man for millenia, within the past 50 years utilization has increased dramatically. However, despite the fact that it is obvious that human activity has drastically modified the wetlands, the extent of this effect is difficult to quantify and evaluate. From the standpoint of coastal erosion in Louisiana the most important human activities have been related to 1. flood control, 2. canal construction, and 3. fluid withdrawal from the subsurface.

1. Flood Control – Dams and Levees: Wetland loss in the Louisiana coastal plain in recent decades can be related in part to the human alteration of natural systems at both local and regional scales. Modifications of the Mississippi River, its tributaries, and their drainage basins far from Louisiana have resulted in a major decrease in the amount of suspended sediment reaching the delta. The major modifications to the system include the construction of dams, levees, and other channel 'improvement' features. One of the major recent modifications is the addition of the Old River Control Structure which now directs 30% of the Mississippi River water through the Atchafalaya Basin (Keown et al 1981). Prior to 1963, the sediment load of the Mississippi River, plus the component diverted through the Atchafalaya River, was 434×10^6 tons/year. The present value is 255×10^6 tons/year, a decrease of 41%. Measurements of bedload show that there has been a shift in grain size to finer fractions, indicating the trapping of the coarser grained sizes by dams upstream. Thus, the sediment that normally serves as the foundation material for delta growth has been greatly reduced in quantity.

The artificial levees that now line the delta's distributaries have altered natural processes in two major ways. First, they direct all of the sediment that reaches the delta out the mouths of the distributaries. Second, artificial levees prevent overbank flooding and thus aid in depriving the delta's wetland areas of the normal addition of sediment that is so important in marsh nourishment and maintenance and in subsidence compensation.

2. Highway and Canal Construction: The construction of highways and the dredging of canals have impacted heavily on the wetlands. Prior to man's heavy use of the area, drainage was determined by the distribution of the bayous, which were mostly abandoned distributaries. Initially, roads were placed on top of natural levees and had only minimal effect on drainage. However, as a transverse (east–west) highway system developed, natural drainage courses were blocked.

Canal construction in the Louisiana wetlands was one of the earliest ways man altered the area. The earliest hunting and trapping canals had little effect on drainage patterns and marsh development (Davis 1973). However, as canals began to be dredged to support general navigation (e.g. the Intracoastal Canal) and to provide access to petroleum exploration and exploitation sites, major modifications occurred. The canal network that has evolved has contributed to wetland loss both directly and indirectly (Johnson and Gosselink 1982, p. 60). Directly, there is the conversion of wetland into waterways, spoilbanks, and the subsequent widening that accompanies canal-bank erosion. Indirectly, canals create an artificial drainage network which alters marsh hydrology as well as inorganic sedimentation with major consequences on marsh maintenance. This hydrographic interruption can change surface water and groundwater flow patterns, thereby restricting nutrient and sediment dispersal to some areas while others are impacted with increased discharge and possibly salt-water intrusion (Turner et al. 1983). Salt-water intrusion results in the death of non-salt-tolerant species, leaving the marsh surface exposed to erosion from surface runoff.

3. Fluid Withdrawal: Depressurization of both shallow and deep aquifers and hydrocarbon reservoirs has resulted in an increase in subsidence over that which would be expected from natural processes only. In Louisiana, subsidence with the loss of pressure in aquifers has been greatest in or

near metropolitan areas where demand for fresh water is high (Keady et al. 1975). Groundwater use and associated subsidence in New Orleans caused a lowering of the ground surface by 0.52 m between 1938 and 1964.

Subsidence in areas of oil and gas fields has been documented at only a few locations (Gilluly and Grant 1949) but well-documented accounts from Louisiana are not readily available. It is likely that some of the fields producing from shallow horizons have undergone subsidence. In the case of those fields situated in the wetlands, even small amounts of subsidence can accelerate salt-water intrusion and marsh deterioration.

Summary and Conclusions

In Louisiana, coastal wetland loss is affected by regional long-term processes, regional short-term processes, and by local short-term processes. Regional long-term processes include 1. down-warping because of sedimentary loading and 2. global sea-level rise. Regional short-term processes include 1. the change in location of delta formation, 2. compaction, dewatering, and oxidation of coastal fine-grained and organic-rich sediments, and 3. human modification of the riverine system. Local short-term processes include 1. those of a catastrophic nature, 2. those of a biologic nature, and 3. those of a human nature such as canal dredging, levee construction, and fluid withdrawal from the subsurface.

The loss of Louisiana's wetlands at a rate of over 100 km^2/year has now been measured. Further, many of the processes that are responsible for this loss have been identified. However, these processes have yet to be quantified in a satisfactory manner. Those data sets that do exist usually reflect multiple interactions and are often in conflict. Therefore, any ranking of the relative importance of the causes of Louisiana's wetland loss must be tentative. With such qualifications in mind, it is here proposed that these causes, in order of decreasing importance, are:

1 change in the depositional site and stage in the delta cycle,
2 compaction and localized differential subsidence,
3 sea-level change and long-term climatic change,
4 human modification of the Mississippi River system,
5 canal dredging,
6 biological degradation,
7 short-term catastrophic events,
8 fluid extraction, and
9 regional geosynclinal down-warping.

References

Boyd, R, and Penland, S., 1981: Washover of deltaic barriers on the Louisiana coast. *Transactions Gulf Coast Assoc. Geol. Soc.*, 31: 243–248.

Chabreck, R. H., 1982: The effect of coastal alteration on marsh plants. In: Boesch, D. F. (ed.): *Proc. Conf. Coastal Erosion and Wetland Modification in Louisiana: Causes, Consequences, and Options*. U.S. Dept. of the Interior, FWS/OBS-82/59: 92–98.

Coleman, J. M., 1976: *Deltas: processes of deposition and models for exploration*. Continuing Education Publ., Champaign, Ill., 102 p.

Coleman, J. M., Roberts, H. H. and Tye, R. P., 1986: Causes of Louisiana Land Loss, Report: Louisiana Mid-Continental Oil and Gas Assn. 28 p.

Craig, N. J., Turner, R. E. and Day, J. W., Jr., 1979: Wetland losses and their consequences in coastal Louisiana. *Zeitschr. für Geomorph.*, 34: 225–241.

Davis, D. W., 1973: *Louisiana canals and their influence on wetland development*. Ph.D. dissertation, Louisiana State University, Baton Rouge, 199 p.

DeLaune, R. D., Patrick, W. H. and Buresh, R. J., 1978: Sedimentation rates determined by Cs^{137} dating in a rapidly accreting salt marsh. *Nature*, 275 (5680): 532–533.

DeLaune, R. D. and Smith, C. J., 1984: The carbon cycle and the rate of vertical accumulation of peat in the Mississippi River deltaic plain. *Southeastern Geology*, 25: 61–69.

Dillon, W. D. and Oldale, R. N., 1978: Late Quaternary sea level curve: reinterpretation based on glaccio-eustatic influence. *Geology*, 6: 56-60.

Dozier, M. D., 1983: *Assessment of change in the marshes of southwestern Barataria Basin, Louisiana, using historical aerial photographs and* a spatial information system. M. S. thesis, Louisiana State University, Baton Rouge, 102 p.

Fisk, H. N., 1944: *Geological investigations of the alluvial valley of the lower Mississippi River.* U.S. Army Engineers Miss. River Comm., Vicksburg, Miss. 78 p.

Fisk, H. N. and McFarlan, E., Jr., 1955: Late Quaternary deltaic deposits of the Mississippi River. *Geol. Soc. America Spec. Paper 62*: p. 279–302.

Frazier, D. E., 1967: Recent deltaic deposits of the Mississippi River, their development and chronology. *Trans. Gulf Coast Assoc. Geol. Soc.*, 17: 287–315.

Gagliano, S. M., 1981: *Special report on marsh deterioration and land loss in the deltaic plain of coastal Louisiana.* Coastal Environments, Inc. Baton Rouge, 6 p.

Gilluly, J. and Grant, U.S., 1949: Subsidence in the Long Beach Harbor Area, California. *Geol. Soc. Amer. Bull.*, 60: 461–530.

Gosselink, J. G., 1984: *The ecology of delta marshes of Coastal Louisiana: a community profile.* U.S. Fish and Wildlife Service, FWS/OBS-84-09, 134 p.

Hatton, R. S., 1981: *Aspects of marsh accretion and geochemistry: Barataria Basin, Louisiana.* M. S. Thesis, Louisiana State Univ., Baton Rouge, 116 p.

Hoffman, J. S., Keys, D. and Titus, J. G., 1983: *Projecting future sea level rise: Methodology, estimates to the year 2100, and research needs.* U.S. Environmental Protection Agency, EPA 230–09–007, 121 p.

Johnson, W. B. and Gosselink, J. G., 1982: Wetland loss directly associated with canal dredging in the Louisiana coastal zone. *In*: Boesch, D. F. (ed.): *Proc. Conf. Coastal Erosion and Wetland Modification in Louisiana: Causes, Consequences, and Options.* U.S. Dept. of Interior, FWS/OBS–82/59: 60–72.

Kahn, J. H., 1980: *The role of hurricanes in the long-term degradation of a barrier island chain: Chandeleur Islands, Louisiana.* Unpublished Master of Science Thesis, Louisiana State University, 98 p.

Kahn, J. H. and Roberts, H. H., 1982: Variations in storm response along a microtidal transgressive barrier-island arc. *Sedimentary Geology*, 33: 129–146.

Keady, D. M., Lins, T. W. and Russell, E. E., 1975: *Status of land subsidence due to ground-water withdrawal along the Mississippi Gulf Coast.* Water Resources Research Inst., Mississippi State Univ., 25 p.

Keown, M. P., Dardeau, E. A. and Causey, E. M., 1981: *Characterization of the suspended sediment regime and bed material gradation of the Mississippi River Basin.* Environmental Laboratory, Potamology Program, USACOE, Vicksburg, Miss. 2 volumes.

Kolb, C. R. and Van Lopik, J. R., 1958: *Geology of the Mississippi deltaic plain-southeastern Louisiana.* USACOE, Waterways Exp. Sta., Tech. Rept. 2, 482 p.

Kraft, J. C., 1971: Sedimentary facies patterns and geologic history of a Holocene marine transgression: *Geological Soc. America Bull.*, 82: 2131–2158.

Milliman, J. D. and Emery, K. O., 1968: Sea-levels during the past 35,000 years. *Science*, 162: 1121–1123.

Nummedal, D., 1983: Future sea level changes along the Louisiana coast. *Shore and Beach*, 51: 10–15.

Nummedal, D., Penland, S., Gerdes, R., Schramm, W., Kahn, J. and Roberts, M., 1980: Geologic response to hurricane impact on low-profile Gulf Coast barriers. *Trans. Gulf Coast Assoc. Geol. Soc.*, 30: 183–195.

O'Neil, T., 1949: *The muskrat in the Louisiana coastal marshes.* Louisiana Department of Wildlife & Fisheries, New Orleans, 152 p.

Penland, S. and Boyd, R., 1981: Shoreline changes on the Louisiana barrier coast. *IEEE Oceans*, 81: 209–219.

Penland, S. and Boyd, R., 1982: Assessment of geological and human factors responsible for Louisiana coastal barrier erosion. *Proceedings of the conference on coastal erosion.* U.S. Department of the Interior, Washington, D.C., 14–38.

Ringold, P. L. and Clark, J., 1980: *The coastal almanac: for 1980 – the year of the coast.* The Conservation Foundation, San Francisco, W. H. Freeman and Company, 172 p.

Roberts, H. H., 1985: *A study of sedimentation and subsidence in the south-central coastal plain of Louisiana:* Summary Report to the New Orleans Branch of the U.S. Army Corps of Engineers, New Orleans, La., 57 pp.

Russell, R. J., 1936: Lower Mississippi River delta: reports on the geology of Plaquemines and St. Bernard Parishes. *Louisiana Department of Conservation Geol. Bull.*, 8, 199 p.

Salinas, L. M., DeLaune, R. D. and Patrick, W. H., Jr., 1986: Changes occurring along a rapidly submerging coastal area: Louisiana, USA. *Journal of Coastal Research*, 2: 269–284.

Scruton, P. C., 1960: Delta building and the deltaic sequence. *In*: Shepard, F. P., et al. (eds.): *Recent Sediments, N. W. Gulf of Mexico*, Amer. Assoc. Petrol. Geologists, Tulsa, Okla., 82–102.

Suter, J. R., 1986: Buried late Quaternary fluvial

channels in the Louisiana continental shelf. *Journal of Coastal Research*, Special issue No. 1, 27–37.

Terzaghi, K., 1943: *Theoretical soil mechanics:* Wiley, New York, 510 p.

Trahan, D. B., 1982: Monitoring local subsidence in areas of potential geopressured fluid withdrawal, southwestern Louisiana. *Trans. Gulf Coast Assoc. Geol. Soc.*, 32: 231–236.

Turner, R. E., McKee, K. L., Sikora, W. B., Sikora, J. P., Mendelssohn, I. A., Swenson, E., Neill, C., Leibowitz, S. G. and Pedrazini, F., 1983: The impact and mitigation of man-made canals in coastal Louisiana. *Symp. Integration of Ecological Aspects in Coastal Engineering Projects*, Rotterdam, The Netherlands, 497–504.

Vail, P. R., Mitchum, R. M., Jr. and Thompson, S., III, 1977: Seismic stratigraphy and global changes of sea level, Part 4: global cycles of relative changes of sea level. *In:* Payton, C. E. (ed.): Seismic Stratigraphy – Applications to Hydrocarbon Exploration. *Amer. Assoc. Petrol. Geologists Memoir* 25: 83–97.

van Beek, J. L. and Meyer-Arendt, K. J., 1982: *Louisiana's eroding coastline: recommendations for protection.* Office of Coastal Zone Management, Baton Rouge, 49 p.

Van Heerden, I. Ll. and Roberts, H. H., 1980: The Atchafalaya delta: rapid progradation along a traditionally retreating coast. *Zeitschr. für Geomorph. N. F.* Supplementband 34: 225–240.

Wells, J. T., Chinburg, S. J. and Coleman, J. M., 1982: *Development of the Atchafalaya River delta.* Coastal Studies Institute, Louisiana State Univ., prepared for USACOE, Waterways Experim. Sta., Vicksburg, Miss. 91 p.

Worzel, J. L. and Burk, C. A., 1979: The margins of the Gulf of Mexico. *Amer. Assoc. Petrol. Geologists Memoir* 29: 403–419.

33

CORAL REEFS – A CHALLENGING ECOSYSTEM FOR HUMAN SOCIETIES*

B. Salvat

Six hundred thousand square kilometres of coral reefs are distributed along the coasts of the Pacific, Indian and Atlantic oceans, mostly between 25° north and south of the Equator, but occasionally up to 30° of latitude (Figure 33.1). They are bathed by warm waters and extend from the surface down to 100 m. Some particularly large reefs occupy continental shelves or shallow seas (eg northeast Australia's 2200 km long Great Barrier Reef), whereas others form protective fringes around high volcanic islands or low atolls. Most of the world's 410 atolls are slowly subsiding and drifting in response to tectonic movements and remain intact solely due to the protection afforded by growing reefs.

Coral reefs constitute the most diverse, complex and productive marine ecosystem. They exhibit a high degree of self-sufficiency as evidenced by the presence of luxurious reefs in very nutrient-poor waters of central areas in the Indian and Pacific oceans. A symbiotic association between coral hosts and *zooxanthellae* (ie unicellular algae) is the key to success of this ecosystem. Corals and calcareous algae form the main framework and all other organisms of the community provide the essential biodetritus.

Slightly more than 100 countries are bordered by coral reefs. The overwhelming majority are developing countries, including many of the world's least developed.[1] Most are archipelagoes that were formerly colonies of European states. Several developed countries also contain coral reefs either within the metropolitan state or overseas dependencies. They include: Australia, Japan (Ryukyu), the USA (Florida, Hawaii, Caribbean dependencies and Micronesia), and France (Caribbean, Indian and Pacific dependencies).

The daily management of most coral reefs is in the hands of developing countries that give high priority to improving the economic well-being of their citizens but lack the requisite expertise to apply scientific management principles. In contrast, coral reef specialists are generally based in developed countries of the northern hemisphere

* Originally published in *Global Environmental Change*, 1992, vol. 2, pp. 12–18.
[1] Bernard Salvat, 'Preservation of coral reefs: Scientific whim or economic necessity – past, present and future', *Proceedings of the Fourth International Coral Reef Symposium*, vol. 1, Manila, 1981, pp. 225–229.

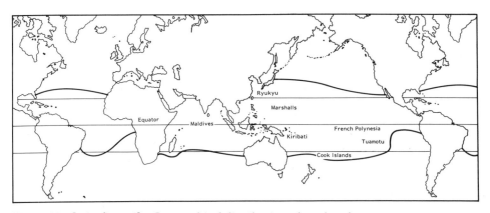

Figure 33.1 [*orig. figure 1*]   Geographical distribution of coral reefs.
Basically, the distribution of scleractinian corals follows the warm ocean currents and the 20°C winter isotherms.

where they place greater emphasis on fundamental research than on applied management.

Human Use of Coral Reefs

Population data are very incomplete but it is believed that several tens of millions of people inhabit coral reef coasts and harvest their resources for subsistence or for sale in local markets. Although the archipelagic countries of the Pacific are more numerous, most of the people who are dependent on coral reefs live in Southeast Asia, especially the Philippines and Indonesia. With one-fifth of the world's coral reefs and two-thirds of the atolls, Pacific ocean states support a little more than two million people on 90 000 km² of land.

Coral reef fishes are not yet exploited by industrialized fishing fleets for ecological and technical reasons. Their potential is believed to be about one-tenth of the world's present marine catches. The sustainable annual fish harvest from coral reefs is about 10–15 tonnes/km².[2] In the South Pacific, for example, apart from subsistence fishing, reef and lagoon fish may constitute 29% of the commercialized local fishery and yield 100 000 tonnes a year.[3] Up to 90% of the animal protein which is consumed on many of the Pacific islands is attributable to marine sources.

Relationships between people and reefs vary in response to factors such as local natural resources, history and culture. Polynesians are much more closely associated with reefs than are Melanesians or African peoples who were transplanted to Caribbean islands to provide labour for sugar cane plantations.

Subsistence economies predominate in the least-developed and developing countries that possess coral reefs. Given limited alternatives and the closed nature of island systems, ancient reef-using societies were often conservative managers. Subsistence production provided a long-term, ecologically secure method of survival on atolls.[4] Penetration by Western lifestyles, laws, and market economies often produce a tragedy of the commons and permitted uncontrolled

[2] John L. Munro and David McB. Williams, 'Assessment and management of coral reef fisheries: biological, environmental and socio-economic aspects', *Proceedings of the Fifth International Coral Reef Congress*, Tahiti, 1985, vol. 4, pp. 545–578.
[3] Bernard Salvat, 'The living resources of the South Pacific: Past, present and future', UNESCO-MAB *Technical Notes*, 1980, vol. 13, pp. 131–148.
[4] Moshe Rapaport, 'Population pressure on coral atolls: Trends and approaching limits', *Atoll Research Bulletin*, no. 340, 1990, pp. 1–33.

overfishing.[5] Places such as Japan remained exceptions to this pattern because reefs were the responsibility of local collective institutions and the status of marine coastal waters was firmly embedded in existing laws.

Today, after at least two decades of mass air transportation and development oriented to sun, sand and sea, the tourist industry is the leading economic sector in many tropical reef areas. Tourism does not generally extract resources from the reefs but uses them for aesthetic and recreational purposes. Nonetheless, use pressures are rapidly growing due to increasing local populations and integration of reef regions into the global market. Reefs are under increasing pressure to provide specimens and raw materials for handicrafts, and to support aquaculture for food.

Effects of Natural Events on Coral Reefs

Since early in the Pleistocene period the most important natural causes of changes to coral reefs have included: rises and falls of sea level (up to 150 m) associated with glaciations, and accompanying fluctuations of sea-surface temperatures that affected the geographic distribution of reefs.

Natural events inflict damage at least one order of magnitude greater than human activities, especially when they occur in regions like French Polynesia where reef structure and functioning are not adapted to such destructive phenomena. At present, the most important natural causes of damage to coral reefs are predator outbreaks and diseases, certain sea-atmosphere processes (eg El Niño), and hurricanes.[6] For example, the El Niño of 1982–83 brought lower sea level and warmer sea-surface temperatures

in the Pacific, both of which contributed to mass coral reef mortality. Hurricanes completely destroy shallow reef areas and outer slopes of barrier reefs by breaking up coral colonies and rolling debris down the seaward faces.[7] Outbreaks of *Acanthaster planci* (the predatory Crown-of-Thorns starfish) brought massive mortality to large areas of coral reef in the Indo-Pacific region during the 1960s. Over the past decade in the Caribbean, disease-bearing pathogens have been a cause of mortality in corals and echinids (*Diadema*). All these natural events cause large-scale destruction of reefs, but they generally occur over long periods. Recovery times may range up to 50 years for the most complex and stable (climax) reef systems.

Human-induced Degradation of Coral Reefs

Anthropogenic impacts on coral reefs have been well documented in many recent reviews.[8] More than 20 categories of stress-causing damage have been identified but some are more important than others. These include: runoff from land clearance; chemical pollution and eutrophication due to domestic and industrial wastes or agriculture; coral and sand mining; overfishing; coastal construction projects; and oil pollution.

The destruction and degradation of reefs is increasing in frequency both spatially and temporally. Before 1840, human impacts were small in scale and involved 'soft' technologies and natural materials in a context of low population pressure. Contemporary activities tend to use high technologies, new materials and chemicals.[9]

Where subsistence economy and small-

[5] Salvat, *op. cit.*, 1981.

[6] Susan Wells, 'Coral reefs of the world', vols I, II, III, *UNEP/IUCN* Cambridge, 1988; Richard W. Grigg and Steven J. Dollar, 'Natural and anthropogenic disturbance on coral reefs', in *Coral Reefs*, ed. Z. Dubinsky, Elsevier, 1990, pp. 439–452.

[7] David Stoddardt, 'Hurricane effects on coral reefs', *Fifth International Coral Reef Congress*, vol. 3, Tahiti, 1985, pp. 349–350.

[8] Bernard Salvat, *Human Impacts on Coral Reefs: Facts and Recommendations*, Antenne Museum EPHE, French Polynesia, 1987, pp. 1–253; Wells, *op. cit.*, Ref 6; Grigg and Dollar, *op. cit.*, Ref 6.

[9] Arthur L. Dahl and Bernard Salvat, 'Are human impacts, either through traditional or contemporary

scale exploitation are still important, the main problem is overfishing due to demographic pressure. For example, between 1966 and 1986 the productivity of coral reefs in the Philippines dropped by one third as the national population doubled.[10] As the increasing proportion of young specimens and juveniles in catches indicates, fish and edible benthic species such as molluscs or crustaceans are being heavily exploited. Nevertheless, because of their larval reproduction process, no coral reef species are known to be threatened with extinction.[11] But stocks are depleted in many areas, often because new harvesting means are used to increase the catch in minimum time, with great damage to habitat. The dynamiting of coral reefs and the use of industrial poisons (pesticides) are examples. Pollution impacts are usually confined to urban areas where they have generally been preceded by overfishing.

In several countries management of coral reefs is complicated by the fact that populations that are unfamiliar with the coast are migrating there from interior zones. For example, in Madagascar and Indonesia many people are faced with a new environment, without the cultural knowledge that would enable them to manage the reefs. Typically, such groups employ all available materials and techniques to secure increased food supplies without providing for the preservation and perpetuation of reef ecosystems.

Anthropogenic or Natural Disturbances?

The long-term variability of coral reefs has increasingly become understood during the past two decades. It is now clear that in any one location, algae, echinoderms (ie echinids, starfish), molluscs, corals and fish show extreme variability of recruitment from one year to another. But there is a general lack of historical information about environmental parameters and population dynamics at specific sites. This makes it extremely difficult to distinguish between natural variability and anthropogenic impacts. This dilemma is well illustrated by two examples: the Crown-of-Thorns population explosion since the 1960s; and the coral bleaching phenomenon since the 1980s.

Acanthaster planci populations exploded at the end of the 1950s and during the following decade, on many sites in the Indian and Western Pacific oceans.[12] On average, a single starfish consumes about 6 m^2 of living coral per year. Fourteen percent of the Great Barrier Reef and 90% of the fringing reefs of Guam were destroyed as a result of this process. Dead corals were replaced by an algae and sea urchin dominated community.

At the time, the causes and effects of outbreaks of *Acanthaster planci* became major items of coral reef research funding. It was found that during outbreaks starfish densities were thousands of times larger than the usual 2–3 per km^2. Most researchers initially expressed the view that such outbreaks were human-induced. Destruction of starfish predators and the effects of pollution on larvae were suspected. Since then, ancient outbreaks have been identified and most analysts regard them as cyclical and natural worldwide events. In other words, data were insufficient for long-term interpretation of population dynamics and reef variability. As a result, human impacts were exaggerated. The experience of this controversy should not be forgotten!

uses, stabilizing or destabilizing to reef community structure?', *Sixth International Coral Reef Symposium*, vol. I, Australia, 1988, pp. 63–69.

[10] Don E. McAllister, 'Environmental, economic and social costs of coral reef destruction in the Philippines', *Galaxea*, vol. 7, 1988, pp. 161–178.

[11] Bernard Salvat, 'Menace et sauvegarde des espèces des récifs corralliens', *Cahiers d'Outre Mer*, Bordeaux, vol. 172, 1990, pp. 489–501.

[12] Robert Endean, 'Population explosions of *Acanthaster planci* and associated destruction of hermatypic corals in the indo-west Pacific region', *Biology and Geology of Coral Reefs*, vol. II, no. 1, 1973, pp. 389–438.

The contemporary bleaching phenomenon is another story that demands close consideration. Bleaching events are taking over from the Crown-of-Thorns plague of the 1960s and 1970s as the pre-eminent disturbance that affects coral reefs worldwide. Major bleachings occurred in 1979–80; 1982–83 (related to the El Niño phenomenon);[13] and 1986–88. These events were not spatially restricted, but occurred throughout large areas, such as archipelagoes and biogeographic provinces. Bleaching involves disruption of the delicate symbiosis between coral hosts and *zooxanthellae*. Stress can be thermal, or due to heavy sedimentation, or chemical pollutants, or other factors. *Zooxanthellae* are ejected by the host with much mucus and the white skeleton of the coral is apparent through the translucent soft tissue. This produces the 'bleaching' phenomenon. Because corals live near their upper lethal temperature, it is possible to bleach them in experimental situations by raising the temperature a few degrees. If the stress is neither strong nor prolonged, corals may recover by multiplication of *zooxanthellae* that remain in their tissue, but more often they die.

During the past decade the Caribbean was much affected by repeated bleaching events that induced extensive mortality of corals and coral reef mass. Temporal and spatial relationships between the bleaching phenomenon and warmer sea-surface temperatures (1–3° above normal summer temperature) have been correlated.[14] It is now quite clear that higher temperature is the cause of, or at least the main factor in, the bleaching events. A worldwide overview of bleaching has been produced from published data and interviews with researchers.[15] This shows that bleaching events were much more frequent than usual in the 1980s, and observers have hypothesized that there are links between bleaching, sea-surface warming and global change. Could the coral reef be a harbinger of global environmental change? At present, we simply do not know. Lack of long-term information on coral reef stability and instability hampers any attempt to discriminate between natural and anthropogenic causes of change. This deficiency argues for setting up a network of coral reef observatories in the next century.[16]

Global Change and Coral Reefs

It is believed that, by the mid-21st century, global air temperature may have risen by 2–4°C, sea-surface temperature may have risen by 1.5–4.5° C, and sea level may be 25–40 cm higher than at present.[17] Similar changes occurred in the geologic past, as recently as two million years ago. Despite the fact that these changes were larger than those that are anticipated by the mid-21st century, coral reefs survived on continental slopes and around tropical islands. However, there are two major differences between the previous and present cases. First, it took thousands of years to effect the earlier change: we face the prospect of rapid change over half a human lifetime – albeit at a rate of sea-level rise that is less than that of the post-glacial transgression (ie 5mm/year

[13] Barbara B. Brown, 'Coral bleaching', *Coral Reefs*, special issue, vol. 8, no. 4, 1990, pp. 153–232; Lucy Bunkley-Williams and Ernest H. Williams, 'Global assault on coral reefs', *Natural History*, 1990, pp. 47–54.

[14] Thomas J. Goreau, 'Coral bleaching in Jamaica', *Nature*, vol. 343, 1990, p. 417; Thomas J. Goreau, Raymond L. Hayes, Jenifer W. Clark, Daniel J. Basta and Craig N. Robertson, *Elevated Satellite Sea Surface Temperatures Correlate with Caribbean Coral Reef Bleaching*, NOAA Technical Report No. 137, National Ocean Survey, Washington, DC, 1991, pp. 1–60.

[15] Ernest H. Williams Jr and Lucy Bunkley-Williams, 'The worldwide coral reef bleaching cycle and related sources of coral mortality', *Atoll Research Bulletin*, no. 335, 1990, pp. 1–71.

[16] Bernard Salvat, 'World coral reef site network project: A long term global plan to monitor coral reefs on indices of natural and anthropogenic changes', *XVII Pacific Science Congress: Abstracts*. Honolulu, 1991, p. 118.

[17] David Hopley, 'Global change and the coastline: environmental, social and economic implications', *Fourth Pacific Congress on Marine Science and Technology: Proceedings*, PACON 90, Tokyo, 1990, pp. 17–22.

compared to 13mm/year). Second, 50% of the world's population is located near coasts and about 300 million people live within 1 m of sea level. They will be seriously affected by sea-level changes, not least because of their impact on coastal ecosystems. Of the two main tropical ecosystems – mangroves and coral reefs – the first will be most endangered by sea-level rise because it is restricted to present intertidal zones, whereas the second already lies below water – between low tide level and 50–100 m below the surface.

The probable first impact of anticipated sea-level rise will be submergence of all atolls. This was the case 1500 years ago when sea level was 1 m higher than today and it may occur again with or without the contributions of potential increased storminess.[18] (From time to time hurricanes and storms completely erode the detrital carbonate that constitutes the emerged parts of low islands in the Marshalls, Kiribati, Tuamotu, Maldives, and Cook islands, among others.)

Coral reefs are especially involved in the atmospheric carbon dioxide cycle and the global carbon cycle because corals lose carbon dioxide through the air-sea interface.[19] However, the aggregate contribution of coral reef communities and the entire coral reef ecosystem is unknown. At present, coral reefs act as an annual sink of 900 million tonnes of calcium carbonate or 111 million tonnes of carbon equivalent. This is about 2% of the present output of anthropogenic carbon dioxide. A sea-level rise of 0.5–1 m will allow corals to colonize the present fringing reefs and reef flats. As a result, corals may gain more shallow surface than they will lose in deepening water and calcium carbonate deposition may double in the next century.

Extension of coral reef surfaces and increase of calcium carbonate deposition due to sea-level rise is a primary effect of global change that may produce overall benefit for the ecosystem. But secondary and tertiary effects, that are more open to debate, do not favour coral reefs.[20] An increase of temperature could be catastrophic for coral reefs, as occurred during the bleaching events of the 1980s. More El Niños may occur, bringing more tropical storms in their wake. These may extend their range into areas where they are rare at present. Finally, increased rainfall may lead to heavier sedimentation and consequent destruction of some coral reefs.

A Challenging Ecosystem

Degradation of coral reefs and impoverishment of their living resources is already well underway throughout the entire ecosystem due to a combination of human and natural factors. Coasts of least developed and developing countries are most affected. Whether the loss of reefs is due to global climate change is not yet possible to determine but it is true that the coral reef ecosystem is very sensitive to slight increases in temperature. It may be that coral reefs will provide indicators of global climate change. Clearly, the management of coral reefs presents major challenges for present and future human societies. At the least, it is now time to monitor coral reefs[21] and to manage tropical coastal ecosystems in light of possible greenhouse effects.[22]

[18] Peter W. Glynn, 'Coral reef bleaching in the 1980s and possible connections with global warming', *Trends in Ecology and Evolution*, vol. 6, no. 6, 1991, pp. 175–178.

[19] Paolo P. Pirazzoli, Lucien F. Montaggioni, Bernard Salvat and Gerard Faure, 'Late Holocene sea-level indicators from twelve atolls in the central and eastern Tuamotu (Pacific Ocean)', *Coral Reefs*, vol. 7, 1988, pp. 57–68.

[20] Hopley, *op. cit.*, Ref 17; Pirazzoli et al, *op. cit.*, Ref 19.

[21] Bunkley-Williams and Williams, *op. cit.*, Ref 15.

[22] Richard Grantham, 'Approaches to correcting the global greenhouse drift by managing tropical ecosystems', *Tropical Ecology*, vol. 30, no. 2, 1989, pp. 157–174.

34

IS DESERTIFICATION A MYTH?*

T. Binns

Most geography students, by the age of 16, are familiar with the term 'desertification'. It appears in GCSE syllabuses and crops up regularly in textbooks, the media and examinations.[1] The common image conjured up by the term is one of an advancing Sahara desert inexorably moving south, smothering villages and destroying farmland and pasture once and for all. By 'A' level, this image has been further reinforced by being portrayed as an environmental catastrophe, the product of a long-term decline in rainfall exacerbated by unwise human practices such as overgrazing, burning and deforestation.

This paper hopes to shed some further light on these views and will argue that this stereotypical impression is inappropriate and, for much of Africa, blatantly inaccurate. A number of recent publications suggest that longstanding perceptions of desertification need to be closely re-examined and school syllabuses and accompanying texts should take note and adapt accordingly.

Much, but by no means all, of the work on desertification focuses on Africa, particularly the Savanna–Sahel zone stretching across the northern part of the continent from Mauritania in the west to Sudan in the east. This region also provides the focus for this paper, since the author has firsthand experience of several countries in this zone, and the Savanna-Sahel has figured prominently in recent discussions on desertification.

Interest in Desertification

Reports of an advancing Sahara are by no means new, and it was at a meeting of the Royal Geographical Society on 4th March 1935 that Professor E. P. Stebbing, Professor of Forestry at the University of Edinburgh, delivered a much-quoted paper entitled, "The encroaching Sahara: the threat to the West African colonies". Stebbing, reporting on a recent trip he had made through West Africa, commented on his observations in northern Nigeria:

And the desert is advancing! How, or how fast, I have yet to learn when I cross the frontier. It is impossible to travel on the road up to Geidam from Kano, followed by us, or back from Geidam via Nguru to Kano, without realizing the serious threat.

* Originally published in *Geography*, 1990, vol. 75, pp. 106–13.
[1] For example, the June 1988 examination of the Southern Examining Group GCSE Geography Syllabus B included a map showing areas of the world affected by 'desertification' and asked candidates, amongst other things, 'What is desertification?', 'Explain one reason for desertification' and 'Suggest how the problems of desertification could be reduced and explain the practical problems likely to arise in doing what you suggest?' (Southern Examining Group, 1988).

I have rarely been more impressed with the seriousness of the position than during the last few days. The people are living on the edge, not of a volcano, but of a desert whose power is incalculable and whose silent and almost invisible approach must be difficult to estimate. But the end is obvious: total annihilation of vegetation and the disappearance of man and beast from the overwhelmed locality (Stebbing, 1935, p. 510).

As one might expect from a Professor of Forestry, he then went on to suggest that, "it should not be beyond the power of man to put up a barrier to this threat," and he proposed two massive forest belts, one through Burkina Faso (then Upper Volta) and northern Nigeria to Lake Chad, and another further south from the Ivory Coast, through Ghana (formerly Gold Coast) to Jebba or Minna in Nigeria's middle belt. Sir Arthur Hill, Director of the Royal Gardens at Kew, who was listening to Stebbing's presentation, estimated (presumably without the aid of a calculator!) that some 15,914,000,000 trees would have to be planted, whereupon Stebbing replied that his 'belts' would develop largely from existing woodland which would be closed and protected from farming, fire and grazing. He recognised that some planting might be needed later.

Fourteen years later, Aubreville (1949) is generally recognized as the first person to use the term 'desertification' to refer to the increasing extent of deserts – dry areas with few plants – usually into semi-arid lands. He said, "these are real deserts that are being born today, under our very eyes, in regions where the annual rainfall is from 700 to 1500 mm" (quoted in Glantz and Orlovsky, 1986).

The events which perhaps did most to popularise the use of the term 'desertification' in recent years were the two droughts in Africa's Savanna-Sahel zone: 1968–1974 and 1979–1984. Following the first drought and the associated media coverage, the United Nations Environment Programme organised a World Conference on Desertification (UNCOD) in Nairobi in 1977 to consider the extent and character of the problem (figure 34.1) and to propose measures to combat it on an international scale (Griffiths and Binns, 1988). UNCOD went further than either Stebbing or Aubreville and referred to desertification as:

the diminution or destruction of the biological potential of the land . . . leading ultimately to desert-like conditions . . . Desertification is a self-accelerating process, feeding on itself and as it advances, rehabilitation costs rise exponentially (UNCOD, 1977).

Since this important conference there have been numerous comments from influential politicians and others perpetuating the concept of desertification. On 14th March 1986, Vice-President George Bush of the United States was being urged to give aid to the Sudan because "desertification was advancing at 9 km per annum" (quoted in Warren and Agnew, 1988, p. 2). Mostafa Tolba, director of the United Nations Environment Programme, after a major UN review of action against deserts in 1984, concluded that, "currently, 35 per cent of the world's surface is at risk . . . each year 21 million ha is reduced to near or complete uselessness" (Forse, 1989). In the same year Stiles suggested that in the Sudano-Sahelian region of Africa, 87 per cent of the population living on rangelands are affected, 78 per cent of the population on rain-fed croplands, and 30 per cent of that on irrigated croplands – altogether, 49.5 million people (Stiles, 1984).

Similar gloomy statements abound in academic and school textbooks. But what is the reality of the situation? Is desertification happening? This begs the question of an appropriate definition. The author's own field observations (Binns, 1986), coupled with discussions with various field experts and the reading of a number of recently published research papers (eg. Forse, 1989; Mortimore, 1989; Nelson, 1988; Warren and Agnew, 1988) suggest that desertification has been grossly over-emphasised, largely because of insufficient investigation on the ground as to how environments, societies and food production systems respond to periods of drought. As Ridley Nelson, in a

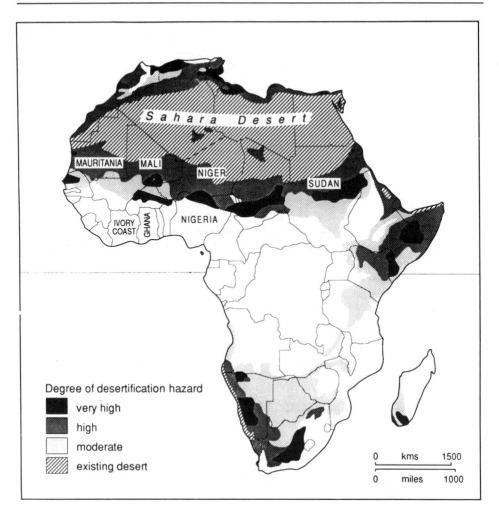

Figure 34.1 [*orig. figure 1*] Desertification in Africa as defined by the United Nations Conference on Desertification, 1977. Adapted from *World Map of Desertification 1:25,000,000*, Nairobi: UNEP.

recent UN Environment Department Working Paper says:

> Contrary to popular belief, the extent of desertification is not at all well known . . . there is extremely little scientific evidence based on field research or remote sensing for the many statements on the global extent of the problem (Nelson, 1988).

Searching for a Definition

There seems to be much confusion between the terms 'desertification' and 'land degra-dation'. The present author would agree with Nelson who believes 'desertification', "is a term that served a purpose in raising public awareness, but it now cloaks the problem in a disguise that obscures its true shape" (Nelson, 1988).

Over the years the definition of desertification has moved from "the expansion of desert-like conditions" to "a process of sustained land (soil and vegetation) degra-dation in arid, semi-arid and dry sub-humid areas, caused at least partly by man", and reducing "productive potential to an extent which can neither be readily reversed by removing the cause nor easily reclaimed without substantial investment" (Nelson, 1988, p. 2).

Desertification therefore, may give the impression of expanding deserts – in Africa the southward advancing Sahara as mentioned by Stebbing – whereas what is more commonly the case is various stages of land degradation. 'Desertification' implies a once-and-for-all process on a massive scale, whereas what seems to happen on the ground is the development of areas of land degradation, in some cases quite severe, but which with care and two or three good rainy seasons might return to their former status. It is not argued here that desertification does not exist, but rather that its extent over space and time is much less than the popular image. The desert fringe might be likened to an ebbing and flowing tide, moving southwards during dry periods and northwards when rainfall is plentiful. Pockets of degraded land may occur at any time for a variety of human and environmental reasons, but the process of degradation is likely to accelerate during dry spells.

Inadequate Evidence?

Nelson (1988) points to two studies which have been quoted frequently to support the existence of desertification in Africa, but he argues that in both cases there are serious flaws. The UNEP study, reported in Mabbutt (1984), estimates that approximately 20 million ha annually are reduced to zero or negative economic productivity, and the number of people inhabiting land being desertified has increased by 35 per cent since 1977. Nelson is critical of information yielded from the questionnaire used in this study and wonders how government staff completing it differentiated between 'moderately desertified' and 'severely desertified' land, particularly when in Africa they were completing it at the height of the drought.

The point is to emphasise that the results, which are by far the most widely quoted evidence on the extent of desertification, have an extraordinarily shaky basis and

have been enormously influenced in Africa by being completed after a long and exceptionally dry period (Nelson, 1988, p. 6).

Nelson is equally critical of another major source, Hugh Lamprey, an ecologist, who conducted a study in Sudan in 1975, comparing ecological boundaries with those mapped by Harrison and Jackson in 1958. Lamprey suggested that the Sahara desert was advancing south at a rate of 5.5 km per year, a total of 90–100 km between 1958 and 1975 (Harrison and Jackson, 1958; Lamprey, 1975). These findings have been disputed by Hellden, a Swedish researcher who undertook a useful remote sensing and ground study of a transect in the same area and concluded:

There was no creation of long lasting desert-like conditions during the 1962–79 period in the area corresponding to the magnitude described by many authors and commonly accepted by the Sudanese Government and international aid organisations . . . The impact of the Sahelian drought was short lasting followed by a fast land production recovery (Hellden, 1984).

In the same area Olsson observed:

no woody species has been eradicated from the area, no ecological zones have been shifted southwards and the boundaries between different vegetation associations appear to be the same now as they were 80 years ago . . . there are also examples of soil and vegetation recovery within the area, which clearly demonstrates the dynamics of the countryside (Olsson, 1984).

These observations in Sudan certainly accord with the present author's observations in Mali at the end of the drought in September 1985, where, in an area to the east of the Bandiagara escarpment and also further north, the Sahel was very green.

There was little evidence of any 'desertification' and the healthy millet crop indicated that even areas which have been desertified' in the past may be 'resurrected' with adequate rainfall (Binns, 1986, p. 250).

However, six months earlier during the dry season, and after several years of well below average rainfall, this and other areas may well have been classified as desertified. The timing of observations is crucial and it appears that many of the studies upon which statements on desertification have been based were conducted during drought periods after which environment and food production systems slowly but significantly recovered. Warren and Agnew (1988, p. 4) argue that because of fluctuations in climate and vegetation, "It would be difficult, if not impossible, to determine whether the condition of the Sahel at any time between 1968 and 1984 was a consequence of long-term degradation or of drought".

A further problem revealed by studies of drought, degradation and desertification is a lack of reliable statistics over any considerable length of time:

Aside from the need for more reliable information than that provided by guesses and estimates in understanding where desertification is worsening or improving, we need data because the credibility of statements on desertification is now in question. Claims that the Sahara is expanding at some horrendous rate are still made despite the absence of evidence to support them. It may have been permissible to say such things ten or twenty years ago when remote sensing was in its infancy and errors could easily be made in extrapolating limited observations. It is unacceptable today (Dregne, 1987).

Warren and Agnew (1988) are equally concerned about the paucity and unreliability of data, saying that data needed to classify land are available for very few areas and for very few years and in most of Africa little is known about range condition, crop yield or the extent of soil erosion.

We must conclude from all this that major questions surround the existence and recognition of desertification. It is probably much less widespread than has often been portrayed in the media and areas have frequently been classified as desertified on the basis of incomplete evidence and statistics. In many, perhaps most, cases the term 'desertification' has been used when 'land degradation' would have been more appropriate.

Resilience: A Key to Better Understanding?

If the term 'desertification' is restricted to mean simply the creation of deserts, as is suggested here, it may be regarded as an extreme form of land degradation occurring where vegetation cover falls below 35 per cent on a long-term basis. Warren and Agnew introduce the term 'resilience' to describe the property of a resource that makes its sustainable use possible. They argue that land degradation occurs when resilience is damaged and that often resilience can only be seen to have been damaged if the system does not recover after a shock, such as drought. Recovery in dry lands is in any case slow because of an absence of water to assist important soil processes and vegetative growth (Warren and Agnew, 1988).

Interestingly, Mortimore (1988; 1989) writing about semi-arid northern Nigeria, uses the concept of resilience in a rather different sense to refer not merely to the land resource but to food production systems which are carefully adapted by farmers and pastoralists to the rigours of environment:

During the 1980s, droughts have been experienced again, and in some areas they equalled or exceeded in severity those of the previous decade (this was true, for example, of south-eastern Niger Republic and the extreme north-east of Nigeria). Yet both farming and livestock systems in the semi-arid zone have survived with, if anything, less hardship. This has been achieved by intensified exploitation of alternative opportunities, and spatial mobility, using to the full adaptive capabilities developed in response to the droughts of the 1970s (which followed a long period of relatively reliable rainfall) (Mortimore, 1988, p. 64).

There is now plenty of accumulated evidence to show how pastoral and farming communities in marginal semi-arid and arid regions have devised, over many generations, mechanisms to cope with harsh environmental episodes such as drought, which is endemic in these regions and is a key element in indigenous folklore and oral histories. Food production systems frequently possess a high degree of structural resilience. Amongst pastoralists, mobility, management of size and composition of herds and complex arrangements for exchanging livestock and livestock products are elements of this in-built resilience. Their knowledge of the location of water, different types of pasture and the incidence of pests such as tsetse fly is often quite impressive (Binns, 1984; Binns and Mortimore, 1989; Swift, 1975, 1977, 1982; Western, 1982).

Farmers, too, frequently display a whole repertoire of adaptive and risk-aversion strategies in areas which are susceptible to drought. The use of drought-tolerant varieties, light tilling and mulching to preserve organic matter in the soil and minimise erosion, and the utilisation of wild plants and tree crops, are some examples. In areas such as northern Nigeria, farmers may take on dry season occupations, often in nearby towns, to provide extra security. Such short-term migration is not a new phenomenon (Prothero, 1959).

Much of the early writing on desertification was critical of the way farmers and pastoralists used their land resources and accelerated the process of degradation. Present at the Royal Geographical Society on the occasion of Stebbing's lecture in 1935, was Mr George Hemmant, Acting Governor of Nigeria from November 1930 to June 1931. Referring to problems of deforestation and run-off, he said, ". . . a country with native peasants is not so fortunate. They have not the intelligence to protect themselves or their country" (Stebbing, 1935,

p. 520). Research on farming and pastoral communities in many parts of the Third World, including the drought-prone Savanna-Sahel zone of West Africa, has shown that such a statement, the like of which is not unfamiliar today, is very far from the truth (see for example, Brokensha et al., 1980; Chambers, 1983; Richards, 1985, 1986).

Conclusion

The concept of desertification has many attendant problems, as we have seen. But frequent references in the media to desertification have played an important role in raising awareness of environmental problems in the world's marginal lands and a significant interest has developed in drylands research.[2] The evidence seems to suggest that the idea of the Sahara desert sweeping southwards on a broad front, as Stebbing postulated in 1935 and others have argued more recently, is no longer tenable. However, land degradation of varying degrees does exist and the causes and possible remedies of this problem must be understood.

Massive anti-deforestation schemes, such as Stebbing's proposed forest belts, are not favoured by recent writers. Instead, the collection of reliable data and more detailed local level studies to understand the interaction between food production systems and environmental resources are increasingly favoured. The search for better systems for indigenous participation and co-operative management of resources is also important. There is an urgent need to examine linkages between management systems and land productivity. Existing projects need to be carefully evaluated and Warren and Agnew (1988) suggest research is needed on the basic processes of desertification and land degradation, together with agricultural and social research into how communities decline, survive and adapt.

[2] During 1988 two important new journals on arid and semi-arid areas were established in the United Kingdom alone. *Haramata* is published by the International Institute for Environment and Development (3 Endsleigh Street, London WC1H 0DD) and *Baobab* comes from the Arid Lands Information Network based at Oxfam (274 Banbury Road, Oxford OX2 7DZ).

It is time to modify our stereotype views of desertification, and teachers at all levels need to take this on board. The important recent contributions of Mortimore, Nelson, Warren and Agnew, Forse and others provide powerful ammunition in the debate over desertification which in spite, or perhaps because, of their work will no doubt continue for some considerable time to come.

The author would like to thank Professor M. J. Mortimore for his constructive comments on an earlier version of this paper.

References

Aubreville, A. (1949) *Climats, Forêts et Desertification de l'Afrique Tropicale*, Société d'Editions Géographiques et Coloniales: Paris.

Binns, J. A. (1984) "People of the six seasons", *Geographical Magazine*, 56, pp. 640–644.

Binns, J. A. (1986) "After the drought: field observations from Mali and Burkina Faso", *Geography*, 71 (3), pp. 248–252.

Binns, J. A. and Mortimore, M. J. (1989) "Ecology, time and development in Kano State, Nigeria", in Swindell, K., Baba, J. M. and Mortimore, M. J. (eds) *Inequality and Development: Case Studies from the Third World*, London: Macmillan, pp. 359–380.

Brokensha, D. W., Warren, D. M. and Werner, O. (eds) (1980) *Indigenous Systems of Knowledge and Development*, Lanham, USA: University Press of America Inc.

Chambers, R. (1983) *Rural Development: Putting the Last First*, London: Longman.

Dregne, H. E. (1987) "Reflections on the PACD", *Desertification Control Bulletin*, (Special tenth anniversary of UNCOD issue), Nov 15.

Forse, W. (1989) "The myth of the marching desert", *New Scientist*, 1650, pp. 31–32.

Glantz, M. H. and Orlovsky, N. S. (1986) "Desertification: anatomy of a complex environmental process", in Dahlberg, K. A. and Bennett, J. W. (eds) *Natural Resources and People: Conceptual Issues in Interdisciplinary Research*, Westview Special Studies in Natural Resources and Energy Management, Boulder, Colorado: Westview Press.

Griffiths, I. L. and Binns, J. A. (1988) "Hunger, help and hypocrisy: crisis and response to crisis in Africa", *Geography*, 73 (1), pp. 48–54.

Harrison, M. N. and Jackson, J. K. (1958) "Ecological classification of the vegetation of

the Sudan", *Forests Bulletin (Khartoum)*, 2.

Hellden, U. (1984) *Drought Impact Monitoring, a Remote Sensing Study of Desertification in Kordofan, Sudan*, Sweden: Lunds Universitets Naturgeografiska Institution in co-operation with Institute of Environmental Studies, University of Khartoum, Sudan.

Lamprey, H. F. (1975) *Report on the Desert Encroachment Reconnaissance in Northern Sudan*, UNESCO/UNEP Consultant Report Oct–Nov., 1975.

Mabbutt, J. A. (1984) "A new global assessment of the status and trends of desertification", *Environmental Conservation*, 11(2) pp. 100–113.

Mortimore, M. J. (1988) "Desertification and resilience in semi-arid West Africa", *Geography*, 73(1), pp. 61–64.

Mortimore, M. J. (1989) *Adapting to Drought: Farmers, Famines and Desertification in West Africa*, Cambridge: Cambridge University Press.

Nelson, R. (1988) "Dryland management: the 'desertification' problem", *Environment Department Working Paper No. 8*, Washington: World Bank.

Olsson, K. (1984) *Long-term Changes in the Woody Vegetation in Northern Kordofan, Sudan*, Sweden: Lunds Universitets Naturgeografiska Institution in co-operation with the Institute of Environmental Studies, University of Khartoum, Sudan.

Prothero, R. M. (1959) *Migrant Labour from Sokoto Province, Northern Nigeria*, Kaduna: Government Printer.

Richards, P. (1985) *Indigenous Agricultural Revolution*, London: Hutchinson.

Richards, P. (1986) *Coping with Hunger*, London: Allen and Unwin.

Southern Examining Group (1988) *General Certificate of Secondary Education, Geography Syllabus B, Paper 1 (6th June)*, Guildford, Surrey: SEG.

Stebbing, E. P. (1935) "The encroaching Sahara: the threat to the West African colonies", *Geographical Journal*, 85, pp. 506–524.

Stiles, D. (1984) "Desertification: a question of linkage", *Desertification Control Bulletin*, 11, pp. 1–6.

Swift, J. (1975) "Pastoral nomadism as a form of land-use: the Twareg of the Adrar n Iforas", in Monod, T. (ed.) *Pastoralism in Tropical Africa*, Oxford, IAI, pp. 443–453.

Swift, J. (1977) "Desertification and man in the Sahel", in *Land Use and Development, African Environment Special Report No. 5*, London, pp. 171–178.

Swift, J. (1982) "The future of African hunter-gatherer and pastoral peoples", *Development and Change*, 13, pp. 159–181.

United Nations Conference on Desertification (UNCOD) (1977) *Desertification: its Causes and Consequences*, Nairobi: Pergamon.

Warren, A. and Agnew, C. (1988) "An assessment of desertification and land degradation in arid and semi-arid areas", *International Institute for Environment and Development, Drylands Programme, Paper 2*, London: IIED.

Western, D. (1982) "The environment and ecology of pastoralists in arid savannas", *Development and Change*, 13, pp. 183–211.

35

LARGE HERBIVOROUS MAMMALS: EXOTIC SPECIES IN NORTHERN AUSTRALIA*

W. J. Freeland

Introduction

Compared to the tropical savannas of Asia and Africa, the northern Australian savannas are depauperate in native species of large herbivorous mammal. In the Northern Territory north of 18°S there are only six species of macropod marsupial that could qualify as large herbivorous mammals. It is unusual to find more than three or possibly four of these species living in the same geographic area, and even then the species appear to exhibit different patterns of habitat choice. When compared to the Asian and African savannas which have at least six to eight species of large herbivore (McKay and Eisenberg, 1974; Eisenberg and Seidensticker, 1976), the northern Australian region is clearly depauperate.

The paucity of large herbivorous mammals in northern Australia is due to extinction of large herbivorous marsupials rather than the habitat being in some way unsuitable for large grazing or browsing mammals. Not only do the northern Australian savannas bear a strong structural resemblance to those from other parts of the

tropics, but over two-thirds of the vascular plant genera present in the Northern Territory between 11 and 16°S have global distributions including other areas of the world's tropics. Only 15% of genera have distributions restricted to Australia (Bowman, Wilson and Dunlop, 1988).

Up until the Pleistocene Australia had a fauna of large herbivorous marsupials, as well as large ratite birds, and large marsupial and varanid predators (Hope, 1984). Man's first arrival on the continent may have caused the extinction either through hunting or indirectly by habitat modifications wrought by the use of fire, or climate change may have played the major role (Horton, 1980). Whatever the cause of the extinctions, it is clear that when European man arrived in northern Australia approximately 150 years ago the habitat was underutilized by herbivorous mammals, and had a flora with a long evolutionary history of herbivory from large mammals.

Since the arrival of European man the extinct megafauna has been replaced by large ungulates introduced from Asia, Europe and Africa. A total of eight species of ungulate (and one size variant, the Timor

* Originally published in the *Journal of Biogeography*, 1990, vol. 17, pp. 445–9.

Pony) have been introduced to various places in the Northern Territory. Some species are now widely spread throughout the region (e.g. *Bos taurus* and *Equus caballus*), others have retained a localized distribution (e.g. *Bos banteng* and *Cervus unicolor*), and some are still expanding their ranges (e.g. *Camelus dromedarius*) (Bayliss and Yeomans, 1989a, b). The species are either entirely feral, primarily feral, or maintain feral populations together with minimally managed harvested stock. Expansion of the feral and the near feral herds has resulted in overgrazing and alteration of native habitats (Letts, Bassingthwaight and deVos, 1979; Braithwaite et al., 1984; Bowman and Panton, 1989).

In northern Australia it is assumed that there is something unusual or undesirable about habitat change caused by the feral herds. Undesirable effects include loss of habitat for native species and soil erosion (e.g. Letts et al., 1979; Bowman and Panton, 1989). It is open to question whether the impact of the feral herds is in any way greater than or in some way different from that which occurs in the savannas of Asia or Africa.

If the impact of the feral herds on savanna vegetation in northern Australia is greater than occurs with native herds, one possible explanation is that feral herds exist at densities greater than those achieved in their native habitats. Data on the densities of introduced herbivorous mammals in the Australasian region are summarized and compared to expectations of population density based on body size (Damuth, 1981), and densities particular species of introduced herbivore achieve in their native habitats. Results are discussed in relation to what is known of factors likely to influence the size of Australian and native populations.

Methods

Data on the population densities of ten species of herbivorous mammal in Australasia, and seven species in both Australasian and native habitats, have been extracted from a variety of sources and are available on request. All species and populations considered have been in their introduced habitats for greater than 100 years. Data for the large mammals are taken primarily from aerial survey results, whereas data for the smaller species are based on capture–recapture studies. Data on the introduced populations are compared to the data set assembled by Damuth (1981) in his study of the relationship between mammalian herbivore body size and population density. Some of the variance in his data set was due to inclusion of data from populations introduced to alien environments. All such data were removed from the data set and body size–population density relationship recalculated for comparison with the data on introduced species. The adjusted body size–population density relationship is described by:

$$\log_{10}\text{Density} = 4.196 - 0.74(\log_{10}W),$$

where Density is km^{-2}, and W=body mass in g (r^2=0.73, N=319).

Results and Discussion

The phenomenon

Herbivorous mammals introduced to Australasia exist at higher population densities than those predicted from knowledge of their body sizes or that occur in the species' native habitats. All ten introduced species for which there are data have ecological densities greater than predicted by the Damuth relationship (table 35.1). This frequency of occurrence of densities greater than predicted is higher than expected by chance (binomial probability=0.00098). As five of the six species for which there are data in both the Australasian and native habitats have observed native densities less than predicted by the Damuth relationship (table 35.1), the observed high densities in Australasia are not due to the species being a biased selection that has higher than usual natural densities. In all cases for which there are data, the mean Australasian density is greater than is observed in the native habitat (table 35.1). Maximum densities over minimum areas of at least 100 km^2 in

Table 35.1 [*orig. table 1*] Ecological densities of herbivorous mammals as predicted from the Damuth relationship, as observed in their native habitats and as observed in Australasia. Densities are in km^{-2}, – indicates missing data, and * indicates species for which there are data from northern Australia. The native density for *E. asinus* is that of *E. heminonus*. Species have been arranged in ascending order of body size.

| Species | Population density | | | |
| | Predicted | Native habitat | Australasia | |
			Mean	Maximum
Mus musculus	1842.00	–	31769.00[1]	–
Rattus exulans	890.00	1073.00[2]	8500.00[3]	18500.00[3]
Rattus rattus	446.00	–	4500.00[4]	6400.00[4]
Oryctolagus cuniculus	59.00	–	135.00[5]	–
*Sus scrofa**	4.78	0.79[6]	6.61[7]	12.61[7]
Capra hircus	3.81	3.10[8]	5.00[9]	–
*Bos banteng**	1.50	0.85[10]	2.46[10]	–
*Equus asinus**	1.35	0.33[12]	2.20[13]	10–15[13]
*E. caballus**	1.10	–	1.40[13,14]	7.20[13,14]
*Bubalus bubalis**	1.08	0.83[6]	3.30[14]	25.20[14]

1=Newsome (1969); 2=Harrison (1969) and Dwyer (1978); 3=Strecker (1962) and Wirtz (1972); 4=Tamarin and Malecha (1971); 5=Dunsmore (1974) and Wood (1980); 6=Eisenberg and Seidensticker (1976); 7=Hone (1986); 8=Damuth (1982); 9=Henzell and McCloud (1984); 10=Hoogerwerf (1970); 11=Bayliss and Yeomans (1989a); 12=Gee (1963) and Wolfe (1979); 13=Graham *et al.* (1982); 14=Bayliss and Yeomans (1989b).

Australasia for species surveyed from the air are 16–45 times densities recorded in native habitats.

Possible explanations for the unusually high population densities of introduced herbivores in the Australasian environment include: (a) an absence of competition from species rich herbivore communities, (b) a paucity of potential predators, (c) a paucity of parasites and diseases, and (d) an absence of allelochemical/physical defences capable of protecting Australasian plants against introduced herbivores.

Competition

Lack of interspecific competition may contribute to the high densities of introduced herbivores in Australia.

Species richness of mammalian herbivores (native and introduced) in northern Australia compares favourably with that for savannas in other parts of the tropics. Large mammal species richness for areas in the Australian tropical savannas from which some of the density data were obtained

range from seven to ten. Examples include the Cobourg Peninsula (nine species) and the Victoria River District (ten species) (table 35.2). These areas were selected because together they embrace the majority of the latitudinal moisture gradient present in the Northern Territory, and because both areas have been used to derive some of the population density estimates. These levels of species richness compare favourably with those of the Asian (six to eight species) and many African savannas (McKay and Eisenberg, 1974; Eisenberg and Seidensticker, 1976).

While species richness may not be a significant factor in any possible absence of competition among the feral and native populations in northern Australia, the combinations of species involved differs greatly from that of natural communities, and this may influence the types and effects of competitive interactions. The major differences between the Australian man-made communities and those of Asia/Africa are: (a) the absence of any extensive period

Table 35.2 [*orig. table 2*]   Species present in two communities of large mammalian herbivores in the Northern Territory, Australia. (a) = horse. (b) = Timor Pony.

Species	Cobourg Peninsula	Victoria River District
Native		
Macropus robustus		*
M. antilopinus	*	*
M. agilis	*	*
Onychogalea ungifera	*	*
Introduced		
Bos banteng	*	
B. taurus		*
Bubalus bubalis	*	*
Cervus unicolor	*	
Camelus dromedarius		*
Equus caballus (a)	*	*
E. caballus (b)	*	
E. asinus		*
Sus scrofa	*	*
Total species	9	10

during which the Australian herbivores could have undergone co-evolution resulting in a minimization of competitive interactions, (b) the absence from Australia of exceedingly large herbivores such as elephants and rhinoceros, and (c) the absence from Australia of browsing species. The first and last factors result in an expectation of more intense competitive interactions in Australian as opposed to natural communities, while any possible effects of the second are difficult if not impossible to predict.

A reduction of more than 50% of the feral donkeys (*E. asinus*) in an area in the Victoria River District appeared to result in a significant increase in the population of feral horses (*E. caballus*) (Freeland and Choquenot, 1990). Competition clearly influences patterns of relative abundance of large herbivorous mammals in northern Australia, but it is not known whether this has any greater or lesser effect in Australian as opposed to native communities. The interaction between donkeys and horses may be a consequence of the haphazard

man-made nature of the community resulting in sympathy of species lacking a co-evolutionary background. Natural communities do not usually have more than one species of wild equid.

Available information on levels of biomass per unit of land area suggests that northern Australia ungulates may experience more inter-specific competition than do ungulate populations in southern Asia. Data from southern Asia (Eisenberg and Seidensticker, 1976) suggest a mean biomass of native ungulates of 1153 kg km^{-2} standard deviation =865, N=8) significantly lower than is found in introduced ungulates in parts of northern Australia (water buffalo, horses, cattle and donkeys only) (mean=2225 kg km^{-2} standard deviation=767, N=6 areas (9000–17,700 km^2) that have been colonized by introduced species for >50 years (Bayliss and Yeomans, 1989b). Because the introduced ungulates of northern Australia have ecological equivalents (in some cases the same species) in Asia, it might be taken that levels of competition could be higher in Australia. This is especially so given the absence of introduced leaf-eating specialists from northern Australia, and the absence of feral pigs and native macropods from the Australian data. Any conclusion drawn from these data requires that soils, rainfall and survey methods have not significantly biased the results.

Australian levels of biomass are in one case equivalent to that of Africa, and in another the Australian biomass is much lower. The pronounced biological differences between the African and introduced Australian faunas make interpretation of these patterns difficult if not impossible. The Australian biomass mean is equivalent to that for ungulates in Rwenzori National Park, Uganda (2696 kg km^{-2}), even though the Australian figure does not include data for native species or feral pigs. The Australian estimate is far less than the reported 8427 kg km^{-2} for Transvaal lowveld (Damuth, 1982). While the Australian figure does not include pigs or the large native herbivores, it seems unlikely that they could account for the difference between Australia and the lowveld. The absence from Australia of functional equivalents to the giraffe (40% of the lowveld biomass), the leaf eating kudu (6% of biomass) and a migratory equivalent to the wildebeeste (26% of biomass) make direct comparisons difficult at best. Migratory species can have 11 or more times the population sizes of non-migratory species (Fryxell, Greever and Sinclair, 1988). These biological differences are likely to have a major impact on any differences between the areas' potentials for interspecific competition. Certainly the levels of biomass recorded from northern Australia are substantial, and given the haphazard nature of the community composition and the absence of browsing species, interspecific competition may be unusually high, rather than unusually low.

Resolution of the problem requires more detailed data on total community biomass (rather than species specific population sizes) and species richness in relation to rainfall and soil fertility, as well as manipulative experiments in the Australian and native environments. Account needs also to be taken of any effects that wildlife management (or its absence) and fencing may play in biasing some of the African results.

Predators and pathogens

Low levels of predation and low pathogen loads are likely to contribute towards the maintenance of high densities of introduced mammalian herbivores in Australia.

Predation can have a marked impact on juvenile survivorship of large herbivorous mammals (e.g. Beasom, 1974), and Australia is depauperate in large predators. The only predator potentially capable of consuming any of the larger species of introduced herbivore is the dingo (Canis familiaris dingo) (i.e. the domestic dog, introduced by man). Although the dingo's effect on prey populations remains unquantified, its effect is unlikely to equate with that of species rich natural communities of predators that have species with much larger body sizes (e.g. lions, tigers, hyaenas).

Introduced ungulates in northern Australia experience a reduced pathogen load. Native Australian herbivores are marsupials and rodents, whereas the introduced species are primarily ungulate. Relatively few of the native pathogens are capable of infesting the introduced species (e.g. Freeland, 1983). Relatively few pathogens accompanied introduction of the herbivores, and virtually none of the major ungulate pathogens is present in Australia (e.g. African horse sickness, anthrax, foot and mouth disease, rinderpest, blue tongue, equine influenza) (Seddon, 1953; Meischke and Geering, 1985). Disease is a major cause of mortality in natural populations of herbivorous mammals (Dobson and May, 1982), and is likely to be less significant among Australia's feral herds.

Plant defences

Plants in the northern Australian savannas are unlikely to lack defences against, or have less impact on mammalian populations than do plants in other savannas. Nor should any such effects be expected given the long plant–mammalian herbivore coevolution of Australian environments, and the great similarity between Australian and pantropical savanna floras (Bowman et al., 1988). What might be expected is that because of reduced impact from predators and pathogens, populations of feral species in Australia exceed bounds imposed in native environments, only to have imposed on them limits not usually observed in natural situations.

Feral donkey populations in northern Australia are limited primarily by poor juvenile survivorship at carrying capacity densities (Choquenot, 1988). Poor juvenile survivorship is linked to the apparent inability of females to lactate when depleted of mineral nutrients (Na, Ca, P) (Freeland and Choquenot, 1990). Females at carrying capacity are depleted of these nutrients relative to females in growing populations. Low mineral nutrient status is caused by a lower level of mineral intake, and fecal loss of mineral nutrients (apparently caused by the high fibre, species poor diet eaten when at carrying capacity density) (Freeland and Choquenot, 1990).

If, as seems likely, predators and pathogens have significant impacts on populations of mammalian herbivores in natural settings, population limitation as observed in feral donkeys may only occur in feral or man-disturbed communities.

Conclusions

Species of large mammalian herbivore introduced to northern Australia exist at densities greater than those achieved in the species' native habitats. These high density populations appear to result from the absence of significant predation and/or significant pathogen loads. In at least one case this has resulted in population regulation imposed by grazing caused changes in food eaten. The northern Australian savannas may be subject to unusually high levels of degradation from large mammals because of the absence of factors that regulate natural populations of ungulates. This expectation is in accordance with reports of severe habitat alteration (e.g. Letts et al., 1979). Conservation of the northern Australian environment requires the control of feral ungulates. This may be achieved at least cost and least environmental disturbance by the introduction of pathogens, rather than physical intervention by man or the introduction of predators.

Acknowledgments

D. Bowman, R. Braithwaite, S. Tidemann, P. Whitehead and J. Woinarski are thanked for their critical comments. J. Damuth is thanked for making his 1981 data set available.

References

Bayliss, P. and Yeomans, K. M. (1989a) Correcting bias in aerial survey population estimates of feral livestock in Northern Australia using the double count technique. *J. Appl. Ecol.* 26, 925–934.

Bayliss, P. and Yeomans, K. M. (1989b) The distribution and abundance of feral livestock in the 'Top End' of the Northern Territory (1985–1986), and their relation to population control. *Aust. Wildl. Res.* 16, 651–676.

Bayliss, P. and Yeomans, K. M. (1989c) The distribution and abundance of Bali Cattle and other feral ungulates on Cobourg Peninsular, Northern Territory, 1985. *Aust. Wildl. Res.* (in press).

Beasom, S. L. (1974) Relationships between predator removal and white-tailed deer net productivity. *J. Wildl. Managemnt,* 38, 854–859.

Bowman, D. M. J. S. and Panton, W. (1989) Banteng (*Bos javanicus*) and pig (*Sus scrofa*) habitat impact, Cobourg Peninsular, Northern Australia. *Aust. J. Ecol.* (in press).

Bowman, D. M. J. S., Wilson, B. A. and Dunlop, C. R. (1988) Preliminary biogeographic analysis of the Northern Territory vascular flora, *Aust. J. Bot.* 36, 503–517.

Braithwaite, W. R., Dudzinski, M. L., Ridpath, M. G. and Parker, B. S. (1984) The impact of water buffalo on the monsoon forest ecosystem in Kakadu National Park. *Aust. J. Ecol.* 9, 309–322.

Choquenot, D. (1988) Feral donkeys in northern Australia: population dynamics and the cost of control. M. Sc. dissertation, Canberra College of Advanced Education.

Damuth, J. (1981) Population density and body size in mammals. *Nature,* 290, 699–700.

Damuth, J. (1982) The analysis of the degree of community structure in assemblages of fossil mammals. Ph.D. dissertation, University of Chicago.

Dobson, A. P. and May, R. M. (1982) Disease and conservation. *Animal disease in relation to animal conservation* (ed. by M. A. Edwards and U. McDonnell), pp. 345–365. Symposia of the Zoological Society of London, No. 50. Academic Press, London.

Dunsmore, J. D. (1974) The rabbit in subalpine southeastern Australia. I. Population structure and productivity. *Aust. Wildl. Res.* 1, 1–16.

Dwyer, P. D. (1978) A study of *Rattus exulans* (Peale) (Rodentia: Muridae) in the New Guinea highlands. *Aust. Wildl. Res.* 5, 221–248.

Eisenberg, J. F. and Seidensticker, J. (1976) Ungulates in Southern Asia: a consideration of biomass estimates for selected habitats. *Biol. Conserv.* 10, 293–308.

Freeland, W. J. (1983) Parasites and the co-existence of animal host species. *Am. Nat.* 121, 223–236.

Freeland, W. J. and Choquenot, D. (1990) Determinants of herbivore carrying capacity: plants, nutrients and *Equus asinus* in northern Australia. *Ecology* (in press).

Fryxell, J. M., Greever, J. and Sinclair, A. R. E. (1988) Why are migratory ungulates so abundant? *Am. Nat.* 131, 781–798.

Gee, E. P. (1963) The Indian Wild Ass (a survey). *Oryx,* 7, 9–21.

Graham, A., Raskin, S., McConnell, M. and Begg, R. (1982) An aerial survey of feral donkeys in the V. R. D. Conservation Commission of the

Northern Territory, Technical Report.

Harrison, J. L. (1969) The abundance and population densities of mammals in Malayan lowland forests. *Malayan Nature J.* 22, 174–178.

Henzell, R. P. and McCloud, P. I. (1984) Estimation of the density of feral goats in part of arid South Australia by means of the Petersen Estimate. *Aust. Wildl. Res.* 11, 93–102.

Hone, J. (1986) An evaluation of helicopter shooting of feral pigs. Report to the Conservation Commission of the Northern Territory.

Hoogerwerf, A. (1970) *Udjung Kulon: the land of the last Javan rhinoceros.* Brill, Leiden.

Hope, J. (1984) The Australian Quaternary. *Vertebrate zoogeography and evolution in Australasia* (ed. by M. Archer and G. Clayton), pp. 69–81. Hesperian Press, Victoria Park.

Horton, D. R. (1980) A review of the extinction question: Man, climate and megafauna. *Archaeol. Phys. Anthropol. Oceania,* 15, 86–97.

Letts, G. A., Bassingthwaight, A. and deVos, W. L. (1979) Feral animals in the Northern Territory: report of a board of inquiry. Northern Territory Government Printer, Darwin.

McKay, G. M. and Eisenberg, J. F. (1974) Movement patterns and habitat utilization of ungulates in southeastern Ceylon. In: *The behaviour of ungulates and its relation to manage-ment* (ed. by V. Geist and F. Walther). IUCN Publ. No. 24.

Meischke, H. R. C. and Geering, W. A. (1985) Exotic animal diseases. *Pests and parasites as migrants* (ed. by A. Gibbs and R. Meischke), pp. 23–27. Cambridge University Press.

Newsome, A. E. (1969) A population study of house mice permanently inhabitating a reedbed in South Australia *J. Anim. Ecol.* 38, 361–377.

Seddon, H. R. (1953) *Diseases of domestic animals in Australia,* Parts 1–6. Commonwealth of Australia, Div. Vet. Hyg. Serv. Publ. No. 1–6.

Strecker, R. L. (1962) Population levels. *Pacific Island Rat Ecology* (ed. by T. I. Storer), pp. 74–79. Bull. Bishop Mus. Honolulu 225, 274 pp.

Tamarin, R. H. and Malecha, S. R. (1971) The population biology of Hawaiian rodents: demographic parameters. *Ecology,* 52, 383–394.

Wirtz, W. O. (1972) Population ecology of the Polynesian rat, *Rattus exulans,* on Kure Atoll, Hawaii. *Pacific Science,* 26, 433–464.

Wolfe, S. L. (1979) Population ecology of the Kulan. *Symposium on the ecology and behaviour of wild and feral equids* (ed. by R. H. Denniston), pp. 205–220. University of Wyoming, Laramie.

Wood, D. H. (1980) The demography of a rabbit population in an arid region of New South Wales, Australia *J. Anim. Ecol.* 49, 55–78.

36

CHLORINATED HYDROCARBONS AND EGGSHELL CHANGES IN RAPTORIAL AND FISH-EATING BIRDS*

J. J. Hickey and D. W. Anderson

New perspectives on the role of chlorinated hydrocarbon insecticides in our environment have come into focus in recent years. Successive discoveries have demonstrated that these compounds are systematically concentrated in the upper trophic layers of animal pyramids (1). Raptorial bird populations have simultaneously suffered severe population crashes in the United States and Western Europe (2, 3, 4). These involve reproductive failures which, at least in Britain, are characterized by changes in calcium metabolism and by a decrease in eggshell thickness resulting in the parent birds' breaking and eating their own eggs (4, 5, 6). Such a derangement of calcium metabolism or mobilization perhaps could result from breakdown of steroids by hepatic microsomal enzymes induced by exposure to low dietary levels of chlorinated hydrocarbons (7).

We have examined the possibility that the eggshell changes reported in Britain (6) have also occurred in the United States and that the raptor population crashes in Europe and North America may have had a common physiological mechanism. The population changes are without parallel in the recent history of bird populations (8). They include the pending extirpation of the peregrine falcon (*Falco peregrinus*) in northwestern Europe, the complete extirpation of the nesting population of this species in the eastern half of the United States, and simultaneous declines among other bird- and fish-eating raptors on both sides of the Atlantic.

We examined 1729 blown eggs in 39 museum and private collections. Shells were weighed to the nearest hundredth of a gram. In 29 percent of these, we were able to insert a micrometer through the hole drilled by the collector at the girth of the shell and to take four measurements of thickness 7 mm from the edge of the blow hole; these were then averaged to the nearest 0.01 mm for each shell. Thickness in each case then represented the shell itself plus the dried egg membranes. Peregrine falcons, bald eagles (*Haliaeetus leucocephalus*), and ospreys (*Pandion haliaetus*) were selected as having one or more regionally declining

* Originally published in *Science*, 1968, vol. 162, pp. 271–2.

populations; golden eagles (*Aquila chrys-aeetos*), red-tailed hawks (*Buteo jamaicensis*), and great horned owls (*Bubo virginianus*) were selected as representative of reasonably stationary populations that may be slowly declining as their habitats are gradually destroyed by man, but for which widespread reproductive failures are currently unknown. In addition, 57 eggs of the herring gull (*Larus argentatus*) were collected from five colonies in 1967. The shells of these were dried at room temperature for 4 months before being measured, and residues of the entire egg contents were analyzed by the Wisconsin Alumni Research Foundation for chlorinated hydrocarbons but not for polychlorinated biphenyls. Analytical procedure followed that outlined by the U.S. Food and Drug Administration (9). Analyses were conducted on a gas chromatograph (Barber Coleman, model GC 5000, and Jarrell-Ash, model 28-700) with electron-capture detectors. The glass column (0.6 cm by 1.2 m) was packed with 5 percent DC 200 (12,500) on Cromport XXX. The column temperature was 210°C, and the nitrogen flow rate was 75 cm³/min. Each portion of the ground and dried samples was extracted for 8 hours or more in a Soxhlet apparatus with a mixture of ether

and petroleum ether (70:170). Portions of the extracts were further purified by putting them through a Florisil column.

In California, where the peregrine falcon population is in "a serious condition" (10), a change of 18.8 percent in shell weight occurred from 1947 to 1952. Ratcliffe (6) found a corresponding decrease of 18.9 percent in Britain. The change in California involved a decrease in shell thickness and had no precedent in the previous 57-year recorded history of the peregrine in that state (figure 36.1). In the eastern United States, where the nesting population of peregrines has now been wiped out (3), fragmentary data indicate that the same change took place (table 36.1). Broken

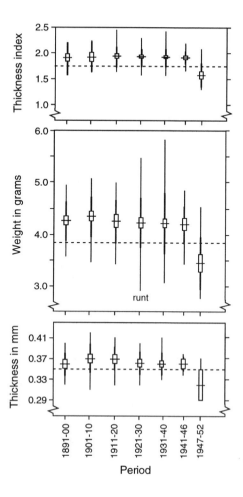

Figure 36.1 [*orig. figure 1*]   Measurements of 614 California peregrine eggshells collected since 1891. The dotted horizontal line is the midpoint between the 95 percent confidence limits for 1947–52 and the lowest of any preceding group. Solid horizontal bars are means; rectangles, 95 percent confidence limits; heavy vertical lines, standard deviations; narrow vertical lines, range in sample. The thickness index, calculated as ten times the weight divided by the product of the length and breadth (in mm) of each egg, was devised by Ratcliffe (6) for the study of museum eggs and appears here to be a meaningful statistic. The sizes of samples for measurement of weight and of thickness index for the periods were 71, 49, 36, 85, 155, 30, and 31, respectively; and the sizes of the samples for measurement of thickness were 24, 31, 29, 23, 37, 7, and 6.

Table 36.1 [*orig. table 1*]   Weights of raptor eggshells in museum and private collections. Citations (*23–25*) refer to the data for the population trend; S. E., standard error of the mean.

Region	Period	No.	Weight (g) Mean ± S.E.	Change (%)	Population trend (or reproduction)
			Red-tailed hawk		
Calif. (*23*)	1886–1937	386	6.32 ± 0.032		
	1943–44	6	6.09 ± 0.237	– 3.6	Stationary
	1953–67	8	6.49 ± 0.214	2.7	Stationary
			Golden eagle		
Calif. (*23*)	1889–1939	278	13.03 ± 0.083		
	1940–46	28	12.70 ± 0.161	– 2.5	Stationary
	1947–65	33	13.41 ± 0.232	2.9	Stationary
			Bald eagle (24a)		
Brevard Co., Fla.	1886–1939	56	12.15 ± 0.127		
	1947–62	12	9.96 ± 0.280	– 18.0	Declining
Osceola Co., Fla.	1901–44	25	12.32 ± 0.240		
	1959–62	8	9.88 ± 0.140	– 19.8	Declining
			Osprey (24b)		
Md.-Va.	1890–1938	152	7.05 ± 0.054		
	1940–46	21	6.91 ± 0.164	– 2.0	Stationary
	1955	3	6.85	– 2.8	Stationary
N.J.	1880–1938	117	7.08 ± 0.069		
	1957	6	5.30 ± 0.446	– 25.1	Declining
			Peregrine (25)		
B.C.	1915–37	–9	4.24 ± 0.061		
	1947–53	15	4.18 ± 0.081	– 1.4	Stationary
Calif. (*23*)	1895–1939	235	4.20 ± 0.031		
	1940–46	49	4.07 ± 0.038	– 3.1	No data
	1947–52	31	3.41 ± 0.084	– 18.8	Declining
N.H. to N.J.*	1888–1932	56	4.38 ± 0.034		
Vt.	1946	3	4.30	– 1.80	Stationary
Mass.	1947	3	3.47	– 20.8	Extirpated
N.J.	1950	3	3.24	– 26.0	Extirpated
			Great horned owl		
Calif. (*23*)	1886–1936	154	4.50 ± 0.033		
	1948–50	12	4.62 ± 0.119	2.4	Stationary

* Including Vermont and Massachusetts

eggshells in a North American peregrine eyrie were observed for the first time in 1947 by J. A. Hagar 60 miles (9.6 km) from the Massachusetts eyrie cited in this table (*11*). They were next inferred in Quebec in 1948 when egg-eating was observed at the same site in 1949 (*12*), and were observed in Pennsylvania in 1949 and 1950 (*13*). Chlorinated hydrocarbon data for this now-extinct regional population are completely absent. For nine surviving adult peregrines

in Canada's Northwest Territories in 1966, the data are reported to have averaged 369 parts per million (ppm) (fresh weight) in fat (*14*). For four adults in another migratory population in northern Alaska, values were even higher (*15*).

For the five other raptorial species we have studied, the data do not permit a precise delineation of the onset of the change in calcium metabolism or mobilization, but the decrease (table 36.1) in shell weight (and hence thickness) has involved only declining populations and not stationary ones. Change in shell thickness occurs in poultry as a result of dietary deficiencies and age (*16, 17*). This phenomenon would probably not occur simultaneously on two continents 1 year after the chlorinated hydrocarbon insecticides came into general usage. Other chemicals affect shell thickness in poultry (*17*), but the finding of high concentrations of chlorinated hydrocarbons in the eggs of wild populations of raptors and the time correlation of shell changes with the introduction of DDT [1,1,1-trichloro-2,2-bis (*p*-chlorophenyl) ethane] tend strongly to suggest that chlorinated hydrocarbons are the major contributing cause, although it is not unlikely that other chemicals could be contributory.

In order to test the hypothesis that these recent changes of thickness in raptor eggshells were the result of differences in exposure to chlorinated hydrocarbons we analyzed 10 to 14 eggs taken in 1967 from each of five colonies of the herring gull (*Larus argentatus*). Mean shell weight and thickness in 55 eggs collected in the same five states prior to 1947 disclosed no geographic gradients or significant differences. The 1967 mean thicknesses for each colony were therefore compared to mean levels of residual DDE [1,1-dichloro-2,2-bis (*p*-chlorophenyl) ethylene] on a fresh-weight basis, with the result shown in figure 36.2, the *r* value being significant, with $P = .001$. The residues of polychlorinated biphenyls (*18*) have not been studied in these ecosystems, but DDE has been consistently high in the Lake Michigan birds, averaging (fresh weight) 1925 ppm (S. E. 274) in the fat of 12 healthy adults collected in 1963–64 (*19*).

Reproduction in these gull colonies was generally normal in 1967 except perhaps in Wisconsin. At the latter colony where an 11 percent mean decrease in shell thickness occurred, some egg breakage and shell flaking was evident in 1967, although not at the frequency seen in previous years. Excessive reproductive failure occurred at this site in 1964 when about 18 percent of the eggs lost about one-third of the shell due to

flaking, when clutch size decreased and embryonic mortality was high, and when DDE residues averaged 202 ppm (S. E. 34) in nine eggs (*20*). (If linear extrapolation of the 1967 values is carried out to 202 ppm, the shell thickness in 1964 could be estimated as having decreased by about 32 percent.) The effectiveness of DDE in the enzymatic metabolism of aminopyrine has been reported by Hart and Fouts (*21*), and our data suggest that this compound, because of its prevalence, has played a major role in inducing the hepatic microsomal metabolism of steroids that in turn resulted in the eggshell changes we have encountered in museum collections. Without doubt DDE is the commonest insecticide or insecticide analog now being found in avian tissues (*22*). In 1966, it was found to average 284 ppm (S. E. 62) in the fat of nine arctic-breeding peregrine falcons (*14*) and about 414 ppm in four others on a wet-weight basis (*15*). Concentrations of this compound and other chlorinated hydrocarbons in the pere-

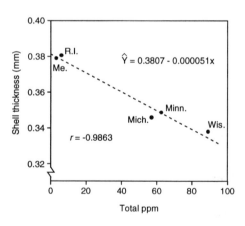

Figure 36.2 [*orig. figure 2*]   Variation in shell thickness and DDE concentrations in the eggs of herring gulls in 1967. The eggs were taken off Block Island, R. I.; Green Island in Penobscot Bay, Me.; Rogers City, Mich., on Lake Huron; near Knife River, Minn., on Lake Superior; and the Sister Islands in Green Bay, Wis. Some polychlorinated biphenyls probably occurred in these eggs, but they have not yet been identified.

grine populations that crashed farther south can be assumed to have been as high – and they may have been much higher.

From the above evidence and that accumulated by others (2, 4, 6, 8), we have reached these conclusions: (i) many of the recent and spectacular raptor population crashes in both the United States and Western Europe have had a common physiological basis; (ii) eggshell breakage has been widespread but largely overlooked in North America; (iii) significant decreases in shell thickness and weight are characteristic of the unprecedented reproductive failures of raptor populations in certain parts of the United States; (iv) the onset of the calcium change 1 year after the introduction of chlorinated hydrocarbons into general usage was not a random circumstance; and (v) these persisting compounds are having a serious insidious effect on certain species of birds at the tops of contaminated ecosystems.

References and Notes

1   E. G. Hunt and A. L. Bischoff, *Calif. Fish Game* 46, 91 (1960); review in R. L. Rudd, *Pesticides and the Living Landscape* (Univ. of Wisconsin Press, Madison, 1964), pp. 248–264.

2   C. Demandt, *Ornithol. Mitt.* 7, 5 (1955); P. Linkola, *Suomen Luonto* 18, 3, 34 (1959); P. Linkola, *ibid.* 19, 20 (1960); P. Linkola, *ibid.* 23, 5 (1964); K. Kleinstäuber, *Falke* 10, 80 (1963); C. Kruyfhooft, in *Working Conference on Birds of Prey and Owls* (International Council Bird Protection. London, 1964). p. 70; J.-F. Terrasse, *ibid.*, p. 73.

3   D. D. Berger, C. R. Sindelar, Jr., K. E. Gamble, in *Peregrine Falcon Populations: Their Biology and Decline*. J. J. Hickey, Ed. (Univ. of Wisconsin Press, Madison, 1968), p. 165.

4   D. A. Ratcliffe, *Bird Study* 10, 56 (1963).

5   ——, *Brit. Birds* 51. 23 (1958).

6   ——, *Nature* 215, 208 (1967).

7   D. B. Peakall, *ibid.* 216, 505 (1967); L. G. Hart, R. W. Shulties, J. R. Fouts, *Toxicol. Appl. Pharmacol.* 5, 371 (1963); A. H. Conney, *Pharmacol. Rev.* 19, 317 (1967); D. Kupfer, *Residue Rev.* 19, 11 (1967).

8   J. J. Hickey, Ed., *Peregrine Falcon Populations: Their Biology and Decline* (Univ. of Wisconsin Press, Madison, 1968).

9   U.S. Food and Drug Administration. *Pesticide Analytical Manual*, vol. 1 [U.S. Dept. of Health, Education, and Welfare, FDA Adm. Publ. (1963, revised 1964 and 1965)].

10   B. Glading, in Hickey (8), p. 96.

11   J. A. Hagar, in Hickey (8), p. 123.

12   G. H. Hall, *Brit. Birds* 51, 402 (1958).

13   J. N. Rice, in Hickey (8). p. 155.

14   J. H. Enderson and D. D. Berger. *Condor* 70, 149 (1968).

15   T. J. Cade, C. M. White, J. R. Haugh, *ibid.*, p. 170.

16   A. L. Romanoff and A. J. Romanoff, *The Avian Egg* (Wiley, New York, 1949), pp. 154–157.

17   T. G. Taylor and D. A. Stringer, in *Avian Physiology*, P. D. Sturkie, Ed. (Cornell Univ. Press, Ithaca, N. Y., ed. 2, 1965), p. 486.

18   D. C. Holmes, J. H. Simmons, J. O'G. Tatton, *Nature* 216, 227 (1967).

19   J. J. Hickey, J. A. Keith, F. B. Coon, *J. Appl. Ecol.* 3 (Suppl.), 141 (1966).

20   J. A. Keith. *J. Appl. Ecol.* 3 (suppl.), 57 (1966).

21   L. G. Hart and J. R. Fouts, *Arch. Exp. Pathol. Pharmakol.* 249, 486 (1965).

22   E. H. Dustman and L. F. Stickel, *Amer. Soc. Agron. Spec. Publ.* 8, 109 (1966).

23   J. B. Dixon, B. Glading, W. C. Hanna, E. N. Harrison, S. B. Peyton, personal communication.

24a A. Sprunt, IV, personal communication.

24b W. A. Stickel, in Hickey (8), p. 337.

25   Except for the California data (23) the data on population trends are given in (8) by F. L. Beebe for British Columbia; by W. R. Spofford for Vermont; by J. A. Hagar for Massachusetts; and by D. D. Berger et al. for New Jersey.

26   Research carried out under contract with the Bureau of Sport Fisheries and Wildlife, Fish and Wildlife Service, U.S. Dept. of Interior. W. H. Drury, J. T. Emlen, and M. E. Slate provided gull eggs for analysis. E. N. Harrison, W. C. Hanna, and many other zoologists greatly facilitated our measurements of egg shells. We thank D. A. Ratcliffe for advice throughout the entire study.

37

FOREST DECLINE AND ACIDIC DEPOSITION*

L. F. Pitelka and D. J. Raynal

Introduction

In the past few years the problems of acidic deposition (acid rain) and unexplained forest decline in Europe and North America have attracted considerable public and scientific interest. As with almost any major environmental problem, discussions of these issues are often characterized by alarming claims and scenarios as well as debate over the extent to which a serious problem actually exists. The discrepancies in reports of the extent and causes of forest decline make it difficult for those not close to the research effort to develop well-informed evaluations. This is almost as true for scientists as it is for the general public. Since this state of affairs typically develops with any major environmental issue, it may not be noteworthy. However, the forest decline and acidic deposition problems, and the research effort that has been organized to study them, should be of particular interest to ecologists. Our purpose in this article is to review the current forest decline phenomenon briefly and to discuss central aspects of the research effort and scientific debate.

Section I: The Current Forest Decline Problem

Concern about the possible effects of acidic deposition on forests has been stimulated by recent reports of forest decline in Europe (Schütt and Cowling 1985) and North America (Johnson and Siccama 1983, Morrison 1984) and by the increasing awareness of the nature of precipitation chemistry and soil acidity (e.g., National Atmospheric Deposition Program 1985, 1986, 1987, National Academy of Sciences 1986). Numerous hypotheses relating potential causal agents to forest decline have been proposed, and these have also stimulated research on effects of atmospheric deposition on forests. The major problem in assessing such effects is that forests are complex, long-lived ecosystems that may respond slowly and subtly to stress. Recognizing abnormal changes in forests and ascribing causes to these changes are often very difficult (Ashmore and Tickle 1987, Franklin et al. 1987, Klein and Perkins 1987, Woodman and Cowling 1987). Nonetheless, establishing the health of forests and measuring responses of forest ecosystems to atmospheric deposition are

* Originally published in *Ecology*, 1989, vol. 70, pp. 2–10.

vitally important as the chemical climate of the earth changes.

Crucial issues regarding supposed instances of forest decline include (1) whether a forest is actually undergoing a significant abnormal decline, and (2) whether the decline might be linked to human activities. It is quite normal for some trees, upon reaching a certain age or stage in stand development, to decline in growth, to appear unhealthy, or to die (Ashmore and Tickle 1987, Franklin et al. 1987, Harcombe 1987, Pastor et al. 1987, Peet and Christensen 1987). Tree populations may experience natural variations in health in response to changes in climate and other environmental factors (LeBlanc et al. 1987a, b, Mueller-Dombois 1987, Waring 1987). Forest decline is not a new phenomenon that is necessarily associated with human activities (Manion 1985, Mueller-Dombois 1986, Klein and Perkins 1987). In fact, plant pathologists recognize a category of tree diseases called "decline" in which no single causal agent is known, but which is caused by interactions among abiotic and biotic factors that produce gradual deterioration (Manion 1981). Climatic cycles, extremes of weather, attacks by insects and pathogens, and natural succession, as well as natural disturbances and those caused by human activities, all influence forest health and can produce major changes in ecosystem structure and function. Numerous examples of historical forest decline have involved species found in Europe, North America, Hawaii, and other Pacific forests (Mueller-Dombois 1987). Establishing the existence of forest decline depends largely on current species surveys, continuous stand inventories, and tree ring measurements. With any of these methods, care must be taken in sampling and in data analysis to ensure that decline can be distinguished from changes associated with succession or stand maturation. At present, the frequent lack of sufficient long-term quantitative information on forest health, as well as contrasting interpretations of current forest condition, contribute to different evaluations of the extent and magnitude of unexplained decline.

When it is determined that a forest is declining and that anthropogenic pollutants are likely to be a primary cause, we are often only part way to an understanding of cause–effect relationships. Forests are often exposed to a wide array of air pollutants derived from many sources. Various pollutants frequently co-occur and different pollutants and pollutant combinations have been implicated in decline. Numerous hypotheses have been proposed to explain the role of air pollutants in causing forest decline (Nihlgard 1985, Schütt and Cowling 1985, Hinrichsen 1986, Klein and Perkins 1987, Woodman and Cowling 1987). Ultimately, if cause–effect and dose-response relationships are to be understood, it is essential to identify the specific pollutants and the mechanisms that are responsible. Amassing additional data for use in judging changes in forest health will provide a more rigorous basis for evaluating the existence and nature of declines in the future.

Forest Decline in Europe

West Germany

Forest decline in West Germany has attracted more attention than in most other regions of the world for the following reasons: (1) early warnings of forest decline originated there (Ulrich et al. 1980), (2) a governmental "forest damage survey" has publicized the problem, and (3) concern has resulted in considerable research on the problem.

Widespread decline of Norway spruce (*Picea abies* (L.) Karst.), silver fir (*Abies alba* Mill.), and European beech (*Fagus sylvatica* L.) in West Germany has been reported. Severity of decline is greatest in fir, although the species is far less important in areal coverage than the other species. Kandler (1985) has reported that historical diebacks in silver fir are well documented and have not been synchronized with adverse climatic change. Further, after analyzing the current dieback, he concluded that there is little evidence for an involvement of atmospheric pollution. Decline in beech, which occurs on both acidic and alkaline soils, is not well understood, although soil acidification and aluminum toxicity may be involved (Ulrich 1983a). Norway spruce, the most important timber species in Germany, appears to be the most widely studied and best understood species that is exhibiting decline symptoms.

Rehfuess (1987) reports that there are several distinct decline or disease syndromes of Norway spruce observed in different regions. In the Ore Mountains of northern Bavaria, decline symptoms are associated with the very high local concentrations of sulfur dioxide. In southern Bavaria, widespread needle reddening and loss that were observed extensively in lowland stands in the early 1980s are apparently unrelated to acidic precipitation, and, possibly, to any air pollutants: these symptoms may result from adverse weather such as frost shock. The importance of weather is indicated by recovery of many stands following drought. In the calcareous Alps of southern Bavaria, deficiencies of manganese and potassium may be involved in decline: effects of acidic precipitation there are unlikely because of the calcareous soil, although ozone and acid mists may exacerbate foliar nutrient deficiencies (Prinz 1985).

The most extensive decline symptom of Norway spruce, and one in which acidic precipitation may play a role, is the needle yellowing and loss that occurs at high elevation on acidic soils in the Bavarian Alps, Black Forest, and the Harz and Hils Mountains. Zöttl and Hüttl (1986) and Rehfuess (1987) report that magnesium deficiency is responsible for these symptoms and that fertilization can ameliorate the condition. Structural changes in yellow needles from spruce in the Black Forest are similar to changes induced under experimental conditions by nutrient deficiencies and are not similar to changes caused by exposure to gaseous pollutants (Fink, *in press*). The influence of atmospheric deposition on soil nutrient loss has not been fully established, although Ulrich (1983b) has proposed a model of acid deposition as a driving force for forest ecosystem destabilization. Soil acidification can be expected to lead to some base cation (nutrient) leaching

(Reuss and Johnson 1986). Research must determine the level of leaching due to natural acidification and that due to anthropogenically caused soil changes. Characterization of interrelationships among acidification, soil and foliar nutrient loss, and nutrient cycling remains a vitally important research task.

One of the more significant features of forest decline in Germany is the synchrony of appearance of decline symptoms in different species and regions. This occurrence is a major reason that acidic precipitation or air pollutants were suspected to be the proximal cause. Further study has revealed that no single factor or cause explains all declines. The occurrence and magnitude of decline are not correlated with any specific pollutant, and some cases of decline have occurred where it is very unlikely that pollutants are involved (Rehfuess 1987). Some scientists believe adverse weather conditions, specifically drought or frost, have contributed to the synchrony of the declines (Prinz 1985, Blank 1986, Rehfuess 1987). Evidence indicates that ozone may increase the sensitivity of Norway spruce to winter cold (Brown et al. 1987, Barnes and Davison 1988).

Other European forests

Among the most severe instances of forest dieback are those reported in eastern Europe (Paces 1985, Mazurski 1986), where mortality is attributed to high sulfur dioxide concentrations from point sources of industrial activity. Poor forest management practices may also contribute to decline (Marzurski 1986). Local forest decline associated with point sources of pollutants is different from regional forest decline in areas more distant from pollutant sources.

Another instance of forest decline attributable to a specific local pollution problem is the ammonium sulfate-related decline of forests in the Netherlands (van Breeman et al. 1982, Boxman et al. 1987). The major problems here are a consequence of excess nitrogen deposition: the source of the nitrogen is ammonia volatilized from animal manure.

The extent of forest decline in other countries in Europe is less clear. Many nations are conducting forest inventories, but few have produced data that compare with those in West Germany. In Great Britain, damage surveys conducted since 1984 reveal a progressive worsening in the condition of forests (Innes and Boswell 1987), but climatic factors may be largely responsible (Redfern et al. 1987). In Scandinavia, forest monitoring has documented reduced vitality and abnormal needle loss (Andersson 1986): however, geographic patterns are not correlated with levels of acidic deposition of gaseous pollutants. Hauhs and Wright (1986) compared forest decline in Scandinavia and Germany; there was little evidence of decline in Norway. Based on experiments involving accelerated acidification of a conifer forest. Abrahamsen et al. (1988) concluded that direct effects of acid precipitation on Norwegian forests are unlikely at current levels.

There is growing evidence that soil acidification has occurred in recent decades (Berden et al. 1987, Tamm and Hallbacken 1988) and that soil nutrient deficiencies are responsible for some instances of forest decline in Europe (Zöttl and Hüttl 1986). Complex interactions between soil and atmospherically mediated changes in the forest environment and between these changes and various biotic processes necessitate experimental approaches to determine cause-and-effect relationships. Current research should help clarify the role of physical and biological factors in forest decline.

Forest Decline in North America

In North America, unexplained declines of red spruce (*Picea rubens* Sarg.), several southern pines (*Pinus taeda* L., *P. elliottii* Engelm., and *P. echinata* Miller), and sugar maple (*Acer saccharum* Marsh.) have been reported and are receiving considerable research attention. We will not discuss the well-documented influence of ozone on ponderosa pine (*Pinus ponderosa* Dougl. ex Laws), Jeffrey pine (*P. jeffreyi* Grev., and Balt.), and eastern white pine (*P. strobus* L.) (e.g., Miller 1983), because acidic deposition

is apparently not a primary causal factor (Chevone and Linzon 1988).

Red spruce

Red spruce is currently in a state of growth decline throughout its natural range, which extends from the southern Appalachian Mountains to the northeastern United States and adjacent Canada (Johnson and Siccama 1983, Hornbeck and Smith 1985, Johnson and McLaughlin 1986, McLaughlin et al. 1987). Red spruce decline is not a new phenomenon. Although decline has occurred since at least the late 19th century (Fox 1895, Weiss et al. 1985, McLaughlin et al. 1987), the current growth decline observed in all age and size classes, and in stands at a variety of elevations with different disturbance histories, may be unprecedented (McLaughlin et al. 1987). Symptoms in inland stands include needle discoloration and loss, twig and branch death, and outer crown thinning (Johnson and McLaughlin 1986). Coastal Maine populations show inner crown thinning and tufted branches, symptoms not unlike those observed in Norway spruce in Europe (Jagels 1986). At Camels Hump Mountain in Vermont, Whiteface Mountain in New York, and Mt. Mitchell in North Carolina, deteriorating spruce stands are particularly obvious (Siccama et al. 1982, Johnson and Siccama 1983, Scot et al. 1984, Vogelmann et al. 1985, LeBlanc et al. 1988). In some plots on Whiteface Mountain there has been 60–70% decline in basal area of red spruce from the early 1960s to the early 1980s (Scott et al. 1984). Dendroecological investigations have documented widespread synchronous growth reduction beginning in the mid 1960s and continuing to the present (Adams et al. 1985, McLaughlin et al 1987, LeBlanc et al. 1988, Raynal et al. 1988). At lower elevations, decline is more subtle than in subalpine forests, but synchronous growth reductions are typical (Hornbeck and Smith 1985, Hornbeck et al., 1986).

Possible explanations of red spruce decline include climatic factors, disease, natural senescence, gaseous pollutants (particularly ozone), acidic precipitation and cloud water, and combinations of these.

Hornbeck et al. (1986) suggest that many stands at lower elevation are declining because of normal stand maturation processes. Logging and spruce budworm infestations that occurred around 1900 resulted in establishment of many even-aged stands. By 1960 these stands had reached an age where normal competitive interactions are expected to cause reduced growth. Although this explanation does not account for all decline instances, especially in stands of mixed age composition or at high elevation, it does emphasize the importance of normal stand development in growth decline. Hamburg and Cogbill (1988) suggest that climate warming may be important in the current decline of spruce.

Current hypotheses that relate acidic deposition and decline and that are receiving substantial research attention include the following: (1) excess nitrogen deposition interfering with winter hardening of buds and foliage, (2) ozone-acid mist causing nutrient leaching from foliage, (3) soil acidification and cation leaching, leading to nutrient deficiency, (4) aluminum toxicity caused by soil acidification, (5) heavy metal toxicity, (6) direct gaseous oxidant effects, including ozone and hydrogen peroxide, and (7) adverse meteorological conditions. At present, there is little evidence to support any of the hypotheses as causal generally. The discoloration of foliage at high elevation may result from winter injury (Friedland et al. 1984); the role of pollutants as contributing factors is poorly established. Experimental studies have shown red spruce seedling growth to be sensitive to soluble aluminum (Hutchinson et al. 1986, Thornton et al. 1987), and the concentration of soluble aluminum may reach the toxicity threshold of red spruce in some soils (Cronan and Goldstein 1987., C. S. Cronan, *personal communication*). Ozone is a prime candidate as a contributing factor because ambient concentrations may cause reductions in photosynthesis and growth of other trees (Reich and Amundson 1985, Wang et al. 1986) and predispose trees to winter injury (Brown et al. 1987, Barnes and Davison 1988). Other cause–effect relationships are being evaluated in current research sponsored by the United States Forest Service–Environmental Protection Agency Spruce–Fir Research Cooperative and the Electric Power Research Institute (EPRI). It is reasonable to project that decline results from multiple stresses and that the relative contribution of any single stress varies from site to site.

Southern pines

Inventories of the growth of yellow pines in the mountain, piedmont, and coastal regions of the Southeast show 30–50% reduction in annual radial growth during the last 20–30yr (Sheffield et al. 1985, Sheffield and Cost 1987). Concern about the possible influences of acidic deposition and gaseous pollutants on growth has promoted coordinated study of both natural and anthropogenic stresses. The role of acidic deposition and air pollutants in growth reductions remains largely speculative at this time. Certainly natural determinants of growth including stand aging, hardwood species competition, weather conditions, and land use changes need to be evaluated if declines are to be understood.

Sugar maple

Sugar maple decline is known in the northeastern United States and Canada. Some have linked the problem to acidic deposition, although there is little direct evidence of cause and effect (McLaughlin et al. 1985, Vogelmann et al. 1985). After a review of more than 100 reports of sugar maple decline, McIlveen et al. (1986) concluded that the importance of anthropogenic stresses in contributing to the present decline has yet to be demonstrated. Raynal et al. (1982a, b, Dustin and Raynal 1988) report that seed germination and early seedling growth of sugar maple are not adversely affected by acidic precipitation at current ambient levels. Reich and Amundson (1985) indicate that the species is not highly sensitive to ozone, although slight reductions in net photosynthesis due to ambient ozone levels are possible. Late summer defoliation may be an important contributing factor in maple decline

(Gregory et al. 1986). Thornton et al. (1986) have shown that sugar maple seedlings are moderately sensitive to soluble aluminum, but concentrations of aluminum in soils in maple forests commonly may not reach the toxicity threshold for the species. It seems likely that many factors, including stand management, insect and pathogen attack, and weather conditions may lead to decline of sugar maple. Although decline of sugar maple undoubtedly occurs, Hornbeck et al. (1988) report evidence of increasing basal area increments between 1950 and 1980 in New England forests.

Section II: The Research Effort

The present research effort to assess the extent and causes of unexplained forest decline in Europe and North America is unprecedented and noteworthy for four reasons. The first is simply the sheer magnitude of the effort in terms of the financial resources, the number of funding agencies, and the number of researchers that are involved. The federal research effort in the United States is organized under the Forest Response Program FRP of the National Acid Precipitation Assessment Program (NAPAP). Several agencies including the United States Environmental Protection Agency and the United States Forest Service fund research that is part of the FRP. The current annual budget for the FRP is ≈$18 million, approximately the same as the entire 1987 budget for the National Science Foundation Ecosystem Studies Program. EPRI is currently spending ≈$5 million per year on forest effects research, while the National Council of the Paper Industry for Air and Stream Improvement (NCASI) has an annual budget of ≈$1 million for research on the problem. Many states and individual utilities and industries are also investing substantial amounts on research. Finally, additional research is supported by federal competitive grants programs outside of the FRP. A conservative estimate is that $25 million per year is being spent on the problem. It is doubtful whether any past environmental problem has been the subject of so much intensive research funded by so many different agencies or groups in the United States. This effort becomes even more impressive when it is recognized that a larger effort is underway in Europe and that Canada also has a substantial research program.

A second important feature of the research effort is the extent to which the different funding agencies are coordinating their efforts and cooperating to achieve mutual objectives. NAPAP itself is composed of representatives from 12 different federal agencies or entities. The agencies involved in the FRP have worked together to develop a single research plan for assessing the effects of acidic deposition on forests. The cooperation and coordination extend well beyond the federal sector. Rather than fund an independent research effort, NCASI supports the FRP and is represented on the management team. Although the EPRI effort is more independent, specific EPRI projects are recognized by the FRP as being sufficient to test certain hypotheses or meet particular scientific objectives. Thus, efforts are not duplicated. There are many instances in which one funding agency has initiated a project and other agencies have funded complementary projects that build on the initial project or make use of the same facilities. There is also communication and often cooperation among those involved in the research efforts in the United States, Canada, and Europe.

A third aspect of this research effort is the extent to which the funding agencies have recognized that basic, long-term mechanistic research will be needed to explain forest decline and to assess the impacts of acidic deposition. The need for this basic focus is largely a consequence of the complexity of forest ecosystems and the recognition that the cause–effect explanations of forest decline cannot be easily determined. Thus, this particular environmental problem involves more than simply assessing the dose-response relationship between a specific suspected cause and an obvious effect (e.g., DDT and eggshell thinning in birds). The complexity of the research problem, the need for a more basic research approach, and the availability of considerable funds to support that research have helped to attract the interest of a broader community of scientists than in the case of most environmental problems. Many scientists who had never been involved in "applied ecological research" have become engaged in this effort.

A final unusual feature of this effort is that, so far, the research is being conducted before passage of any legislation to control acidic deposition. Consequently, there is a real chance that the research results can have an impact on the policy-making process. Some may feel that we do not need more results before acting, but having more information clearly is desirable.

The perceived threats from acidic deposition are a major reason to mobilize such extensive resources to address one environmental problem. To a greater degree than for almost any past environmental problem, acidic deposition is a regional phenomenon that could have wide-ranging effects on many types of organisms and ecosystems. This contrasts with other environmental threats that involve more local effects or impacts on a narrower range of organisms. The regional or global extent of acidic deposition and other types of air pollution means that many ecosystems, including national parks and nature preserves, may be affected. The apparent effects of acidic deposition on aquatic systems and the early observations of forest decline led many to assume that a catastrophe was imminent. Although it is likely that acidic deposition has some effects on forest ecosystems, research to date has not documented significant direct effects on plants. In any event, the perception that extensive areas of forest and other pristine ecosystems are threatened by acidic deposition was probably the major factor that attracted the interest of scientists, government agencies, and the public and led to the current research effort.

Obvious questions concerning a research effort of this magnitude are, "Are we making progress, has the investment been worth it, and has the cooperation and coordination contributed to a more effective effort?" There seems little question that the answers are affirmative. The interpretations

in the NAPAP Interim Assessment (NAPAP 1987) have been loudly criticized, but the research itself has not. Many good research projects have been funded by NAPAP and other funding agencies, and significant findings are emerging. This success is due to the careful planning and peer review of the research plans, as well as the involvement of many competent scientists in the effort. Continuation of the present level of effort is likely to resolve many questions concerning forest decline and the impacts of acidic deposition, even if it does not produce a definitive evaluation of the role of acidic deposition in the current forest declines.

The research on acidic deposition and forest decline may provide results that will be useful in policy making, but in addition, there will be other benefits and lessons learned. Ecology and forestry will benefit from the research funds that have and will continue to be invested in understanding basic aspects of ecosystem processes and tree physiology, genetics, and ecology. One of the reasons that the present forest declines have proved so difficult to explain is that too little research focusing on forest ecosystem processes had been conducted. The consequence was an inadequate understanding of tree response to different stresses or of natural variation in forest health. We hope that the current influx of funds will result in significant advances in our understanding of forest ecology.

A related issue is the importance of long-term ecological research and monitoring (Likens 1983). If more long-term studies of forests and of environmental conditions in forests had been established before the first reports of forest decline began appearing, we would have been in a much better position to distinguish normal natural variability from significant unnatural or unprecedented changes in forest condition. As Likens (1983) has noted, there are many problems and challenges involved both in designing effective long-term studies and in obtaining sustained support. However, the current forest decline problem illustrates the importance of such studies.

The current research effort also provides a good model for investigating certain other multidisciplinary environmental problems. The cooperation among funding agencies, the careful planning of an integrated federal research program, the attention paid to focusing on only those scientific questions that address the important policy questions, and the recognition of the need for more basic mechanistic research may be some of the more useful lessons for future efforts.

Section III: The Role of Science and the Scientist

The phenomenon of unexplained forest decline is an intriguing scientific problem at the most basic level, and some scientists undoubtedly became interested in it purely because of the intellectual challenge of a new research problem. Concurrently, forest decline as it relates to acidic deposition is also a highly charged political and environmental issue. This has certain consequences for both the science and the scientists involved in investigating the problem. It is not possible, even if individuals wanted to do so, to detach the science from the political debate, and therefore it is important for scientists to be aware of the relationships between the scientific effort and the political debate.

One obvious implication for the scientific effort is that it is driven largely by the politics of the environmental problem. Only because forest decline and acid rain are major political issues, have public and private funding agencies devoted such large financial resources to the problem. A likely consequence is that much of the funding will disappear if the problem lessens as a major political issue or if policy decisions to control acidic deposition are made, even if all the important scientific questions have not been resolved. This is of course possible because data from scientific studies represent only one of many inputs to the formulation of policy (Bernabo 1986); decisions are often made before definitive data are available. It is interesting that lobbying efforts, public fears based on unsubstantiated claims or loose speculation about the threats posed by acidic deposition, or loss of public interest in the problem, could influence the decision-

making process more than solid scientific evidence. Ironically, it has been the concerns over the threats from acidic deposition that have created the demand for the large research effort. Thus the political forces that helped to foster the well-funded research effort may also work to diminish the relative importance of objective scientific input to the debate.

Given that the forest decline and acid rain problem is a major political issue, what role should scientists, particularly ecologists, attempt to play? Should they attempt to influence the decision-making process? Some scientists may stick strictly to conducting scientific investigations and providing rigorous answers to relevant scientific questions, without becoming at all involved in the public debate or without attempting to influence the decision makers. One could argue that this is the appropriate role for the scientist: that scientists compromise their objectivity and scientific credibility when they become advocates for a particular position in such a debate. However, as already noted, purely scientific data or reports may not attract much attention from decision makers. In addition, it is only natural for ecologists to become actively involved in debates on environmental problems because they have relevant expertise and because their interests and concerns lead them to do so. Bliss (1984) argued convincingly the need for ecologists to become more involved in environmental issues.

The challenge for scientists who do become involved in environmental debates is to distinguish clearly in their advocacy between statements that are based on personal views and those that are based on scientific evidence. Ecologists are likely to have strong opinions on environmental problems. But as scientists they have a responsibility to maintain appropriate standards of objectivity and to apply sound scientific criteria when evaluating evidence. Bliss (1984) noted that ecologists may have trouble maintaining their objectivity, especially when their investigations show "no effect" from some human-caused disturbance to natural systems. Some ecologists make "... up their minds in advance that a [particular disturbance is] wrong prior to the studies" (Bliss 1984). It is also important to distinguish between scientific and policy questions. When technical experts offer their opinions on questions such as "Do we know enough to mandate emission controls", their answers represent policy statements from laymen, not technical answers from expert scientists (Bernabo 1986). In the long run, it does not enhance the credibility of ecologists or the place of sound scientific evidence in the policy process to misstate risks simply to promote a particular position or cause.

Discussion of current forest declines and the possible role of acidic deposition in these declines has been characterized by considerable unsupported claims and speculation in the popular press and even semiscientific publications. One example is the use of inappropriate photographs or descriptions to portray the devastation supposedly caused by acidic deposition. Thus, photographs of fir-waves or Fraser fir (*Abies fraseri* (Pursh) Poir.) killed by the balsam wooly adelgid are sometimes presented as examples of forests possibly killed by acid rain. Similarly, there are locations where air pollutants (SO_2 in particular) have killed forests around smelters (e.g., Bormann 1982) or in very heavily industrialized regions, but these are significantly different situations from the regional cases of forest decline in areas farther from point sources of specific pollutants. Yet photographs or descriptions of these areas have been used by some to build an image of widespread devastation linked specifically to acid rain.

Another example of how information has been used inappropriately is the use of correlation to imply cause and effect. Many of the worst cases of forest-decline occur in areas now known to receive high acidic deposition, leading some to conclude a cause–effect relationship. However, the correlation is often not strong, and other natural or anthropogenic stresses can also be correlated with patterns of decline. Most scientists not actively involved in research on the problem rely on non-peer-reviewed sources for information on environmental issues such as forest decline and acid rain. Although inappropriate uses of data and descriptions can serve to advance a particular position, they clearly do not contribute to informed discussions or to a rational, objective resolution of the problem.

It is reasonable to suspect that acidic deposition plays some role in some of the current unexplained instances of forest decline. Acidic deposition can have effects on soil chemistry, and there is a growing consensus in Europe that soil nutrient deficiencies are responsible for much of the forest decline observed there. However, there is presently little compelling evidence to indicate that acidic deposition alone is a major cause of decline. Extensive investigations have failed to document effects that many scientists probably assumed would be detectable (e.g., direct effects on plant physiological processes and growth). When entering the policy debates on forest decline and acidic deposition, ecologists have a responsibility to distinguish clearly between rigorous scientific evidence, anecdotal information, and speculation. Policy makers can then weigh the available information from scientific studies against other concerns in determining whether any action should be taken.

Acknowledgments

We thank D. J. Leopold, P. D. Manion, and F. A. Pitelka for their comments and suggestions on an earlier version of this manuscript. This paper reflects our own views, however, and not necessarily those of our colleagues.

Literature Cited

Abrahamsen, G., B. Tweite, and A. O. Stuanes. 1988. Wet acid deposition effects on soil properties in relation to tree growth. *In* D. P. Lavender, editor, *Woody Plant Growth in a Changing Physical and Chemical Environment*. University of British Columbia, Vancouver, British Columbia, Canada, *in press*.

Adams, H. S., S. L. Stevenson, T. J. Blasing, and D. N. Duvick. 1985. Growth-trend declines of spruce and fir in mid-Appalachian subalpine forests. *Environmental and Experimental Botany* 25:315–325.

Andersson, F. 1986. Acidic deposition and its

effects on forests of Nordic Europe. *Water, Air, and Soil Pollution* 30: 17–29.

Ashmore, M., and A. Tickle. 1987. Acid rain: never pure and rarely simple. *Trends in Ecology and Evolution* 2:58–59.

Barnes, J. D., and A. W. Davison. 1988. The influence of ozone on the winter hardiness of Norway spruce (*Picea abies* (L.) Karst.). *New Phytologist* 108:159–166.

Berden, M., S. I. Nillson, K. Rosen, and G. Tyler. 1987. Soil acidification: extent, causes and consequences. Report 3292 National Swedish Environmental Protection Board. Solna, Sweden.

Bernabo, J. C. 1986. Science and policy: notes from a former congressional fellow. EOS. *Transactions of the American Geophysical Union.* February 1986 67(7):82.

Blank, L. W. 1986. Deterioration, but some recovery. *Nature* 319:529.

Bliss, L. C. 1984. Address of the past president: ecologists need to increase their involvement in society. *Bulletin of the Ecological Society of America* 65:439–444.

Bormann, F. H. 1982. The effects of air pollution on the New England landscape. *Ambio* 11:338–346.

Boxman, A. W., H. F. G. van Dijk, and J. F. M. Roelofs. 1987. Some effects of ammonium sulphate deposition on pine and deciduous forests in the Netherlands. Pages 680–687 *in* R. Perry, R. M. Harrison, J. N. B. Bell, and J. N. Lester, editors. *Acid Rain: Scientific and Technical Advances.* Selper, London, England.

Brown, K. A., T. M. Roberts, and L. W. Blank. 1987. Interaction between ozone and cold sensitivity in Norway spruce: a factor contributing to the forest decline in central Europe? *New Phytologist* 105:149–155.

Chevone, B. I., and S. N. Linzon. 1988. Tree decline in North America. *Environmental Pollution* 50:87–99.

Cronan, C. S., and R. A. Goldstein. 1987. ALBIOS: a comparison of aluminum biogeochemistry in forested watersheds exposed to acidic deposition. *In* D. Adriano, editor. *Case Studies on Acid Precipitation.* Springer-Verlag, New York, New York, USA, *in press.*

Dustin, C. D., and D. J. Raynal. 1988. Effects of simulated acid rain on sugar maple seedling root growth. *Environmental and Experimental Botany* 28:207–213.

Fink, S. *In press.* Pathological anatomy of conifer needles subjected to gaseous air pollutants or mineral deficiencies. *Aquilo Ser Botanica.*

Fox, W. F. 1895. The Adirondack black spruce.

1894 Annual report of the New York State Forest Commission, Albany, New York, USA.

Franklin, J. F., H. H. Shugart, and M. E. Harmon. 1987. Tree death as an ecological process. *BioScience* 37:550–556.

Friedland, A. J., R. A. Gregory, L. Karenlampi, and A. H. Johnson. 1984. Winter damage to foliage as a factor in red spruce decline. *Canadian Journal of Forest Research* 14:963–965.

Gregory, R. A., M. W. Williams, B. L. Wong, and G. J. Hawley. 1986. Proposed scenario for dieback and decline of *Acer saccharum* in northeastern USA and southeastern Canada. *International Association of Weather Analysts Bulletin* 7:357–369.

Hamburg, S. P., and C. V. Cogbill. 198. Historical decline of red spruce populations and climatic warming. *Nature* 331:428–431.

Harcombe, P. 1987. Tree life tables. *BioScience* 37:557–568.

Hauhs, M., and R. F. Wright. 1986. Relationship between forest decline and soil and water acidification in Scandinavia and Northern Germany. Pages 15–26 *in* B. G. Blackmon and R. S. Beasley, editors. *Proceedings of the Mid-South Symposium on Acid Deposition.* University of Arkansas, Little Rock, Arkansas, USA.

Hinrichsen, D. 1986. Multiple pollutants and forest decline *Ambio* 15:258–265.

Hornbeck, J. W., C. A. Federer, and R. B. Smith. 1988. Using tree rings to evaluate acid deposition and other causes of forest decline. Pages 346–350 *in* A. Carey, R. Blair, and C. Saint, editors. *Forest Response Program Annual Meeting Project Status Reports.* Atmospheric Impacts Research Program, North Carolina State University, Raleigh, North Carolina, USA.

Hornbeck, J. W., and R. B. Smith. 1985. Documentation of red spruce growth decline. *Canadian Journal of Forest Research* 15:1199–1201.

Hornbeck, J. W., R. B. Smith, and C. A. Federer. 1986. Growth decline in red spruce and balsam fir relative to natural processes. *Water, Air, and Soil Pollution* 31:425–430.

Hutchinson, T. C., L. Bozix, and M. Munoz-Vega. 1986. Response of five species of conifer seedlings to aluminum. *Water, Air, and Soil Pollution* 31:283–294.

Innes, J. L., and R. C. Boswell. 1987. Forest health surveys 1987, part 1: results. *Forestry Commission Bulletin* 74. Her Majesty's Stationery Office, London, England.

Jagels, R. 1986. Acid fog, ozone and low elevation spruce decline. *International Association of Weather Analysts Bulletin* 7:299–307.

Johnson, A. H., and S. B. McLaughlin. 1986. The nature and timing of the deterioration of red spruce in the northern Appalachian Mountains. Pages 200–230 in Acid Deposition, Long-term Trends. National Academy Press, Washington, D.C., USA.

Johnson, A. H., and T. G. Siccama. 1983. Acid deposition and forest decline. Environmental Science and Technology 17:294A–306A.

Kandler, O. 1985. Waldschaden – Theorie and Praxis auf der Suche nach Antworten. R. Oldenbourg Verlag, Munich, Germany.

Klein, R. M., and T. D. Perkins. 1987. Cascades of causes and effects of forest decline. Ambio 16:86–93.

LeBlanc, D. C., D. J. Raynal, and E. H. White. 1987a. Acidic deposition and tree growth: I. The use of stem analysis to study historical growth patterns. Journal of Environmental Quality 16:325–333.

LeBlanc, D. C., D. J. Raynal, and E. H. White. 1987b. Acidic deposition and tree growth: II. Assessing the role of climate in recent growth declines. Journal of Environmental Quality 16:334–340.

LeBlanc, D. C., D. J. Raynal, E. H. White, and E. H. Ketchledge. 1988. Comparative historical growth patterns of Adirondack conifers. In D. P. Lavender, editor. Woody Plant Growth in a Changing Physical and Chemical Environment. University of British Columbia. Vancouver, British Columbia, Canada, in press.

Likens, G. E. 1983. Address of the past president: a priority for ecological research. Bulletin of the Ecological Society of America 64:234–243.

Manion, P. D. 1981. Tree Disease Concepts. Prentice-Hall, Englewood Cliffs, New Jersey, USA.

——,1985. Prepared discussion. Journal of the Air Pollution Control Association 35:919–922.

Mazurski, K. R. 1986. The destruction of forests in the Polish Sudetes Mountains by industrial emissions. Forest Ecology and Management 17:303–315.

McIlveen, W. D., S. T. Rutherford, and S. N. Linzon. 1986. A historical perspective of sugar maple decline within Ontario and outside of Ontario. Report Number ARB-141-86-Phyto, Ministry of the Environment, Ontario, Canada.

McLaughlin, S. B., D. J. Downing, T. J. Blasing, E. R. Cook, and H. S. Adams, 1987. An analysis of climate and competition as contributors to decline of red spruce in high elevation Appalachian forests of the eastern United States. Oecologia (Berlin) 72:487–501.

McLaughlin, D. L., S. N. Linzon, D. E. Dinna, and W. D. McIlveen. 1985. Sugar maple decline in Ontario. Report Number ARB-141-85-Phyto, Ministry of the Environment, Ontario, Canada.

Miller, P. R. 1983. Ozone effects in the San Bernardino National Forest. Pages 161–198 in D. D. David, A. A. Miller, and L. Dochinger, editors. Air Pollution and the Productivity of the Forest. Isaac Walton League of America, Arlington, Virginia, USA.

Morrison, I. K. 1984. Acid rain, a review of literature on acid deposition effects in forest ecosystems. Forestry Abstracts 45:483–506.

Mueller-Dombois, D. 1986. Perspectives for an etiology of stand-level dieback. Annual Review of Ecology and Systematics 17:221–243.

——, 1987. Natural dieback in forests. BioScience 37:575–585.

National Academy of Sciences. 1986. Acid Deposition, Long-term Trends. National Academy of Sciences Press. Washington, D.C., USA.

National Acid Precipitation Assessment Program. 1987. Interim assessment. The Causes and Effects of Acidic Deposition. United States Government Printing Office, Washington, D.C., USA.

National Atmospheric Deposition Program. 1985, 1986, 1987. NADP/NTN annual data summary. Precipitation Chemistry in the United States. NADP/NTN Coordinator's Office, Natural Resource Ecology Laboratory, Colorado State University, Fort Collins, Colorado, USA.

Nihlgard, B. 1985. The ammonium hypothesis – an additional explanation to the forest dieback in Europe. Ambio 14:2–8.

Paces, T. 1985. Sources of acidification in central Europe estimated from elemental budgets in small basins. Nature 315:31–36.

Pastor, J., R. H. Gardner, V. H. Dale, and W. M. Post. 1987. Successional changes in nitrogen availability as a potential factor contributing to spruce declines in boreal North America. Canadian Journal of Forest Research 17:1394–1400.

Peet, R. K., and N. L. Christensen. 1987. Competition and tree death. BioScience 37:586–595.

Prinz, B. 1985. Prepared discussion. Environmental Science and Technology 35:913–915.

Raynal, D. J., D. C. LeBlanc, B. T. Fitzgerald, E. H. Ketchledge, and E. H. White. 1988. Historical growth patterns of red spruce and balsam fir at Whiteface Mountain, New York. In Effects of Atmospheric Pollution on Spruce and Fir Forests in the Eastern United States and the Federal Republic of Germany. United States Forest Service, Washington, D.C., in press.

Raynal, D. J., J. R. Roman, and W. M. Eichenlaub. 1982a. Response of tree seedlings to acid precipitation. I. Effect of substrate acidity on seed germination. *Environmental and Experimental Botany.* 22:377–383.

Raynal, D. J., J. R. Roman, and W. M. Eichenlaub. 1982b. Response of tree seedlings to acid precipitation. II. Effect of simulated acidified canopy throughfall on sugar maple seedling growth. *Environmental and Experimental Botany* 22:385–392.

Redfern, D. B., S. C. Gregory, J. E. Pratt, and G. A. MacAskill. 1987. Foliage browning and shoot death in Sitka spruce and other conifers in northern Britain during winter 1983–84. *European Journal of Forest Pathology* 17:166–180.

Rehfuess, K. E. 1987. Perceptions on forest diseases in central Europe. *Forestry* 60:1–11.

Reich, P. B., and R. G. Amundson. 1985. Ambient levels of ozone reduce net photosynthesis in tree and crop species. *Science* 230:566–570.

Reuss, J. O., and D. W. Johnson. 1986. *Acid Deposition and the Acidification of Soils and Waters.* Springer-Verlag, New York, USA.

Schütt, P., and E. B. Cowling. 1985. Waldsterben, a general decline of forests in Central Europe: symptoms, development and possible causes. *Plant Disease* 69:548–58.

Scott, J. T., T. G. Siccama, A. H. Johnson, and A. R. Breisch. 1984. Decline of red spruce in the Adirondacks, New York. *Bulletin of the Torrey Botanical Club* 111:438–444.

Sheffield, R. M., and N. D. Cost. 1987. Behind the decline. *Journal of Forestry* 85:29–33.

Sheffield, R. M., N. D. Cost, W. A. Bechtold, and J. P. McClure. 1985. Pine growth reductions in the Southeast. *United States Forest Service Resource Bulletin* SE-83.

Siccama, T. G., M. Bliss, and H. W. Vogelmann. 1982. Decline of red spruce in the Green Mountains of Vermont. *Bulletin of the Torrey Botanical Club* 109:162–168.

Tamm, C. O., and L. Hallbacken. 1988. Changes in soil acidity in two forest areas with different acid deposition: 1920s to 1980s. *Ambio* 17:56–61.

Thornton, F. C., M. Schaedle, and D. J. Raynal. 1986. Effect of aluminum on the growth of sugar maple in solution culture. *Canadian Journal of Forest Research* 16:892–896.

Thornton, F. C., M. Schaedle, and D. J. Raynal. 1987. Effects of aluminum on red spruce seedlings in solution culture. *Environmental and Experimental Botany* 27:489–498.

Ulrich, B. 1983a. Soil acidity and its relations to acid deposition. Pages 127–146 *in* B. Ulrich and J. Pankrath, editors. *Effects of Accumulation of Air Pollutants in Forest Ecosystems.* D. Reidel, Boston, Massachusetts, USA.

———, 1983b. A concept of forest ecosystem stability and of acid deposition as driving force for destabilization. Pages 1–29 *in* B. Ulrich and J. Pankrath, editors. *Effects of Accumulation of Air Pollutants in Forest Ecosystems.* D. Reidel, Boston, Massachusetts, USA.

Ulrich, B., R. Mayer, and T. K. Khanna. 1980. Chemical changes due to acid precipitation in a loess-derived soil in central Europe. *Soil Science* 130:193–199.

van Breeman, N., P. A. Burrough, E. J. Velthorst, H. F. van Dobben, T. de Wit, T. B. Ridder, and H. F. R. Reijnders. 1982. Soil acidification from atmospheric ammonium sulphate in forest canopy throughfall. *Nature* 299:548–550.

Vogelmann, H. W., J. D. Badger, M. Bliss, and R. M. Klein. 1985. Forest decline on Camels Hump, Vermont. *Bulletin of the Torrey Botanical Club* 112:274–287.

Wang, D., D. F. Karnosky, and F. H. Bormann. 1986. Effects of ambient ozone on the productivity of *Populus tremuloides* Michx. grown under field conditions. *Canadian Journal of Forest Research* 16:47–54.

Waring, R. H. 1987. Characteristics of trees predisposed to die. *BioScience* 37:569–574.

Weiss, M. J., L. R. McCreery, I. Millers, M. Miller-Weeks, and J. T. O'Brien. 1985. Cooperative survey of red spruce and balsam fir decline and mortality in New York, Vermont and New Hampshire. United States Forest Service Report NA-TP-11.

Woodman, J. N., and E. B. Cowling. 1987. Airborne chemicals and forest health. *Environmental Science and Technology* 21:120–126.

Zöttl, H. L., and R. F. Hüttl. 1986. Nutrient supply and forest decline in southwest Germany. *Water, Air, and Soil Pollution* 31:449–462.

38

THE BIODIVERSITY CHALLENGE: EXPANDED HOT-SPOTS ANALYSIS*

N. Myers

Introduction

This paper reviews the findings of an expanded analysis of 'hot-spot' areas, these being areas that (a) feature exceptional concentrations of species with high levels of endemism, and (b) face exceptional threats of destruction. This analytic approach can serve as a key contribution to conservation strategies.

To demonstrate the value of this approach, an earlier appraisal (Myers, 1988) examined the case of 10 areas in tropical forests, viz., Madagascar, the Atlantic coast of Brazil, western Ecuador, the Colombian Choco, the uplands of western Amazonia, eastern Himalayas, peninsular Malaysia, northern Borneo, the Philippines and New Caledonia. It found the areas' expanse of 292,000 km² comprised only 3.5 percent of primary tropical forests, yet these contained endemic plant species amounting to 27 percent of all tropical forests' plant species, together with a much greater proportion (albeit undocumented) of animal species. At a higher scale of calculation, it revealed that 0.2 percent of the Earth's land surface contains 13.8 percent of Earth's plant species, for an 'area/species index' of 69 (13.8 divided by 0.2). Moreover, all these 10 areas are losing their forests so fast that it is unlikely they will retain as much as 10 percent of original tree cover by the year 2000. This means, according to the theory of island biogeography (for brief review, see Myers, 1988), they will then be unable to support more than 50 percent of their original species. So in these 10 areas we can expect the early demise of 17,000 plant species and at least 350,000 animal species (perhaps several times more animal species). This in itself would amount to a spasm of species extinctions unparalleled in human history.

Furthermore, since the hot-spots approach identifies key localities of biotic richness under acute threat, it enables conservationists to determine their priorities in a more informed and methodical manner than has often been the case to date. Of course, it is not the only 'silver bullet' strategy available; it should be viewed in complementary accord with the 'Megadiversity Countries Strategy' developed by Mittermeier (1988).

* Originally published in *The Environmentalist*, 1990, vol. 10, no. 4, pp. 243–56.

Table 38.1 [*orig. table 1*]   Tropical forests: additional hot-spot areas.

Area (km²)	Extent of forest (km²)		Present plant species	Species endemic to area	(%)
	Original	Present (primary*)			
Ivory Coast	160,000	4,000**	2,770	200	(7)
Tanzania	31,000	6,000	1,600	535	(33)
Western Ghats in India	50,000	8,000	4,050	1,600	(40)
Southwestern Sri Lanka	16,000	700***	1,000	500	(50)
Totals	257,000	18,700	9,420	2,835	(30)

In *summary*, almost 3.8 percent of Earth's plant species (30 percent of them endemic) occur in 0.013 percent of Earth's land surface.

*   Some, though not many, primary forest species can survive in secondary forests.
**   Of these 4,000 km² of remaining primary forest, 2,000 km² are in the Tai Forest National Park, supporting 870 plant species (31 percent of the total), 150 of them (17 percent) being endemic.
***   Of these 700 km² of remaining primary forest, 56 km² occur in the Sinharaja Forest Reserve, supporting 75 percent of Sri Lanka's endemic tree species.
Sources: numerous, as cited under the area reviews in the texts.

Analytic Approach

For each hot-spot area, the analysis depends on establishing three prime factors: 1) the number of plant species originally extant; 2) the number of species remaining today; and 3) the number of species likely to survive into the opening part of the next century.

Plant species are chosen as an indicator category of species, rather than mammals, butterflies or some other taxon, because they are the best known of any category with large number of species. We can be pretty sure the 250,000 species identified include virtually all plant species extant; it is estimated that only another 10,000 species are still to be discovered. We have a better idea of their distributions and ranges than is the case with other categories except for mammals and birds, which total only 13,000 species. In the main, we have also a clearer understanding of plants' conservation status – much clearer than for any other category except for mammals and birds.

That is not to say, of course, that other categories might not reveal different sets of hot-spots. One of the original 10 hot-spots, New Caledonia, contains relatively few mammals and other vertebrate species. But one can reasonably suppose, even though the situation is almost entirely undocumented, that the rich flora has led to the evolution of a similarly rich invertebrate fauna. Since invertebrates make up the great bulk of all animal species, and since the hot-spots analysis is concerned with the mass extinction overtaking Earth's biotas generally, it is considered that plant species serve as adequate indicator species, at least for the purpose of an exploratory appraisal. So the focus of this paper is confined to the best known category of organisms, vascular plants (or 'higher' plants – of which flowering plants comprise 88 percent), henceforth referred to simply as 'plant species'.

This is not to leave animal species entirely out of account. True, we know all too little about animal species in hot-spot areas. In tropical forests, for instance, there could well be 30 million, and conceivably as many as 80 million, insect species (Adis, 1990; Erwin, 1988; Stork, 1988). Yet to date we have identified little more than 0.5 million of all invertebrate species. We can assert, however, in light of species–area inventories in several sectors of the tropics, that there are at least 20 animal species for every one plant species – albeit a calculation based upon a minimum planetary total of 5 million species. This reflects the fact that terrestrial life depends ultimately for its existence on some 250,000 higher plant species, these being the great bulk of organisms capable of photosynthesis. These plant species contrast with the tens of millions of heterotrophic species. So the extinction of one of the photosynthetic species should lead, in a food-web or energetic sense alone, to an average of between 30 and 250 extinctions of other kinds of organisms (Raven, 1976). Obviously a rare herb would not affect as many other organisms, but the disappearance of a widespread tree could affect many times more. For the purposes of this paper, however, and in order to remain consistent with the earlier analysis (Myers, 1988), also to be cautiously conservative, it will be assumed that one plant species is matched by only 20 animal species.

The data base: strengths and weaknesses

In many instances the statistical information is considered accurate to within 5 percent or better. In many others it is sufficiently accurate to rank as sound support for 'working estimates'. In some again, it is regarded as qualitatively correct, even though the quantitative data are sorely deficient: indicative information and best-judgement appraisals have their role to play in a paper such as this, provided their constraints are explicitly recognised. In a few further instances, which are identified as such, the information base is so poor that it is nothing better than an 'educated assessment', or a rough-and-ready estimate, albeit carefully conservative. The author believes this overall approach, uneven as it is, is justified in an analytical exercise that seeks to delineate the conservation challenge of a mass-extinction episode in its full scope.

Moreover, in cases where there is inadequate documentation of quantitative sort, one should not be preoccupied with what can be counted to the detriment of what also counts. After all, to decide that an area should not be evaluated because it lacks a conventional degree of accurate data, is effectively to decide that its conservation needs cannot be evaluated either – in which case its cause tends to go by default. As in other policy-and-planning realms where uncertainty is a salient factor, it will be better

for us to find we have been roughly right than precisely wrong.

Hot-Spot Areas: Additional Sites

Tropical forests

During the course of the 'first cut' analysis of hot-spot areas in tropical forests (Myers, 1988), it became apparent that a second-order list of areas warrants analysis. They are described below, four in all (for a tabulated review, see table 38.1).

First, a brief point of clarification. In many instances, a patch of forest may be grossly disrupted without being destroyed outright. Or, when a deforested area regenerates, the result is secondary forest, with species complements that differ markedly from those of the original primary forest. This paper confines itself to primary forest, on the grounds that only primary forest contains the high species diversity that characterises the tropical-forest biome. Moreover there appears to be little overlap in species composition between primary forest and secondary forest. In certain sectors of the Brazilian Amazonia, for instance, secondary vegetation as much as 15 years old shows little regeneration of primary forest species, even though the new vegetation is structurally similar to the original forest (Lisboa et al., 1989).

(1) *Southwestern Ivory Coast.* The southwestern sector of the Ivory Coast features part of a centre of diversity that straddles the border with Liberia, and coincides with the so-called Upper Guinea Pleistocene Refuge. Its forest area supports 2770 of the country's 4700 plant species, with perhaps 200 of these being endemic (Guillaumet, 1984; Hamilton, 1976; Myers, 1982).

Of Ivory Coast's original forest cover totalling 160,000 km², or 50 percent of the country, 118,000 remained as recently as 1956. But today there are only 16,000 km² at most, or 10 percent of the original expanse; and of this, only 4,000 km² can be classified as primary forest (Maitre, 1987; Myers, 1989; Spears, 1985). The remnants are being cut at a rate of up to 2,000 km² per year and if present trends continue, there will surely be next to nothing left outside protected areas by the mid-1990s.

This makes the Tai National Park all the more important, as the only large tract of original forest left, this contains 870 plant species, around 150 of them endemic (17 percent) (Assi, 1988; Bousquet, 1978; Dosso et al., 1981; Roth, 1984). Nominally the park covers 3,300 km^2, but it is being steadily reduced by logging (both legal and illegal), by gold prospecting, and most of all by encroaching cultivation. Less than 2,000 km^2, or 61 percent, are reported to be in primary-forest form, comprising half of the remaining primary forests in the country.

The main cause of deforestation lies with slash-and-burn farmers, who in turn reflect recent immigration patterns. Following the Sahel droughts, Ivory Coast has received 1.4 million foreigners from countries to the north, and by the mid-1980s every fifth person in the country was a foreigner (Ahonzo, 1984). Given the continuing drought situation in the Sahel, it can be anticipated that a persistent influx of environmental refugees will maintain the country's annual population growth rate at more than 4 percent.

(2) *The Eastern Arc Forests of Tanzania.* Ranging along central-eastern Tanzania is an arc of montane forests. They include, from north to south, the Pare, Usambara, Nguru, Uluguru, Ukaguru, Rubeho, Uzungwa, Mahenge and Matengo forests. Their expanse amounts to 15,000 km^2 out of an original expanse thought to have totalled some 31,000 km^2. Moreover, less than half, and probably only about 40 percent, i.e. 6,000 km^2 is considered to be primary forest. Being located on ancient hills and mountains, these forests have mostly enjoyed lengthy periods of isolation, leading to much speciation and high levels of endemism. In the East Usambara mountains, for instance close to 80 percent of millipedes, gastropods and amphibians are endemic. Some biologists consider the region is of paramount importance at global level for its potential insights into biogeography and evolution. (This review is based on Brenan, 1978; Hamilton, 1976; Hawthorne, 1984; Lovett, 1987 and 1989; Myers, 1982; Polhill, 1989;

Rodgers and Homewood, 1982; Rodgers et al., 1986).

Altogether these forests contain an estimated total of 1,600 plant species, endemics totalling 535 species (33 percent). They thus contain almost 15 percent of Tanzania's 11,000 plant species. Further, they contain a greater wealth of endemic species than the forest expanse of West Africa covering 150,000 km^2, and they feature 1,600 of the 30,000 plant species in all of tropical Africa with its 20 million km^2, i.e. over 5 percent of species on 0.075 percent of the land area.

Population pressures on the forests are high and rising rapidly, expressed in the form of agricultural encroachment, fuelwood gathering and commercial logging (Lovett, 1989; Rodgers and Hall, 1986). The formerly continuous forest of the East Usambaras is believed to have lost 70 percent of its original expanse in just the last 30 years, and has been fragmented into a patchwork of some 30 remnants of mixed primary and secondary forest. The Usambara mountains feature over 300 people per km^2, one of the highest densities in the world for an agriculture-based economy. Much the same story can be told of most of the forest tracts listed.

All the more regrettable is that certain of the plant species offer marked potential for economic development (Lovett, 1988). For instance, the African violet, an ornamental flower popular in affluent nations, enjoyed a retail trade of $30 million in 1983; of 20 known species of the violet, 18 are confined to Tanzania's montane forests. Only three species have so far been exploited for cultivation, but others could lend themselves to development, including a hardy variety that grows at 2,000 m altitude on the Uluguru mountains where it has become adapted to occasional frosts. Tanzania's forests also contain 16 species of wild coffee, 10 of them endemic; only three have been exploited as commercial crops. The worldwide trade in coffee in 1981 was worth $12 billion.

As recently as 1987 the protected-areas system in Tanzania had almost entirely collapsed, as a result of the country's declining economy and extreme poverty (average GNP per head today, $220, one of the lowest in the world).

(3) *The Western Ghats of India.* Of India's 15,000 plant species with 5,000 endemics (33 percent), there are 4,050 plants with 1,600 endemics (40 percent) in a 17,000 km² strip of forest along the seaward side of the Western Ghats mountains in Maharashtra, Karnataka, Tamil Nadu and Kerala States (Collins et al., 1990; Dayanandan, 1983; Jain and Sastry, 1983; Kendrick, 1989; Nair et al., 1986; Nayar and Sastry, 1987; World Conservation Monitoring Centre, 1988). Forest tracts up to 500 m elevation, comprising one fifth of the entire forest expanse, are mostly evergreen, while between 500 and 1,500 m the forest becomes semi-evergreen. There are two main centres of diversity, the Agastyamalai Hills and the Silent Valley/New Amarmbalam Reserve basin.

Little has recently been documented about the status of the forest cover, except that it seems to have declined between 1972 and 1985 at a rate paralleling that for India as a whole, viz. a loss of over 2.4 percent per year (Munni and Naidu, 1988). If we extrapolate from 1986 to 1989, this means a total loss for 1972/1989 of almost 34 percent. Still worse is the decline of the primary forest: the amount remaining seems to be no more than 8,000 km². All but isolated pockets of original forest have been opened up by shifting cultivation, allowing a take-over by deciduous species and bamboo among other forms of 'degenerate' vegetation.

(4) *Southwestern Sri Lanka.* The 65,610 km² island featured forest cover in more than 50 percent of its expanse as recently as 1950. Today, this residual forest had been reduced by over half, and much of it amounts to scattered fragments. Worse, of the island's richest biodiversity zone, the wet sector in the southwest, there remains only 670 km² of montane forest and 740 km² of lowland forest, a total of 1,410 km² (roughly half being primary forest), or a mere 9 percent of original forest covering almost 16,000 km². (This review is based on Ashton and Gunatilleke, 1987; Balasubramaniam, 1985; Collins et al., 1990; Erdelen, 1988; Geiser and Sommer, 1982; Gunatilleke and Ashton, 1987; Gunatilleke and Gunatilleke, 1985; Ishwaran and Erdelen, 1990; Kendrick, 1989; Nanayakkara, 1982; Senanayake et al., 1977; Sumithraarachchi, 1989; Werner, 1982).

Of Sri Lanka's 3,365 species, at least 1,000 species (30 percent) are endemic to the island. Of these endemics, 497 (50 percent) are confined to the lowland wet forest in the southwestern sector of the island; they constitute half of the 1,000-plus species (rough estimate, probably on the low side) found in the lowland forest. The only large remaining patch of this lowland forest is the 89 km² Man and Biosphere Reserve of the Sinharaja Forest (now a World Heritage site), where more than 70 percent of tree species are endemic (in some families of trees, e.g. the dipterocarps, endemism is greater than 90 percent); indeed the Reserve harbours 75 percent of Sri Lanka's endemic tree species. Of bird species endemic to Sri Lanka, 95 percent have been recorded in Sinharaja, plus 58 percent of mammals and 51 percent of butterflies.

The wet forest offers insights into the floristics of all Southern Asia. It is truly primeval, with an ancestry that traces back to the Deccan flora when an unbroken stretch of forest covered both Peninsular India and Sri Lanka (today's forests contain many more relict forms than the whole of southern India). The endemic plants are mostly phylogenetic hangovers, with closest relatives in Malaysia, Sumatra, Madagascar and South Africa. Hence the exceptional significance of the wet forest in general, and the Sinharaja Forest in particular, for ecological and biogeographic reasons as well as because of the sheer numbers of plant species. Moreover the relict status of many species leaves them confined to very small localities such as a single hilltop; 195 species are considered to be very rare, thus subject to summary extinction.

While the Sinharaja sector is thought to have once covered at least 150 km², today the reserve amounts to only 89 km² (59.3 percent), and only 56 km² (62.9 percent) rank as primary forest. During the 1970s the State Timber Corporation engaged in logging, and there are still threats from a major plywood complex that faces shortages of raw materials. The reserve is situated in the most densely populated sector of the island, surrounded by settlements of 5,000 people or more, many of them practising various forms of shifting cultivation. At least 8 percent of this populace is entirely dependent upon forest products.

Summary of tropical-forest areas

In a brief review (see table 38.1) of these further four hot-spots in tropical forests, an aggregate area of 18,700 km² of primary forest harbours an estimated total of 9,420 plant species, of which 2,835 (30 percent) are deemed to be endemic. In other words, 0.013 percent of Earth's land surface harbours 1.13 percent of the Earth's plant species, for an area/species index of 87.

If, as seems all too likely, these areas lose 90 percent of their original forest within the foreseeable future, they will then be able, according to island biogeography theory, to support only 50 percent of their original complements of species. This implies the impending demise of almost 1,420 plant species. If there is an associated stock of endemic animal species, in a conservative ratio of 20 animals to one plant (probably a good deal higher), there would be an accompanied extinction of at least 28,400 of these animal species.

These figures compare with those for the leading 10 hot-spots areas in tropical forests, identified earlier (Myers, 1988): 34,400 endemic plant species in an aggregate forest expanse of 292,000 km², i.e., 13.8 percent of Earth's species in 0.2 percent of Earth's land surface, for an area/species index of 69.

In addition to these four further tropical-forest areas, a number of others could qualify as subsidiary hot-spot localities. They include: The Mosquitia Forest of Honduras and Nicaragua; the Darien Gap of Panama; the Mount Nimba area of Liberia; lowlands forests of southeastern Nigeria and northwestern Cameroon; eastern Zaire; the Ethiopian highlands; the Andamans/Nicobar Islands of the Indian Ocean; the uplands of northern Burma; southern Thailand; and certain sectors of Sumatra and Sulawesi in Indonesia. They are not included here since they do not appear to be quite so exceptional in species richness, nor in degrees of threat, as the fourteen tropical-forest areas already considered.

Mediterranean-type Areas

Mediterranean-type zones, located in the sub-tropics, are exceptional in terms of their species abundance and diversity, and the degree of destruction overtaking them. The

Table 38.2 [orig. table 2]   Mediterranean-type hot spots: summary.

Area (km²)	Extent of former habitat (km²)	Extent of present habitat (km², and % of former habitat)	Present plant species	Species endemic to area (% of plants total)	Known to be threatened or rare (% of total flora)
Cape Floristic Province of South Africa	134,000	89,000 (66)	8,600	6,300 (73)	1,500 (17)
Southwestern Australia	112,260	54,700 (49)	3,630	2,830 (78)	860 (24)
California Floristic Province	324,000	246,000 (76)	4,450	2,140 (48)	1,140 (26)
Central Chile	140,000	46,000 (33)	2,900	1,450 (50)	580 (20)
Totals	710,260	435,700 (61)	19,580	12,720 (65)	4,080 (21)
In summary, almost 7.8 percent of Earth's plant species (65 percent of them endemic) occur in 0.3 percent of Earth's land surface.					
Mediterranean Basin as generalised mega-zone	n/a	n/a	25,000	12,500 (50)	3,000 (12)

Sources: numerous, as cited under the area reviews in the texts.

principal areas are: the Cape Floristic Province of South Africa, southwestern Australia, the California Floristic Province and central Chile. For a tabulated view, see table 38.2. Note that the Mediterranean Basin itself obviously qualifies as well, but it is far too large to rank as a hot-spot area (see below).

Insofar as the four areas identified are widely separated geographically, they are dissimilar in their floras and faunas, as in their evolutionary histories (Cody, 1986; Cody and Mooney, 1978). Their characteristic vegetation is dry grassland and shrub, known locally as chaparral, fynbos, maquis, etc. They feature high levels of endemism, especially narrow endemism: in the South African Cape, 80 percent of plant species have distributional ranges of less than 100 km^2 (Cody, 1986), while in southwestern Australia 12 percent are similarly limited (Pate and Beard, 1984). Moreover the areas mostly feature large human populations, hence their natural environments have been much disrupted.

(1) *The Cape Floristic Province of South Africa.* The Cape flora is so remarkable that it has been accorded a floristic kingdom of its own, one of six in the world. In this 89,000 km^2 area, with almost 78,000 km^2 of fynbos heathlands, there occur 8,600 plant species, over 6,300 (73 percent) of them endemic; and in the southwestern tip of the Cape, only 18,000 km^2 in extent, there are about 6,000 plant species, at least 4,200 (70 percent) of them endemic. (This review is based on Bond and Goldblatt, 1984; Cowling and Roux, 1982; Gibbs-Russell, 1985 and 1987; Goldblatt, 1978; Hall, 1987; Jarman, 1986; Moll and Bossi, 1984; Moll and Jarman, 1984; Rutherford and Westfall, 1986; Tansley, 1988; Werger, 1987).

Those species known to be rare or threatened (plus 26 extinct) total around 1,500 (over 17 percent), or more than the entire flora of the British Isles. The true total could well be more like 2,000 species, or as many threatened plants as in all the United States. One third of the original fynbos has already been lost to agriculture among other forms of development, or to invasion of exotic plants; and much of the remaining two-thirds suffers accelerating attrition. Much wildland habitat is broken up into a patchwork of relicts dispersed among farmlands and urban areas. Most species occur in areas of less than one square kilometre, and many feature fewer than the 500 individuals often regarded as a minimum for genetic viability (Soule, 1987).

If one anticipates that original habitats will shortly be reduced to 10 percent of their original extent, and if one applies the findings of island biogeography to the total of 6,300 endemic plant species, then one must expect the early extinction of 3,150 species in this area. There would be numerous extinctions too of associated animal species, though whether one can here apply the ratio of one plant to 20 animal species is debatable and unascertainable. All one can assert is that animal extinctions would surely total tens of thousands.

In summation, this area qualifies as one of the hottest of all hot-spots, both in terms of its plant diversity and its extreme threats on habitat destruction. The area is exceptional in one further respect: there is a large scientific presence, there are numerous conservation bodies, and the government may be starting to show more readiness to tackle the challenges at issue.

(2) *Southwestern Australia.* With its 54,700 km^2 of remnant 'kwongan' heathlands and associated Mediterranean-type vegetation (out of a former expanse of 112,260 km^2), this area features 3,630 plant species. At least 2,830 species (78 percent) are endemic; and at least 430 species have geographical ranges of less than 100 km^2. More than 860 species (24 percent) are classified as rare or threatened, and the true total should surely exceed 1,000 species. (This short summary is based on Hopper, 1979; Hopper and Muir, 1984; Pate and Beard, 1984; Rye, 1982; Rye and Hopper, 1981).

The main forms of habitat depletion lie with agricultural clearing of wildlands, mining, wild fires, and spread of weeds and fungal pathogens – most of which are growing worse rapidly. Fortunately the government, at both state and federal levels, seems to be taking greater interest in the conservation challenge.

(3) *California Floristic Province.* The California Floristic Province, originally comprising 324,000 km², possesses at least 4,450 plant species, 2,140 (48 percent) of them endemic. The Province is as large as the State of California with 411,000 km²; and 10 percent of it extends into southern Oregon and northern Baja, California. But 4,119 of the Province's species (93 percent) occur within California. The Province's flora is equivalent to roughly one quarter of all plants in the United States and Canada combined. The 1.5 million km² of the northeastern United States feature only 5,500 plant species, only a few hundred of them being restricted to the region. (This review is based on Barbour and Major, 1988; California Department of Forestry and Fire Protection, 1988; Elias and Nelson, 1987; Jones and Stokes Associates, 1987; Keeley, 1988; Raven and Axelrod, 1978; Smith, 1987; Smith and Berg, 1988; Stebbins, 1980; Stebbins and Major, 1965).

By 1980 the Province had lost at least 78,000 km² of its former expanse to urbanisation and irrigation agriculture, leaving an undisturbed area of 246,000 km² (O'Neill, 1990). A good deal more was being disrupted by stock grazing, logging and mining. Moreover, California is projected to have an extra 5 million people by the year 2000. As a result the flora has been much depleted. Plant species known to be rare or threatened (including those already extinct) number around 1,140 (26 percent). Another 500-plus (11 percent) would probably qualify as threatened if they were checked in detail. The problem lies not only with habitat destruction through human activities. It derives too from the localised distribution of many species: at least 163 species are confined to 100 km², and 73 are limited to 13 km² or less. As a measure of the general disregard that has been directed at the issue, consider that a human community with one of the finest scientific concentrations in the world remains so uninformed about its threatened biodiversity that another 10 plant species, mostly endemics, are still being discovered each year.

(4) *Central Chile.* In central Chile there remain roughly 46,000 km² of Med-iterranean-type habitat, or 33 percent of an original expanse of 140,000 km². The area contains more than 2,900, probably well over 3,000, of the country's 5,200 plant species, i.e. 56 percent of the national flora in 6 percent of national territory. Of these species, endemics total at least 1,450, or 50 percent of the area's total. Because of the exceptional amount of habitat diversity and 'border effects', the true figure could well be over 2,000 species, or 69 percent; after all, the endemism level for the entire country is 57 percent, and central Chile is biotically the richest sector in terms of species numbers. But for the sake of caution, the figure accepted here is the one documented (albeit a strictly minimum figure), 50 percent. Endemism is even more remarkable at generic level: at least 10 percent, by contrast with less than 4 percent in the California Floristic Province. (This short account is based on Arroyo, 1990; Corporacion Nacional Forestal, 1989; Fuenzalida, 1984; Gajardo, 1983; Marticorena, 1980; Pizarro, 1977; Poblete, 1985; and Ramirez, 1984).

This is the most densely inhabited part of Chile, containing over half the national population. Rural families rely heavily on natural vegetation for fuelwood and fodder. It seems as if wildland environments will continue to be depleted if only because of the rate of population growth, 1.7 percent per year, which means that human numbers double in just over 40 years. It is estimated that one fifth, or 580, of the plant species can already be characterised as rare or threatened – though again, the true figure could be a good deal higher.

Summary of Mediterranean-type areas.

In summary, of these four Mediterranean-type hot-spots areas, their 435,700 km² (0.3 percent of Earth's land surface) contain 19,580 plant species, or 7.8 percent of the Earth's plant species. Of these hot-spot species, 12,720 (65 percent; 5.1 percent of Earth's total of plant species) are endemics. This means an area/species index of 26. At least 4,080, or 21 percent, are already designated as officially rare or threatened (the true figure would probably be more like 5,000, 26 percent) (table 38.2). It is these at-risk species that dominate national or regional lists of threatened plant species: in the United States, California's 1,140 rare or threatened species make up almost 57 percent of the national threatened species total. But in terms of island-biogeography constraints we must consider that for longer-run purposes, the total threatened with ultimate extinction must surely be more like one half, viz. over 6,360 species.

There is also the Mediterranean Basin itself. Within its 2.3 million km² occur at least 25,000 plant species (Gomez-Campo, 1985; Henly, 1977; Le Houerou, 1977), or 28 percent more than in all the four areas described above, and 10 percent of all plant species on Earth in 1.6 percent of Earth's land surface. Roughly half of these species are endemic to the basin, and 3,583 are endemic to individual countries (excluding Turkey, Syria and Lebanon). Most of the basin's vegetation has been transformed from its native state, and all too few species appear to be part of the region's climax formations. Habitat disruption persists apace: the Mediterranean's 40,000 km of coastline now support 360 million people in littoral zones, plus 100 million tourists each year. Rare or threatened plant species already total about 3,000 or 12 percent of the flora; the number can be expected to rise swiftly in the foreseeable future.

This figure of rare or threatened species in the Mediterranean Basin increases the total for Mediterranean-type areas described by 42 percent. But clearly the Mediterranean Basin is so extensive that it cannot rank in itself as a single hot-spot area. Rather it would be necessary to identify key sectors that warrant special treatment, such as southern Greece, Cyprus and the Atlas Mountains of Algeria.

On top of present habitat disruptions, there is a further threat to two of the Mediterranean-type areas. While global warming will prove detrimental to species communities in many parts of the world (Peters, 1991), it could well be specially damaging to those in the Cape Region of South Africa and Southwestern Australia. As temperature bands move away from the equator, vegetation bands will try to follow them (Oppenheimer and Boyle, 1990; Schneider, 1989). This will be difficult enough in most parts of the world because plants will be trying to migrate through lands that amount to a 'development desert'. But in the case of the two hot-spot areas mentioned, there will not even be any land available: plant communities movement would be into the sea. They face the prospect of wipeout.

Summation and Conclusion

This second hot-spot analysis shows that another eight areas – four in tropical forests and four in Mediterranean-type zones – deserve consideration for urgent conservation support (table 38.3). The four tropical-forest areas contain at least 2,835 endemic plant species in 18,700 km² of primary forest, or 1.1 percent of Earth's plant species in 0.013 percent of the Earth's land surface (for an area/species index of 87). Taken together with the original 10 hot-spot areas in tropical forests, this means that at least 37,235 endemic plant species occur in 310,700 km², or 14.9 percent of Earth's plant species occur in 0.2 percent of the Earth's land surface (an index of 74.5). If, as is likely, in the absence of much more vigorous conservation measures, these forest areas lose 90 percent of their original forest cover within the next decade or shortly thereafter, we shall witness the imminent demise of at least half of these plant species, i.e., 18,600 species in these areas alone.

The four Mediterranean-type areas harbour at least 1,720 endemic plant species in 435,700 km², or 5.1 percent of Earth's plant species in 0.3 percent of the Earth's land surface (for an index of 17). These areas too seem set to lose 90 percent of their original habitat within the foreseeable future, involving the extinction of at least 6,360 plant species.

Considering these 18 areas together, they support a total of at least 49,955 endemic plant species, or at least 20 percent of Earth's plant species, in 746,400 km², or 0.5 percent of the Earth's land surface (table 38.4). (This translates into one fifth of all plant species in half of one percent of the Earth's land

Table 38.3 [*orig. table 3*]   Eight additional hot-spot areas: overview.

Zone	Extent of present habitat (km²)	Total plant species	Endemic plant species (% of total plant species)
Tropical forests (4 areas)	18,700	9,420	2,835 (30)
Mediterranean-type areas (4 areas)	435,700	19,580	12,720 (65)
Totals	454,400	29,000	15,555 (54)

In summary, 11.6 percent of Earth's plant species (54 percent of them endemic) occur in 0.3 percent of Earth's land surface.

Sources: numerous, as cited under the area reviews in the texts.

Table 38.4 [*orig. table 4*]   Eigtheen hot-spot areas: composite review.

Zone	Extent of present habitat (km²)	Total plant species	Endemic plant species
Tropical forests			
Original 10 areas	292,000	*	34,400
Further 4 areas	18,700	9,420	2,835
Mediterranean-type areas, 4 of them	435,700	19,580	12,720
Totals	746,400		49,955

In summary, almost 500,000 endemic plant species, or one fifth of Earth's plant species, occur in 0.5 percent of Earth's land surface.

* It has been unrealistic to sum total figures for total plant species in the original 10 hot-spot areas, on the grounds that there is some overlap between adjacent hot-spot areas e.g., some plants in Peninsular Malaysia, Northern Borneo and the Philippines.
Sources: numerous, as cited under the area reviews in the texts.

surface). It also means that insofar as all these areas are threatened with imminent destruction of their natural vegetation, the elimination of at least 25,000 plant species, or 10 percent of Earth's plant species, could be witnessed in these areas alone, unless greatly increased conservation measures are implemented forthwith. In addition there will surely ensue the extinction of at least 500,000 endemic animal species (probably several times more), in these areas.

Fortunately this biological debacle can still be forestalled in major measure, provided that suitable-scale conservation initiatives are undertaken during the last decade of this century – preferably during the next few years, since any delay will mean that much more action will have to be undertaken at much greater cost while achieving only a fraction as much success.

Acknowledgements
I am happy to thank several persons for critical comments on an early version of this article, and/or for other forms of exceptional backup help; notably Mark Collins and his colleagues at the World Conservation Monitoring Centre, Cambridge, UK; Steve Davis, Christine Leon and their colleagues at the Threatened Plants Unit, Royal Botanic Gardens, Kew, London, UK;

C. V. S. and I. U. N. Gunatilleke of the University of Peradeniya, Sri Lanka; Angus Hopkins, Principal Policy Officer to the Minister for the Environment, Perth, Western Australia; Ghillean T. Prance of the Royal Botanic Gardens, Kew, London, UK; and Peter H. Raven of the Missouri Botanical Garden in St Louis, USA.

References

Adis, J. 1990. 30 million anthropod species – too many or too few? *Journal of Tropical Ecology*, 6. 116–118.

Ahonzo, J. 1984. *La Population de la Cote d'Ivoire*. Direction de la Statistique. Abidjan, Ivory Coast.

Arroyo, M. T. K. 1990. Personal communication, letter of July 30th, 1990. Universidad de Chile, Santiago, Chile.

Ashton, P. S. and Gunatilleke, C. V. S. 1987. New light on the plant geography of Ceylon. I. Historical plant geography. *Journal of Biogeography*, 14, 249–285.

Assi, A. L. 1988. Especes rares et en voie d'extinction de la flore de la Cote d'Ivoire. In: Goldblatt, P. and Lowry, P. P. (eds), *Modern Systematic Studies in African Botany*, 461–463. Missouri Botanical Garden, St Louis, Missouri, USA.

Balassubramaniam, S. 1985. Tree flora of Sri Lanka. In: Jayatilleke, A. (ed.), *Proceedings of International Conference on Timber Technology*. 58–67, Moratuwa, Sri Lanka.

Barbour, M. and Major, J. (eds). 1988. *Terrestrial Vegetation of California*. California Native Plants Society, San Francisco, California, USA.

Bond, P. and Goldblatt, P. 1984. Plants of the Cape Flora. *Journal of South African Botany*, Supplement 13, 1–455.

Bousquet, B. 1978. Un parc de foret dense en Afrique: le parc national de Tai (Cote d'Ivoire). *Bois Forets Tropicales*, 179, 27–46.

Brenan, J. P. M. 1978. Some aspects of the phytogeography of tropical Africa. *Annals of the Missouri Botanical Garden*, 65, 437–478.

California Department of Forestry and Fire Protection. 1988. *California's Forests and Rangelands: Growing Conflict Over Changing Uses*. California Department of Forestry and Fire Protection, Sacramento, California, USA.

Cody, M. L., and Mooney, H. A. 1978. Convergence *versus* nonconvergence in Mediterranean-climate ecosystems. *Annual Review of Ecology and Systematics*, 9, 265–321.

Cody, M. L. 1986. Diversity, rarity and conservation in Mediterranean-climate regions. In: Soule, M. E. (ed.), *Conservation Biology*, 122–152. Sinauer Associates, Sunderland, Mass, USA.

Collins, N. M., Sayer, J. A. and Whitmore, T. (eds). 1990. *Conservation Atlas of Tropical Forests – Asia and the Pacific*. Macmillan, London, UK for International Union for Conservation of Nature and Natural Resources, Gland, Switzerland, and the World Conservation Monitoring Centre, Cambridge, UK.

Corporacion Nacional Forestal (de Chile). 1989. *Redbook of Terrestrial Flora of Chile*. Corporacion Nacional Forestal (de Chile), Santiago, Chile.

Cowling, R. M. and Roux, P. W. (eds). 1982. The Karoo Biome: A preliminary synthesis. *South African National Scientific Programme Report*, 142, Pretoria, South Africa.

Dayanandan, P. 1983. Conserving the flora of the Peninsular Hills. *Bulletin of Botanical Survey of India*, 23, 250–253.

Dosso, H., Guillaumet, J. L. and Hadley, M. 1981. The Tai Project: land use problems in a tropical rain forest. *Ambio*, 10, 120–125.

Elias, T. S. and Nelson, J. (eds). 1987. *Conservation and Management of Rare and Endangered Plants*. California Native Plants Society, San Francisco, California, USA.

Erdelen, W. 1988. Forest ecosystems and nature conservation in Sri Lanka. *Biological Conservation*, 43, 115–135.

Erwin, T. L. 1988. The tropical forest canopy: the heart of biotic diversity. In: Wilson, E. O. (ed.), *Biodiversity*, 105–109. National Academy Press, Washington, DC, USA.

Fuenzalida, M. 1984. *Evaluation of Native Forest Destruction in the Andes of South Central Chile: Conservation Alternatives*. Comite Nacional pro Defensa de la Fauna Y Flora. Santiago, Chile.

Gajardo, R. 1983. *Sistema Basico de Clasificacion de la Vegetacion Nativa Chilena*. Universidad de Chile/Corporacion Nacional Forestal, Santiago, Chile.

Geiser, U. and Sommer, M. 1982. Up-to-date information on Sri Lanka's forest cover. *Loris*, 16, 66–69.

Gibbs-Russell, G. E. 1985. Analysis of the size and composition of the South African flora. *Bothalia*, 15, 613–629.

Gibbs-Russell, G. E. 1987. Preliminary floristic analysis of the major biomes in Southern Africa. *Bothalia*, 17, 213–227.

Goldblatt, P. 1978. An analysis of the flora of Southern Africa: its characteristics, relationships and origins. *Annals of the Missouri Botanical Garden*, 65, 369–436.

Gomez-Campo, (ed.). 1985. *Plant Conservation in the Mediterranean Area*. W. Junk, Dordrecht, Netherlands.

Guillaumet, J. L. 1984. The vegetation: an extraordinary diversity. In: Jolly, A. et al., (eds), *Key Environments: Madagascar*, 27–54. Pergamon Press, Oxford, UK.

Gunatilleke, C. V. S. and Ashton, P. S. 1987. New light on the plant geography of Ceylon II. The ecological biogeography of the endemic lowland flora. *Journal of Biogeography*, 14, 295–327.

Gunatilleke, C. V. S. and Gunatilleke, I. U. N. 1985. Phytosociology of Sinharaja – Contribution to rain forests conservation in Sri Lanka. *Biological Conservation*, 31, 21–40.

Hall, A. V. 1987. Threatened plants in the Fynbos and Karoo Biomes, South Africa. *Biological Conservation*, 40, 29–52.

Hamilton, A. C. 1976. The significance of patterns of distribution shown by forests plants and animals in tropical Africa for the reconstruction of Upper Pleistocene and palaeoenvironments: a review. In: Van Zinderen Bakker, E. M. (ed.), *Palaeoecology of Africa, the Surrounding Islands, and Antarctica*, pp. 63–97. Balkema Press, Cape Town, South Africa.

Hawthorne, W. D. 1984. Ecological and biogeographical patterns in the coastal forests of East Africa, PhD Thesis, University of Oxford, Oxford, UK.

Henly, P. 1977. The Mediterranean: a threatened microcosm, *Ambio*, 6, 300–307.

Hopper, S. D. 1983. Applied plant systematics: case studies in the conservation of rare Western Australian flora. *Australian Systematics Botanical Society Newsletter*, 35, 1–6.

Hopper, S. D. and Muir, B. G. 1984. Conservation of the Kwongan. In: Pate, J. S. and Beard, J. S. (eds), *Kwongan – Plant Life of the Sandplain*, pp. 253–266. University of Western Australia Press, Nedlands, Western Australia.

Ishwaran, N. and Erdelen, W. 1990. Conserving Sinharaja – an experiment in sustainable development in Sri Lanka. *Ambio*, 19, 237–244.

Jain, S. K. and Sastri, A. R. K. 1983. *A Catalogue of Threatened Plants of India*. Department of the Environment, Government of India, New Delhi, India.

Jarman, M. L. 1986. Conservation priorities in the lowland regions of the Fynbos Biome. *South African National Scientific Programmes Report No 87*. Cooperative Scientific Programs, Council for Scientific and Industrial Research, Pretoria, South Africa.

Jones and Stokes Associates. 1987. *Sliding Toward Extinction: the State of California's Natural Heritage*. California Nature Conservancy, San Francisco, California, USA.

Keeley, J. E. (ed.). 1988. *Bibliographies on Chaparral and the Phyto-ecology of other Mediterranean Systems*. California Water Resources Centre. University of California, Berkeley, California, USA.

Kendrick, K. 1989. India. In: Campbell, D. G. and Hammond, H. D. (eds), *Floristic Inventory of Tropical Countries*, pp. 133–140. New York Botanical Garden, Bronx, NY, USA.

Le Houerou, H. N. 1977. Man and Desertisation in the Mediterranean Region. *Ambio*, 6, 363–365.

Lisboa, P. L. B., Maciel, U. N. and Prance, G. T. 1989. *Some Effects of Colonisation on the Tropical Flora of Amazonia: a Case Study from Rondonia*. Royal Botanic Gardens, Kew, Richmond, UK.

Lovett, J. C. 1987. Endemism and affinities of the Tanzanian montane forests. *Monograph in Systematic Botany*. Missouri Botanical Garden, St. Louis, Missouri, USA.

Lovett, J. C. 1988. Practical aspects of moist forest conservation in Tanzania. In: Goldblatt, P. and Lowry, P. P. (eds), *Modern Systematic Studies in African Botany. Monographs in Systematic Botany*, 25, 491–496.

Lovett, J. C. 1989. Tanzania. In: Campbell, D. G. and Hammond, H. D. (eds), *Floristic Inventory of Tropical Countries*, pp. 232–235. New York Botanical Garden, Bronx, NY, USA.

Maitre, H. 1987. Natural forest management in Cote d'Ivoire. *Unasylva*, 39, 53–60.

Marticorena, C. 1980. *Threatened Plants and Areas of Chile*. Universidad de Concepcion, Santiago, Chile.

Mittermeier, R. A. 1988. Primate diversity and the tropical forest: case studies from Brazil and Madagascar and the importance of the megadiversity countries. In: Wilson, E. O. (ed.), *Biodiversity*, pp. 145–154. National Academy Press, Washington, DC, USA.

Moll, E. J. and Bossi, L. 1984. Assessment of the extent of the natural vegetation of the Fynbos Biome of South Africa. *South African Journal of Science*, 80, 355–358.

Moll, E. J. and Jarman, M. L. 1984. Clarification of the term Fynbos. *South African Journal of Science*, 80, 351–352.

Munni, N. V. and Naidu, K. S. M. 1988. *Monitoring Changes in Forest Vegetation Cover in the Western Ghats through Satellite Remote Sensing*. National Remote Sensing Agency. Balanagar, Hyderabad, India.

Myers, N. 1982. Forest refuges and conservation in Africa – with some appraisal of survival prospects for tropical moist forests throughout the biome. In: Prance, G. T. (ed.) *Biological Diversification in the Tropics*, pp. 658–672. Columbia University Press, New York, USA.

Myers, N. 1988. Threatened biotas: 'hot-spots' in tropical forests. *The Environmentalist*, 8, 187–208.

Myers, N. 1989. *Deforestation Rates in Tropical Forests and their Climatic Implications*. Friends of the Earth, London, UK.

Nair, K. S. S., Gnanaharan, R. and Kedharnath, S. (eds). 1986. *Ecodevelopment of the Western Ghats*. Kerala Research Institute, Peechi, Kerala, India.

Nanayakkara, V. R. 1982. Forests – policies and strategies for conservation and development. *Sri Lanka Forester*, 15, 75–79.

Nayar, M. P. and Sastry, A. R. K. (eds). 1987. *Red Data Book of Indian Plants*. International Union for Conservation of Nature and Natural Resources, Gland, Switzerland.

O'Neill, C. J. 1990. Personal communication, letter of August 22nd, 1990, citing a number of references. UCI Arboretum, University of California, Irvine, California, USA.

Oppenheimer, M. and Boyle, R. H. 1990. *Dead Heat: the Race Against the Greenhouse Effect*. Basic Books Inc, New York, USA.

Pate, J. S. and Beard, J. S. (eds). 1984. *Kwongan – Plant Life of the Sandplain*. University of Western Australia Press, Nedlands, Australia.

Peters, R. L. (ed.) 1991. *Consequences of the Greenhouse Warming to Biodiversity*. Yale University Press, New Haven, Connecticut, USA (in press).

Pizarro, M. 1977. Threatened and endangered species of plants in Chile. In: Prance, G. T. and Elias, T. S. (eds), *Extinction is Forever*, pp. 267–282. New York Botanical Garden, Bronx, New York, USA.

Poblete, I. C. (ed.) 1985. *Flora Nativa Arborea y Arbustiva de Chile Amenazada de Extincion*. Ministry of Agriculture, Santiago, Chile.

Polhill, R. M. 1989. East Africa (Kenya, Tanzania and Uganda). In: Campbell, D. G. and Hammond, H. D. (eds), *Floristic Inventory of Tropical Countries*, pp. 217–231. New York Botanical Garden, Bronx, NY, USA.

Ramirez, C. 1984. *Bibliografia Vegetacional de Chile*. Universidad Austral, Valdivia, Chile.

Raven, P. H. 1976. Ethics and attitudes. In: Simmons, J. B., Beyer, R. I., Brandham, P. E., Lucas, G. L. and Parry, U. T. H. (eds), *Conservation of Threatened Plants*, pp. 155–179. Plenum Press, New York, USA.

Raven, P. H. and Axelrod, D. I. 1978. Origin and relationships of the California flora. *University of California Publications in Botany*, 72, 1–134.

Rodgers, W. A. and Hall, J. B. 1986. Pole cutting pressure in Tanzanian forests. *Forestry Ecology and management*, 14, 133–140.

Rodgers, W. A., Mziray, W. and Shishira, W. 1986. *The Extent of Forest Cover in Tanzania using Satellite Imagery*. Department of Natural Resources, University of Dar es Salaam, Tanzania.

Rodgers, W. A. and Homewood, K. M. 1982. Species richness and endemism in the Usambara mountain forests, Tanzania. *Biological Journal of the Linnean Society*, 18, 197–242.

Roth, H. H. 1984. We all want the trees: resource conflict in the Tai National Park, Ivory Coast. In: McNeely, J. A. and Miller, K. R., (eds), *National Parks Conservation and Development: the Role of Protected Areas in Sustaining Society*, pp. 127–129. Smithsonian Institution Press, Washington Press, USA.

Rutherford, M. C. and Westfall, R. H. 1986. Biomes of Southern Africa: an objective categorisation. *Memoirs of the Botanical Survey of South Africa*, 54, 1–98.

Rye, B. L. 1982. *Geographically Restricted Plants of Southern Western Australia*. Department of Fisheries and Wildlife of Western Australia, Perth, Australia.

Rye, B. L. and Hopper, S. D. 1981. *A Guide to the Gazetted Rare Flora of Western Australia*. Department of Fisheries and Wildlife of Western Australia, Perth, Australia.

Schneider, S. H. 1989. *Global Warming: are we Entering the Greenhouse Century?* Sierra Club Books, San Francisco, California, USA.

Smith, J. P. 1987. California: leader in endangered plant protection. *Fremontia*, 15, 3–7.

Smith, J. P. and Berg, K. (eds). 1988. *Inventory of Rare and Endangered Vascular Plants of California*. California Native Plants Society. Sacramento, California, USA.

Soule, M. E. (ed.). 1987. *Viable Populations for Conservation*. Cambridge University Press, New York, USA.

Spears, J. 1985. *Malaysia Agricultural Sector Assessment Mission: Forestry Subsector Discussion paper*. The World Bank, Washington, DC, USA.

Stebbins, G. L. 1980. Rarity of plant species: a synthetic viewpoint. *Rhodora*, 82, 77–86.

Stebbins, G. L. and Major, J. 1965. Endemism and speciation in the California flora. *Ecological Monographs*, 35, 1–35.

Stork, M. E. 1988. Insect diversity: facts, fiction and speculation. *Biological Journal of the Linnean Society*, 35, 321–337.

Sumithraarachchi, D. B. 1989. Sri Lankan forests: diversity and genetic resources. In: Holm-Nielsen, L. B., Nielsen, I. C. and Balslev, H. (eds), *Tropical Forests: Botanical Dynamics,*

Speciation and Diversity, pp. 253–258. Academic Press, New York.

Tansley, S. A. 1988. The status of threatened Proteaceae in the Cape Flora, South Africa and the implications for their conservation. *Biological Conservation,* 43, 227–239.

Werger, M. J. A. (ed.). 1987. *Biogeography and Ecology of Southern Africa.* W. Junk Publishers, The Hague, Netherlands.

Werner, W. L. 1982. The upper montane rain forests of Sri Lanka. *Sri Lanka Forester,* 15, 119–135.

World Conservation Monitoring Centre. 1988. *India: Conservation of Biological Diversity.* World Conservation Monitoring Centre, Cambridge, UK.

PART VI

Conclusion

INTRODUCTION

Many themes have run through this collection of papers. One is that the human impact has been long continued and that even in prehistoric and historic times humans were capable of achieving a great deal (chapters 4 and 9). Secondly, with developments in technology, the number of ways in which humans are affecting the environment is proliferating, and many problems, e.g. the ozone hole, are of relatively recent occurrence (chapters 3, 11, 16 and 36). Thirdly, some changes are reversible and some environments can recover when particular stresses are removed (chapters 5 and 18). Fourthly, some environments may be particularly sensitive, vulnerable, or significant (chapters 1, 33 and 38). Fifthly, in many cases change is the result of both natural and anthropogenic causes and it is often difficult to disentangle the two (chapters 9, 19, 32 and 37). Sixthly, and following on from that, there is still great uncertainty about future impacts, and this was made clear in Goudie (1993, pp. 372–3):

The debate about the relative power of human and natural influences is but one of many components of environmental uncertainty. In almost all attempts to predict the future major uncertainties arise. One can provide various explanations for this. First, environmental change is the result of very complex interactions between several closely coupled non-linear systems. The complexity creates problems for both modelling and comprehension, while the non-linearity means that the dimensions of a response are not by any means necessarily directly proportional to the size of the stimulus that promotes change. Secondly, prediction of environmental change depends on models and many models are imperfect because of the gross assumptions they employ (see the discussion of General Circulation Models for an exemplification of this problem). Thirdly, nature always has surprises in store (e.g. various types of extreme event or catastrophe)and these will not cease in the future. They may work to counteract or increase the consequences of various human actions. Fourthly, there are bound to be factors that we have ignored, which turn out to be highly important. For example, who several decades ago could have predicted the role of CFCs in creating the ozone hole? Fifthly, identification of future trends requires some knowledge of background or natural levels. Often we lack the necessary long-term data to enable us to ascertain whether observed trends have happened before and whether they are or are not cyclical. Sixthly, some key issues are less easy to predict than others and therefore preclude accurate prediction of phenomena that depend on them. For example, without a clear idea of precipitation patterns in a warmer world it is well nigh impossible to predict the response of rivers and biota. Seventh, we may often be able to predict that change will probably occur, but we find it much less easy to predict the speed of response. Ice caps and glaciers may well melt in a warmer world, but how fast? Finally, there are problems of definitions. Without clear definitions of phenomena, measurement is difficult, 'results' can be meaningless and trends or changes difficult to identify. We find it difficult to define such phenomena as desertification and deforestation.

For the most part, the papers in this volume have not specifically addressed the future. The limitations of space have been severe, but some background to the future is given in Goudie (1993, chapter 8).

The book concludes with an elegant overview by Sir Crispin Tickell (chapter 39), an environmental adviser to successive British prime ministers, a former President of the Royal Geographical Society, and Warden of Green College, Oxford. He asks the question 'Has the human species been a suicidal success?' Time will tell.

Reference

Goudie, A. S. 1993, *The Human Impact on the Natural Environment* (4th edition) Oxford: Blackwell Publishers.

39

THE HUMAN SPECIES: A SUICIDAL SUCCESS?*

C. Tickell

I want first to put us in our place. I remember an occasion at the Royal Institution at which the principal speaker was the famous anthropologist Louis Leakey. The audience was formally dressed, and the officers of the Institution were in tail coats, boiled shirts and white ties. At the stroke of nine o'clock the doors opened, and in strode Dr Leakey in a bush shirt. He looked up and down and said: 'Animals: that is what you are, Animals'.

Indeed we are animals, a member of the class, or taxon, of mammals, a reasonably successful group until we remember that there are some 4000 mammal species while insect species run to millions. Within the mammal taxon we belong to the order of primates which includes the super family of hominoids. We are of the genus *homo*, which we share with chimpanzees (only 1.6 per cent of our DNA is different from that of the chimpanzee).

Yet if Darwin was right in measuring the success of a species by its numbers, we have done amazingly well since the last Ice Age ended. We were less than 10 million in number 10,000 years ago, but by 1930, the year of my birth, we were 2 billion. Now we are 5.4 billion and the human population is increasing by around 93 million every year,

of which Africa and South East Asia account for more than half (UNPFA, 1992). That is a quarter of a million people every day and the equivalent of a new China every ten years.

Viewed from far enough away, we have the look of an assemblage of sociable insects. As Lewis Thomas once said of ants, 'They are so much like human beings as to be an embarrassment. They farm fungi, raise aphids as livestock, launch armies for wars, use chemical sprays to alarm and confuse their enemies, capture slaves. The families of weaver ants engage in child labour, holding their larvae like shuttles to spin out the threads that sew the leaves together for their fungus gardens. They exchange information ceaselessly. They do everything but watch television' (Thomas, 1974).

A prime reason for our success is our flexibility as a switcher predator and scavenger. We are consummately adaptable, able to switch from one resource base – grasslands, forests, or estuaries – to another, as each is exploited to its maximum tolerance or used up. Like other successful species we have learned to adapt ourselves to new environments. But, unlike other animals, we made a jump from being successful to being a runaway success. We have made this jump because of our ability to adapt environments

* Originally published in *Geographical Journal*, 1993, vol. 159, pp. 219–26.

for our own uses in ways that no other animal can match. We now consume – or abuse – 40 per cent of total photosynthetic production on land. We live in all climates from the jungles to the poles, and have changed the face of the land. We have coped with most infectious diseases, thereby greatly increasing life expectancy, even in poor countries. We have changed the genetic inheritance of other animals and plants to meet our food tastes. We face no challenge from other animal predators except for bacteria, viruses and retroviruses.

Since the industrial revolution began 250 years ago, those who have benefited from it have amazingly increased their living standards. Economic wealth on the normal – highly misleading – definition has risen at an almost incredible rate during this century. Global gross net product was around US$ 600 billion in 1900. By 1960 it was US$ 5 trillion and by 1988 US$ 17 trillion (of which 14.7 trillion came from the industrial countries, accounting for less than a quarter of the world's population, and 2.3 trillion from the rest of the world, accounting for over three-quarters of it) (World Resources Institute, 1992). This imbalance is becoming more marked. In industrial countries per capita income rose by 2.4 per cent in the 1980s. Elsewhere it rose by 1.5 per cent (World Resources Institute, 1992).

How can all this have happened? How have we removed or circumvented the restraints or obstacles which normally limit the increase of a species? The answer is that our cultures have evolved much faster than the biological systems on which they depend.

At the end of the eigtheenth century Malthus identified the relationship between resources and population, and incidentally greatly influenced Darwin. Malthus argued that unless profligate childbearing were checked, preferably through abstinence, hunger and famine would follow. Malthus was wrong in that he did not anticipate the enormous potential of advancing technology to raise land productivity. He was, after all, writing before Mendel had formulated the basic principles of genetics, and before Leibig had demonstrated that all

nutrients taken from the soil by plants could be returned in mineral form. He also grossly underestimated human ingenuity. But Malthus was right in identifying the difficulty of expanding food output as fast as population growth. The first explosive increase in such growth came with the introduction of agriculture, the creation of cities, and the specialization of human functions. Demand for labour was a spur to population increase. The second spur came with the industrial revolution, based on new resource exploitation, in particular fossil fuel. Again demand for labour was a spur to population increase. A new feature of the last few decades has been the introduction of such new devices as solid state electronics and biotechnology which have enabled us to do and produce much more from much less. In those parts of the world affected, supply of labour now generally exceeds demand and this trend is likely to accentuate.

It is an ingrained belief that human progress, with one or two blips, has been upwards and onwards. In fact few trends go in this fashion. They jump backwards as well as forwards. All previous human civilizations have collapsed. The proximate reasons are various. But all are subject to three variables: population, resources and environment. They are not always easy to distinguish in their effects. The interaction between them is different in different times and circumstances.

Let me give some examples, some better known and more definite than others. Perhaps the earliest was the fate of the Harappa culture in the Indus valley some 3500 to 4500 years ago. The destruction of forest cover and removal of topsoil probably led to the creation of dust clouds, up to 10,000 metres, which maintained a downward flow of cool air and prevented the rise of moisture, even in summer. With sharply-diminishing rainfall and declining fertility of soils, Harappan society may possibly have lost its natural resource base and simply collapsed (Lamb, 1982). The same could have happened in the valley of the Tigris and Euphrates; around Petra in the Wadi Araba; and in parts of the Sahel belt across Africa today. Growing aridity and exhaustion of soils in what is now central Mexico from the seventh to the tenth centuries probably contributed to the collapse of Teotihuacan and the other classic civilizations of pre-Columbian America.

Perhaps the best example of what humans can do is what befell Easter Island (Diamond, 1991). A small group of Polynesians discovered it around 400 AD. It was then covered in rich forest. By 1500 the population was over 7000 divided into little nations, with virtually no forest but over 1000 statues. Then came civil war, fighting over resources, with not a single tree left to be made into a canoe to permit escape. The Dutch explorer Roggerveen arrived in 1722. By then the population had fallen to less than 500. All were in a miserable condition. They could not even remember their own history or what the statues around them were for.

In looking at the effects of human population increase we should not forget the parallel increase in the animals and plants we have somewhat arbitrarily chosen for our nourishment, support and enjoyment: among animals, cattle, goats, horses, sheep, pigs and dogs; and among plants wheat, rice, maize, cassava and potatoes. The effects of human population increase on resources and the environment cannot be dissociated from those of our involuntary companions in the processes of life.

But the prime engine of the recent dizzy-making rise in the human population and change generally is the industrial revolution. We have the misfortune to be perhaps the first generation in which the magnitude of the global price to be paid is becoming manifest.

There are many aspects. First comes land use. The most significant change has not been the spread of brick, stone, concrete and urban sprawl, but an acceleration of the destruction of forests worldwide and declining fertility of soils. According to a report by the World Resources Institute published in 1992, some 10 per cent of the vegetation-bearing surface of the earth is suffering from moderate to extreme degradation, an area the size of China and India combined. Although forest cover is slightly increasing in industrial countries, its destruction elsewhere, with accompanying loss of species, is on a scale to change the global ecosystem. Enough productive topsoil to cover the whole of France washes away or degrades beyond reasonable use every year. In addition to degradation of land by farming practices, outside forces are also taking a toll on agriculture. Air pollution has already reduced US crop production by five to ten per cent, according to US Government estimates (Brown, 1991). It is probably having a still worse effect in Eastern Europe and in China.

On a Malthusian reckoning we should have had a major crisis in food products from 1950 onwards. We were of course saved by the green revolution which led to the establishment of new and more productive strains of plants, in particular, wheat and rice. Such strains are genetically vulnerable and are often dependent on abundant water and fertilizer. Between 1950 and 1984 there was a ninefold increase in fertilizer use. The question we have to consider for the future is whether the application of similar techniques can save us again on the scale likely to be required. There are enormous possibilities for biotechnology, even in semi-arid areas. So far, the best agricultural technology has been able to meet a growth in demand of two to three per cent against a likely growth in demand of three to five per cent. Unfortunately there are indications that some new technology has been at the cost of future growth; and the penalties of use of fertilizers and pesticides – and dependence on them – are well known.

Fresh water is a particular problem. The global demand for water doubled between 1940 and 1980, and is expected to double again by the year 2000 (WRI, 1992). More people need more water. Many countries already suffer severe shortages and droughts. Competition for water was a prime source of conflict in the past and will be in the future: for example over the Nile, which flows through nine states, each with its own interests and demands; over the Euphrates and the Jordan which nourish Turkey, Syria, Iraq, Jordan and Israel; and over the Colorado, now no more than a sickly salty stream when it finally reaches the sea.

Then there are the direct effects of industrialization: pollution and recent accidents have demonstrated the international character of industrial hazards. Within the vast landmass of the former Soviet Union, some 16 per cent (or 3 577 998 square kilometres) was recently judged an ecological disaster area by the Academy of Sciences. Chemical accidents may be limited in their effects, but disposal of chemical wastes is a world-wide problem. Nuclear war could damage the world as a whole but even nuclear accidents are serious enough: the fallout from Chernobyl was some 50 times that of Hiroshima. More insidious but more real for the ordinary citizen is the growing problem of waste disposal. No part of the world is now exempt from the wastes produced by industrial activity.

Next comes the atmosphere. Acid precipitation is a problem for those downwind of industry; but it is essentially local in character and can be solved if there is political will to solve it. Depletion of the ozone layer is more serious. The miracle molecules known as chlorofluorocarbons and halons (for use as refrigerants, deodorants, fire extinguishers, etc.) have been depleting the protective screening which prevents short-wave ultraviolet radiation reaching the surface of the earth. Damage to the human metabolism (melanoma, etc.) may seem alarming to us, but the more fundamental problem could be the effects on other organisms, not least phytoplankton in the oceans.

Global warming through enhancing the natural – and indispensable – greenhouse effect could affect every aspect of human society. The main conclusions of the Intergovernmental Panel on Climate Change of 1990, updated in 1992, represented a broad scientific consensus (Houghton et al., 1990). On the assumption that we continue to pump carbon dioxide, methane, chlorofluorocarbons and nitrous oxide into the atmosphere, there is likely to be a rise of global mean temperature of around 1.0°C by 2025, and around 3.0°C by the end of that century. Compare this with a drop of around 5.0°C during the last glacial episode.

There would be markedly different results in different places. Temperature change is always more pronounced over land. So the northern hemisphere varies more than the southern. In the north the most marked increases have so far been in minimum night-time temperatures rather than maximum day-time temperatures, and in warmer winters rather than warmer summers. Southern Europe and North America would have less summer precipitation and lower soil moisture. But in general there would be more precipitation, and more extreme and irregular rainfall (as we have seen so vividly in 1992), particularly in areas subject to the monsoon. Arid regions such as southern Africa are likely to see more prolonged and severe droughts. There would be a general redistribution of weather patterns with drastic local effects. Sea levels might rise by over half a metre between now and the end of the next century. The long lag time between cause and effect is the result of the stabilizing effect of the oceans.

There are of course many uncertainties, among them variations in solar radiation; the behaviour of clouds and the hydrological cycle; oceanic exchanges with the atmosphere; mechanisms of the carbon cycle; and the behaviour of polar ice sheets. Most important is the sheer speed of change. According to the Intergovernmental Panel, change in temperature will be greater than has occurred naturally in the last 10,000 years and the rise in sea level will be three to six times faster than in the last 100 years (IPCC, 1992).

Lastly come the consequences of the industrial revolution on other forms of life. The destruction of species can be compared to earlier disasters in the history of the earth: for example the elimination of around 90 per cent of species at the end of Permian times 250 million years ago, or the famous extinction of the dinosaur family and many less famous ecosystems at the end of Cretaceous times 65 million years ago. Current calculations suggest that perhaps a quarter of the earth's remaining biodiversity is now at serious risk (WRI, 1992). It is a crisis with two aspects: mass extinction of species, and gross depletion of genetic variability within given species as a large proportion of a population is wiped out. The destruction of one species can profoundly change a whole ecosystem. Some changes could be quick, like the evolution of new viruses and bacteria; others could be very slow. Once a species is destroyed it has gone forever. As has been well said 'Death is one thing; an end to birth is something else'.

So now we have an accumulation and combination of formidable hazards, each driven to a greater or lesser degree by human population increase. In the past, societies have responded to change, whether caused by depletion of their resource base or by deterioration of their environment, by collapse or dispersal or both. The process was painful. It was usually accompanied by drastic reductions in population. Sometimes it was accelerated by external as well as internal shocks. Sometimes it was slow and lingering. Often most people did not understand what was happening, and failed to link causes with effects. Even if a few did understand, they felt powerless to do anything about it. Their inability to cope was matched by lack of vision of any alternative way to go.

We are supremely adaptable and ingenious mammals. If our brains get us into a mess, our brains should at least try to get us out of it. They will not be able to do so until they recognize what is happening and what will happen. In short, we need eyes to see. There are some obvious points of high significance.

In an overpopulated world any important changes in weather patterns could affect hundreds of millions of people. We have seen beginnings in Africa and Asia. The same goes world-wide for even a small rise in sea levels. Around half the world's population lives by the sea or along rivers and estuaries. Egypt, Bangladesh and the Netherlands are particularly vulnerable. Crude figures for population growth do not convey the magnitude of the accompanying growth of cities: over 80 per cent of population growth is likely to be in the urban areas. This is the equivalent of ten cities the size of

London every year. Already there is increasing dependence of poor countries on food imported from a few rich countries. Africa is particularly affected, where we see a combination of political instability, breakdown of internal administration and often civil war.

Ethiopia is a good example of what can go wrong. The fertile upland areas, which have long had the bulk of the population (88 per cent) have been losing fertile topsoils at a rate of between 1.5 and 3.5 billion tons a year (WRI, 1992). Population increased from around 20 million in 1950 to 31 million in 1970, and is almost 50 million now. The resulting economic strain, when combined with small shifts in patterns of rainfall, produced the famines of 1984/5 when at least 5 million people faced starvation and some 15 million people remained undernourished. Yet in terms of Ethiopia's total population, the million or so who died during the famine were replaced in just over six months.

In the last 15 years there has been a steep rise in human displacement: there were less than five million refugees on a strict political definition in 1978, but over 17.3 million in 1990. If we add a further 10 million environmental refugees and economic migrants, we find that more than 27 million people today are displaced in one way or another. I recall my lecture on this subject to the Society in 1989, and will not go into the details again now. Since then the prospects have become even more serious, and barriers against refugees have been going up throughout the industrial world.

We must also expect changes in patterns of disease. Temperature and moisture are determining factors for biological agents in the human environment: in water, food, air and soil. Variations in both affect the ability of viruses, bacteria and insects to multiply and prosper. We are already seeing a remarkable return of certain diseases whose agents have become resistant to modern drugs. History is full of examples of societies and civilizations brought down by diseases to which local populations had no immunity. The Black Death reduced the medieval population of Europe by between one-third and a half, and smallpox and measles that of the indigenous population of the Americas by over three-quarters. Populations already debilitated for other reasons will be particularly vulnerable. Thus we could see the spread of such non-parasitic diseases as yellow fever, dengue, poliomyelitis, cholera, dysentery, encephalitis, tuberculosis and pneumonia and of such parasitic diseases as malaria, leishmaniasis, schistosomiasis, hook-worm, tape-worm and other helminthic afflictions. We must also reckon with problems arising from drainage and sewage disposal, algal blooms from nitrate pollution, salinization and aluminium toxicity.

I mention two examples. During an exceptionally hot, rainy period in northern Australia in the summer of 1991 there were four epidemics of Ross River fever, an arbovirus infection usually confined to the northernmost parts of the Northern Territory (*The Lancet*, 1992). This coincided with the many anomalies of temperature and moisture associated with the El Niño phenomenon in the Pacific. The same year the population of the Pacific coast of Central and South America was struck by cholera through many vectors: fish, molluscs, crustacea and impure water. The bacillus can coexist with a wide variety of algae and plankton. Under unfavourable conditions it hibernates, re-emerging with warmth and nutrients. Phytoplankton and algae are fertilized by increased levels of carbon dioxide and could harbour large populations of cholera bacilli ready to erupt at warmer temperatures.

For a moment it is worth standing back and asking whether there is a sustainable human carrying capacity of our planet. Obviously any answer must be somewhat subjective and only brave people attempt one. One calculation from the World Hunger Project at the University of Rhode Island is that if we all had a vegetarian diet and shared our food equally, the biosphere could support around 5.5 billion people; if 15 per cent of our calories came from animal products (and again food were shared equally), the figure would come down to 3.7 billion people; if 25 per cent of our calories came

from animal products, then it would fall to 2.8 billion; and if 35 per cent of our calories came from animal products, as in North America today, then it would fall to a little over 2 billion (Kates, 1992). As we are already at 5.5 billion and will be over eight billion in the next quarter century, this calculation is scarcely cheering.

This brings me back to the centre of the argument; the relationship between the three variables of population, resources and the environment. Robert McNamara in his Rafael Salas lecture in December 1991 put it in mathematical terms:

Ed (environmental damage) = P (population) multiplied by C (consumption per capita) multiplied by D (environmental damage per unit of consumption).

And Paul and Anne Ehrlich, in their book *Healing the Planet* write of:

I (impact on the environment) = P (population) multiplied by A (per capita affluence) multiplied by T (damage done by technologies supplying each unit of consumption).

Whether expressed as Ed = P×C×D, or as I = P×A×T, the results as we can now foresee them are ultimately the same: catastrophe, whether in fast or easy stages.

But if there is one principle in human affairs to which we should cling, it is this. Nothing is inevitable. It ain't necessarily so. My family motto is *Facta Ficta*. There is something dubious about any particular selection of facts which this or that generation accepts as revealed truth. We need new flexibility of mind. This is painful. It means abandoning assumptions, changing hallowed habits, creating new models of thought, accepting different values, and seeing the world through other eyes, and still knowing how little we know.

We need to recast our vocabulary. Words are not only a means of expression but also the building blocks of thought. The instruments of economic analysis are blunt and rusty. Such words as 'growth', 'development', 'cost benefit analysis' and 'gross national product', all require redefinition.

We need to realize that conventional

wisdom is often a contradiction in terms. There are no universal answers to anything. The closer to those who exercise power, the better the realization that they have little idea what they are doing.

We need to change the culture. Many have lamented the division between the cultures of science and the arts. They are right to do so. But neither is now in charge. Our real bosses are the business managers, and they are not known for their ability to think long.

We need to recast our educational system to promote better understanding of broad issues and lateral thinking between them. Specialization is the bane of wisdom.

We need a value system which enshrines the principle of sustainability over generations. Sustainable development may mean different things to different people, but the idea itself is simple. We must work out models for a relatively steady state society, with population in broad balance with resources and the environment.

So let us go back to population (P), resource consumption (R) and environment (E). Projections of population increases vary widely. Some suggest that after reaching around ten billion in the middle of the next century (in other words around double present levels), population will become steady and then decline. Others foresee a rise to 12.5 billion and a continuing rise to 20 billion sometime in the twenty-second century. Much will depend on what happens in the next ten years (UNPF, 1992).

But neither the present nor the past is a sure guide to the future. The reasons for increase – lower child mortality, longer lives, better health and so on – are well known. There is a rough, but only rough, correlation between lower rates of increase and improved living standards. No wonder that population levels in most industrial countries are broadly stable whereas those elsewhere are not. But improvement of living standards means higher resource consumption and greater impact on the environment; and the highest rates of population increase are in areas low in resources with fragile environments. In such areas current rates of population increase far exceed those of the industrial countries at the time of their expansion in the last century. In the 1950s and 1960s the global rate rose rapidly. It diminished in the 1970s, but there was no further fall in the 1980s. Declines in fertility in East and South East Asia in the 1970s were not repeated in South Asia or some parts of Latin America. Brazil and Mexico are among the few industrializing countries where fertility declined in the 1980s.

Sooner or later 'Nature' will of course take care of us. Lack of resources, environmental degradation, famine and disease will in the painful fashion known by our ancestors cut our species back. AIDS is the obvious example of a way in which to do it. According to the computer predictions at Imperial College London, conditions already exist in several African countries for the virus to kill more people than are being born. When the proportion of women of childbearing age who are infected exceeds 30 per cent, even the highest growth rates (and some countries have rates exceeding three per cent per annum or a population doubling in a little over 20 years) would be cancelled out. This could lead to further disaster: the most creative and active, and the very young, would be the worst affected. The longer-term impact of AIDS across the world is uncertain. In our age of rapid travel it can spread easily. With its incubation period of as much as ten years or even more, AIDS is not a boom-and-bust infection like the Black Death. Unchecked it could move on a time-scale of 200 rather than 20 years. But the effects could be as devastating.

This is not the way most of us would like things to happen. Peering ahead we realize that demography has much in common with economics: it is a combination of social studies, algebra and necromancy. Some obviously believe there is a case for letting 'Nature' rip. In such places as Africa, it is ripping already, with results that we have seen on our television screens. It is a case implicitly favoured by the Pope of Rome, by those engaged in racial competition between peoples and tribes, by those all too used to high infant mortality, by those who are looking for care in their old age, and by millions who have never known anything else and have no sure means of limiting family size. Fertility anyway rises and falls for reasons not fully understood. Mysteriously sperm-counts the world over seem to be dropping (*New Scientist*, 1992). It could be the product of some so far unidentified pollution. Perhaps the factors which operate in limiting the increase of such animals as rats may affect us too.

But this is a doctrine not only of despair but also of profound irresponsibility. In many parts of the world people are already taking control of their destiny. There are many ways of doing so. At present the Chinese, whose population is now 1.2 billion and will rise inevitably to 1.5 billion, allow one birth per family in the town, one or two in the country and freedom for their minority peoples. State power is used to discriminate against those who do not respect the rules, and the effects are already evident in a generally disciplined society.

Another way is open persuasion using the apparatus of State power and propaganda for the purpose, and the provision of cheap or free contraceptive devices wherever possible. The efforts of the National Family Planning Programme in Thailand helped to reduce the average fertility rate from 6.5 in 1969 to 2.1 in 1989.

A further way is to try and cope with the underlying cultural problems. Most important is to raise the status of women, and give them more power of choice so they are no longer regarded as baby-making machines or second class citizens. It is also necessary to reduce infant and maternal mortality; protect the old and ensure they are supported and cared for; and provide access to information about family planning and services, and better education, particularly for girls and women, whereby ignorance about sex is reduced.

It is often, and rightly, said that the industrial countries, which inadvertently created the conditions for current population increase, should do a lot more to help others cope with it. Obviously they have to proceed with circumspection. It is a delicate issue: for a few industrial countries, especially the United States, abortion is an internal political issue; while for poor countries advocacy

of population restraint is sometimes interpreted as a form of racism. At the Earth Summit in Rio, there was a tacit conspiracy whereby the biggest single environmental issue was almost ignored. This may be righted at a meeting of scientific academies on the subject in Delhi in October 1993 and by the United Nations conference on Population and Development in September 1994. Industrial countries should increase the proportion of their Aid programmes devoted to population issues, and give more support to the United Nations Population Fund, the increasing work of the World Bank in this area, and the efforts of non-governmental organizations.

In the last resort population policies have to come from individuals, families and communities in the countries themselves. Some significant successes are in Thailand, Swaziland, Namibia and even Bangladesh. A particularly interesting example is in India where the five northern states of Rajasthan, Uttar Pradesh, Madhya Pradesh, Bihar and Orissa, which make up 40 per cent of the Indian population, have the steepest population growth and the greatest poverty. They are an exemplar of disaster for humans, resources and environment alike. By contrast Kerala in the south has low infant and maternal mortality, high use of contraceptives, high literacy rates especially among women, and diminishing poverty. The fertility rate is now down to 2.3 (UNPF, 1992). The most significant factor is the status of women who have the inheritance of land. Perhaps it is no accident that there is also much better care of the environment.

In 1986 the US National Research Council made a simple calculation based on computer modelling: that a one per cent reduction in the growth of the labour force would increase growth of per capita income by 0.5 per cent a year or 16 per cent over thirty years. In 1992 in its annual report – *The State of World Population* – the United Nations Population Fund noted a central difference between groups of countries with high economic growth rates in the 1960s and 1970s. Of 19 countries whose populations were growing faster than 2.5 per cent between 1965 and 1980, only 10 achieved positive income growth in the 1980s (and four of them only just). Of 17 whose populations were growing less than 1.5 per cent during the same period, 15 had positive income growth in the 1980s.

Let me turn briefly to Resources (R) and the world's consumption of them. It should go without saying that the more people there are, the more resources will be consumed. Most resources are renewable, and even those that are non-renewable – for example fossil fuels – can sometimes be replaced. Pressure on resources was the principal theme of *Limits to Growth* published in 1972 (Meadows, 1972). Those who questioned it may care to have a look at its successor *Beyond the Limits* published by the same author, earlier this year (Meadows, 1992).

The problem today is that pressure of consumption can easily render renewable resources unrenewable, or renewable only after long stretches of time. Any price system which does not attach values to such things as clean air, clean water, the hydrological service rendered by forests, or the waste-absorbing capacity of the atmosphere, is bound to create increasingly dangerous distortions. Renewable resources under threat include the fertility of land, animals and plants (in particular trees).

At present there is a gross imbalance in consumption between people in industrial countries and those in the rest of the world. The economic power of industrial countries enables them to take both renewable and non-renewable resources from others as well as exploiting their own. There is little that the others, often in debt, dependent on industrial country markets and striving desperately to meet unrealistic expectations of economic progress, can do about it. This issue was also judged too delicate for proper discussion at the Earth Summit at Rio.

Last I come to Environment (E). Here too there is a gross imbalance between the damagers and the damaged. The primary responsibility for the accelerating changes we are seeing rests directly or indirectly with the industrial countries. To take one example relating to the greenhouse effect, over 70 per cent of current carbon emissions into the atmosphere come from them and 23

per cent from the United States alone. In 1988 average per capita emissions per year from industrial countries was about 3.5 tons, and from the rest of the world less than 0.5 ton. But the poor are much more vulnerable than the rich. They lack resources to react to change: and have neither the skills nor the administrative structures capable of organizing adaptation to new circumstances.

This issue was of course discussed at the Earth Summit. At least it was exposed, and the world became more aware of it. Through agreements reached at Rio, in particular the Climate Convention, frameworks will be set in place for dealing with global problems which many only 20 years ago scarcely knew existed. But the shortfall between the size of the problems and the money offered to cope with them, between the rhetoric and the action, was stark. For example, on carbon emissions, the industrial countries, some sheltering behind the United States, made no specific commitments, or even targets, to reduce them. The failure of the big polluters to give an example on this, as on resource consumption and other issues, rendered unconvincing their efforts to persuade the rest of the world to take seriously the nexus of population/resource/environment problems which may affect every member of the human species.

In my view there are five principles on which all governments of whatever complexion should act singly and together. First, they should now do what makes sense for reasons other than any one environmental factor. For example, global warming might suggest the conservation of forests and the creation of new ones to draw carbon out of the atmosphere, and using fuels which do not add to atmospheric carbon dioxide. There are equally good reasons for not running down too fast our non-renewable resources of coal and oil. Next, they should look at their systems of value and attempt to introduce environmental pricing to take account of intergenerational equity. This would make it easier for them to take out insurance policies against disaster, and pay the necessary premiums in terms of precautionary investment: for example attaching a proper price to waste disposal in all its aspects, and anticipating changing patterns of rainfall and thus availability of fresh water. They should retarget relevant scientific research and coordinate the results on a global basis. At a global level we need much better monitoring of changes in land, sea and air. They should work out an international strategy which sets a framework for collective action, takes good account of equity and is founded on national as well as international interest. Last, they should always see and deal with environmental issues together. Isolated measures to cope with one of them can sometimes make others worse.

Yet, when all this is said and some of it is done, we still have to ask ourselves whether our actions are nearly enough and whether we are not dealing with symptoms rather than the disease. At least such actions would represent a start, and once started other actions would follow. They would indeed by driven by events. The critical issue then would be whether we had some control over events, or whether events took control of us.

I sometimes reflect idly on what would happen if we, like nearly all species that have ever lived, were simply to be eliminated. At the beginning of this century rabbits were introduced to Lisianki Island west of Hawaii. Within a decade they had, in the words of Jared Diamond, 'eaten themselves into oblivion by consuming every plant on the island except two morning glories and a tobacco patch' (Diamond, 1991). Could we do likewise to ourselves on this small planet?

There is a book by Dougal Dixon entitled *After Man*, a zoology of the future, in this case a future 50 million years ahead. The author invents some formidable creatures developing from those we recognize as most successful today: for example rats, squirrels, baboons and rabbits. If I have a reproach, it is that he concentrates unduly on mammals, a class of organism, which might no more flourish in the future than did the dinosaurs after the cataclysm 65 million years ago. It then took up to 10 million years for the mammals to occupy the niches left vacant. Next time it could be the insects or the crustaceans who inherit the earth.

My own day-dream is shorter term. If we perished more or less together – say over 50 rather than 50 million years – what would become of the earth? How long would it take for the cities to fall apart, for the earth to regenerate, for the animals and plants we have selected for ourselves to find themselves a more normal place in nature, for the waters and seas to become clean, for the chemistry of the air to return to what it was before we disturbed it?

Life itself is so robust that the human experience could soon become no more than a tiny episode. Nature is not fragile. But we are. I like to recall that throughout his life Darwin focused his powerful attention on one of the most vital creatures of regeneration, the earthworm. He calculated there were 53 767 worms to the acre on his estate. Whatever he or others might do to the land, this army of worms was always at work re-establishing its fertility. For him, as for us, it is a kind of consolation.

References

Brown, L. 1991. The new world order. In *State of the World*, 1991. London: Earthscan.

Diamond, J. 1991. *The Rise and Fall of the Third Chimpanzee*. London: Hutchinson.

Houghton, J. T., Jenkins, C. J. and Ephraums, J. J. (eds). 1990. *Climate Change. The IPCC Scientific Assessment*. Cambridge: CUP.

Houghton, J. T., Callander, B. A. and Varney, S. K. (eds) 1992 *Climate Change, 1992*. The supplementary report to the IPCC scientific assessment. Cambridge: CUP.

Lamb, H. H. 1982. *History, Climate and the Modern World*. London: Methuen.

McNamara, 1991. Rafael Salas lecture, UNFPA, New York, December.

Meadows, D. 1972. *Limits to Growth*. London: Earthscan.

——, 1992. *Beyond the Limits*. London: Earthscan.

New Scientist, 17 October 1992. London.

The Lancet, 30 May, 1992. Editorial. London: BMA.

Thomas, L. 1974. *Societies as Organisms*. London: Viking Press.

United Nations Population Fund, 1992. *The state of world population*, 1992. New York: UNPF.

World Resources Institute, 1992. *World resources*, 1992–93. New York: WRI/Oxford University Press in collaboration with UNDP and UNEP.

INDEX